Lecture Notes in Computer Science 1673
Edited by G. Goos, J. Hartmanis and J. van Leeuwen

Springer

Berlin
Heidelberg
New York
Barcelona
Hong Kong
London
Milan
Paris
Singapore
Tokyo

Patrick Lysaght James Irvine
Reiner Hartenstein (Eds.)

Field Programmable Logic and Applications

9th International Workshop, FPL'99
Glasgow, UK, August 30 - September 1, 1999
Proceedings

Volume Editors

Patrick Lysaght
James Irvine
Department of Electronic and Electrical Engineering
University of Strathclyde
George Street, Glasgow G1 1XW, Scotland, UK
E-mail: p.lysaght@eee.strath.ac.uk
j.m.irvine@strath.ac.uk

Reiner Hartenstein
Computer Science Department, University of Kaiserslautern
P.O. Box 30 49, D-67653 Kaiserslautern, Germany
E-mail: hartens@rhrk.uni-kl.de

Cataloging-in-Publication data applied for

Die Deutsche Bibliothek - CIP-Einheitsaufnahme

Field programmable logic and applications : 9th international workshop ; proceedings / FPL '99, Glasgow, UK, August 30 - September 1, 1999. Patrick Lysaght ... (ed.). - Berlin ; Heidelberg ; New York ; Barcelona ; Hong Kong ; London ; Milan ; Paris ; Singapore ; Tokyo : Springer, 1999
(Lecture notes in computer science ; Vol. 1673)
ISBN 3-540-66457-2

CR Subject Classification (1998): B.6-7, J.6

ISSN 0302-9743
ISBN 3-540-66457-2 Springer-Verlag Berlin Heidelberg New York

Typesetting: Camera-ready by author
SPIN: 10705539 06/3142 – 5 4 3 2 1 0 Printed on acid-free paper

Preface

This book contains the papers presented at the 9th International Workshop on Field Programmable Logic and Applications (FPL'99), hosted by the University of Strathclyde in Glasgow, Scotland, August 30 – September 1, 1999. FPL'99 is the ninth in the series of annual FPL workshops.

The FPL'99 programme committee has been fortunate to have received a large number of high-quality papers addressing a wide range of topics. From these, 33 papers have been selected for presentation at the workshop and a further 32 papers have been accepted for the poster sessions. A total of 65 papers from 20 countries are included in this volume.

FPL is a subject area that attracts researchers from both electronic engineering and computer science. Whether we are engaged in research into soft hardware or hard software seems to be primarily a question of perspective. What is unquestionable is that the interaction of groups of researchers from different backgrounds results in stimulating and productive research.

As we prepare for the new millennium, the premier European forum for researchers in field programmable logic remains the FPL workshop. Next year the FPL series of workshops will celebrate its tenth anniversary. The contribution of so many overseas researchers has been a particularly attractive feature of these events, giving them a truly international perspective, while the informal and convivial atmosphere that pervades the workshops have been their hallmark. We look forward to preserving these features in the future while continuing to expand the size and quality of the events.

A great many people have contributed to making FPL'99 a success and we wish to thank all of them for their efforts. These include the many authors who submitted their work to the workshop, the keynote speakers, and the members of the steering group and the programme committee. In addition to the many staff members at the University of Strathclyde who have contributed to FPL'99, we would like to recognise the assistance of the teams at the University of Kaiserslautern and Imperial College who are veteran organisers of previous FPL workshops. We thank the members of the local organising committee, especially Aaron Ferguson, David Robinson, John MacBeth, and David Downey, and also Tariq Durrani, Morag Aitken, Elaine Black, and Bob Stewart.

Finally we gratefully acknowledge the contribution of Thomas Hoffman at Springer-Verlag in publishing this book, and those companies who have made generous grants to the workshop to ensure its continued success. We are pleased to record our gratitude (alphabetically) to Altera Ltd., Nallatech Ltd., System Level Integration Ltd., and Xilinx Inc.

June 1999

Patrick Lysaght
James Irvine
Reiner Hartenstein

General Chairman

Patrick Lysaght, University of Strathclyde, Scotland

Honorary Programme Chairman

Reiner Hartenstein, University of Kaiserslautern, Germany

Programme Committee

Peter Athanas, Virginia Tech, USA
Samary Baranov, Ben Gurion University of the Negev, Israel
Gorden Brebner, Edinburgh University, Scotland
Stephen Brown, University of Toronto, Canada
Klaus Buchenrieder, Siemens AG, Germany
Steven Casselman, VCC Corp., USA
Bernard Courtois, TIMA Laboratory, France
Carl Ebeling, University of Washington, USA
Norbert Fristacky, Slovak Technical University, Slovakia
Manfred Glesner, Technical University Darmstadt, Germany
John Gray, Xilinx, UK
Herbert Gruenbacher, Vienna University, Austria
Brad Hutchings, Brigham Young University, USA
Udo Kebschull, University of Tuebingen, Germany
Andres Keevallik, Tallinn Technical University, Estonia
Wayne Luk, Imperial College, UK
Toshiaki Miyazaki, NTT Laboratories, Japan
Will Moore, Oxford University, UK
Wolfgang Nebel, University of Oldenburg, Germany
Paolo Prinetto, Politecnico di Torino, Italy
Jonathan Rose, University of Toronto, Canada
Zoran Salcic, University of Auckland, New Zealand
Eduardo Boemo Scalvinoni, University of Madrid, Spain
Hartmut Schmeck, University of Karlsruhe, Germany
Marc Shand, Compaq, USA
Stephen Smith, Altera, USA
Steve Trimberger, Xilinx, USA
Roger Woods, The Queen's University of Belfast, Northern Ireland

Steering Committee

Reiner Hartenstein (Honorary Programme Chairman)
James Irvine (Publications Chairman)
Wayne Luk (General Co-Chair, FPL '97)
Patrick Lysaght (General Chairman)

Local Organising Committee

David Downey, Aaron Ferguson, James Irvine, Patrick Lysaght,
John MacBeth, David Robinson

Table of Contents

Novel Architectures

Machine Applications

Short Papers

Reconfigurable Processors for High-Performance, Embedded Digital Signal Processing*

Paul Graham and Brent Nelson

459 Clyde Building, Brigham Young University, Provo UT 84602, USA
grahamp@ee.byu.edu, nelson@ee.byu.edu

Abstract. For high-performance, embedded digital signal processing, digital signal processors (DSPs) are very important. Further, they have many features which make their integration with on-chip reconfigurable logic (RL) resources feasible and beneficial. In this paper, we discuss how this integration might be done and the potential area costs and performance benefits of incorporating RL onto a DSP chip. For our proposed architecture, a reconfigurable coprocessor can provide speed-ups ranging from 2-32x with an area cost of about a second DSP core for a set of signal processing applications and kernels.

1 Introduction

For high-performance, embedded digital signal processing, digital signal processors (DSPs) are very important, but, in some cases, DSPs alone cannot provide adequate amounts of computational power. As shown in [1, 2], reconfigurable logic (RL), specifically, FPGAs, can profitably be used for signal processing applications which require large amounts of computation and outperform existing high-performance DSPs despite the weaknesses of current commercial FPGAs in performing arithmetic. Unfortunately, FPGAs cannot effectively handle applications requiring high-precision arithmetic or complex control. As a compromise, we have been exploring DSP-RL hybrid architectures which enjoy the flexibility and precision of DSPs while experiencing performance improvements due to RL.

In this paper, we introduce a hybrid DSP-RL processor which tightly couples the DSP core to reconfigurable logic for greater performance and flexibility with digital signal processing applications without altering the DSP's embeddable nature. First, we will discuss the benefits and drawbacks of such an architecture. Following this discussion, we will describe many of the possible hybrid architectures and provide our performance and silicon-area estimates for one specific architecture. The conclusion to the paper provides some summary comments on this hybrid and a few words regarding on-going and future work in this area.

* Effort sponsored by the Defense Advanced Research Projects Agency (DARPA) and Rome Laboratory, Air Force Materiel Command, USAF, under agreement number F30602-97-1-0222. The U.S. Government is authorized to reproduce and distribute reprints for Governmental purposes notwithstanding any copyright annotation thereon.

2 Why a DSP Hybrid?

A large number of projects, including [3–8], have explored the use of processor-RL hybrids for "general-purpose" computing applications. These projects have generally involved coupling a RISC, often a MIPS, processor core with reconfigurable logic and, by doing so, have shown promising speed-ups for a range of applications, including DES encryption, the SPECint92 benchmarks, and image processing. The combination of the processor core with either reconfigurable function units or coprocessors enables the host processor to communicate through high-bandwidth, on-chip interconnection to the RL resources rather than the relatively slow interconnection made through an interface with an external system bus. With reconfigurable resources, these processors were able to compute with greater efficiency than software alone, having less software overhead due to address generation, branching, and function calls and exploiting more parallelism than is possible with the processor's normal data path.

One hybrid processor-RL system called Pleiades [9, 10] has specifically targeted ultra-low power, embedded digital signal processing. The approach fuses an ARM core with a combination of reconfigurable logic, reconfigurable data path, and/or reconfigurable data-flow resources. In this case, due to power and performance constraints, the RISC core is used mainly for administrative tasks such as programming the reconfigurable resources and not for computation. According to the results in the above-cited papers, the architecture has proven its power-performance advantage over a number of common technologies such as DSPs and FPGAs for a few digital signal processing applications.

Unfortunately, RISC processors, despite their computation power, are not well suited for many embedded high-performance applications. These processors are often power hungry, require a number of support ICs, and exhibit behavior which is hard to deterministically characterize for real-time applications. These qualities often make DSPs better choices for embedded applications. Further, a handful of DSPs offer glueless multiprocessing and provide a computing density per board advantage over multiple embedded RISC processors—a metric very important to embedded system designers.

Though DSPs may be good candidates for embedded processors, they also have some disadvantages resulting from their design as efficient embeddable processors. For instance, programming DSPs is laborious relative to programming RISC processors since the related tools are less sophisticated and less efficient. Much of this follows from the emphasis on making DSPs both power and memory efficient as opposed to being compiler friendly. Further, DSPs' explicit parallelism must be managed efficiently, which is not always an easy task. Also, due to the emphasis on low-power and low-latency design, DSPs tend to operate at lower clock frequencies than RISC processors.

Despite these drawbacks, DSP cores have several existing features which make them good candidates for supporting reconfigurable resources. First, many DSPs have either multiple on-chip memory banks or otherwise provide multiple memory accesses per cycle. Thus, the memory architecture of the processor does not have to be drastically modified to provide many independent memory ports to the reconfigurable resources. Second, the parallel execution of the DSP core and the RL resources can often be expressed as simple extensions of DSP instruction sets, which already directly express some level of concurrency. Third, the on-chip DMA controllers which many DSPs already have may prove useful for configuring the RL without burdening the processor with the task. Moreover, since the execution of many DSP applications are quite deterministic, configuration pre-fetching scheduled at compile time [11] should be a useful technique. The expense of configuration management hardware does not appear to be justified for a DSP-RL hybrid processor if only a few kernels are deterministically loaded during application execution.

DSPs can benefit in several ways from using reconfigurable logic for on-chip processing. First, DSPs are often asked to perform tasks which require a large amount of bit-level data manipulation such as error control coding—functions for which they are ill-suited but can perform. Second, the parallelism which is inherent in many DSP applications often cannot be exploited well by a processor which can, at most, perform a few arithmetic operations per cycle. Some recently announced DSPs such as the Analog Devices' Hammerhead and Tiger SHARCs use SIMD techniques to boost the performance of the processors for some applications, but, unfortunately, these techniques can only be used for a portion of all DSP applications. The parallelism that the on-chip reconfigurable resources can exploit is not limited to SIMD techniques. Another interesting, but minor benefit that DSPs can experience by using RL resources on-chip is the ability to generate application-specific address streams with RL. In [1], we discussed an application, delay-sum sonar beamforming, which required address generation for which the usual DSP hardware address generators were inefficient. As we show in Sect. 4, RL used just for address generation can provide a three-times speed-up for this application's kernel at a very small silicon area cost. Lastly, RL can lead to more efficient use of memory bandwidth by performing computations on data as they are streamed to and from memory.

The DSP core also provides the hybrid processor with capabilities which a strictly reconfigurable architecture cannot offer. First, due to the fast reconfigurability of the DSP core from instruction to instruction and the size of its instruction store, the processor is better suited for complex, control-heavy portions of applications. Second, depending on the processor core used, the core can provide higher precision arithmetic than may be practical in the RL resources, meaning that the processor core and RL can play complementary roles in applications. The processor core can deal with operations which require high precision while the RL can provide speed-ups for functions which require large amounts of computation on lower-precision data objects.

3 Hybrid DSP Architectures

For our architectural studies, we have decided to use the Analog Devices' SHARC DSP family as a model because of its features and performance in DSP applica-

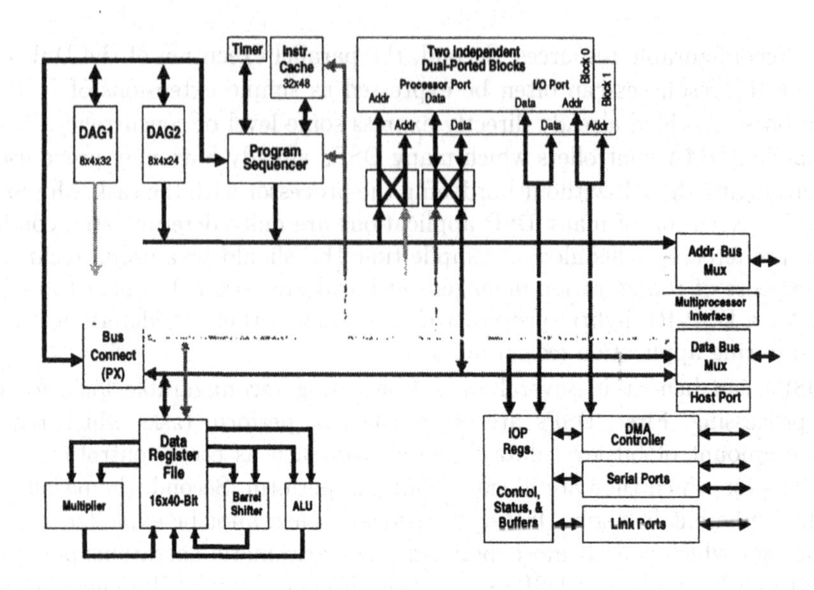

Fig. 1. SHARC Architecture

tions [12]. A block diagram of the architecture is provided in Fig. 1. Among the features which influenced our choice of the SHARC are:

1. its large on-chip memory (512 KB) organized as two banks of dual-ported memory, allowing for concurrent off-chip I/O and processing;
2. a data path capable of performing up to 3 arithmetic operations per cycle (a multiply with subtract and add);
3. several on-chip I/O processor peripherals including DMA controllers;
4. and, the support of glueless multiprocessing.

We chose this DSP despite the fact that it was a floating-point DSP and the applications and kernels in the benchmarks, described in Sect. 4, are implemented in fixed-point. Our rationale for this choice was that no fixed-point DSP currently has features equivalent to those listed above, all of which greatly improve the processor's performance for large applications. Second, considering that the DSP core is only 12% of the total die area, we do not believe the area comparisons would be greatly affected if the core was a fixed-point processor.

The reconfigurable portion of a DSP-RL architecture can be interfaced with the DSP core in several ways. First, the reconfigurable logic can act much like a function unit in the data path of the processor, having access only to the processor's register file. This approach is problematic since the memory bandwidth to the reconfigurable logic is effectively restricted to the one or two memory ports which service the register file. As was shown in [1, 13] and other work, restricting memory ports to only one or two greatly limits the performance of the architecture and causes the memory subsystem to be the bottleneck.

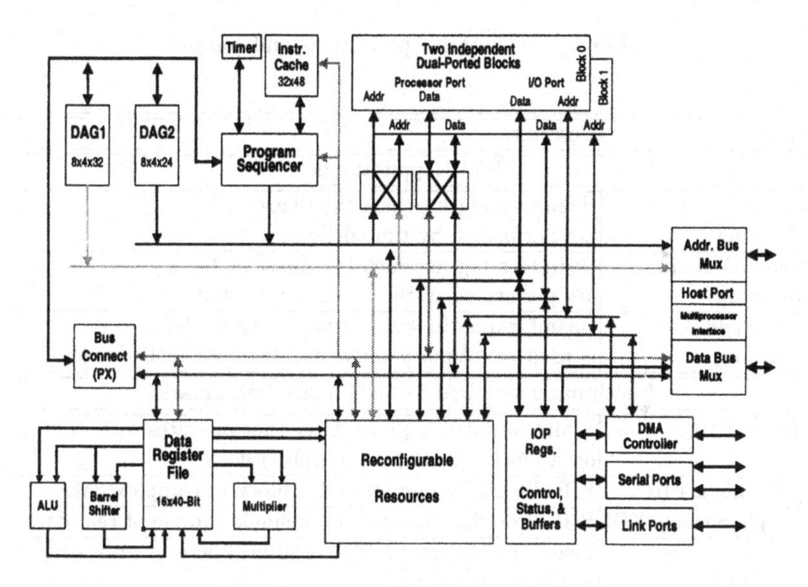

Fig. 2. DSP with a Reconfigurable FU/Coprocessor Combination

Another alternative is to treat the reconfigurable portion of the processor as a coprocessor. If the reconfigurable logic is treated as another peripheral such as the I/O processor of the SHARC, it again has only two memory ports and must compete with the I/O processor for memory, but the RL can operate concurrently with the processor core without disturbing the core's operation. Of course, this does not alleviate the lack of memory ports.

Another variation of this architecture is to modify the memory interfaces to provide a total of 4 memory ports for the reconfigurable processor; in the SHARC architecture, this means sharing both the DSP core's memory ports and the I/O processors' memory ports, i.e., having access to all of the memory ports of the two dual-ported, on-chip memories. Though the DSP cannot run concurrently with the reconfigurable coprocessor when all of the memory ports are being used by the coprocessor, the accessibility of the independent memory ports can enable the reconfigurable coprocessor to speed up applications beyond what the DSP core itself can provide. With the proper, careful assignment of memory ports, concurrent operation should still be possible for the DSP and reconfigurable coprocessor with this structure.

Among the other possibilities, another variation is to provide the reconfigurable resources both access to the processor register file and the four on-chip memory ports; this is illustrated in Fig. 2. This makes the reconfigurable resources a cross between a function unit and a coprocessor. This approach provides flexibility and reduces the requirement of using memory as the means of communication between the two data paths, further reducing the memory bandwidth burden and allowing tighter cooperation between the DSP core and the RL.

Table 1. Benchmark Applications and Kernels

Function	Description
BYU Applications	
Delay-Sum Beamforming	Beamforming in which the phase shifts are performed by time delays
Frequency-Domain Beamforming	Beamforming in which the phase shifts are performed in the frequency domain
Matched Field Beamforming	Beamforming in which a multi-ray model is used to estimate angle and distance of the target.
Benchmarks Inspired by BDTImark benchmark	
Real Block FIR	Finite impulse response filter that operates on a block of real (not complex) data.
Complex Block FIR	FIR filter that operates on a block of complex data.
Real Single-Sample FIR	FIR filter that operates on a single sample of real data.
LMS Adaptive FIR	Least-mean-square adaptive filter; operates on a single sample of real data.
IIR	Infinite impulse response filter that operates on a single sample of real data.
Vector Dot Product	Sum of the point-wise multiplication of two vectors.
Vector Add	Point-wise addition of two vectors, producing a third vector.
Vector Maximum	Find the value and location of the maximum value in a vector.
Convolutional Encoder	Apply convolutional forward error correction code to a block of bits.
Finite State Machine	A contrived series of control (test, branch, push, pop) and bit manipulation operations.
256-Point, Radix-2, In-Place FFT	Fast Fourier transform converts a normal time-domain signal to the frequency domain.

4 Area and Performance Results

As a starting point for our work, we have been evaluating the architecture of Fig. 2 for performance potential and silicon area considerations. As a way of benchmarking the performance, we have chosen a collection of applications and kernels which represent typical digital signal processing computations; these are listed in Table 1. In the mix of applications and kernels, we have included a few sonar beamforming applications [1, 14] with which we are very familiar. The other kernels are modeled after the collection of kernels in the BDTImark benchmark defined by Berkeley Design Technology Inc., who have identified these kernels as important to many DSP applications[15].

Our initial studies assume that the reconfigurable logic looks much like the Xilinx 4000 family of FPGAs. Though we expect coarser-grain RL architectures such as CHESS [16], CFPA [17], and the Garp RL array [6] would be better for the hybrid, we use this RL architecture because it is well known and understood. Also, it provides a nice upper bound for the area-performance trade-offs in the hybrid architecture since the Xilinx 4000 architecture will generally be more costly for implementing data-path operations than the coarse-grain field programmable architectures mentioned above. In fact, we expect these coarse-grain architectures to require only about 40% to 55% of the silicon area of the Xilinx 4000 architecture for the same data-path functionality. To its credit, though, the Xilinx 4000 architecture will often be more flexible for control hardware and bit-level operations due to its finer granularity.

From [12], we learn that the SHARC, using a .6 micron, two-layer metal process, requires about $3.32x10^9\,\lambda^2$ in area. Due to the large on-chip memories of the architecture, about $1.86x10^9\,\lambda^2$, or 56% of the die, is devoted to SRAM, while about 12% of the die area, or $3.99x10^8\,\lambda^2$, is the DSP core and data path. The remaining die area is devoted to on-chip peripherals, the I/O pads, and interfacing logic. With this information and area estimates for Xilinx 4000 CLBs with interconnect drawn from [18], we can estimate the relative area increases due to the reconfigurable logic for each of the applications in the benchmark suite. For instance, from [18], we learn that a CLB has a cost of approximately $1.26x10^6\,\lambda^2$, so a design requiring 100 CLBs would require about $1.26x10^8\,\lambda^2$ in silicon area, which is only a 3.79% increase in total die area.

Table 2 describes the performance per area increase for the hybrid architecture for several of these benchmarks. The entries are sorted in decreasing speed-up per silicon area change. As an explanatory note, we should point out that the speed-up numbers do not take into account reconfiguration time for the kernel or application because our analysis assumes that reconfiguration is infrequent. Also, the RL implementations account for bit growth due to arithmetic operations, though simple scaling is essential in some cases. You should further note that only the reconfigurable logic portion of the hybrid is executing the benchmarks, except in a few cases that are noted in the table where the RL and the DSP core are operating cooperatively. In this study, we also assume that the reconfigurable coprocessor operates at the same clock frequency as the DSP. Though, in general, we would not expect this from the Xilinx 4000 architecture for higher-frequency DSPs (100–200 MHz), we expect that coarser-grain, data-path-oriented field-programmable architectures such as CHESS, CFPA, or the Garp RL array would be able to support these frequencies. For instance, the Garp RL array has been estimated to execute at a minimum of 133 MHz[6]. Lastly, as mentioned before, the reconfigurable implementations of the benchmarks are fixed-point implementations, not floating point.

With a reconfigurable logic budget of about 1000 CLBs, a good portion of the kernels and applications can be accelerated without frequent reconfiguration of the RL. This accounts for about a 40% increase in chip area. From our estimates and the published work on other field-programmable architectures, we believe that the cost of this amount of reconfigurable logic can be as little as 16–20% with computational fabrics such as CHESS or CFPA—an area just a

Table 2. Performance and Area of Hybrid for Benchmark Circuits

Application	Implementation	Pipe-lining	Speed-Up	Area (CLBs)	Total Area	Speed-Up / Δ Area
Convolutional Encoder	For V.32 Modem, with differential encoding	Full	34	5	1.0019	17929
Vector Add	Cooperative w/ DSP	N/A	2	40	1.0152	131
Vector Max.	16b data	Full	2	48	1.0182	110
Delay Sum BF	Coop. w/ DSP (Addr. Gen.)	N/A	3	100	1.0379	79.1
Delay Sum	Full PE	Full	3	246	1.0933	32.2
IIR Biquad	4 8b ×, 4 20b +	1/2	4	360	1.1365	29.3
FIR	32 tap, 16b KCM, 40b +	1/2	32	2976	2.1287	28.4
FIR	16 tap, 16b KCM, 40b +	1/2	16	1488	1.5643	28.4
FIR	2 tap, 16b KCM, 40b +	1/2	2	186	1.0705	28.4
FFT	CORDIC-based butterfly	1/2	4	380	1.1441	27.8
Matched Field BF	Sub-voxel, memory sensitive	1/2	8	928	1.3520	22.7
Matched Field BF	Sub-voxel, memory sensitive	Full	8	1073	1.4070	19.7
Vector Dot	Coop. w/ DSP, 16b ×, 40b +	1/2	2	308	1.1168	17.1
Matched Field BF	Sub-voxel, memory intensive	1/2	6	928	1.3520	17.0
Matched Field BF	Sub-voxel, memory intensive	Full	6	1073	1.4070	14.7
Freq. BF	Reported in [1]	1/2	4	856	1.3247	12.3
FIR	2 tap, 16b ×, 40b +	1/2	2	552	1.2094	9.55
IIR Biquad	4 16b ×, 4 24b +	1/2	4	1200	1.4551	8.79
IIR Biquad	5 16b ×, 4 24b +	1/2	4	1488	1.5643	7.09

little larger than the size of the DSP core. In addition, since the computation density per board is often one of the most important characteristics for embedded system designers, silicon area increases of 16%–40% are acceptable if the total computational density per board is increased by a factor of 2 or more.

5 Conclusions and Future Work

We have demonstrated that a reconfigurable architecture which incorporates a reconfigurable coprocessor into a DSP can have performance benefits for a reasonable increase in chip area. In addition, DSPs have many architectural features which make a combination with reconfigurable logic feasible and beneficial. Despite the raw clock rate disadvantage of DSPs when compared with general-purpose microprocessors, DSPs serve an important role in high-performance, embedded computing; a reconfigurable coprocessor on-chip can help DSPs exploit more of the parallelism found in digital signal processing applications, thus improving the processor's overall performance.

A large amount of work still remains to be performed and many outstanding questions exist. For instance, the amount of concurrent DSP-RL execution which can be expected for DSP algorithms should be determined—a process which may require the use of automated programming tools and the mapping of many DSP algorithms to the architecture. If our current experience is representative, there may not be many situations in which concurrent operation is beneficial or possible, indicating that the connection of the RL array to the DSP's register file may not be needed.

Another unanswered question is whether the DSP and RL can truly serve complementary roles in applications—the DSP performing the control-heavy and high-precision arithmetic portions of the algorithm while the RL performs highly parallel operations on lower-precision data items. Again, a large amount of application mapping to the architecture may be required to answer this question, though adaptive signal processing algorithms may provide a fruitful set of applications for this study.

Other on-going and future work include a refinement of the programming methodology for the DSP hybrid processor as well as the many issues of interfacing the RL with the DSP. We expect that programming methodologies for the architecture will use a library-based design approach for the reconfigurable logic in which the designer simply describes the interconnection of the library modules using an assembly-like language. The intention is to make the programming task resemble more of a software creation problem than a hardware design exercise, a requirement crucial in making the architecture usable by DSP programmers and not just hardware designers. Future work will also quantify the effects of frequent reconfiguration on application performance.

References

1. P. Graham and B. Nelson. FPGA-based sonar processing. In *Proceedings of the Sixth ACM/SIGDA International Symposium on Field-Programmable Gate Arrays (FPGA '98)*, pages 201–208. ACM/SIGDA, ACM, 1998.
2. P. Graham and B. Nelson. Frequency-domain sonar processing in FPGAs and DSPs. In *Proceedings of the IEEE Symposium on FPGAs for Custom Computing Machines (FCCM '98)*. IEEE Computer Society, IEEE Computer Society Press, 1998.
3. Rahul Razdan. *Programmable Reduced Instruction Set Computers.* PhD thesis, Harvard University, Cambridge,MA, May 1994.
4. R. D. Wittig and P. Chow. OneChip: An FPGA processor with reconfigurable logic. In J. Arnold and K. L. Pocek, editors, *Proceedings of IEEE Workshop on FPGAs for Custom Computing Machines*, pages 126–135, Napa, CA, April 1996.
5. Scott Hauck, Thomas W. Fry, Matthew M. Hosler, and Jeffery P. Kao. The Chimaera reconfigurable functional unit. In Kenneth L. Pocek and Jeffery Arnold, editors, *Proceedings of the IEEE Symposium on FPGAs for Custom Computing Machines*, pages 87–96. IEEE Computer Society, IEEE Computer Society Press, April 1997.
6. John R. Hauser and John Wawrzynek. GARP: A MIPS processor with a reconfigurable coprocessor. In J. Arnold and K. L. Pocek, editors, *Proceedings of IEEE Workshop on FPGAs for Custom Computing Machines*, pages 12–21, Napa, CA, April 1997.

7. Charle R. Rupp, Mark Landguth, Tim Garverick, Edson Gomersall, Harry Holt, Jeffery M. Arnold, and Maya Gokhale. The NAPA adaptave processing architecture. In Kenneth L. Pocek and Jeffrey M. Arnold, editors, *Proceedings of the IEEE Symposium on FPGAs for Custom Computing Machines (FCCM '98)*, pages 28–37. IEEE Computer Society, IEEE Computer Society Press, April 1998.
8. Jeffery A. Jacob and Paul Chow. Memory interfacing and instruction specification for reconfigurable processors. In *Proceedings of the Seventh ACM/SIGDA International Symposium on Field-Programmable Gate Arrays (FPGA '99)*, pages 145–154. ACM/SIGDA, ACM, 1999.
9. Arthur Abnous and Jan Rabaey. Ultra-low-power domain-specific multimedia processors. In *Proceedings of the IEEE VLSI Signal Processing Workshop*. IEEE, IEEE, October 1996.
10. Arthur Abnous, Katsunori Seno, Yuji Ichikawa, Marlene Wan, and Jan Rabaey. Evaluation of a low-power reconfigurable DSP architecture. In *Proceedings of the Reconfigurable Architectures Workshop*, March 1998.
11. Scott Hauck. Configuration prefetch for single context reconfigurable coprocessors. In *Proceedings of the Sixth ACM/SIGDA International Symposium on Field-Programmable Gate Arrays (FPGA '98)*, pages 65–74. ACM/SIGDA, ACM, 1999.
12. Joe E. Brewer, L. Gray Miller, Ira H. Gilbert, Joseph F. Melia, and Doug Garde. A single-chip digital signal processing subsystem. In R. Mike Lea and Stuart Tewksbury, editors, *Proceedings of the Sixth Annual IEEE International Conference on Wafer Scale Integration*, pages 265–272, Piscataway, NJ, 1994. IEEE, IEEE.
13. Csaba Andras Moritz, Donald Yeung, and Anant Agarwal. Exploring optimal cost-performance designs for Raw microprocessors. In Kenneth L. Pocek and Jeffery M. Arnold, editors, *Proceedings of the IEEE Symposium on FPGAs for Custom Computing Machines (FCCM '98)*, pages 12–27. IEEE Computer Society, IEEE Computer Society Press, 1998.
14. Paul Graham and Brent Nelson. FPGAs and DSPs for sonar processing—inner loop computations. Technical Report CCL-1998-GN-1, Configurable Computating Laboratory, Electrical and Computer Engineering Department, Brigham Young University, 1998.
15. Garrick Blalock. The BDTImark: A measure of DSP execution speed. Technical report, Berkeley Design Technology, Inc., 1997. Available at http://www.bdti.com/articles/wtpaper.htm.
16. Alan Marshall, Tony Stansfield, Igor Kostarnov, Jean Vuillemin, and Brad Hutchings. A reconfigurable arithmetic array for multimedia applications. In *Proceedings of the Seventh ACM/SIGDA International Symposium on Field-Programmable Gate Arrays (FPGA '99)*, pages 135–143. ACM/SIGDA, ACM, 1999.
17. Alireza Kaviani, Daniel Vranesic, and Stephen Brown. Computational field programmable architecture. In *Proceedings for the IEEE Custom Integrated Circuits Conference (CICC '98)*, pages 12.2.1–12.2.4. IEEE, IEEE, 1998.
18. André DeHon. *Reconfigurable Architectures for General-Purpose Computing*. PhD thesis, Massachusetts Institute of Technology, September 1996.

Auditory Signal Processing in Hardware: A Linear Gammatone Filterbank Design for a Model of the Auditory System

M. Brucke, A. Schulz, W. Nebel

University of Oldenburg, Computer Science Department
D-26111 Oldenburg, Germany
Brucke@Uni-Oldenburg.DE,
WWW home page: http://eis.informatik.uni-oldenburg.de/~silco

Abstract. A digital hardware implementation of an algorithm modelling the "effective" signal processing of the human auditory system is presented in this paper. The model can be used as a preprocessor for speech, for example in automatic speech recognition, objective speech quality measurement, noise reduction and digital hearing aids.
A direct hardware implementation of the floating point software version of the model has many disadvantages. The main problem when converting floating point arithmetic to fixed point arithmetic is the determination of the necessary numerical precision which implies the wordlength of internal number representation. The necessary internal wordlength for the linear parts of the system can be assessed in a straight-forward way because the filters are linear time invariant systems, where classical numerical parameters like SNR can be applied. For the realization of the nonlinear parts this procedure is not applicable. One application of the model (objective speech quality measurement) was used to determine the necessary internal precision. By observing the degradiation of the performance with decreasing internal precision it shows that necessary internal wordlength can be reduced while the performance almost stays the same. To validate the approach a prototype of the design was implemented on a Xilinx XC40200XV-09.

1 Introduction

The Medical Physics Group at the University of Oldenburg has been working in the field of psychoacoustical modelling, speech perception and processing, audiological diagnostics and digital hearing aids for several years. One particular aspect of the work done is the development of a psychoacoustical preprocessing model [1] and the demonstration of its applicability as a preprocessing algorithm for speech, for example in automatic speech recognition [2], objective speech quality measurement [3], noise reduction and digital hearing aids [4]. The model describes the "effective" signal processing in the human auditory system and provides the appropriate internal representation of acoustic signals. In interaction with the "Graduate School in Psychoacoustics" at the University of Oldenburg the processing model has been improved and optimized.

A transfer of the model into hardware leads to the positive effect that more ressources of a DSP or host CPU remain for a given application which uses the internal representation of an acoustic signal as an input. By this real-time implementations of complex auditory-based speech processing algorithms will become possible, because the chipset performs substantial parts of the necessary calculations on a dedicated hardware.

The VLSI group at the University of Oldenburg and the IMA group at the University of Hamburg are now working on a transformation of this signal processing algorithm into a set of integrated digital circuits. It is planned to produce the ASICs in a 0.25 μm technology. The FPGAs are used as prototypes to validate the approach.

2 The preprocessing algorithm

Figure 1 outlines the structure of the psychoacoustical perception model. The model combines several stages of processing simulating spectral properties of the human ear (spectral masking, frequency-dependent bandwidth of auditory filters) as well as dynamical effects (nonlinear compression of stationary or dynamic signals and temporal masking). The appropriateness of this approach was shown in several psychoacoustical experiments by Dau [1]. The gammatone filterbank represents the first processing stage and simulates the frequency-place transformation on the basilar membrane [5]. It is build up from 30 bandpass filters with center frequencies from 73 Hz to 6,7 kHz equidistant on the ERB scale, which is approximately logarithmic. The bandwidth of the filters grows with increasing frequency. The output of each channel of the gammatone filterbank is halfwave rectified and lowpass filtered at 1 kHz to preserve the envelope of the signal for high carrier frequencies. This is motivated by the limited phase locking of auditory nerve fibers at higher frequencies. The adaption loops model the dynamic compression of the input signal. Stationary signals are compressed almost logarithmically whereas fast fluctuating signals are transformed linearly [6]. The adaptation stage is a chain of five consecutive feedback loops with different time constants range from 5 to 500 ms. Each feedback loop consists of a divider and a lowpass filter. The divisor determines the “charging state” of the capacitor low pass. Thus the system has some kind of “memory” determining the compression of the current signal level based on the signal history which accounts for the ability to correctly model temporal masking effects. In the next stage the signal is filtered by an 8 Hz first order modulation lowpass. This accounts for decreasing modulation detection at higher frequencies found for many broadband carrier signals.

3 Implementation of the model

The design was partitioned to split up the necessary computations on two chips. The first chip (Chip 1) being developed in Oldenburg contains the gammatone filterbank, the halfwave rectification, the 1 kHz lowpass and the threshold which

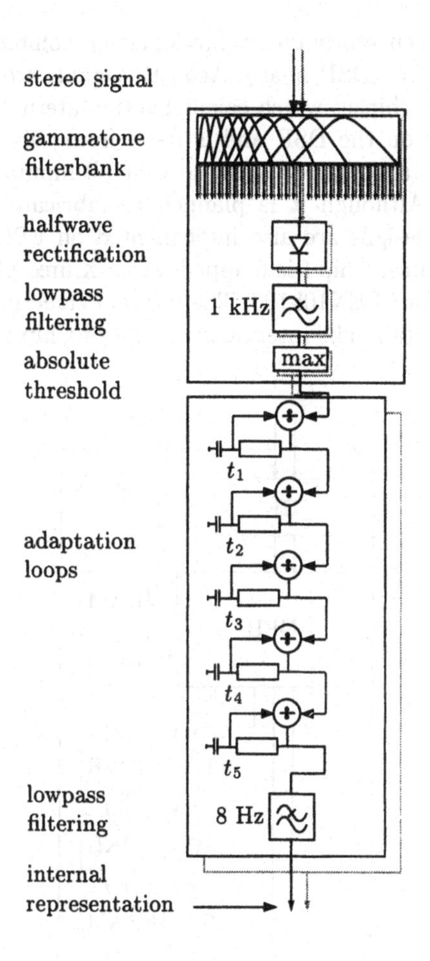

Fig. 1. Signal processing scheme of the binaural perception model.

work in stereo (i.e., two independent channels are used for binaural signal processing). Furthermore a module is incorporated which calculates the phase difference and the amplitude quotient between the two stereo channels for each auditory filter. The second chip (Chip 2) being developed in Hamburg processes the output data of Chip 1 and applies the dynamic compression of the adaptation loops and the modulation lowpass filtering on the signal. Due to its complexity it works monaurally and has to be used twice in the target environment where the data streams of the chipset (Chip 1 + 2*Chip 2) are combined to a resulting data stream with a throughput of about 12 Mbps. This data stream contains the "internal representation" for each input sample and the phase difference and amplitude quotient between left and right stereo channel for each of the 30 filters of the gammatone filterbank.

An addon board is currently under developement combining the chipset with a Texas Instruments C6x DSP board. Acoustic signals are sampled by a 16 bit ADC and fed into the chipset which calculates the internal representation. The user application runs on the DSP and is freed from this task. For debugging purposes the board can be reconfigured so that all inputs and outputs can be routed to the DSP. Although it is planned to fabricate the chips in digital VLSI (0.25 μm), the designs are also implemented on FPGAs to test the performance and algorithms. Chip 1 is mapped into a Xilinx XC40200XV, Chip 2 is mapped into an Altera FLEX10k100. The internal structure of Chip 2 is shown in [7] or [8] in more detail. The internal structure of Chip 1 is shown in figure 2.

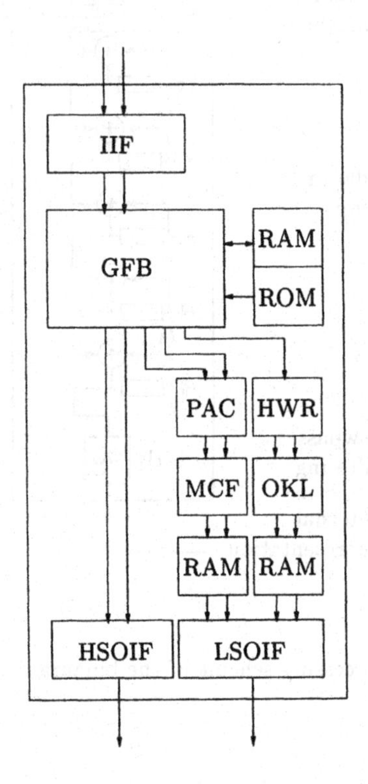

Fig. 2. Internal structure of Chip 1

Besides the serial data input interface (IIF) Chip 1 contains the 30-channel binaural filterbank (GFB). The ouptut data of the filterbank is converted into a serial data stream by the High Speed Output Interface (HSOIF) which leads to a data rate of ≈ 30 Mbps. Parallel to that it is processed by two other datapaths. The first one – consisting of the halfwave rectification (HWR) and the 1 kHz lowpass (OKL) – calculates the envelope of the signal, whereas the second one computes the sliding mean (Mean Calculation Filter MCF) of phase differences and magnitude quotients (Phase and Amplitude Calculation PAC) for each channel. The Low Speed Output Interface (LSOIF) combines these two data streams to a resulting serial output with a rate of ≈ 12 Mbps.

The gammatone filterbank consists of a 4th order IIR bandpass filter (Figure 3) which is multiplexed through all 30 channels and twice for the left and the right side signal. Including a six-stage pipelined multiplier and one adder/subtractor this kernel is realized by a quad cascade of a single stage complex IIR filter (See figure 4). This is primarily motivated by the focus on area reduction for an ASIC implementation and saves chip ressources but requires a 50 MHz system clock. A RAM unit saves temporary data, the filter constants are stored in the ROM block.

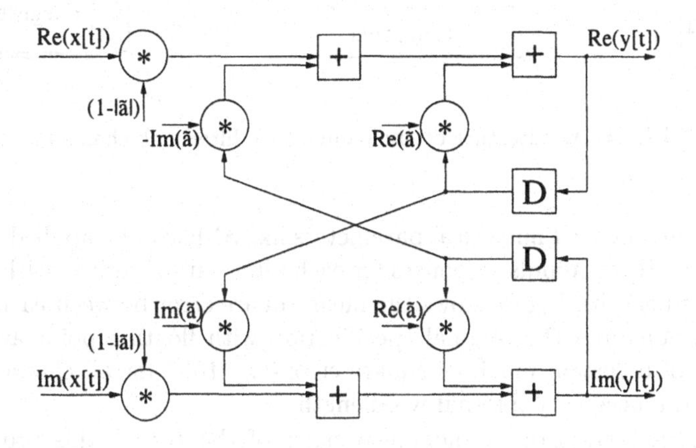

Fig. 3. One stage of the IIR bandpass filter for a gammatone filterbank channel (iterated 4 times).

4 Transformation from floating point to fixed point arithmetic

A direct implementation of the floating point version of the model is not possible due to limitations of area and power consumption (ASIC) and ressources (FPGA). The main problem when converting floating point arithmetic to fixed point arithmetic is the determination of the necessary numerical precision. This implies the wordlength of internal number representation. Therefore the perception model was recoded in C++ using a self-developed scalable data type. This data type takes the internal wordlength as a parameter and saves the values exactly in the same format as they would be saved in a register. So numerical effects of imprecise arithmetic can be simulated.

The necessary internal wordlength for the gammatone filterbank can be assessed in a straight-forward way, because the filters are linear time invariant

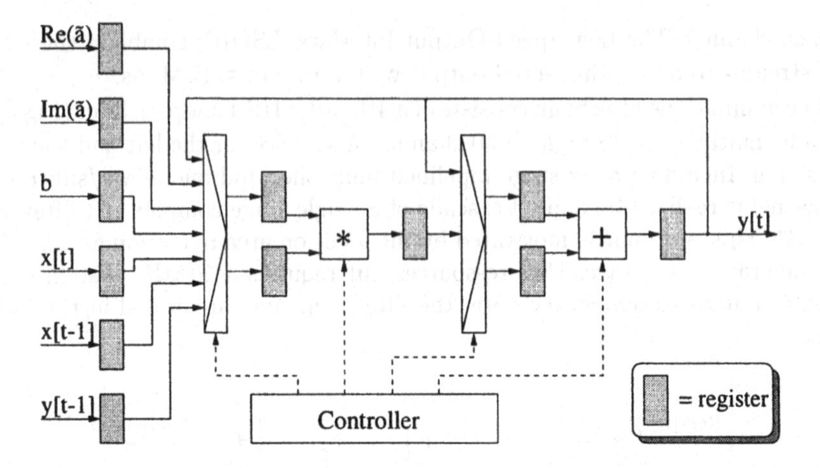

Fig. 4. Architecture of one gammatone filterbank channel

systems where classical numerical parameters like SNR can be applied. It is sufficient to record the impuls responses for each filter parameterized with different internal wordlengths. Fig. 5 shows the mean square error between each of these implementations and the original specification with floating point arithmetic. The choice of a certain maximal square error (e.g. 10^{-3} for all channels) leads directly to the necessary internal wordlength.

For the realization of the nonlinear parts of the system this procedure is not applicable. The only method to determine the optimal wordlengths and to validate the correct function of the model is to simulate various implementations with different wordlengths in a given target application and to observe the influence of the wordlength on the performance of the application.

One possible application of the model is the use as a preprocessing stage for the measurement of objective speech quality [9]. In this application, distorted speech signals are generated by low-bit-rate speech coding-decoding devices ("codecs") such as used in mobile telephony. These codecs produce a speech signal that is fully intelligible and allows almost normal speaker identification, compared to standard telephony. However, they exhibit a clearly reduced speech quality due to their highly nonlinear and/or time-variant algorithms. In listening experiments carried out by the research center of Deutsche Telekom, the speech quality has been subjectively rated by test subjects.

In objective speech quality measurement, the application of psychoacoustical preprocessing models is motivated by the assumption, that subjects are able to judge the quality of a test speech signal by comparing the "internal perceptual representation" of the test sound with that of a reference sound [9–11]. This representation is thought of as the information that is accessible to higher neural stages of perception. It should contain the perceptually relevant features of the incoming sound. Differences in this "internal representation" of input and output signal are expected to correspond to perceivable differences of the two

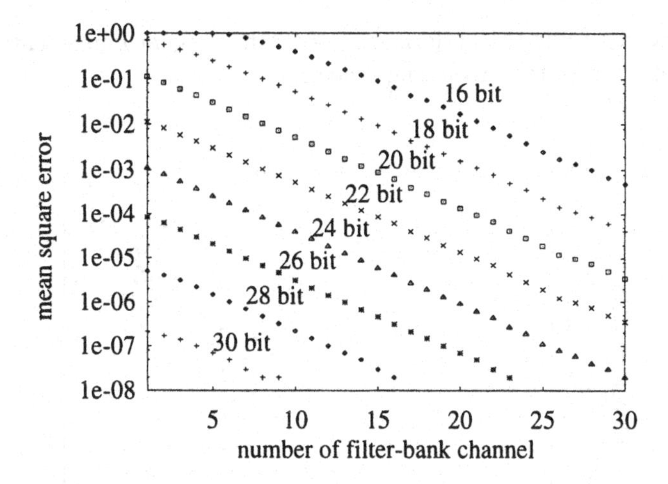

Fig. 5. Mean square error between implementations with different wordlengths and the floating point implementation for each channel of the gammatone filterbank. Note, that due to increased analysis bandwidth the error for a given wordlength decreases with increasing frequency / channel number.

signals and thus to indicate a decreased speech quality of the output signal. In subjective listening tests, typically sentences of different speakers are encoded using the same codec-condition and are rated individually by the subjects. For the objective method, these sentences of different speakers were concatenated and their average mean opinion score (MOS) were calculated in order to reduce the variability of the MOS due to the different voices of the speakers. The original and the distorted signal are then aligned with respect to overall delay and overall RMS. Both signals are then transformed to their internal representations. The objective-subjective speech quality data can be fitted by a monotonic function with only small deviation. A high correlation coefficient is achieved and, in particular, no clusters of different codec types occur. This indicates that the individual signal degradations introduced by the different types of codecs are transformed in a perceptually "correct" way into the internal representations. Therefore, this method can be used to analyze effects of internal wordlength by observing the degradiation of the correlation when decreasing internal arithmetical precision.

Fig. 6, 7 and 8 show the results of the ETSI tests for different combinations of wordlengths. Based on this information it is easy to decide whether the performance of the application is sufficient for a given wordlength. The results are determined at middle signal levels. The derived wordlengths were validated by simulations with input data at different signal levels. Using wordlength N_{opt} (Fig. 7) only small deviations to a floating point implementation occurred whereas for N_{bad} substantial deviations are noticed (Fig. 8). This indicates that the deviations from the original floating point algorithm (Fig. 6) can be assessed quite well by observing the performance of the fixed point implementation in a typical application of the processing scheme.

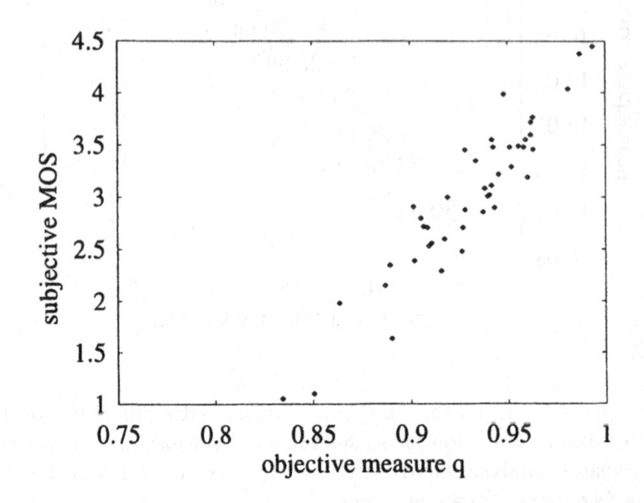

Fig. 6. Results of the objective speech quality measurement with the auditory preprocessing model: Subjective quality (MOS) versus objective measure q for the ETSI Halfrate Selection Test: Floating point version of the model.

5 Results and future work

The design is currently implemented in a XILINX XC40200XV-09. It contains the computation of all necessary data for Chip 2 (IIF, GFB, RAM, ROM, HSOIF, HWR, OKL and LSOIF). Currently the calculation of the amplitude quotient and phase difference (PAC) is not implemented because these computations are not part of the model and the data is not used in algorithms for speech recognition. For hearing aid applications (noise reduction, source detection) they will be added. The current implementation utilizes about 2800 CLBs so that an implementation of the PAC will be possible (the XC40200XV provides about 7000 CLBs).

The design is modelled in VHDL and build up of about 50 entities. Most resources are used by the onchip RAM and the multiplier. Push button synthesis already leads to a maximum system clock of 36 MHz. Detailed analysis shows that the main reason for the delay is the complexity of the finite state machine which controls the filterbank channel. This will have to be optimized in the future.

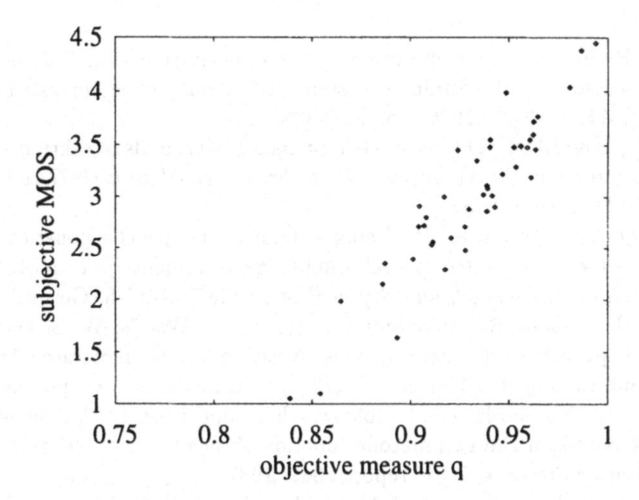

Fig. 7. Results of the objective speech quality measurement with the auditory preprocessing model: Subjective quality (MOS) versus objective measure q for the ETSI Halfrate Selection Test: Fixed Point version of the model with "optimal" wordlength N_{opt}.

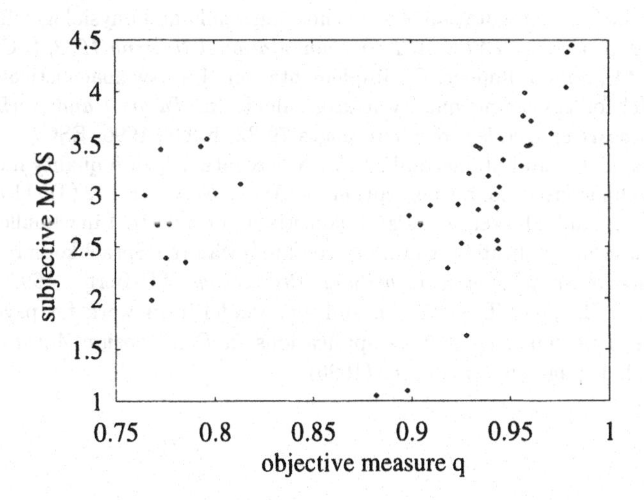

Fig. 8. Results of the objective speech quality measurement with the auditory preprocessing model: Subjective quality (MOS) versus objective measure q for the ETSI Halfrate Selection Test: Fixed Point version of the model with "insufficient" wordlength N_{bad}.

References

[1] Dau, T., Pschel, D. and Kohlrausch, A.: A quantitative model of the "effective" signal processing in the auditory system I. *Journal of the Acoustical Society of Amerika (JASA) 99 (6)*, 1996, pp. 3633-3631.

[2] Tchorz, J., Wesselkamp, M. et al: Gehörgerechte Merkmalsextraktion zur robusten Spracherkennung in Störgeräuschen *Fortschritte der Akustik-DAGA96*, pp532-533, DEGA, Oldenburg, Germany

[3] Hansen, M., and Kollmeier, B.: Using a quantitative psychoacoustical signal representation for the objective speech quality measurement *Proc. ICASSP-97, Intl. Conf. on Acoustics, Speech and Signal Proc.*, p1387, Munich, Germany

[4] Wittkop, T., Albani, S., Hohmann, V., Peissig, J., Woods, W. S. and Kollmeier, B.: Speech processing for hearing aids: Noise reduction motivated by models of binaural interaction *Acustica united with acta acustica*, vol. 83, pp. 684-699, 1997.

[5] Patterson, R., Nimmo-Smith, I, Holdsworth, J, and Rice, P.: An efficient auditory filterbank based on the gammatone function *Appendix B of SVOS Final Report: The auditory Filterbank*, APU report 2341, 1987

[6] Kohlrausch, A, Püschel, D., and Alphei, H.: Temporal resolution and modulation analysis in models of the auditory system *Mouton de Groyer*, pages 85-98, 1992

[7] Schwarz, A., Mertsching, B., Brucke, M, Nebel, W., Hansen M. and Kollmeier, B.: Silicon Cochlea: A digital VLSI implementation of a psychoacoustically and physiologically motivated speech preprocessor. *To appear in "Psychophysics, physiology and models of hearing"*. Proceedings of the symposium "Temporal processing in the auditory system: Psychophysics, physiology and models of hearing", Bad Zwischenahn, 1998

[8] Brucke, M., Nebel, W., Schwarz, A., Mertsching, B., Hansen, M., Kollmeier, B.: Digital VLSI-implementation of a psychoachustically and physiologically motivated speech preprocessor *NATO ASI on Computational Hearing 1998*, Il Ciocco, Italy

[9] Hansen, M., and Kollmeier, B.: Implementation of a psychoacoustical preprocessing model for the sound quality measurement. *In: Tutorial and workshop on the auditory basis of speech perception*, pages 79-82, Keele, 1996, ESCA

[10] Beerends, J. G., and Stemerdink, J. A.: A perceptual speech quality measure based on a psychoacoustic sound perception. *J. Audio Eng. Soc.*, 42 (3): 115-123, 1994

[11] Berger, J., and Merkel, A.: Psychoakustisch motivierte Einzelmaße als Ansatz zur objektiven Qualitätsbestimmung von ausgewählten Sprachkodiersystemen. *In: Elektronische Sprachsignalverarbeitung, Proceedings*, TU-Berlin, 1994

[12] Kollmeier, B., Dau, T. et al: An auditory model framework for psychoacoustics and speech perception - and its applications In *Proc. Forum Acusticum*, volume 82, Suppl. 1, page 89, Antwerpen (1996)

SONIC - A Plug-In Architecture for Video Processing

Simon D. Haynes[1], Peter Y. K. Cheung[1], Wayne Luk[1], and John Stone[2]

[1] Imperial College of Science, Technology and Medicine, London, England
{s.d.haynes,p.cheung}@ic.ac.uk,wl@doc.ic.ac.uk
[2] Sony Broadcast & Professional Europe, Basingstoke, England
john.stone@adv.sonybpe.com

Abstract. This paper presents the SONIC reconfigurable computing architecture and the first implementation, SONIC-1. SONIC is designed to support the software plug-in methodology to accelerate video image processing applications. SONIC differs from other architectures through the use of plug-in Processing Elements (PIPEs) and the Application Programmer's Interface (API). Each PIPE contains a reconfigurable processor, a scaleable router that also formats video data, and a frame-buffer memory. The SONIC architecture integrates multiple PIPEs together using a specialised bus structure that enables flexible and optimal pipelined processing. SONIC-1 communicates with the host PC through the PCI bus and has 8 PIPEs. We have developed an easy to use API that allows SONIC-1 to be used by multiple applications simultaneously. Preliminary results show that a 19 tap separable 2-D FIR filter implemented on a single PIPE achieves processing rates of more than 15 frames per second operating on 512 x 512 video transferred over the PCI bus. We estimate that using all 8 PIPEs, we could obtain real-time processing rates for complex operations such as image warping.

1 Introduction

Reconfigurable platforms are often designed with little consideration for *how* the board will be used and for *what* purpose. The resulting platforms can be very general [1,2], but performance in specific domains can often be compromised in favour of overall modest acceleration. General platforms are also inherently more difficult to integrate into the software environment efficiently, requiring very general APIs. We recognise that perhaps the biggest issue in custom computing machines is one of software integration: developers will not use a platform, no matter how good it is, if the software interface is poor.

Our architecture, SONIC, is specifically targeted for video image processing tasks. Focusing the application domain in this way means that greater acceleration can be achieved than would be the case for a more general architecture. This also simplifies the Application Programmer's Interface (API).

The SONIC architecture uses the software plug-in architecture. The use of plug-in architectures for reconfigurable processing is not new[3], but the novelty of SONIC lies with the fact that the SONIC architecture was designed specifically for this programming methodology.

When designing SONIC our starting point was the software model. The SONIC architecture has also been developed to simplify the interface between the software and hardware. First we give the software designer a simple, easy to understand software model. Second, the designer of the accelerating hardware is given as much abstraction from the detail of the implementation as possible.

2 Requirements of Video Image Processing

In order to develop a reconfigurable architecture suited to video image processing, it is first necessary to have an understanding of the requirements of typical video image processing tasks. Video image processing in this context means tasks such as image warping, in addition to more typical examples of image processing, such as filtering, and edge detection.

It is well known that image processing, particularly video image processing, is a suitable candidate for hardware acceleration. This is due to two reasons; (i) the large amounts of parallelism, and (ii) the relatively simple nature of the operations required.

Video image processing is typified by high data rates (187.5 Mbytes/sec for real time HDTV), making an efficient method of data transferral between host and platform important. The memory system must also be able to cope with the high data rates. Other work, such as the P^{3}I [4], has emphasised the need for clean efficient memory system when handling video images.

Many video image processing tasks can be decomposed into pipelined sub-operations. For example, a separable 2-D FIR filter can be implemented using two 1-D FIR filters, turning the image through 90° in-between. [5] also shows that pipelined processing and specialised datapaths are important architectural features for image processing. To give good performance, the SONIC architecture should be able to exploit this kind of pipelined, stream-based processing, as platforms like Splash-2 [1] do.

Video image processing tasks can also often be separated into two distinct information paths: One is a high bandwidth datapath, the other a low bandwidth path. The high bandwidth path performs simple operations on the stream of image data, such as interpolation, the low bandwidth path provides the parameters for the operations (normally originating from user input). For operations such as rotation or filtering the parameters can simply be one or two numbers: for more complex operations, such as image warping, they can be a number of vectors. The parameters may change from frame to frame, or over a single image. The generation of the parameters often requires floating point and other complex operations best suited to general purpose microprocessors.

3 The SONIC Software Model

Software plug-ins are widely used in applications such as Adobe® Photoshop® and Adobe® Premiere®. Plug-ins are used to enable third-party vendors to extend applications. Plug-ins such as filters and transformations can be invoked by the application as required. The application further benefits from the more structured style of code development that the architecture encourages.

In order to make software acceleration a practical proposition for application developers, it is beneficial to disguise the fact a reconfigurable platform is being used. Hardware acceleration can be embedded within software plug-ins, without the application designer ever knowing. Indeed, hardware acceleration can be used *after* the application has been written.

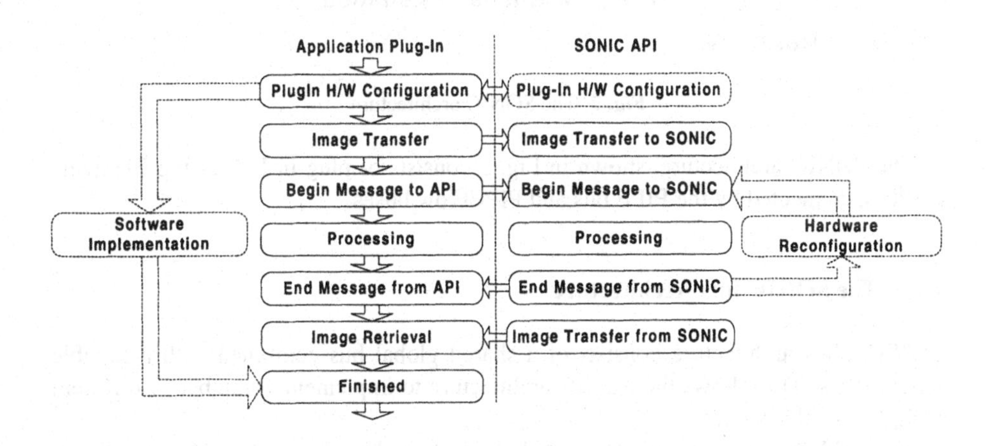

Fig. 1. Software Model for SONIC

Fig. 1 shows the SONIC architecture's software model. The plug-in contains a pure software implementation (used if the board is busy or not present), and a hardware description file. The hardware description file encapsulates the hardware description of the plug-in. It contains the configuration data for one or more FPGAs.

The API includes functions that handle PIPE resource allocation and scheduling in a transparent way. For example, PIPE caching implements the virtual hardware, that minimises the overhead required for configuration.

We believe that the actual hardware design for the plug-in will usually have to be done by a hardware designer. This is because the hardware generated from a software description tends to be inefficient at present [6]. We certainly do not preclude automatic hardware generation from a software description; it is simply that we recognise the need for making SONIC simple to understand for anyone designing plug-in hardware.

4 Architecture of the SONIC Platform

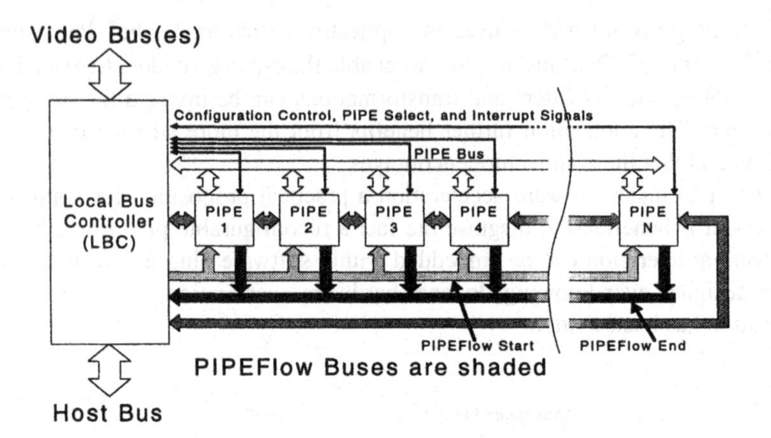

Fig. 2. The SONIC architecture

The SONIC architecture, shown in Fig. 2 consists of plug-in Processing Elements (PIPEs), connected by the PIPE bus and PIPEFlow buses.

4.1 The SONIC Bus Architecture

SONIC's bus architecture consists of a shared global bus combined with a flexible pipeline bus. This allows the SONIC architecture to implement a number of different computational schemes.

The PIPE bus is a synchronous, global bus that should be matched to the bandwidth of the host bus. This bus is used for *fast image transfer* to the memory on the PIPEs, *PIPE Parameter access* (run-time data required by the PIPEs), *control of the routing* of the PIPEFlow bus through the PIPEs, and *configuration* of the PIPEs (if a high configuration bandwidth is required).

Each PIPE has a number of unique control signals that are used for configuration control, interrupt signalling and device selecting.

The SONIC architecture implements pipelined operation using the PIPEFlow buses. Data passes along the pipeline using the PIPEFlow buses connecting adjacent PIPEs. The PIPEFlow-Start bus can be used to get data to the start of the pipeline, and the PIPEFlow-End bus to retrieve the data from the end. Data are sent over these buses using a pre-defined 'raster-scan' protocol.

4.2 The PIPE (Plug-In Processing Element)

The PIPEs are the most important part of the SONIC architecture; they perform the processing. Each PIPE consists of the three conceptual parts shown in Fig. 3: the PIPE Router (PR), PIPE Engine (PE), and PIPE Memory (PM).

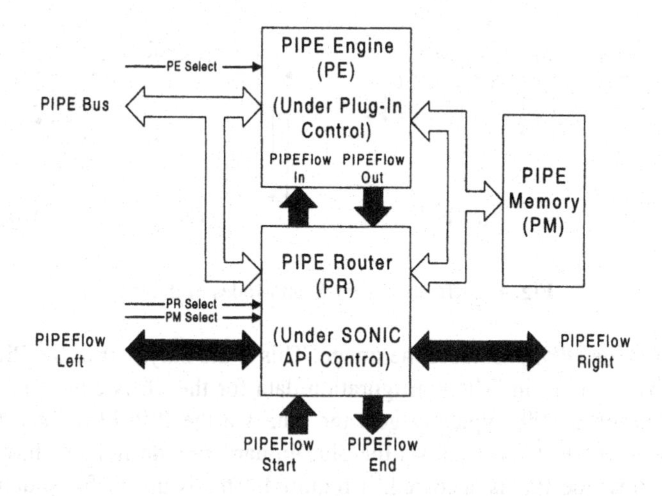

Fig. 3. Architecture of the PIPE

The architecture of the PIPEs means that computation, handled by the PE, is separated from the movement and formatting of the image data, which the PR handles. The PE is controlled by the plug-in; the PR by the API. It is the PR, and the way that it is used, which makes SONIC unique.

The PIPE Router (PR) provides a flexible, scaleable solution to routing and data formatting for the SONIC architecture. The PR is responsible for three tasks: (i) Accessing of the PM by PIPE Bus, (ii) Generating the PIPEFlow-In data for the PE, and (iii) Handling the PIPEFlow Out data from the PE.

The PR is used to present the image data to the PE in the format in which the plug-in hardware expects it. There are three elements to this:

Data Format - The PR is responsible for ensuring that the data are in the correct format for the plug-in in the PE. For example, if the plug-in in the PE is designed to operate with YCrCb components and is pipelined to the previous PIPE that generates RGB components, the PR must perform the YCrCb to RGB conversion. The PR could also support conversions from formats such as HSV.
Data Locations - The PR must route the data from the correct place. Not only can the PR route the data from one of the PIPEFlow buses, but PIPEFlow data could also be routed to or from the PM. This means that precisely the same plug-in can be used either as a single entity, or form part of a larger chain of PIPEs with the data coming from the previous PIPE.
Data Access - The PR is capable of supplying the data to the PE in a variety of ways, as shown in Fig. 4. Simple operations, such as gamma correction, can be carried out using the normal horizontal scan mode. The normal vertical scan mode allows designers to easily implement two-pass algorithms. The more complicated 'stripped' accessing greatly simplifies the design of 2-D Filters, and block processing algorithms.

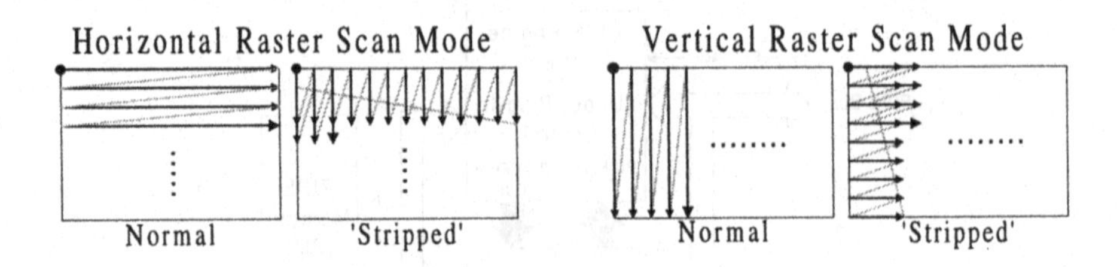

Fig. 4. Different Raster-Scan Modes of the PR

The PIPE Engine (PE) processes the image. This is the only part of the PIPE directly configured by the plug-in. The configuration data for the PE is contained within the plug-in. Although the PE typically gets the data via the PIPEFlow bus, the PE has direct access to the PM. This allows the plug-in hardware designer to have complete control over how the PM is accessed, if required. This is useful for situations where the image must be accessed randomly (explosion effects, for example).

The PIPE Memory (PM) can be used for image storage and manipulation. If the plug-in hardware designer does not use the PM, the SONIC architecture allows the PR to use the PM for image storage by the API.

4.3 Different Implementations of the PIPE

The actual implementation of the PIPE could take many forms. First, despite conceptually consisting of three parts (the PR, PE and PM), the implementation of the PIPE could consist of just one device or even many devices. Second, although the original intention is clearly to use reconfigurable logic, the PE and/or PR could be implemented with a DSP processor, or customised ASIC.

5 Integration of the SONIC Architecture with the Software Model

Fig. 5 shows how the SONIC platform architecture complements our software model. In this example the SONIC platform contains three plug-ins. PIPE 1 contains plug-in

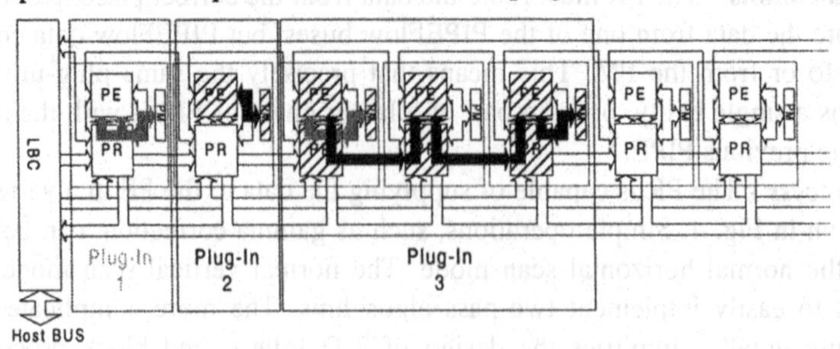

Fig. 5. SONIC Platform with three plug-ins

1, which is currently unused (although plug-in 1 is unused the PIPE still remains configured; so if plug-in 1 should be re-used then there is no need for the API to reconfigure the PIPE). Plug-in 2 is implemented in PIPE 2. The PE for plug-in 2 is accessing the PM directly. Plug-in 3 shows how a larger plug-in can be implemented using multiple PIPEs, in this instance passing data via the PIPEFlow buses. More complex plug-ins can easily be designed by cascading smaller plug-ins which use the PIPEFlow buses. The remaining PIPEs are free to be used to implement further plug-ins.

6 Implementation of the SONIC Architecture

A photograph of our implementation of the SONIC architecture, SONIC-1, is shown in Fig. 8. We implement the PIPEs using daughter board modules that are inserted into the 200 pin DIMM sockets on the main board. The modularity of the design is beneficial for several reasons: (i) easier development, (ii) improved device density on the board, (iii) easier testing (a board with headers for a logic analyser was made which could be inserted in place of a PIPE), and (iv) allowing for future expansion, by allowing different devices to be used in the PIPEs

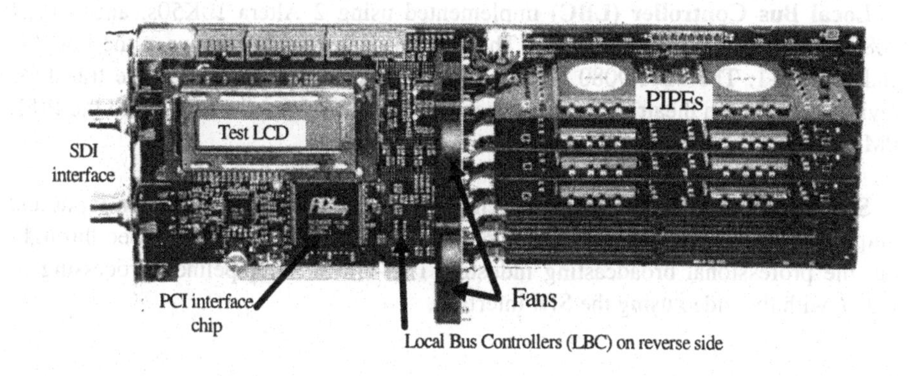

Fig. 6. SONIC-1 platform, with four PIPEs

Implementation of the PIPE

Since we use Altera parts, which cannot be partially reconfigured, it was necessary to place the PR and PE in separate devices (a FLEX10K70 for the PE and a FLEX10K20 for the PR). The 10K70 can be clocked at 33 or 66Mhz.

The PM consists of 4Mbytes of SRAM arranged as 1M x 32 bits. The bandwidth of the PM is 132MB/s, which matches that of the PIPE Bus and the PCI bus, and is twice that of the PIPEFlow bus.

There are also 22 bit bi-directional connections between the PEs of adjacent PIPEs (utilising the remaining device pins), which can be used when multiple PIPEs are combined together.

Implementation of the buses

The PIPE Bus is implemented as a 32 bit multiplexed address/data bus (plus 4 control signals). It is capable of matching the maximum bandwidth of the PCI bus (132MB/s).

The PIPEFlow bus are 19 bits in width (16 bit data + 3 control bits) and operate at (66MB/s). This bandwidth is half that of the PM, so it is possible to read *and* store PIPEFlow data to the same PM. Pin availability on the PIPE and the PR placed the limitation on the size of this bus. Because 8-bit RGBα data is typically used, this bus is time multiplexed between RG & Bα components.

SONIC-1 contains hardware dedicated to smoothly interfacing the PIPEs to the SONIC API through the host PCI bus. It also contains a SDI interface that can be used as an image data stream interface independently from the PCI bus. The elements of the main SONIC board are:

Local Bus Controller (LBC) implemented using 2 Altera 10K50s, and a PLX 9080 to interface with the PCI bus. The PCI bus transfers data between the host PC and SONIC-1. The PLX 9080 PCI interface chip can support burst mode transfers, giving a maximum theatrical transfer speed of 132MB/s from the host PC to the PIPE PM.

Serial Digital Interface (SDI) which can be used simultaneously as an input and output for video independently of the PCI bus. This interface is widely used throughout the professional broadcasting industry. This allows for pipelined processing of video, with the video using the SDI interface.

7 Example: 19 Tap Separable 2-D FIR Filter for Adobe® Premiere®

A 2-D separable filter can be implemented using two 1-D FIR Filters, processing occurring once in the horizontal direction, and once in the vertical direction. Rather than use two filters, we use a single 1-D FIR Filter twice.

The important point is how little the hardware designer needs to know about the SONIC platform; they will receive a stream of data through the PIPEFlow- In port, and must send the processed data out using the PIPEFlow-Out port.

The fragment of the 'C' code for the filter can be seen below. *Precisely* the same hardware design for the PE can be used for both the horizontal and vertical passes, since it is the PR that accesses the data differently.

```
Sonic_Conf(&hPIPE, "SEP_2D_FIR_FILTER.RBF");//Configure PIPE
Sonic_PR_ImageSize_Write(hPIPE, Width,Height); //Setup PR
Sonic_PR_Route_Write(hPIPE, PR_TO_AND_FROM_PM);
Sonic_PM_Write(hPIPE, pSrcImage); //Write source Image
Sonic_PR_ImageMode_Write(hPIPE,PR_HORIZONTAL_RASTER);
Sonic_PR_Pipeflow_Write(hPIPE,PR_PROCESS);  //Start
//Process the frame using horizontal raster scan
do {
    Sonic_PR_Pipeflow_Read(hPIPE,Done);
} until (Done & FINISHED); //Wait for PE to finish

//Setup PR for the vertical pass
Sonic_PR_ImageMode_Write(hPIPE,PR_VERTICAL_RASTER);
Sonic_PR_Pipeflow_Write(hPIPE,PR_PROCESS);

do {
    Sonic_PR_Pipeflow_Read(hPIPE,Done);
    } until (Done & FINISHED);

Sonic_PM_Read(hPIPE,pDstImage); //Read back filtered image
Sonic_Unlock_PIPE(hPIPE); //Free PIPE
```

Software Code Fragment implementing a separable 2D filter using SONIC

The usefulness of the API only configuring the PE when strictly necessary is emphasised by this plug-in. Adobe® Premiere® loads the plug-in in for each *frame*. However, because the API leaves the PIPE configured when the plug-in finishes, the PIPE is only configured once.

We ran Adobe® Premiere® on a 300MHz Pentium II machine using a sequence of fifty 576x461 frames. The results can be seen in Table 1.

Table 1. Perfomance of the Seperable 2-D Filter

	Processing Time (PT)	Framework Time (FT)	Total Time PT+FT
SONIC-1	3.8s	21.6s	35.4s
Software	117.4s	21.6s	139.0s
Processing Speed-Up	**30.9x**		
Total Speed-Up			**5.5x**

The time taken to process the sequence can be split into two times: *Processing Time* - the time actually spent in the plug-in, and *Framework Time* - the time which Premiere requires to prepare each frame (since the frames are stored in a compressed format). The processing speed up is 30 times, although the Premiere framework overhead has reduced the total speedup to 5.5 times. When using the SONIC platform in other applications where no such compression takes place, we would expect to see ≈30 times speedup overall.

Another important fact is that this plug-in only used a single PIPE. Assuming a linear speed up as more PIPE are used, a more complicated plug-in using all 8 available PIPEs would expect to achieve speedups of around 250 times.

8 Conclusions

The uniqueness of the SONIC architecture is due to the PIPE Router (PR) and the API. The API enables resource allocation and scheduling which is invisible to the API user, whilst conforming to a simple software model. The PR simplifies the reconfigurable hardware design process by carrying out the image transferral and conversion necessary to give the PIPE Engine (PE) the correct data in the correct format. The SONIC architecture also demonstrates:

- The advantages of having a reconfigurable platform with a good software model.
- That software plug-ins are particularly suited to reconfigurable platforms.
- Simple simultaneous use of reconfigurable hardware by multiple applications.
- That PIPEs can be pipelined together to create complex plug-ins.
- Good flexibility and expandability.

We have demonstrated that our implementation (SONIC-1) gives impressive performance, and have used the software plug-in methodology to accelerate Adobe® Premiere®. The development of plug-ins for other software is underway.

Future work which we would like to carry out includes: building a library of hardware components which can be used by the designers for the basis of new designs, developing more benchmark plug-ins for various applications, refining of the design flow, and improving the API to enable more sophisticated scheduling of the reconfiguration of the PIPEs.

We gratefully acknowledge the support of the UK Engineering and Physical Sciences Research Council, and Sony Broadcast & Professional Europe.

References

1. Athanas, P.M. and Abbott, A.L.: Real-Time Image Processing on a Custom Computing Platform, IEEE Computer, Vol. 28, Issue 2, (Feb 1995) pp. 16-24
2. Mackinlay, P.I., Cheung, P.Y.K, Luk W. and Sandiford, R.D.: Riley-2: A flexible platform for codesign and dynamic reconfigurable computing research, Field-Programmable Logic and Applications, W. Luk, P.Y.K. Cheung and M. Glesner (editors), LNCS 1304, Springer (1997) pp. 91-100
3. Singh, S. and Slous, R.: Accelerating Adobe Photoshop with Reconfigurable Logic, IEEE Symposium on FPGAs for Custom Computing Machines, April 15th - 17th (1998) pp. 236 - 244
4. Colaïtis, M.J., Jumpertz, J.L., Guérin, B., Chéron, B., Battini, F., Lescure, B., Gautier, E. and Geffroy, J-P.: The Implementation of P^{3}I, a Parallel Architecture for Video Real-Time Processing: A Case Study, Proceedings of the IEEE, Vol. 84, No. 7, July (1996) pp 1019-1037
5. Kung, H.T.: Computational Models For Parallel Computers, Scientific applications of Microprocessors, R.J. Elliot & C.A.R. Hoare, Prentice Hall (1989) pp 1-15
6. Hudson, R.D., Lehn, D.I and Athanas, P.M.: A Run-Time Reconfigurable Engine for Image Interpolation, IEEE Symposium on FPGAs for Custom Computing Machines, April 15th - 17th (1998) pp. 88 - 95

DRIVE: An Interpretive Simulation and Visualization Environment for Dynamically Reconfigurable Systems*

Kiran Bondalapati and Viktor K. Prasanna

Department of Electrical Engineering Systems
University of Southern California
Los Angeles, CA 90089-2562, USA
{kiran, prasanna}@ceng.usc.edu
http://maarc.usc.edu/

Abstract. Current simulation tools for reconfigurable systems are based on low level simulation of application designs developed in a High-level Description Language(HDL) on HDL models of architectures. This necessitates expertise on behalf of the user to generate the low level design before performance analysis can be accomplished. Most of the current simulation tools also are based on static designs and do not support analysis of dynamic reconfiguration.
We propose a novel interpretive simulation and visualization environment which alleviates these problems. The **D**ynamically **R**econfigurable systems **I**nterpretive simulation and **V**isualization **E**nvironment(**DRIVE**) framework can be utilized for performance evaluation and architecture and design space exploration. *Interpretive* simulation measures the performance of an application by executing an abstract application model on an abstract parameterized system architecture model. The simulation and visualization framework is being developed in *Java* language and supports modularity and extensibility. A prototype version of the **DRIVE** framework has been implemented and the complete framework will be available to the community.

1 Introduction

Reconfigurable systems are evolving from rapid prototyping and emulation platforms to a general purpose computing platforms. The systems being designed using reconfigurable hardware range from FPGA boards attached to a microprocessor to systems-on-a-chip having programmable logic on the same die as the microprocessor. Reconfigurable systems have been utilized to demonstrate large speed-ups for various classes of applications. Architectures are being designed which support partial and dynamic reconfiguration. The reconfiguration overhead to change the functionality of the hardware is also being diminished by

* This work was supported by the DARPA Adaptive Computing Systems Program under contract DABT63-96-C-0049 monitored by Fort Hauchuca.

the utilization of configuration caches and multiple contexts on the same device. Compilation of user level programs onto reconfigurable hardware is also being explored.

The general purpose computing area is the most promising to achieve significant performance improvement for a wide spectrum of applications using reconfigurable hardware. But, research in this area is hindered by the absence of appropriate techniques and tools. Current design tools are based on ASIC CAD software and have multiple layers of design abstractions which hinder high level optimizations based on reconfigurable system characteristics. Existing frameworks are either based on simulation of HDL based designs [1, 11, 13] or they are tightly coupled to specific architectures [5, 9, 14](See Section 1.1). It is also difficult to incorporate dynamic reconfiguration into the current CAD tools framework. Simulation tools provide a means to explore the architecture and the design space in real time at a very low resource and time cost. The absence of mature design tools also impacts the simulation environments that exist for studying reconfigurable systems and the benefits that they offer. System level tools which analyze and simulate the interactions between various components of the system such as memory and configurable logic are limited and are mostly tightly coupled to specific system architectures.

In this paper we present a novel interpretive simulation and visualization environment based on modeling and module level mapping approach. The **D**ynamically **R**econfigurable systems **I**nterpretive simulation and **V**isualization **E**nvironment (**DRIVE**) can be utilized as a vehicle to study the system and application design space and performance analysis. Reconfigurable hardware is characterized by using a high level parameterized model. Applications are analyzed to develop an abstract application task model. *Interpretive* simulation measures the performance of the abstract application tasks on the parameterized abstract system model. This is in contrast to simulating the exact behavior of the hardware by using HDL models of the hardware devices.

The **DRIVE** framework can be used to perform interactive analysis of the architecture and design parameter space. Performance characteristics such as total execution time, data access bandwidth characteristics and resource utilization can be studied using the **DRIVE** framework. The simulation effort and time are reduced and systems and designs can be explored without time consuming low level implementations. Our approach reduces the semantic gap between the application and the hardware and facilitates the performance analysis of reconfigurable hardware. Our approach also captures the simulation and visualization of dynamically reconfigurable architectures. We have developed the Hybrid System Architecture Model(HySAM) of reconfigurable architectures. This model is currently utilized by the framework to map applications to a system model.

An overview of our framework is given in Section 2. Various aspects of the simulation and visualization framework including our Hybrid System Architecture Model(HySAM) are described in detail in Section 3. Conclusions and future work are discussed in Section 4.

1.1 Related Work

Several simulation tools have been developed for reprogrammable FPGAs. Most tools are device based simulators and are not system level simulators. The most significant effort in this area has been the Dynamic Circuit Switching(DCS) based simulation tools by Lysaght et.al. [13]. Luk et.al. describe a visualization tool for reconfigurable libraries [11]. They developed tools to simulate behavior and illustrate design structure. CHASTE [5] was a toolkit designed to experiment with the XC6200 at a low level. There are other software environments such as CoDe-X [9], JHDL [1], HOTWorks [7], Riley-2 [14], etc.

These tools study the dynamically reconfigurable behavior of FPGAs and are integrated into the CAD framework. Though the simulation tools can analyze the dynamic circuit behavior of FPGAs, the tools are still low level. The simulation is based on CAD tools and requires the input design of the application to be specified in VHDL. The parameters for the design are obtained only after processing by the device specific tools. Most of the software frameworks do not support system level analysis and are utilized for for low level hardware design and evaluation.

2 DRIVE Overview

Figure 1 shows an overview of our framework. The system architecture can be characterized to capture the parameter space which affects the performance. The implementations of various optimized modules can be encapsulated by characterizing the performance of the module with respect to the architecture. This characterization is partitioned into the *capabilities* of the system and the actual *implementations* of these *capabilities*. The application is not mapped onto a low level design but is analyzed to develop an application task model. The application model can exploit the knowledge available in the form of the system *capabilities* provided by the module characterization. Algorithmic techniques are utilized to map the application task model to the system models, to perform interpretive simulation and obtain performance results for a given set of parameter values.

Interpretive simulation is performed on the system model which permits a higher level abstract simulation. The application does not need to be actually executed by using device level simulators like HDL models of the architectures. The performance measures can be obtained in terms of the application and model parameters and system characteristics. An interpretive simulation framework will permit design exploration in terms of the architectural choices, application algorithm options, various mapping techniques and possible problem decomposition onto the system components. Development of all the full blown designs which exercise these options is a non-realizable engineering task. Simulation, estimation and visualization tools can be designed to automate this exploration and obtain tangible results in reasonable time.

The abstractions and the techniques that are developed are enclosed in the dashed box in Figure 1. Verification of the models, mapping techniques and

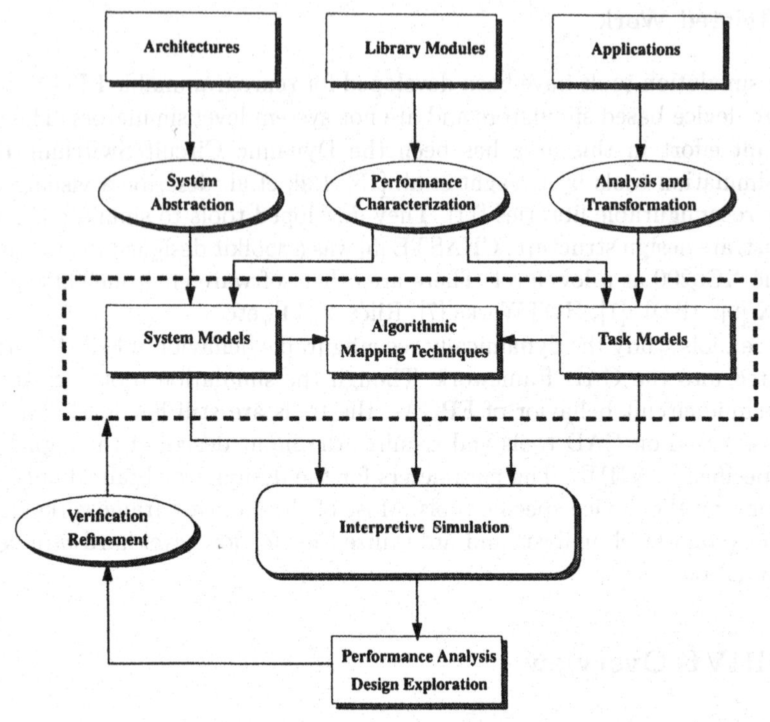

Fig. 1. DRIVE framework

simulation framework can be performed by mapping some designs onto actual architectures. This verification process can be utilized to expand on the abstraction knowledge and refine the various models and techniques that are developed. The verification and refinement process completes the feedback loop of the design cycle to result in final accurate models and efficient techniques for optimal designs.

3 Simulation Framework

The simulation framework consists of abstractions and algorithmic techniques as discussed in Section 2(Fig. 1). A high level model of reconfigurable hardware is needed to abstract the low level details. Existing models supplied by the CAD tools have either multiple abstraction layers or are very device specific. We have developed a parameterized model of configurable computing system, which consists of configurable logic attached to a traditional microprocessor. Our model cleanly partitions the *capabilities* of the hardware from the *implementations* and presents a very clean interface to the user. The algorithmic techniques for mapping are not the focus of this paper. Some algorithms for mapping based on the HySAM model are described in our prior work [3, 4].

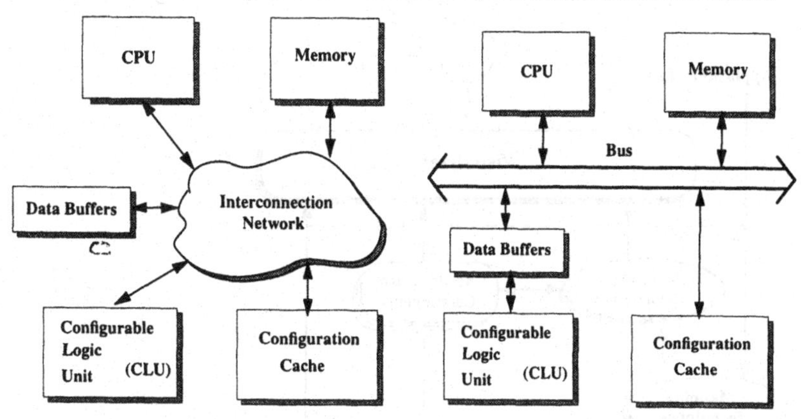

Fig. 2. Hybrid System Architecture and an example architecture

3.1 Hybrid System Architecture Model(HySAM)

The *Hybrid System Architecture Model* is a general model consisting of a traditional microprocessor with additional Configurable Logic Unit(CLU). Figure 2 shows the architecture of the HySAM model and an example of an actual architecture. The architecture consists of a traditional microprocessor, standard memory, configurable logic, configuration memory and data buffers communicating through an interconnection network.

We outline some of the parameters of the Hybrid System Architecture Model (HySAM) below.

- F : Set of functions $F_1 \dots F_n$ which can be performed on configurable logic. (*capabilities*)
- C : Set of possible configurations $C_1 \dots C_m$ of the Configurable Logic Unit. (*implementations*)
- A_{ij} : Set of attributes for implementation of function F_i using configuration C_j.
- R_{ij} : Reconfiguration cost in changing configuration from C_i to C_j.
- G : Set of generators which abstract the composition of configurations to generate more configurations.
- B : Bandwidth of the interconnection network (bytes/cycle).
- N_c : The number of configuration contexts which can be stored in the configuration cache.
- k_c, K_c : The cost of accessing configuration data from the cache and external memory respectively (cycles/byte).
- k_d, K_d : The cost of accessing data from the cache and external memory respectively (cycles/byte).

The functions F and configurations C have a *many-to-many* relationship. Each configuration C_i, can potentially contain more than one function F_j. In the HySAM model, only function can be active in a configuration at any given time. Each function F_i can be executed by using any one configuration from a subset of the configurations. The different configurations might be generated by different tools, libraries or algorithms. These configurations might have different area, time, reconfiguration cost, precision, power, etc. characteristics.

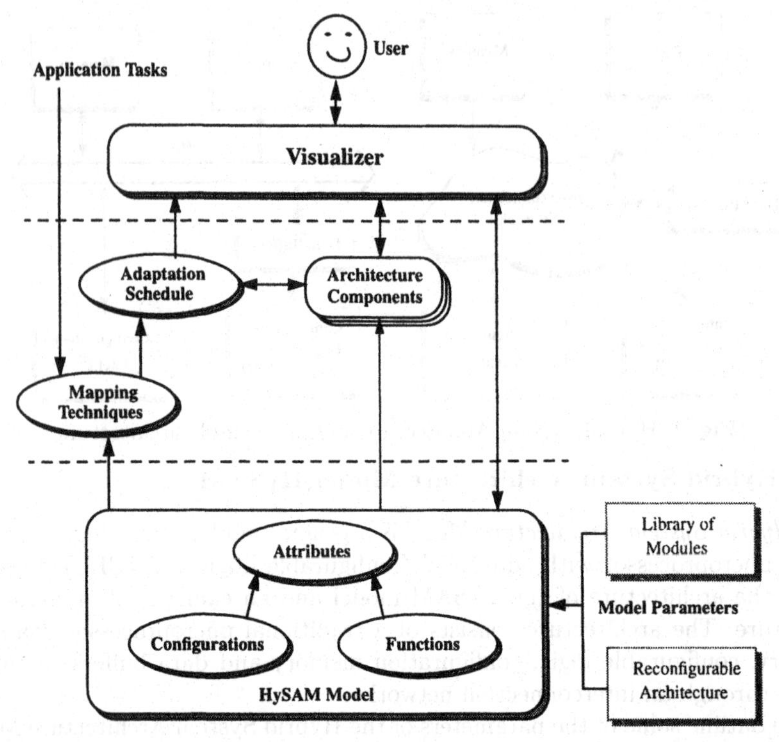

Fig. 3. Major components in the **DRIVE** framework and the information flow

The attributes A define the relationship between the functions and the configurations. The attributes define values such as the execution time and the data accessed during execution of a function in a configuration etc. For example, the different execution times and the different data input patterns when a multiplier is implemented as a bit parallel versus a bit serial multiplier are defined by the attributes. The reconfiguration costs R define the costs involved in changing the configuration of the CLU between two configurations. This cost can be statically evaluated based on the configuration information for different configurations. The cost can also be computed dynamically when the configurations are constructed dynamically.

3.2 DRIVE Framework Implementation

An overview of the major components in the **DRIVE** framework and their interactions is given in Figure 3. The framework utilizes high level models of reconfigurable hardware. The current prototype uses the HySAM model described in Section 3.1.

The main input requirements to the **DRIVE** framework are the model parameters and the application tasks. The model parameters supply information about the Functions, Configurations, Attributes and the Reconfiguration costs. The user can visualize and update any of the instantiated parameters to explore the design space. For a given model parameters, performance results can be obtained for any set of application tasks with various algorithmic mapping techniques.

The high level model partitions the description of the hardware into two components: the Functions(*capabilities*) of the hardware and the Configurations(*implementations*). For example, ability of the hardware to perform multiplication is a capability. The implementations are the different multiplier designs available with varying characteristics such as area, time, precision, structure, etc. Components from a library or modules form the *implementations* in the model and can be determined for different architectures. Vendors and researchers have developed parameterized libraries and modules optimized for a specific architectures. The proposed framework can exploit the various efforts in design of efficient and portable modules [6,12,15]. The framework can incorporate such knowledge as the parameters for the HySAM model.

The user only needs to have a knowledge of the *capabilities*. The application task model consists of specification of the application in terms of the Functions(*capabilities*). The input to the framework consists of a directed acyclic graph of the application tasks specified with the Functions as the nodes of the graph. The edges denote the dependencies between the tasks. This technique reduces the effort and expertise needed on the part of the user. The application need not be implemented as an HDL design by the user to study the performance on various reconfigurable architectures. Automatic compilation efforts [2] can be leveraged to generate the Functions from high level language application programs.

Algorithmic mapping techniques are then utilized to map the application specification to actual implementations. These techniques map the *capabilities* to the *implementations* and generate a sequence of configuration, execution, and reconfiguration steps. This is the *adaptation schedule* which specifies how the hardware is adapted during the execution of the application. The schedule contains a sequence of configurations($C_1 \ldots C_q$) where each configuration $C_i \in C$. This *adaptation schedule* can be computed statically for some applications by using algorithmic techniques. Also, the simulation framework can interact with the model and the mapping algorithms to determine the *adaptation schedule* at run-time.

The interpretive simulation framework is based on module level parameterization of the hardware. The user can analyze the performance of the architecture for a given application by supplying the parameters of the model and the application task. Typically the architectural parameters for the model are supplied by the architecture designer and the library designer. But, the user can modify the model parameters and explore the architecture design space. This provides the ability to study design alternatives without the need for actual hardware. The simulation and the performance analysis are presented to the user through a Graphical User Interface. The framework supports incorporation of additional information in the configurations(C) which can be utilized for actual execution or simulation. It can contain configuration bitstreams or class descriptions which can be utilized to perform actual configuration of hardware or simulation using low level models. Using this information, it is possible to link the abstract definitions to actual implementations to verify and refine the abstract models.

The parameters and attributes of the model can also be evaluated and adapted at run-time to compute the required information for scheduling and visualization. For example, reconfiguration costs can be determined by computing the difference in the configuration information and configurations can even be generated dynamically by future integration of tools like JBits [10]. It is assumed currently that the attributes for configurations are available a priori. It is easy to integrate simulation tools which evaluate the attributes such as execution time by performing simulations as in various module generators [1, 6, 15]. These simulations are based on module generators which do not require mapping using time consuming CAD tools. Once the attribute information for low level modules are obtained by initial simulations and implementations, the attributes for higher level modules can be simulated or computed without the intervention of CAD tools.

The **DRIVE** framework has been designed using object-oriented methodology to support modification and addition to the existing components. The framework facilitates addition of new architectural models, algorithmic mapping techniques, performance analysis tools, etc. in a seamless manner. The framework can also be interfaced to existing tools such as parameterized libraries(Xilinx XBLOX, Luk et. al. [12]), module generators(PAM-Blox [15], Berkeley Object Oriented Modules [6], JHDL [1]), configuration generators(JBits [10]), module interfaces(FLexible API for Module-based Environments [8]), etc. The components of the framework will be made available to the community to facilitate application mapping and modular extensions.

3.3 Visualization

The visualizer for the framework has been developed using the *Java* language AWT toolkit. A previous version of the visualizer was developed using Tcl/Tk. The C programming language was utilized for implementing the simulation engine. The current prototype has been developed in *Java* to utilize the object oriented framework and make the framework modular and easily extensible. Implementing the visualizer and the interpretive simulation in the same language provides for a clearer interface between the components. *Java* is becoming the language of choice for several research and implementation efforts in hardware design and development [1, 6, 10]. Incorporating the results and abstractions from other research efforts is simplified using the current version.

The visualizer acts as a graphical user interface to support the full functionality of the framework. It is implemented as a separate *Java* class communicating with the remaining classes. Any component of the simulation or visualizer framework can be completely replaced with a different component supporting the same interface. The visualizer is oblivious of the algorithmic techniques and implementation details. It accesses information from the different components in the simulation framework on an event by event basis and displays the state

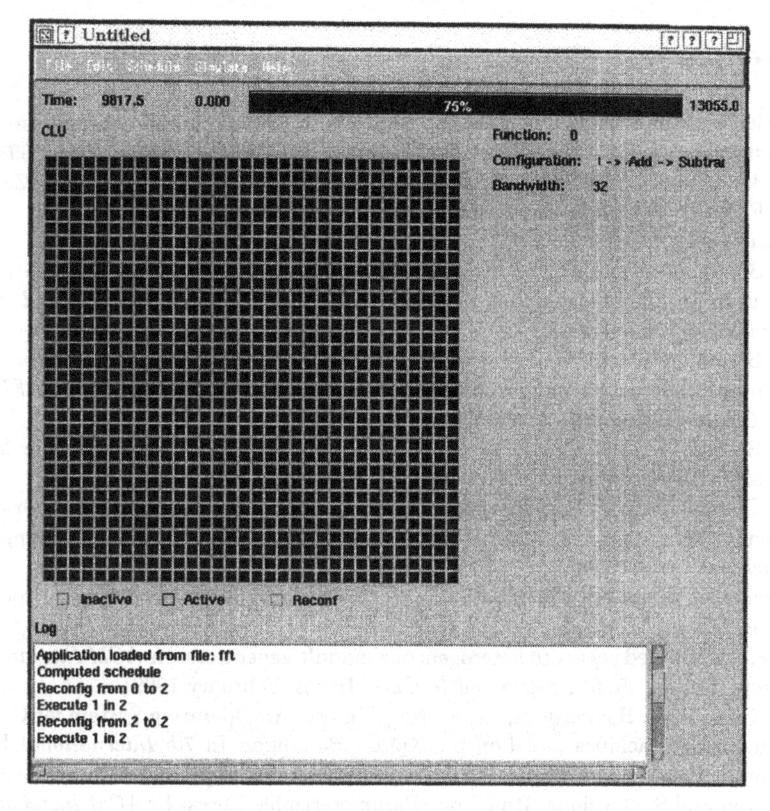

Fig. 4. Sample **DRIVE** visualization

of the various architecture components and the performance characteristics. A sample view of the visualizer is shown in Figure 4.

4 Conclusions

Software tools are an important component of reconfigurable hardware development platforms. Simulation tools which permit performance analysis and design space exploration are needed. The utility of current tools for reconfigurable hardware design is limited by the required user expertise in multiple domains. We have proposed a novel *interpretive* simulation and visualization environment which supports system level analysis. The **DRIVE** framework supports a parameterized system architecture model. Algorithmic mapping techniques have been incorporated into the framework and can be extended easily. The framework can be utilized for performance analysis, design space exploration and visualization. It is implemented in the *Java* language and supports flexible extensions and modifications. A prototype version has been implemented and is currently available. The USC Models, Algorithms and Architectures project is developing algorithmic techniques for realizing scalable and portable applications using configurable computing devices and architectures. Details on **DRIVE** and related research results can be found at http://maarc.usc.edu.

References

1. P. Bellows and B. Hutchings. JHDL - An HDL for Reconfigurable Systems. In *IEEE Symposium on Field-Programmable Custom Computing Machines*, April 1998.
2. K. Bondalapati, P. Diniz, P. Duncan, J. Granacki, M. Hall, R. Jain, and H. Ziegler. DEFACTO: A Design Environment for Adaptive Computing Technology. In *Reconfigurable Architectures Workshop, RAW'99*, April 1999.
3. K. Bondalapati and V.K. Prasanna. Mapping Loops onto Reconfigurable Architectures. In *8th International Workshop on Field-Programmable Logic and Applications*, September 1998.
4. K. Bondalapati and V.K. Prasanna. Dynamic Precision Management for Loop Computations on Reconfigurable Architectures. In *IEEE Symposium on FPGAs for Custom Computing Machines*, April 1999.
5. G. Brebner. CHASTE: a Hardware/Software Co-design Testbed for the Xilinx XC6200. In *Reconfigurable Architectures Workshop, RAW'97*, April 1997.
6. M. Chu, N. Weaver, K. Sulimma, A. DeHon, and J. Wawrzynek. Object Oriented Circuit-Generators in Java. In *IEEE Symposium on FPGAs for Custom Computing Machines*, April 1998.
7. Virtual Computer Corporation. Reconfigurable Computing Products, http://www.vcc.com/.
8. A. Koch. Unified access to heterogeneous module generators. In *ACM International Symposium on Field Programmable Gate Arrays*, February 1999.
9. R. Kress, R.W. Hartenstein, and U. Nageldinger. An Operating System for Custom Computing Machines based on the Xputer Paradigm. In *7th International Workshop on Field-Programmable Logic and Applications*, pages 304–313, Sept 1997.
10. D. Levi and S. Guccione. Run-Time Parameterizable Cores. In *ACM International Symposium on Field Programmable Gate Arrays*, February 1999.
11. W. Luk and S. Guo. Visualising reconfigurable libraries for FPGAs. In *Asilomar Conference on Signals, Systems, and Computers*, 1998.
12. W. Luk, S. Guo, N. Shirazi, and N. Zhuang. A framework for developing parametrised FPGA libraries. In *Field-Programmable Logic, Smart Applications, New Paradigms and Compilers*, 1996.
13. P. Lysaght and J. Stockwood. A Simulation Tool for Dynamically Reconfigurable FPGAs. *IEEE Transactions on VLSI Systems*, Sept 1996.
14. P.I. Mackinlay, P.Y.K. Cheung, W. Luk, and R. Sandiford. Riley-2: A Flexible Platform for Codesign and Dynamic Reconfigurable Computing Research. In *7th International Workshop on Field-Programmable Logic and Applications*, September 1997.
15. O. Mencer, M. Morf, and M.J. Flynn. PAM-Blox: High Performance FPGA Design for Adaptive Computing. In *IEEE Symposium on FPGAs for Custom Computing Machines*, April 1998.

Modelling and Synthesis of Configuration Controllers for Dynamically Reconfigurable Logic Systems Using the DCS CAD Framework

David Robinson and Patrick Lysaght

Dept. Electronic and Electrical Engineering
University of Strathclyde
204 George Street
Glasgow, G1 1XW
United Kingdom

Fax: +44 (0) 141 552 4968
e-mail: d.robinson@eee.strath.ac.uk

Abstract. The overheads contributed to a dynamically reconfigurable logic (DRL) design by its configuration controller can be prohibitive. Not only are the resource requirements and execution delays of the dynamic design adversely effected, but the time to redesign and test a configuration controller for every design iteration can be significant. We present a generic model of a configuration controller. The model is sufficiently complex to be useful in many design scenarios, and is customisable via parameterisation and user defined blocks. A new tool, DCSConfig, has been created within the DCS CAD framework to provide partial automation of configuration controller design. The new CAD tool provides initial estimates of latency and complexity overheads. VHDL models of the configuration controller for a particular dynamic design are produced for simulation and hardware synthesis.

1 Introduction

Proponents of dynamically reconfigurable logic (DRL) cite increased circuit speeds and reduced hardware requirements as two major benefits of the technique [1]. These benefits are extremely susceptible to the overheads associated with the control unit used to manage the reconfigurable aspects of the design. Some increase in physical resource requirements and decrease in system execution speed are inevitable, but the extent of these overheads depends on whether the configuration controller is implemented in hardware, software or a combination of both.

Designing a configuration controller can be a complex process, and like most DRL design tasks, is best delayed until the benefits of employing the technique in the design are established. However, without an accurate model of the configuration controller at an early stage in the design cycle, the benefits of using dynamic reconfiguration for a particular design cannot be properly evaluated [2]. Early estimation of the overheads associated with a design's configuration controller, and automatic synthesis of the controller are required to improve the research and development of DRL designs. A basic model encapsulating the core requirements of a configuration controller was reported by Shirazi *et al* in [3]. They presented a three-stage model containing a monitor, a loader and a configuration store. The monitor recognises requests for configuration and the loader configures the FPGA with data from the configuration store. The configuration store is a structured memory containing information required to configure the device. The configuration store can include a transform agent to modify the configuration data if required. We present a more detailed and generic model of a hardware based configuration controller, allowing both system modelling and automatic synthesis. The model is sufficiently complex to be useful in many designs. Many aspects of the model are customisable by the designer, and standard data transfer protocols and generic interfaces allow the inclusion of user defined blocks to tailor behaviour of the configuration controller for individual designs.

DCS (dynamic circuit switching) [4] is a VHDL based CAD framework for DRL, that integrates with existing CAD tools and standards. The framework comprises of a method of capturing reconfiguration information, and performing design simulation and hardware synthesis. DCS has been extended to allow the impact of the configuration controller to be considered in a design. A new tool, DCSConfig, has been added to supply VHDL models and performance estimates of a design's configuration controller.

This paper is organised into five further sections. Section 2 defines the terminology used in this paper. Sections 3 and 4 introduce configuration controllers and the configuration controller model. Section 5 describes DCSConfig and section 6 concludes the paper.

2 Definitions

Throughout the paper we have used to following terminology extensively. Some of these terms have appeared in previous papers whilst some are introduced here.

A *dynamically reconfigurable* FPGA can have a subset of the user logic reconfigured while the remainder continues to operate. Configuration data is loaded into an FPGA via its *configuration port*, which encompasses all the FPGA hardware resources required to configure the device. A hardware design unit is called a *task*. A task represents a grouping of device primitives and can be of arbitrary complexity. An inverter is an example of a simple task while a microprocessor represents a complex task. The set of tasks that are currently mapped to physical resources is known as the *active set* and the set of tasks that are not is known as the *inactive set*. A *static task* is always present in the active set whilst a *dynamic task* may enter and leave the active set. The *reconfiguration latency* [5] of a dynamic task is the time required to transfer the dynamic task from the inactive set to the active set.

A *reconfiguration condition* is a predefined condition that must be satisfied before a dynamic task may be reconfigured. There are two reasons for reconfiguring a dynamic task: *activation* or *deactivation*. Activation moves the dynamic task from the inactive set to the active set. Deactivation moves the dynamic task from the active set to the inactive set. Dynamic tasks are deactivated to ensure that all allocated resources are left in a safe state to prevent conflict with other members of the active set. This is achieved by loading a safe configuration that matches the dynamic task's footprint exactly. This may not be necessary if the footprint of the new dynamic task completely covers that of the dynamic task that it is replacing [6]. Removing a dynamic task from the active set prematurely to make resources available for a dynamic task with a higher priority is referred to as task *pre-emption*.

3 Configuration Controllers

Configuration Controller Functionality

The configuration control unit for a DRL design has three principal operational requirements [3]:

- Recognising the reconfigurations conditions
- Retrieving configuration data

Using the configuration data to configure the device

A number of further desirable operations can be identified:

• A queuing mechanism for reconfiguration conditions	For storing reconfiguration conditions that occur during periods of high dynamic activity
• Priority levels for dynamic tasks	Priority levels to allow certain dynamic tasks to reconfigure ahead of others and pre-empt dynamic tasks already resident on the FPGA
• Decoding of configuration data at run-time	To allow designers to store the data in an encrypted or compressed format
• Run-time transformation of dynamic task layout	To improve the efficiency of resource usage by adapting dynamic task layouts to match the current operating conditions

Configuration Controller Model

Fig. 1 shows a model of a configuration controller that provides generic support for the functionality described in the previous section. The model is block based with data transfer between blocks controlled by handshaking protocols. Parameterisable blocks provide a range of functionality, and the model is user extensible, either through the addition of new blocks or through customisation of existing blocks. The interfaces to each block have been designed to be generic, de-coupling them from the internal function of the block. Device and system independence is achieved by encapsulating all dependencies in hardware specific blocks. Blocks that are not relevant to a particular design can be removed without effecting the model.

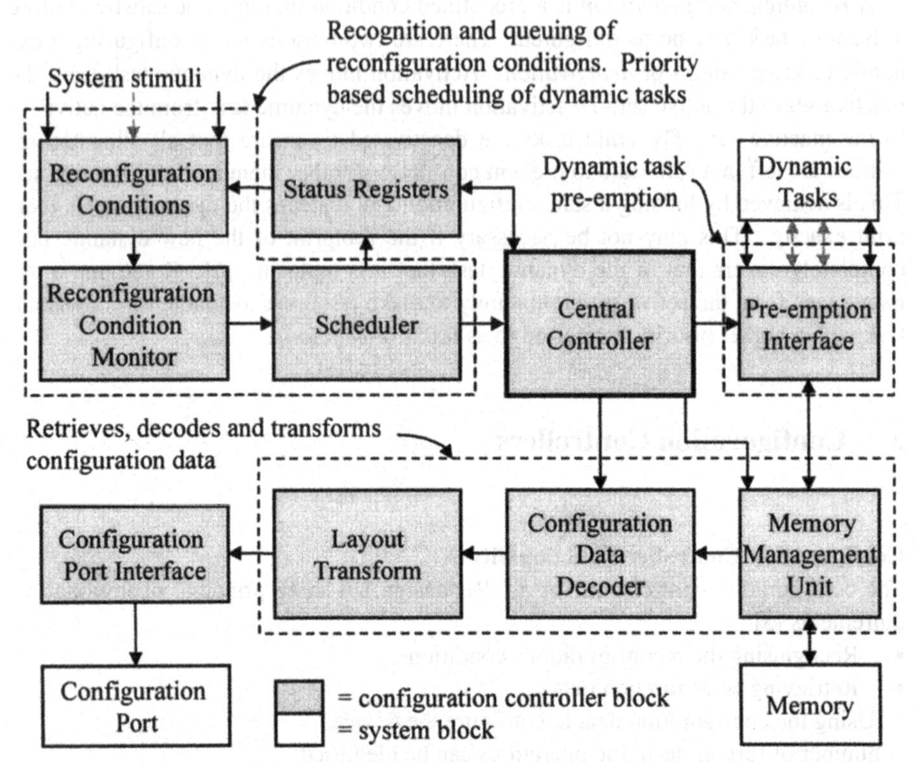

Fig. 1. Block diagram of configuration controller

Configuration controller operation begins with the reconfiguration conditions, shown at the top left of Fig. 1. Active reconfiguration conditions are detected by the reconfiguration condition monitor and are stored in a queue. Dynamic tasks are selected from this queue by the scheduler according to their priority. The central controller synchronises the operation of the remaining blocks to reconfigure the dynamic task, pre-empting other dynamic tasks if required. Configuration data is retrieved by the memory management unit and processed by the configuration data decoder and the layout transform blocks if necessary. The configuration port interface actually configures the FPGA.

Configuration Port Interface and Memory Management Unit

The configuration port interface is a device specific block that interfaces the configuration controller to the configuration port. The memory management unit provides a generic interface to the system memory. This is used by the configuration controller to obtain configuration data and by dynamic tasks for pre-emption purposes. The memory can be a mixture of local memory and different types of external memory. Both of these blocks are almost completely device or design dependent.

Central Controller

The central controller is the block that controls the activation and deactivation of dynamic tasks. This process begins when a dynamic task identifier and a flag appear at the output of the scheduler. The flag indicates whether the dynamic task is to be activated or deactivated. The central controller activates the new dynamic task, pre-empting if necessary any dynamic tasks that need to be removed. When deactivating a dynamic task, the central controller restores any dynamic tasks that were pre-empted as a result of the activation of this one. The central controller ensures the consistency of the status registers throughout this process.

Reconfiguration Conditions

Each dynamic task has two reconfiguration conditions, corresponding to its activation and deactivation. Designers can specify the number of times a reconfiguration condition must occur, whether it must occur on consecutive clock cycles and the number of clock cycles to delay the activation. This level of control enables the designer to specify with great accuracy the reconfiguration schedule of the design.

Reconfiguration Condition Monitor

The reconfiguration condition monitor is responsible for detecting and storing satisfied reconfiguration conditions for processing by the central controller. The reconfiguration condition monitor is structured as a series of independent priority monitors, Fig. 2.

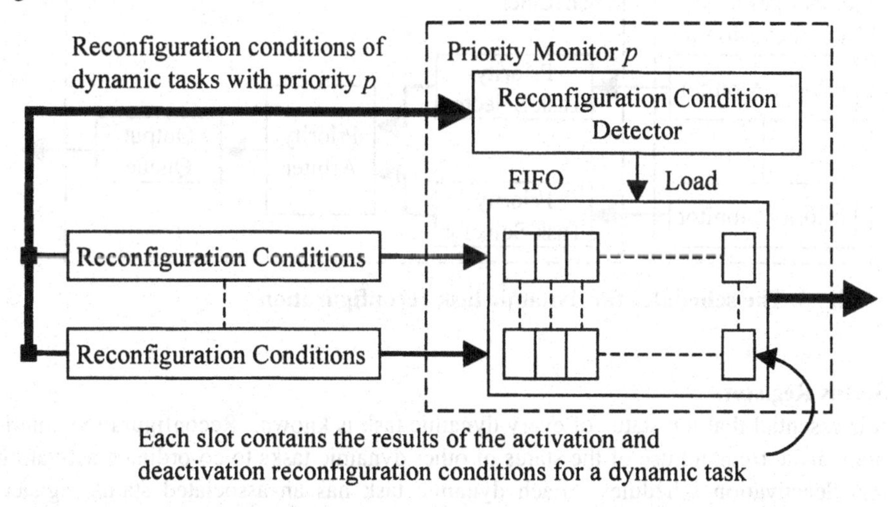

Fig. 2. A priority monitor from the reconfiguration condition monitor block

Each priority monitor *p*, tracks the reconfiguration conditions of dynamic tasks with the priority *p*. A priority monitor is physically implemented as a FIFO, that captures information when a reconfiguration condition activates, and a reconfiguration condition detector. An entry in the FIFO is structured as a series of slots. Each slot is assigned to a dynamic task and contains the reconfiguration conditions that activate or deactivate the dynamic task. Without this scheme, active reconfiguration conditions would be lost if a dynamic task was already being reconfigured. The depth of the FIFO is user defined and should be large enough to cope with peak periods of reconfiguration activity.

Scheduler

The scheduler, shown in Fig. 3, serialises parallel requests for reconfiguration from the reconfiguration conditions. The scheduler receives data from a series of priority queues in the reconfiguration condition monitor. These are processed to create a single output queue, containing a dynamic task identifier and a flag for each reconfiguration condition. The flag specifies whether the dynamic task is to be activated or deactivated. Two schemes are used to select a dynamic task for processing. The first is a simple circular queue used to select a reconfiguration condition from within a priority monitor. The output of a priority monitor is structured as a series of slots, each slot representing a dynamic tasks. Slots are selected sequentially from first to last and the contents converted to an identifier and flag. These are loaded into the output queue. When the last slot is processed, control of the output queue is relinquished. The second scheme used to select the priority of the dynamic task to be processed next, is a user definable circular queue. Priorities are selected in an order specified by the designer. For instance, $p_1 \rightarrow p_1 \rightarrow p_2 \rightarrow p_1 \rightarrow p_1 \rightarrow p_3$ gives priority one (p_1) maximum importance without blocking the other priority levels.

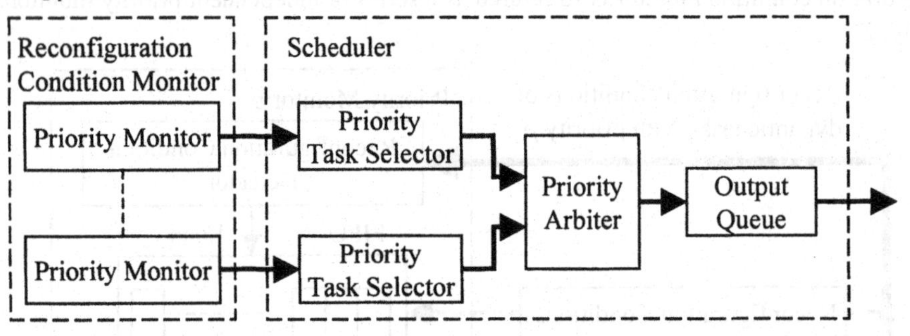

Fig. 3. The scheduler for dynamic task reconfiguration

Status Registers

It is essential that the status of every dynamic task is known. Reconfiguration conditions make frequent use of the status of other dynamic tasks to co-ordinate activation and deactivation schedules. Each dynamic task has an associated status register, irrespective of whether it is on-chip or not. The minimal set of status flags supported by the model is shown below, but the designer can provide support for more if required.

- *present* specifies whether the dynamic task is configured onto the array
- *pre-empted* specifies whether the dynamic task has been pre-empted

- *reconfiguring* specifies whether the task is undergoing a reconfiguration
- *activation scheduled* specifies whether the dynamic task has been scheduled for activation
- *deactivation scheduled* specifies whether the dynamic task has been scheduled for deactivation

Pre-emption Interface

When a dynamic task is pre-empted by one of higher priority, it is removed from the array and is only replaced when the higher priority one has executed. A pre-empted dynamic task must resume operation, either from initialisation or from the point of pre-emption. The latter requires the ability to save the state of the dynamic task to memory and restore it when the dynamic task is reinserted into the active set. The dynamic task is given control of the memory controller and issued with a pre-empt or restore command. The dynamic task then proceeds to perform internal control routines to save or restore enough information for it to resume operation from the point of pre-emption. It then signals completion to the central controller. The pre-emption scheme employed by a dynamic task must be specified and implemented by the designer. Adding the pre-emption scheme automatically is an area of current research.

Configuration Data Decoder

All configuration data for the design is loaded by the memory management unit and passed to the configuration data decoder. If this data is not encoded then it is passed unmodified through the configuration data decoder. If the data is encoded, then the configuration data decoder includes user defined circuits to convert the data into valid configuration data. The configuration data could be encoded in a variety of ways, for example it can be compressed or encrypted. Compression may provide a simple method of reducing the configuration latency for a dynamic task, or reduce the amount of memory required to store the data. Dynamic tasks that have highly regular structures consist of many cells with identical configuration data [7]. In these cases, storing the common cell data and all the cell addresses is an efficient compression scheme where the decoder simply appends the data to each address as it is injected into the system [3]. Encryption of configuration data increases design security by increasing the complexity of reverse engineering the design. The many possible methods of encoding data make this block extremely design dependent.

Layout Transform

Delaying the decision on the final position and footprint of a dynamic task until run-time allows the configuration controller to make maximum use of available array resources [7]. Changing the destination of a dynamic task [8] is feasible in current technology whereas footprint modification is a more futuristic application as it requires run-time placement and routing. The layout transform block exists in the configuration controller model as an empty block consisting of only the data handshaking protocol. User defined algorithms can be implemented between the input and output of the block to transform the data.

4 Overhead Estimation

Design space exploration demands access to design estimates at the earliest opportunity, to exclude designs that show no benefit from using DRL from further attention. Each block in the configuration controller has an overhead profile consisting of resource requirements and execution latency. Although the models are not yet targeted at any specific architecture, the overhead profiles are valid design aids. Post processing or designer experience can convert the data into device specific estimates.

Resource Requirements

Resource usage reports the number of flip-flops and components required by a block. This data can be targeted at a specific device by a simple mapping process. This process may be a simple rule-of-thumb, such as only considering flip-flops in a look-up-table based device, or actual optimisation and mapping software. Routing requirements are not estimated, but at this level of design abstraction they are not critical. A design that fails to meet resource usage constraints will not improve when routing requirements are added. In many cases, a designer's experience will allow a routing resource estimate to be added.

Many blocks in the model consist of circuitry that is either pre-designed or predictably generated based on the reconfiguration information. Resource usage estimates provide detailed requirements for FSMs and custom components, but relies on interrogating libraries for the requirements of standard components, such as counters and FIFOs. This approach allows specific FPGA devices to be targeted with higher accuracy, as custom component implementations can be used, i.e. components that use specific device features to reduce resource usage.

Execution Delay

Adding and removing a dynamic task from the active set incurs a time overhead known as the reconfiguration latency. The reconfiguration latency for a dynamic task can be broken down into constituent latencies [5]. Of these latencies, T_{ACK} and $T_{CONTROL}$[1] are introduced by the configuration controller.

T_{ACK} corresponds to the time taken by the configuration controller to acknowledge a reconfiguration condition has occurred and start processing the request. T_{ACK} has the following constituent latencies:

- The delay of the reconfiguration condition monitor t_{RCM}
- The delay of the scheduler t_S

$T_{CONTROL}$ is the latency of the control algorithms employed by the configuration controller after a dynamic task identifier has been received by the central controller. $T_{CONTROL}$ has the following constituent latencies:

- The delay of the central controller t_{CC}
- The delay of the configuration data decoder t_{CDD}

---

[1] Note that in [5] these latencies were represented using t instead of T

- The delay of the pre-emption interface t_{PEI}
- The delay of the layout transform t_{LT}
- The delay of the memory management unit t_{MMU}
- The delay of the configuration port interface t_{CPI}

T_{ACK} is non-deterministic as t_{RCM} and t_S depend on the number and priority of the other dynamic tasks already scheduled for reconfiguration. $T_{CONTROL}$ is non-deterministic as t_{PEP} and possibly t_{MMU}, have variable execution times. The difficulty in estimating $T_{CONTROL}$ is also compounded by the inclusion of device dependent and user definable blocks. The layout transform and the configuration data decoder blocks are user definable and the configuration port interface block is device dependent. The basic latencies of these blocks due to protocols and common functions introduced by the model can be calculated. The designer is prompted for delay estimations of the remaining circuitry.

Execution delay estimates are specified in clock cycles and it is assumed that the longest routing delay will be shorter than the period of the clock. If exact prediction of execution latencies of a particular block cannot be calculated, both best case and worst case estimates are supplied to the designer.

5 DCSConfig

DCSConfig analyses the reconfiguration information and modifies the reconfiguration conditions to prevent resource contention. The order of activation and deactivation between certain dynamic tasks is critical if the dynamic tasks are assigned to the same physical resources. These dynamic tasks must be mutually exclusive with respect to each other in time to prevent resource contention. Such dynamic tasks are assigned to a mutually exclusive group (mutex group)[4]. DCSConfig adjusts the reconfiguration conditions for every dynamic task in such a group to include the *present* and *deactivation scheduled* status bit of every other dynamic task in the group. This prevents a dynamic task from activating until all others in the group have been deactivated. Dynamic task pre-emption is also defined by the mutex groups. To prevent the requirement of a costly run-time analysis, dynamic tasks can only pre-empt others in the same mutex group. Dynamic tasks do not share resources with those in other mutex groups; therefore, pre-emption is not required across groups. This information is incorporated into the central controller when it is synthesised.

Based on the reconfiguration information, DCSConfig selects and parameterises, or synthesises the standard blocks of the configuration controller. User defined blocks are included and the output is produced. Depending on the mode DCSConfig is used in, this can be a VHDL model of the configuration controller or a report detailing the overheads introduced by the configuration controller.

6 Conclusions

We have presented a general model of a configuration controller for dynamically reconfigurable logic systems. The controller is sufficiently generic to be useful in a large number of design scenarios. Although we have only discussed hardware implementation in this paper, the model can also be implemented in software or a combination of both without any loss of generality.

The model should be viewed as a useful starting point for many designs and not necessarily the final solution. It is impossible to predict and implement every potential situation that a controller must deal with. For this reason, we have designed extensibility into the model. The separation of a block's function from its interface allows functional customisation without breaking its interface to the remainder of the model. The configuration data decoder and layout transform are examples of blocks that require such customisation.

Partial automation of configuration controller synthesis is provided by DCSConfig, but certain design dependent tasks remain for the designer. Creation of the configuration port interface and the memory management unit are essential. Optionally, if pre-emption is required then dynamic tasks must have support added and any custom decoder and transform algorithms must be supplied. Automating the production of these blocks provides an excellent opportunity for further research.

References

[1] M. J. Wirthlin and B. Hutchings, "Improving Functional Density Through Run-Time Constant Propagation", In *ACM/SIGDA International Symposium on FPGAs*, pp. 86-92. 1997

[2] P. Lysaght, G. McGregor and J. Stockwood, "Configuration Controller Synthesis for Dynamically Reconfigurable Systems", in *IEE Colloquium on Hardware - Software Cosynthesis for Reconfigurable Systems*, HP Labs, Bristol, UK. Feb. 1996

[3] N. Shirazi, W. Luk and P.Y.K. Cheung, "Run-Time Management of Dynamically Reconfigurable Designs", in *Field Programmable Logic and Applications*, pp 59-68, Hartenstein, R. and Keevallik, A. (Eds), Tallinn, Estonia, Sept. 1998

[4] D. Robinson, P. Lysaght and G. McGregor, "New CAD Framework Extends Simulation of Dynamically Reconfigurable Logic", in *Field Programmable Logic and Applications*, pp 1-8, Hartenstein, R. and Keevallik, A. (Eds), Tallinn, Estonia, Sept. 1998

[5] P. Lysaght, "Towards an Expert System for *a priori* Estimation of Reconfiguration Latency in Dynamically Reconfigurable Logic, in *Field Programmable Logic and Applications*, pp 183-192, Luk, W., Cheung, P., & Glesner, M. (Eds), Sept. 1997

[6] P. Lysaght and J. Stockwood, "A Simulation Tool for Dynamically Reconfigurable Field Programmable Gate Arrays", in *IEEE Transactions on VLSI Systems*, Vol.4, No. 3, pp. 381-390, September, Sept. 1996

[7] G. Brebner, "A virtual hardware operating system for the Xilinx XC6200", In *Field-Programmable Logic and Applications*, pp 327-336, Hartenstein, R. W. and Glesner, M., (Eds), Darmstadt, Sept. 1996

[8] O. Diessel and H. ElGindy, "Run-time compaction of FPGA designs", In *Field-Programmable Logic and Applications*, pp 131-140, Luk, W., Cheung, P., & Glesner, M. (Eds) Sept 1997

Optimal Finite Field Multipliers for FPGAs

Captain Gregory C. Ahlquist, Brent Nelson, and Michael Rice

459 Clyde Building, Brigham Young University, Provo UT 84602 USA
ahlquist@ee.byu.edu, nelson@ee.byu.edu, mdr@ee.byu.edu

Abstract. With the end goal of implementing optimal Reed-Solomon error control decoders on FPGAs, we characterize the FPGA performance of several finite field multiplier designs reported in the literature. We discover that finite field multipliers optimized for VLSI implementation are not optimized for FPGA implementation. Based on this observation, we discuss the relative merits of each multiplier design and show why each does not perform well on FPGAs. We then suggest how to improve the performance of many finite field multiplier designs[1].

1 Introduction

Since its advent in 1960, efficient and practical Reed-Solomon (RS) decoder design has been an important research topic. RS decoders decode transmitted data using Reed-Solomon linear cyclic block codes to detect and correct errors. Applications include error correction in compact disk players, deep space communications, and military radios. Traditionally, RS decoder design exploits Application Specific Integrated Circuit (ASIC) technology to realize low processing times and small areas. Nevertheless, ASICs achieve these advantages at the expense of flexibility. General Purpose Processors (GPPs), Digital Signal Processors (DSPs), or hybrid ASICs can provide flexibility, but do so at the expense of processing time and area. FPGAs, however, promise to realize RS decoders with time and area performance similar to ASICs but with flexibility similar to GPPs. This is intriguing as it opens the door to dynamically changing error control coding based on the current characteristics of the transmission channel. Indeed, FPGA-hosted RS decoders already exist but little work has been reported on realizing FPGA-based dynamic error control or on simply characterizing the optimal FPGA implementation of RS decoders. Thus, we have focused our research in these areas. As finite field multipliers are the most numerous building blocks of RS decoders, our initial efforts seek to characterize the optimal FPGA finite field multiplier implementation. The rewards could be significant. We recently implemented a RS decoder for the RS(15,9) code on a Xilinx XC4062. This relatively small decoder required 130 finite field multipliers using 800 Configurable Logic Blocks(CLBs) and ran at 15 Megabits(Mb) per second. We estimate a larger decoder, such as for the industry standard 30 Mb RS(255,233) code [1], would require approximately 15,000 multipliers using 270,000 CLBs; equivalent to 118 Xilinx XC4062 Chips! There are a number a RS decoder design parameters we can adjust to reduce the resource requirement, increase speed, and make

[1] Distribution A: Approved for public release; distribution unlimited

FPGA-based finite field decoders more feasible. Nevertheless, attempting to optimizing FPGA-based finite field multipliers in terms of both speed and area is an important first step.

2 Finite Fields

Reed-Solomon codes are based on finite field arithmetic which involves defining closed binary operations over finite sets of elements. Unfortunately, a full review of finite fields is beyond the scope of this presentation but an excellent treatment of the topic is found in [2]. As a brief overview, we will start with the simplest example of a finite field which is the binary field consisting of the elements {0,1}. Traditionally referred to as GF(2) [1], the operations in this field are defined as integer addition and multiplication reduced modulo 2. We can create larger fields by extending GF(2) into vector space leading to finite fields of size 2^m. These are simple extensions of the base field GF(2) over m dimensions. The field GF(2^m) is thus defined as a field with 2^m elements each of which is a binary m-tuple. Using this definition, we can group m bits of binary data and refer to it as an element of GF(2^m). This in turn allows us to apply the associated mathematical operations of the field to encode and decode data. For our purposes, we will limit our discussion to the finite field GF(16). This field consists of sixteen elements and two binary operations; addition and multiplication. There are two alternate (but equivalent) representations for the field elements. First, all nonzero elements in GF(16) may be represented as powers of a primitive field element α [2] (i.e. each nonzero element is of the form α^n for $n = 0, 1, \dots 14$). Second, each element has an equivalent representation as a binary 4-tuple. While the α^n representation has great mathematical convenience, digital hardware prefers the binary 4-tuple representation. These representations are illustrated in Table 1.

Table 1. Canonical Representation of Finite Field $GF(16)$

Element	0	α^0	α^1	α^2	α^3	α^4	α^5	α^6
Representation	0000	0001	0010	0100	1000	0011	0110	1100
Element	α^7	α^8	α^9	α^{10}	α^{11}	α^{12}	α^{13}	α^{14}
Representation	1011	0101	1010	0111	1110	1111	1101	1001

To understand how finite field addition and multiplication work, it is essential to view each field element as describing the coefficients of a binary polynomial. For example, element α^7 whose binary representation is 1011 represents the polynomial $\alpha^3 + \alpha^1 + \alpha^0$. You may gain added insight by noting that the element

[1] The GF stands for Galois Field which is another name for a finite field. The term honors the French mathematician Eviriste Galois who first formalized finite field concepts.

α^7 is in fact the linear summation of α^3, α^1, and α^0. Under this representation, finite field addition and multiplication become polynomial addition and multiplication where the addition operation occurs modulo 2. To illustrate, consider GF(16) elements A and B represented as follows:

$$A = a_3\alpha^3 + a_2\alpha^2 + a_1\alpha^1 + a_0\alpha^0 \quad (1)$$

$$B = b_3\alpha^3 + b_2\alpha^2 + b_1\alpha^1 + b_0\alpha^0 \quad (2)$$

Adding elements A and B becomes a simple bit-wise modulo 2 addition.

$$A + B = (a_3 + b_3)\alpha^3 + (a_2 + b_2)\alpha^2 + (a_1 + b_1)\alpha^1 + (a_0 + b_0)\alpha^0. \quad (3)$$

Multiplication is a little more complicated. We begin by carrying out the polynomial multiplication.

$$\begin{aligned} A * B = {} & a_3 * b_3\alpha^6 + (a_3 * b_2 + a_2 * b_3)\alpha^5 \\ & + (a_3 * b_1 + a_2 * b_2 + a_1 + b_3)\alpha^4 \\ & + (a_3 * b_0 + a_2 * b_1 + a_1 * b_2 + a_0 * b_3)\alpha^3 \\ & + (a_2 * b_0 + a_1 * b_1 + a_0 * b_2)\alpha^2 \\ & + (a_1 * b_0 + a_0 * b_1)\alpha^1 + a_0 * b_0\alpha^0 \end{aligned} \quad (4)$$

The result has seven coefficients which we must convert back into a 4-tuple to achieve closure. We do this by substituting α^6, α^5, and α^4 with their polynomial representations and summing terms.

$$\begin{aligned} A * B = {} & (a_3 * b_3 + a_3 * b_0 + a_2 * b_1 + a_1 * b_2 + a_0 * b_3)\alpha^3 \\ & + (a_3 * b_3 + a_3 * b_2 + a_2 * b_3 + a_2 * b_0 + a_1 * b_1 + a_0 * b_2)\alpha^2 \\ & + (a_3 * b_2 + a_2 * b_3 + a_3 * b_1 + a_2 * b_2 + a_1 + b_3 + a_1 * b_0 + a_0 * b_1)\alpha^1 \\ & + (a_3 * b_1 + a_2 * b_2 + a_1 + b_3 + a_0 * b_0)\alpha^0 \end{aligned} \quad (5)$$

Equation (5) is often expressed in matrix form.

$$\begin{pmatrix} a_0 & a_3 & a_2 & a_1 \\ a_1 & a_3 + a_0 & a_3 + a_2 & a_1 + a_2 \\ a_2 & a_1 & a_3 + a_0 & a_3 + a_2 \\ a_3 & a_2 & a_1 & a_3 + a_0 \end{pmatrix} \begin{pmatrix} b_0 \\ b_1 \\ b_2 \\ b_3 \end{pmatrix} = \begin{pmatrix} c_0 \\ c_1 \\ c_2 \\ c_3 \end{pmatrix} \quad (6)$$

Because GF(16) is an extension of the binary field, The multiplies in equation (5) can be implemented as logical ANDs and the additions as logical XORs. Thus, the expression requires only sixteen ANDs and nine XORs to implement. Nevertheless, a number of hardware designs to perform finite field multiplication have been devised [3–10]. Each new design strives to reduce the overall silicon area and time required for the operation.

3 Finite Field Multipliers

The dependence on finite field arithmetic means that efficient RS decoders are dependent on efficient finite field adders and multipliers. Addition is easy. It equates to a bit-wise XOR of the m-tuple and is realized by an array of m XOR gates. The finite field multiplier is, by comparison, much more complicated and is the key to developing efficient finite field computational circuits. We have conducted an extensive survey of finite field multiplier designs and have characterized their performance on an FPGA. For each design, we created a hardware description using the Brigham Young University developed hardware description language JHDL [11]. This language easily allows us to model, simulate, and netlist our multiplier designs. Using JHDL, we verified the correct operation of each design and created an associated netlist. There were seven designs in all.

3.1 Linear Feedback Shift Register Multiplier

The Linear Feedback Shift Register (LFSR) GF(16) Multiplier design is considered by many the standard or canonical design for finite field multipliers. An excellent and detailed description of this multiplier can be found in [3]. The LFSR GF(16) multiplier has parallel input and output (i.e. inputs and outputs occur in 4-bit blocks) but possesses a 4 clock cycle latency which is not pipelined. Thus, the LFSR multiplier produces one 4-bit result every 4 clock cycles. As Figure 1 illustrates, the hardware consists of a latch and a shift register that feed arguments into a central LFSR . Appropriate combinational logic is appended to the LFSR to combine together the arguments, feedback, and the previous partial product.

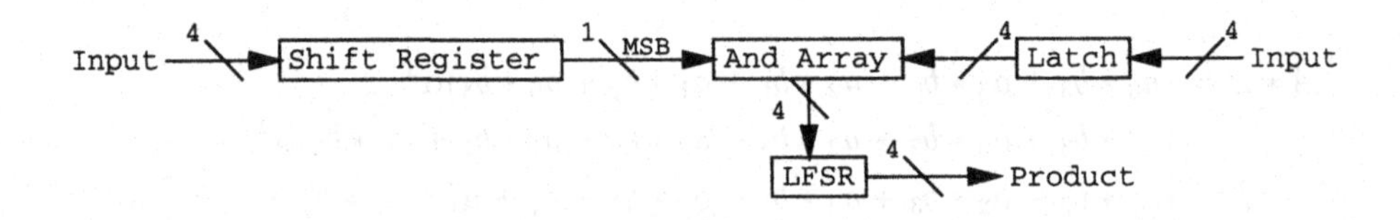

Fig. 1. Block Diagram of the Canonical LFSR Finite Field Multiplier.

3.2 Mastrovito Multiplier

In his doctoral dissertation, Mastrovito [4] provides a completely combinatorial finite field multiplier design. Mastrovito's multiplier has parallel inputs and outputs and no clock cycle latency. Thus, this multiplier produces one 4-bit result every clock cycle. Mastrovito's design mirrors the matrix form of the finite field multiplication equation (6) and consists of two stages. The first stage computes all the multi-variable A matrix elements and feeds these to the second stage. The second stage performs the matrix multiplication and outputs the product. All operations are performed completely in combinatorial logic.

3.3 Massey-Omura Multiplier

The Massey-Omura Multiplier as reported by Wang et. al. [5] operates on GF(16) elements represented in the normal basis. The normal basis merely uses a different binary representation for each GF(16) element. The benefit is that, under this basis, equation (5) becomes highly uniform and facilitates a regular VLSI design. The Massey-Omura multiplier has parallel inputs but produces the product serially. There is no clock cycle latency between argument presentation and results, but it does take 4 clock cycles to produce all 4 bits of the product. As Figure 2 shows, the multiplier loads the arguments into two cyclic shift registers. These, in turn, feed a block of combinatorial logic that produces one bit of the result. On each subsequent clock cycle, the register contents shift and the combinatorial logic produces another bit of the product.

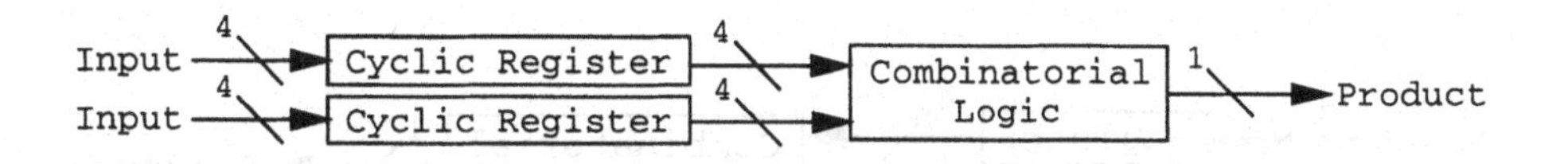

Fig. 2. Block Diagram of Massey-Omura Finite Field Multiplier.

3.4 Hasan-Bhargava Multiplier

Hasan and Bhargava [7] devised a bit-serial finite field multiplier implemented as a systolic array. The multiplier streams bits sequentially through eight small processing units with the product flowing bit-serially from the last processor. Each processing unit consists of two to three registers and a small amount of combinatorial logic. The first four processing units calculate the elements of the A matrix shown in equation (6). The second four units perform the matrix calculation. This design possesses an eight clock cycle latency and produces one 4-bit product every 4 clock cycles thereafter.

3.5 Paar-Rosner Multiplier

Paar and Rosner [8] present a novel multiplier design using a composite basis representation of GF(16). The composite basis essentially divides the GF(16) element into upper and lower bit pairs where each pair represents an element of GF(4). This allows Paar and Rosner to effect the multiplication of two GF(16) elements by performing simpler GF(4) multiplications and additions. The overall effect is to recast equation (5) in combinatorial logic in a very efficient manner. The Paar-Rosner multiplier has parallel inputs and outputs and no clock cycle latency.

3.6 Morii-Berlekamp Multiplier

A multiplier proposed by Morii et. al. [9] and based on a similar multiplier developed by Berlekamp [10] uses a dual basis representation of GF(16) elements. The dual basis is again just another way to represent the elements of GF(16) as binary 4-tuples. The advantage of using the dual basis is it converts the A matrix of equation (6) into a simpler form that can be realized by a simple LFSR. Like the Massey-Omura multiplier, the Morii-Berlekamp multiplier has parallel inputs but produces the output bit-serially. As Figure 3 shows, the multiplier loads one argument in a register and another into an LFSR. On each clock cycle, the LFSR computes a row of the simplified A matrix which is then fed into a block of combinatorial logic. The combinatorial logic block combines this input with the other argument to compute one bit of the result. Although there is no latency between argument presentation and results, the design does require 4 clock cycles to produce a complete 4-bit product.

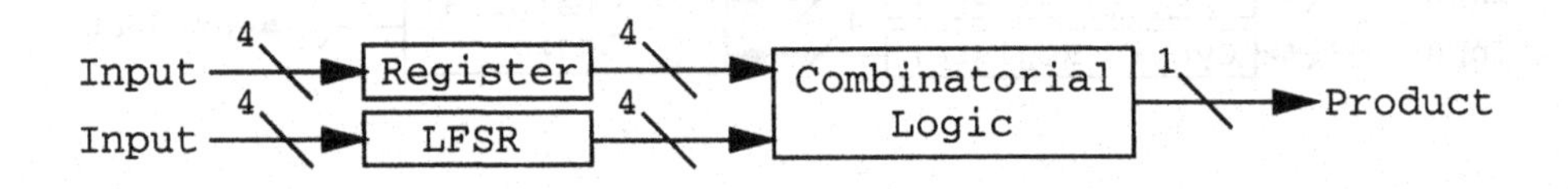

Fig. 3. Block Diagram of Morii-Berlekamp Finite Field Multiplier.

3.7 Pipelined Combinatorial Multiplier

This multiplier is a simple, pipelined implementation of equation (5) which we designed. The multiplier has parallel inputs and outputs and consists of two blocks of combinatorial logic separated by a bank of registers. The first combinatorial logic block breaks equation (5) into eleven parts representing eleven partial products. The register bank latches the partial products and feeds them into the second logic block which combines the partial products to obtain the desired result. The multiplier possesses an initial 2 clock cycle latency but produces a result every clock cycle thereafter. This design is, in effect, yet another recasting of equation (5)

4 Finite Field Multiplier FPGA Performance

Using simulation tools available in our lab, we characterized the performance of each finite field multiplier on the Xilinx XC4062 FPGA. We used two metrics for comparing these designs. The first metric is the area-time product combining the number of Xilinx CLBs and the critical path delay measured in nano-seconds (ns). As we ideally want our designs to take negligible resources and time, a lower number in the CLB*ns column represents better performance. The second metric is like the first except it reports results in units of CLB*ns^2. This simple change adds emphasis to faster designs. Table 2 shows resource usage and clock speed data for all designs. The time value indicates the delay of each multiplier. For example, The Massey-Omura multiplier produces a complete 4-bit result every 40.016 ns.

Table 2. Finite Field $GF(16)$ Multiplier Comparison Results

Design	CLBs	Time(ns)	CLB*ns	CLB*ns^2
Massey-Omura	15	40.016	600.240	24019.4
Hasan-Bhargava	19	24.952	474.088	11892.4
LFSR	13	26.092	339.196	8850.3
Morii-Berlekamp	13	23.780	309.140	7351.3
Paar-Rosner	5	12.318	61.590	758.7
Mastrovito	6	9.784	58.704	574.4
Pipelined Combinatorial	7	6.336	44.352	281.4

5 Performance Analysis

Our study of these multipliers leads us to an interesting observation: finite field multipliers optimized specifically for VLSI are not necessarily optimized for FPGAs. To illustrate our claim, compare results of the literature reported designs with our simple pipelined combinatorial implementation of a finite field multiplier. The simple, straightforward design performs much better on the Xilinx XC4062 FPGA than the more complicated VLSI-optimized designs. This result is very much counter-intuitive to our original predictions and we attempted to explain why. On closer inspection, we observed that the VLSI optimized designs all possess architectures that do not map well to the underlying XC4062 and lead to poor performance. Three significant flaws in comparison to the pipelined combinatorial design emerged.

- Multi-clock cycle operation
- Long unregistered datapaths
- Underutilized logic elements

Specifically, we claim the pipelined combinatorial design outperforms all other designs because it produces a result every clock cycle (after an initial 2 clock cycle latency), registers all logic element outputs, and maximizes the resources of each logic element. Producing a result every clock cycle and registering all logic element outputs leads to a pipeline like architecture which shortens the clock cycle and increases throughput. Maximizing logic element resources minimizes the overall number of logic elements required. The result is a fast, small, pipelined combinatorial circuit that outperforms all other designs in terms of CLB*ns and CLB*ns^2. In comparison, LFSR, Wang, Hasan-Bhargava, and Morii all perform worse than pipelined combinatorial primarily because they require 4 clock cycles to produce a complete 4-bit result. LFSR and Wang also suffer because they have unregistered datapaths that span more than one CLB. This greatly increases the signal propagation delay which in turn increases clock cycle time. Hasan-Bhargava requires an inordinately large number of resources because its small processing elements underutilize the processing capability of the Xilinx CLBs. LFSR, Wang, and Morii suffer from this as well but to a lesser extent. Mastrovito and Paar-Rosner produce complete 4-bit results every clock cycle and make excellent use of available logic resources. Nevertheless, they possess long, unregistered datapaths which increase their processing time. Table 3 summarizes each multiplier's detractions when compared to the pipelined combinatorial design.

Table 3. Finite Field $GF(16)$ Multiplier Architecture Comparison

Design	Multi-Cycle Processing	Unregistered Datapaths	Underutilized Logic Elements
Massey-Omura	√	√	√
Hasan-Bhargava	√	√	√
LFSR	√	√	√
Morii-Berlekamp	√	√	√
Paar-Rosner		√	
Mastrovito		√	
Pipelined Combinatorial			

6 Performance Improvements

Recognizing these performance limitations puts us in the position to improve the performance of each design relative to the Xilinx XC4062 FPGA. Thus, we modified each design along the following guidelines.

- Eliminate multi-clock cycle operation
- Eliminate unregistered datapaths
- Maximize use of each logic element

Our strategy to eliminate multi-clock cycle operation was to pipeline the effected designs such that, after some initial latency, the multipliers produced a complete 4-bit result every clock cycle. The strategy to eliminate unregistered datapaths simply involved registering every signal as it emerged from a CLB logic processing element. The strategy for maximizing the use of each logic element involved eliminating CLBs used simply to register signals or that computed very simple logic results. In most cases, these strategies proved to be at odds with one another and our endeavor quickly turned into a classic time versus area tradeoff. Eliminating multi-clock cycle operation and unregistered datapaths necessitated the use of more Xilinx CLBs; primarily due to the need for more registers. Conversely, attempts to maximize the logic processing capability of each Xilinx CLB often led to multi-level datapaths. In the end, each design modification became an exercise in decreasing processing time while minimizing resource growth. Table 4 shows the performance of the designs listed in table 2 after modifications. A quick comparison with table 2 shows that we achieved some improvement in all designs with the exception of the Mastrovito multiplier. In the Mastrovito multiplier, one block of combinatorial logic feeds another and results in several signals spanning more than one CLB. In an attempt to reduce the length of these datapaths we registered all results from the first block. This introduced a timing problem which required us to register both 4-bit arguments; something we previously did not have to do. Although we were able to shorten the processing time for this multiplier, the associated cost in CLB resources due to the increased number of registers wiped out any gain.

We note with some emphasis that Paar-Rosner-II and Morii-Berlekamp-II either approach or slightly exceed the performance of our pipelined combinatorial design. We achieved these gains by breaking down the Paar-Rosner and Morii-Berlekamp designs into their basic governing logic equations and implementing them in the same fashion as our combinatorial pipelined design. The results imply that the Paar-Rosner and Morii-Berlekamp approaches are just as good as our pipelined combinatorial multiplier design so long as the respective multipliers are implemented in a pipelined combinatorial fashion.

Table 4. Modified Finite Field $GF(16)$ Multiplier Comparison Results

Design	CLBs	Time(ns)	CLB*ns	CLB*ns^2
Hasan-Bhargava-II	17	6.457	109.769	708.8
LFSR-II	17	5.981	101.677	608.1
Mastrovito-II	12	6.481	77.772	504.0
Massey-Omura-II	8	6.723	53.784	361.6
Morii-Berlekamp-II	7	6.338	44.366	281.2
Pipelined Combinatorial	7	6.336	44.352	281.0
Paar-Rosner-II	7	6.258	43.806	274.1

7 Conclusions

Our experiments with finite field GF(16) multipliers taught us the following lessons. First, we cannot simply implement any finite field multiplier design reported in the literature and expect optimal FPGA performance. We must take into account the underlying architecture of the target FPGA. Specifically, optimal performance is dependent on single clock cycle throughput, registered datapaths, and maximum use of each logic element. Applying these principles to optimize finite field multiplier designs is a classic time-area tradeoff. We must be careful that design changes that decrease processing time do not increase resource usage such that any gains are nullified. Our data suggests that the optimal design is a simple, pipelined, combinatorial architecture that implements the governing logic equation of the multiplier. This is true regardless of the basis used to represent the elements of GF(16). Our data also raises a number of unanswered questions. Do these conclusions hold true as the size of the multiplier grows to handle elements of GF(32) on up to GF(256)? How much impact does the multiplier size and speed have on the overall design of RS encoders and decoders? How can we better design FPGAs to facilitate these kind of devices? Our future research will endeavor to address these and other questions.

References

1. J. L. Politano and D. Deprey. A 30 mbits/s (255,223) Reed-Solomon decoder. In *EUROCODE 90, International Symposium on Coding Theory and Applications*, pages 385–391. EUROCODE 90, Udine Italy, Nov 1990.
2. S. B. Wicker. *Error Control Systems for Digital Communication and Storage.* Prentice Hall, 1995.
3. S. Lin and D. J. Costello. *Error Control Coding: Fundamentals and Applications.* Prentice-Hall, Englemwood Cliffs, NJ, 1983.
4. E. D. Mastrovito. *VLSI Architectures for Computations in Galois Fields.* PhD thesis, Linköping University, Dept. Electr. Eng., Linköping, Sweden, 1991.
5. C. A. Wang, T. K. Truong, H. M. Shao, L. J. Deutsch, J. K. Omura, and I. S. Reed. VLSI architectures for computing multiplications and inverses in GF(2^m). *IEEE Transactions on Computers*, 34(8):709–717, Aug 1985.
6. J. L. Massey and J. K. Omura. Computational method and apparatus for finite field arithmetic. U.S. Patent Application, 1981.
7. M. A. Hasan and V. K. Bhargava. Bit-serial systolic divider and multiplier for finite fields GF(2^m). *IEEE Transactions on Computers*, 41(8):972–980, Aug 1992.
8. C. Paar and M. Rosner. Comparison of arithmetic architectures for Reed-Solomon decoders in reconfigurable hardware. *IEEE Transactions on Computers*, 41(8):219–224, Aug 1997.
9. M. Morii, M. Kasahara, and D. Whiting. Efficient bit-serial multiplication and the discrete-time Wiener-Hopf equation over finite fields. *IEEE Transactions on Information Theory*, 35(6):1177–1183, November 1989.
10. E. Berlekamp. Bit-serial Reed-Solomon encoders. *IEEE Transactions on Information Theory*, IT-28(6):869–874, November 1982.
11. B. Hutchings, P. Bellows, J. Hawkins, S. Hemmert, B. Nelson, and M. Rytting. A CAD suite for high-performance FPGA design. In K. Pocek and J. Arnold, editors, *IEEE Symposium on Field-Programmable Custom Computing Machines (FCCM 99)*, page TBA. IEEE Computer Society, IEEE Computer, April 1999.

Memory Access Optimization and RAM Inference for Pipeline Vectorization*

Markus Weinhardt and Wayne Luk

Department of Computing, Imperial College, London, UK
{mw8, wl}@doc.ic.ac.uk

Abstract. This paper describes memory access optimization in the context of pipeline vectorization, a method for synthesizing hardware pipelines in reconfigurable systems from software program loops. Since many algorithms for reconfigurable coprocessors are I/O bound, the throughput of the coprocessor is determined by the external memory accesses. Thus access optimizations directly improve the system's performance. Two kinds of optimizations have been studied. First, we consider methods for reducing the *number* of accesses based on saving frequently-used data in on-chip storage. In particular, recent FPGAs provide on-chip RAM which can be used for this purpose. We present *RAM inference*, a technique which automatically extracts small on-chip RAMs to reduce external memory accesses. Second, we aim to minimize the *time* spent on external accesses by scheduling as many accesses in parallel as possible. This optimization only applies to architectures with multiple memory banks. We present a technique which allocates program arrays to memory banks, thereby minimizing the overall access time.

1 Introduction

The reconfigurable computing community is increasingly interested in high-level compilers which translate a software program into both machine code for a host processor and configuration bitstreams for a field-programmable accelerator [1–3]. This work has shown that loops are the most promising candidates for acceleration. The coprocessors must also exploit the hardware's parallelism to achieve a sufficient speedup. Our approach, called *pipeline vectorization* [1, 4, 5], achieves this by vectorizing loops and synthesizing pipeline circuits. Many image processing and DSP applications exhibit no or little dependences which limit pipelining. Consequently, provided that enough FPGA resources are given, only memory accesses limit the throughput of the pipelines: the coprocessors are I/O bound. We present techniques that reduce these access times, thus directly improving system performance. One of the techniques, the use of shift registers, was first suggested in [6] for pipelines in reconfigurable systems. However, the

* This work is supported by a European Union training project financed by the Commission in the TMR programme, the UK Engineering and Physical Sciences Research Council, Embedded Solutions Ltd., and Xilinx Inc.

registers were declared explicitly and not synthesized automatically as in our approach.

Since FPGA-based accelerators typically organize external memory as one or more fast local SRAM banks accessible from the host and the FPGAs [7–9], we also address the allocation of data to memory banks. Gokhale and Stone [10] independently developed an allocation algorithm based on a technique called implicit enumeration. This algorithm is slightly faster, but less general than our approach.

The remainder of this paper first reviews pipeline vectorization. Then, Section 3 defines access equivalence and explains how shift registers synthesized for equivalence classes reduce the number of memory accesses. Next, Section 4 introduces RAM inference, which automatically extracts small on-chip RAMs by analyzing access equivalence in nested loops. Section 5 presents a technique to allocate program arrays on multiple memory banks. Finally, performance results are given in Section 6 before we conclude the paper in Section 7.

2 Pipeline Vectorization

We outline pipeline vectorization using a computationally expensive running example, image skeletonization [11]. It iteratively performs erosion, dilation, difference and OR operators on an input image, thereby producing its "skeleton". Figure 1 outlines the program. Note that successive versions of the input image and of the resulting skeleton are copied back and forth between `image1/skeleton1` and `image2/skeleton2` during the iterations of the outer `while` loop.

Pipeline vectorization consists of three main steps: candidate loop selection, dataflow graph generation, and pipelining. In the first step, candidates for hardware acceleration are selected. These are inner loops which do not contain non-synthesizable functions and are normalized so that their index variables are initialized to zero and incremented by one. In the example in Figure 1, the innermost `for` loop (line 12 to line 23) is a candidate. Note that the given form of the loop was generated by a transformation [1] which merged the originally independent operators so that they are combined in one loop and can be synthesized into a single coprocessor. The loop contains the functions `Fmin` and `Fmax` (not elaborated in Figure 1) which perform the bulk of the program's computation: they respectively compute the minimum or maximum of 21 pixels of an image (a 5×5 window without the corners). Loops spanning these 5×5 windows are unrolled by another transformation [1] to remove the internal loops in the functions.

Next, a dataflow graph is generated for the loop by analyzing the data dependences within the loop body. For conditionally assigned variables, multiplexers are inserted to select the correct values. We allocate scalar input and output variables to FPGA registers. Array variables are stored in local (off-chip) memory, since it is expensive to store large data arrays, such as image data, on an FPGA. Figure 2 shows the result for the example. The internal structure of the blocks `Fmin` and `Fmax`, the conditions guarding some operations, and the logic

```
 1 char image1[N][N], image2[N][N], skeleton1[N][N], skeleton2[N][N];
 2 char *im_in, *im_out, *skel_in, *skel_out; ...
 3 ... /* read input image to image1 and reset skeleton1 */
 4 pixel_on = true; count = 0;
 5 while (pixel_on) {
 6   pixel_on = false; count++;
 7   im_in    = (count % 2 == 1) ? image1    : image2;
 8   im_out   = (count % 2 == 1) ? image2    : image1;
 9   skel_in  = (count % 2 == 1) ? skeleton1 : skeleton2;
10   skel_out = (count % 2 == 1) ? skeleton2 : skeleton1;
11   for (x=0; x<N+2; x++)
12     for (y=0; y<N+1; y++) { /*** candidate loop ***/
13       if (x<N && y<N)
14         im_out[x,y] = (y<2 || y>=N-2 || x<2 || x>=N-2) /* A */
15                       ? 0 : Fmin(im_in,x,y);
16       if (x>=2 && y>=1) {
17         filter      = (y<3 || y>=N-1 || x<4 || x>=N)   /* B */
18                       ? 0 : Fmax(im_out,x-2,y-1);
19         pixel = im_in[x-2,y-1] - filter;
20         skel_out[x-2,y-1] = skel_in[x-2,y-1] | pixel;
21         if (pixel==255) pixel_on = true;
22       }
23     }
24 }
```

Fig. 1. Skeletonization program.

for updating pixel_on are omitted for clarity. Fmin and Fmax each contain 20 comparators and multiplexers connecting the inputs.

Before coprocessor operation, scalar input variables must be loaded into the FPGA registers, and input arrays copied to the local memory. During operation, values from the arrays whose index depends on the loop index (which we call vector inputs) must be read, and vector outputs written in every iteration. In our example, 42 values have to be read and two written for each iteration computing a pixel. We refer to these inputs and outputs collectively as *vector ports* of the circuit. Though the dataflow graph is acyclic and thus could be pipelined and produce an output in every clock cycle, the performance of the circuit is very low because of the memory accesses.

Additionally, dependences carried by the candidate loop can prevent legally pipelining the dataflow graph and must therefore be analyzed. Our example loop contains such dependences from statement A which writes im_out[x,y] (line 14 in Figure 1) to statement B (line 18) which reads this value in subsequent iterations (as im_out[x,y-1] and im_out[x,y-2]). However, these are regular dependences (with constant dependence distances) which do not prevent pipelining but require some changes to the dataflow graph. They will be explained for our example in the next section.

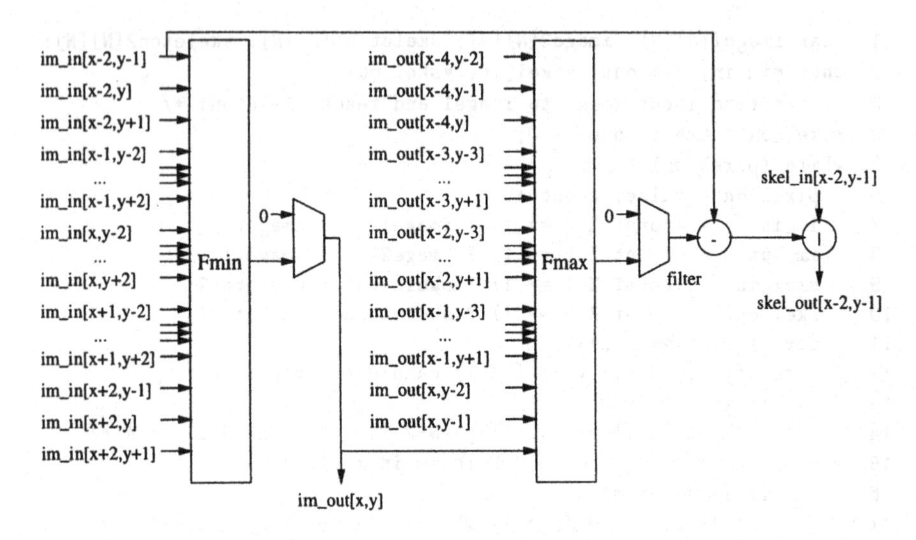

Fig. 2. Dataflow graph for Figure 1.

In contrast to classical vectorization, we do not explicitly generate vector instructions. Rather all instructions of the loop body are vectorized implicitly by pipelining the dataflow graph.

3 Access Equivalence and Shift Registers

It is obvious that the vector inputs in Figure 2 access many pixels of `im_in` and `im_out` repeatedly in successive loop iterations. These redundant reads can be avoided by combining vector ports and storing reused pixels in FPGA registers. To do this automatically, we define classes of *equivalent vector accesses* which access the same array element in different loop iterations.

Definition 1 (Equivalence of Vector Accesses) *Two vector accesses differing only in one index are* equivalent with respect to index variable I *iff the indices have the form $C_1 + S \times I$ and $C_2 + S \times I$, where S, C_1 and C_2 are constants,* and *C_1 mod $S = C_2$ mod S. In other words, the indices are linear functions of I with the same "stride" S, differ only in the constant term, and the constants are in the same residue class modulo S. The accesses'* distance *is $D = (C_1 - C_2)/S$. All equivalent vector accesses form an* equivalence class.

For instance, input ports `im_in[x-2,y-1]`, `im_in[x-2,y]` and `im_in[x-2, y+1]` in Figure 2 differ only in the second index and have the form $C + S \times I$ (with $S = 1$, $I =$ y and $C = -1$, 0 and 1, respectively). Since they are all equal *mod* 1, they form an equivalence class with respect to the candidate loop's index variable `y`. In contrast, accesses `A[2*I]` and `A[2*I-1]` are not equivalent with respect to `I`. The index has the form $C + S \times I$ for both accesses (with $S = 2$), but the modulo condition is not met. This shows that the expressions can never access the same array element.

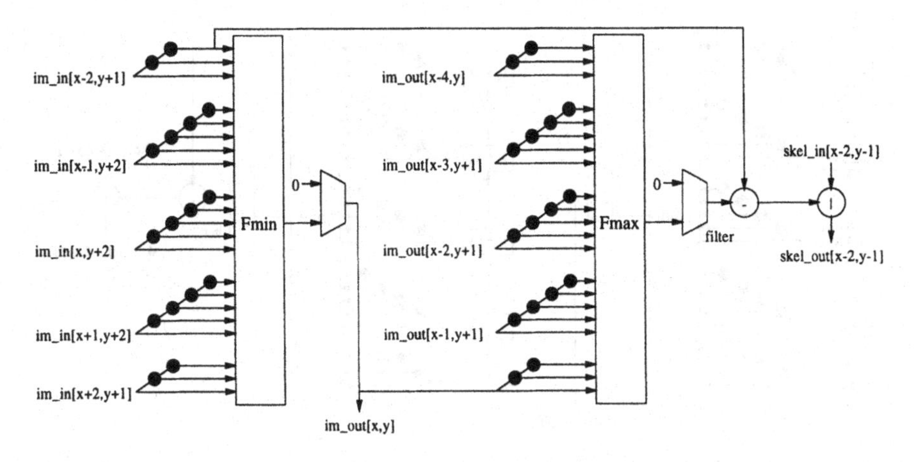

Fig. 3. Circuit with shift registers.

Ports for equivalent vector accesses with respect to the candidate loop's index variable can be combined. Thus only one vector port is necesssary for an entire access equivalence class. The vector input with the largest C (for $S > 0$) or the smallest C (for $S < 0$) for each class is retained and connected to the input of a shift register. Its length is chosen as the largest distance D to any other input in its class. Then every other input in the class is substituted by an access to the value delayed by the distance D for that input in the shift register.

In Figure 2, input port `im_in[x-2,y+1]` is retained and connected to a shift register of length two, since the largest distance D is two (between `im_in[x-2,y-1]` and `im_in[x-2,y+1]`). Input ports `im_in[x-2,y]` and `im_in[x-2,y-1]` are substituted by connections to the shift register. Figure 3 shows the resulting circuit. Registers are represented as black dots. All inputs to the other image lines are combined similarly; and the inputs for `im_out[x]` even use the values produced by `Fmin`. So no vector input for `im_out[x]` is necessary at all, and the dependences mentioned in Section 2 which prohibited pipelining the dataflow graph are removed. Thus memory reads, the actual computation, and memory writes can be pipelined. Since the circuit is acyclic, the combinational delay could be reduced to the delay of a single logic cell or FPGA look-up table by pipelining. However, the throughput is still limited by the time required for the remaining 10 vector read and two vector write accesses per loop iteration. Obviously, additional time is needed for filling the shift registers and the pipeline stages. As long as we have moderately long loops, the overhead is outweighed by the reduced time for each loop iteration. We have implemented the combination of input registers in an earlier prototype compiler [4, 5].

4 RAM Inference

The use of shift registers presented in the previous section greatly reduces the number of memory accesses. However, the method is restricted to data reuse

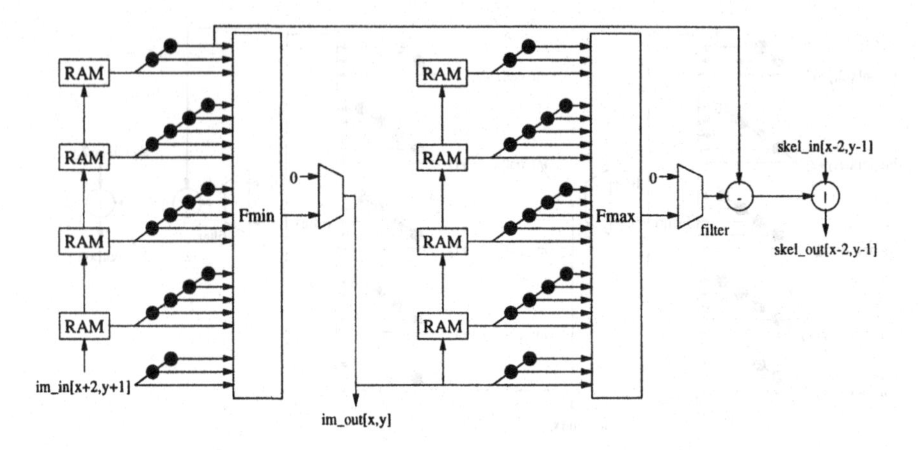

Fig. 4. Circuit with shift registers and on-chip RAMs.

within one pipeline execution, namely within the candidate loop. Further reductions are possible if data accessed in a previous pipeline execution can be reused. We have to store entire slices of an array on-chip. Since this is infeasible with shift registers built from ordinary FPGA registers, we utilize on-chip RAM available on newer FPGA architectures like Xilinx XC4000, Virtex or Altera FLEX10K instead.

Analysis of the next outer loop of a candidate determines which redundant accesses can be removed. We call this process *RAM inference* since it synthesizes small RAMs from a program like HDL synthesis systems use register inference to determine which variables need to be stored in registers. RAM inference is only legal if non-candidate code in the next-outer loop does not alter the arrays in question. If this condition is met, we determine equivalence classes according to Definition 1, but this time with respect to the next-outer loop's index variable. In every equivalence class only one access is necessary. The others are substituted by accesses to new on-chip RAMs which store all values accessed during an entire pipeline execution. The transformation is analoguous to the synthesis of shift registers, but instead of a register an entire small on-chip RAM bank is synthesized. In our example program, the outer `for` loop (index `x`) does not alter any arrays outside the candidate loop and carries no dependences, and all accesses to `im_in` and `im_out` are in respective equivalence classes. Thus only `im_in[x+2,y+1]` needs to be retained. All older image lines are stored in separate on-chip RAM banks which can be accessed concurrently; the RAM banks serve as delay lines. All accesses to `im_out` can directly use stored output values from `Fmin` as well. Thus the number of vector inputs is further reduced to two. Figure 4 shows the resulting circuit. The RAM control logic is omitted for clarity.

Though RAM inference can provide big speedups, it also uses many FPGA resources, especially if the candidate loops are long and consequently large RAMs are needed. Thus a time/area trade-off arises. A hardware estimater must determine the number of RAMs of the required size which fit on the given resources, and select an appropriate implementation.

5 Array Allocation

The previous sections introduced techniques to reduce the required number of external memory accesses. However, the pipeline's throughput is actually determined by the *time* required for these accesses. On architectures with a single memory bank, this time is equivalent to the number of vector accesses or ports. For architectures with multiple memory banks, the throughput can be increased by scheduling as many accesses as possible in parallel. By distributing the ports over the memory banks, the maximal number of accesses required per iteration in one single bank can be minimized.

Since we do not attempt to distribute arrays over several memory banks, all accesses to the same array necessarily have to be sequentialized. We therefore aim to allocate a program's arrays to memory banks such that the resulting accesses are as evenly distributed as possible. The banks must also be long enough to hold the allocated arrays.[1] This is represented by the following problem specification. It applies to a single candidate loop.

Problem 1 (Simple Array Allocation) *Allocate m arrays $A_1, ..., A_m$ of lengths $l_1, ..., l_m$ and with $p_1, ..., p_m$ ports (accesses in one candidate loop iteration) to n memory banks $B_1, ..., B_n$ of size $s_1, ..., s_n$ such that the maximal number of ports allocated to a bank (via the arrays) is minimal and the arrays fit on the banks, that is find a mapping $M = [1:m] \to [1:n]$ which mimimizes*

$$maxp = \max_{i \in [1:n]} \left(\sum_{j \in [1:m], M(j)=i} p_j \right)$$

and fulfills the following constraints:

$$\forall i \in [1:n] : \sum_{j \in [1:m], M(j)=i} l_j \leq s_i$$

Unfortunately this simple problem specification is inaccurate for many programs since accesses can refer to different arrays in different instances of a candidate loop, which we call *situations*. In the program in Figure 1, for instance, accesses to `im_in` can refer to either `image1` or `image2`. Counting both references would greatly overestimate the access numbers and would miss the information that `image1` can either serve as `im_in` or `im_out`, but never as both at the same time. So it is useful to distinguish situations whenever possible, as those for odd and even values of `count` in the example program. We extend problem 1 accordingly.

[1] We do not consider differing memory widths. They could easily be modelled as additional constraints for the allocation.

Problem 2 (Simultaneous Array Allocation for Situations) *Allocate m arrays to n memory banks as in Problem 1. Let p_j^k be the number of ports for array A_j in* situation S_k, *for situations $S_1, ..., S_o$. A simultaneous allocation for all situations must minimize maxp defined as follows:*

$$maxp = \max_{k\in[1:o]} \left(\max_{i\in[1:n]} \left(\sum_{j\in[1:m],M(j)=i} p_j^k \right)\right)$$

We formulate the more general Problem 2 as an *integer linear program (ILP)* and use a standard ILP solving algorithm to find an optimal solution. Although this algorithm is potentially exponential, it works efficiently for the problem at hand since the number of memory banks, arrays and situations found in practical applications is typically less than five. The ILP determines the maximum access number *maxp* and the 0-1 variables $x_{j,i}$ for $j \in [1:m]$ and $i \in [1:n]$ which determine the mapping M: $x_{j,i} = 1 \Leftrightarrow M(j) = i$. The cost function is simply *maxp* which must be minimized. The solution is subject to the following constraints:

$$\forall k \in [1:o], \forall i \in [1:n] : \sum_{j\in[1:m]} p_j^k \times x_{j,i} \le maxp \quad (1)$$

$$\forall j \in [1:m] : \quad \sum_{i\in[1:n]} x_{j,i} = 1 \quad (2)$$

$$\forall i \in [1:n] : \quad \sum_{j\in[1:m]} l_j \times x_{j,i} \le s_i \quad (3)$$

Constraints (1) ensure that *maxp* is met for all banks and situations. Constraints (2) guarantee that M is a function, while (3) represent the memory length constraints.[2]

The example circuit with shift registers (Figure 3) provides the following input values: $p_{image1} = 5$, $p_{image2} = 5$, $p_{skeleton1} = 1$, $p_{skeleton2} = 1$ for both situations (`count % 2 = 1` and `count % 2 = 0`). Given that two memory banks of sufficient size ($s \ge 2 \times N \times N$) are available, `lp_solve` [12] allocates `image1` and `skeleton1` to B_1 and `image2` and `skeleton2` to B_2. $maxp = 6$, thus the 12 accesses are evenly distributed in both situations.

This method can be extended to find a single array allocation for an entire program by combining constraints for the set L of all loops moved to coprocessors in a single ILP. The cost function must be extended to $\sum_{l\in L} w_l \times maxp_l$ where w_l is the estimated iteration count for loop l during the entire program execution. This approach is taken in [10]. A combined allocation has the advantage that data need not be rearranged between coprocessor executions, but it may not be optimal for every single coprocessor. However, if host access to the data is required between the executions, the data have to be copied back and forth anyway. Hence optimal independent allocations should be used in some cases, depending on a dataflow analysis. To make the situation worse, the set L cannot

[2] Note that dual-ported memory can be easily modelled. Constraints (1) must be replaced by $\frac{1}{2}\sum_{j\in[1:m]} p_j^k \times x_{j,i} \le maxp$ if B_i is dual-ported, that is accommodates twice as many accesses as single-ported memory.

Implementation	A - no optim. (Figure 2)	B - shift reg.s (Figure 3)	C - shift reg.s + RAMs (Figure 4)
Memory accesses	44	12	4
Access cycles with 1 bank	44	12	4
Access cycles with 2 banks	22	6	2
Access cycles with 4 banks	21	5	1

Table 1. Performance for skeletonization program.

be determined before array allocation since it depends on the achieved performance for a loop. Hence the dataflow analysis and the estimates for w_l required for optimal allocation have to be combined with hardware/software partitioning [13]. This is beyond the scope of this paper.

6 Results

Table 1 summarizes the performance characteristics of the skeletonization program in Figure 1. There are three implementations: without optimizations (A), with shift registers (B), and with shift-registers and small RAMs (C) on systems with one, two and four memory banks. Obviously implementations without any optimizations (A) are infeasible because they require over 20 access cycles in any case. Shift registers reduce the number to between 5 and 12 cycles, depending on the number of memory banks. RAM inference yields another cycle reduction — and hence execution speedup — by a factor three (for one or two banks) or five (for four banks). Even with optimal allocation of the arrays, the number of access cycles does not linearly fall with the number of memory banks, because accesses with the same arrays are bound to one bank. This is why moving from two to four banks for implementation B only speeds up the coprocessor by 17 % rather than 50 % as for implementation C.

We implemented versions B and C on an RC1000-PP [9] board with two memory banks in Handel-C [14]. For a circuit running at 20 MHz, we measured 6 ms for one iteration of the outer `while` loop on a 128 × 128 image for implementation B, and 2 ms for implementation C — a speedup factor of three as expected. We measured 65 ms per iteration for software execution on a 300 MHz Pentium PC. Thus implementation C with two memory banks is about 32 times faster than software, and the anticipated speed for four banks (1 ms) is 65 times faster. However, there is a small overhead per image for transferring the data (2 ms). Since a typical image needs around 15 to 30 skeletonization iterations, the impact of this overhead is relatively small. Configuring the Xilinx XC4085 FPGA takes 780 ms on the RC1000-PP; only one configuration is required during the entire program execution.

7 Conclusion and Future Work

We have discussed the crucial impact of external memory accesses on the performance of I/O bound reconfigurable coprocessors. We presented methods to synthesize shift registers and small on-chip RAMs to reduce significantly the number of accesses. A technique to allocate arrays to memory banks has also been developed, which minimizes the overall memory access time. We used an extended example to demonstrate the effects of our optimizations. Our approach has been validated by a prototype implementation on the RC1000-PP board.

We are currently including the methods described here in our pipeline compiler based on the SUIF compiler framework [15]. Our methods can be used to extend hardware compilers such as Handel-C as well. Future work includes evaluating different types of on-chip memory in the latest FPGA families. Further investigations on the area/time trade-offs of RAM inference is also required to automatically select an optimal implementation.

References

1. M. Weinhardt and W. Luk. Pipeline vectorization for reconfigurable systems. In *Proc. FCCM'99.* IEEE Computer Society Press, 1999.
2. G. Haug and W. Rosenstiel. Reconfigurable hardware as shared resource in multipurpose computers. In *Proc. FPL'98.* Springer, 1998.
3. M. B. Gokhale and J. M. Stone. NAPA C: compiling for a hybrid RISC/FPGA architecture. In *Proc. FCCM'98.* IEEE Computer Society Press, 1998.
4. M. Weinhardt. Compilation and pipeline synthesis for reconfigurable architectures. In *Reconfigurable Architectures Workshop RAW'97*, 1997.
5. M. Weinhardt. *Übersetzungsmethoden für strukturprogrammierbare Rechner (Compilation techniques for structurally programmable computers, in German).* PhD thesis, Universität Karlsruhe, July 1997.
6. S. A. Guccione and M. J. Gonzalez. A data-parallel programming model for reconfigurable architectures. In *Proc. FCCM'93.* IEEE Computer Society Press, 1993.
7. S. Nisbet and S. A. Guccione. The XC6200DS development system. In *Field Programmable Logic and Applications*, LNCS 1304, 1997.
8. D. A. Buell, J. M. Arnold, and W. J. Kleinfelder. *Splash 2 - FPGAs in a Custom Computing Machine.* IEEE Computer Society Press, 1996.
9. Embedded Solutions Limited. *RC1000-PP Product Information Sheet.* `http://www.embedded-solutions.ltd.uk/ProdApp/RC1000PP.htm`.
10. M. B. Gokhale and J. M. Stone. Automatic allocation of arrays to memories in FPGA processors with multiple memory banks. In *Proc. FCCM'99.* IEEE Computer Society Press, 1999.
11. H. R. Myler and A. R. Weeks. *Computer Imaging Recipes in C.* Prentice Hall, 1993.
12. M. Berkelaar. Unix manual page of `lp_solve`. Eindhoven University of Technology, Design Automation Section, 1992.
13. M. Weinhardt. Integer programming for partitioning in software oriented codesign. In *Proc. FPL'95.* Springer, 1995.
14. Embedded Solutions Limited. *Handel-C Reference Manual*, 1998.
15. The Stanford SUIF Compiler Group. Homepage `http://suif.stanford.edu`.

Analysis and Optimization of 3-D FPGA Design Parameters

Silviu M.S.A. Chiricescu and M. Michael Vai

Northeastern University, ECE Department
Boston, MA 02115, USA
{schirice,vai}@ece.neu.edu

Abstract. This paper describes an experimental approach developed to determine the design parameters of a 3-D FPGA. The FPGA architecture created with these design parameters has high performance while the requirement of balancing the areas of its constituent layers is satisfied.

1 Introduction

This paper proposes an alternative FPGA architecture that uses a 3-D VLSI technology[1]. The work presented on this paper represents a further development of our earlier research efforts [2, 3].

It has been shown that a 3-D FPGA has a smaller footprint and a faster routing structure than a 2-D FPGA counterpart [2, 4]. A straightforward extension of a 2-D FPGA was used for the design of Rothko [2]. Rothko is a sea-of-gates type 3-D FPGA in which every cell can be connected to its immediate neighbors in the planes above and below, and to its immediate neighbors in its own plane. We have observed that, when one maps complex designs onto Rothko, a relatively large number of routing and logic blocks (RLBs) is used for routing purposes only. In order to alleviate this problem, we have decided to split the routing and logic resources into two layers. One cannot completely separate the routing structure from the logic structure without wasting silicon estate since the routing structure area is much larger than the logic area. The challenge is to determine how to distribute the routing structure among these two layers. Besides the routing layer and the routing and logic block (RLB) layer, we also include a memory layer that stores extra configuration bits for both the RLB and routing layers. The FPGA can switch from one configuration to another without having to access an external memory.

2 FPGA Architecture

The general architecture of our FPGA is presented in Fig. 1. It consists of three layers: the RLB layer, the routing layer and the memory layer. As mentioned before, we have separated the logic layer from the routing layer. However, this is not a complete separation. Instead, we have decided to implement in the RLB layer part of the routing structure that will be used for routing short nets.

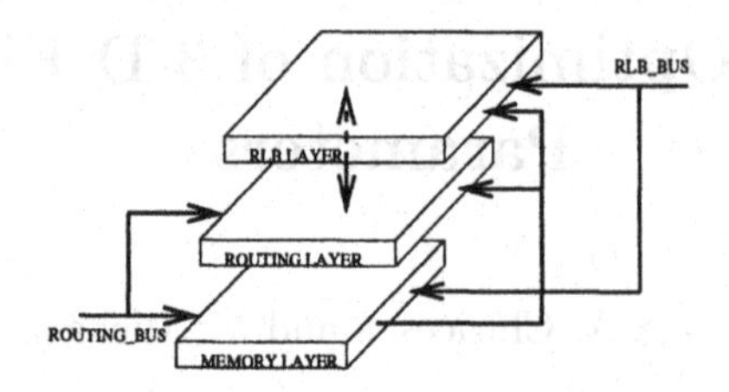

Fig. 1. General structure of the 3-D FPGA

2.1 RLB Layer

The RLB layer consists of RLBs that have direct connections with their north, east, south and west neighbors. Each RLB can be configured to implement a D-type register and an arbitrary logic function of up to four variables. Each RLB, in addition to implementing logic functions, can perform some simple routing tasks (e.g., provide a connection between two of its neighbors). The RLB's ability to perform limited routing tasks allows a large percentage of the connections to be routed in the RLB layer. The RLB layer is organized into clusters. For a general discussion, we consider each cluster as an array of $m \times m$ RLBs. Each cluster is associated with a switch-box in the routing layer (RL). The interconnections between one of the RLBs in a cluster and the cluster's associated switch box are shown in Fig. 2. Each RLB can send R_o signals to the RL (directly into the switch-box) and can receive R_{in} signals from the RL (directly from the switch-box).

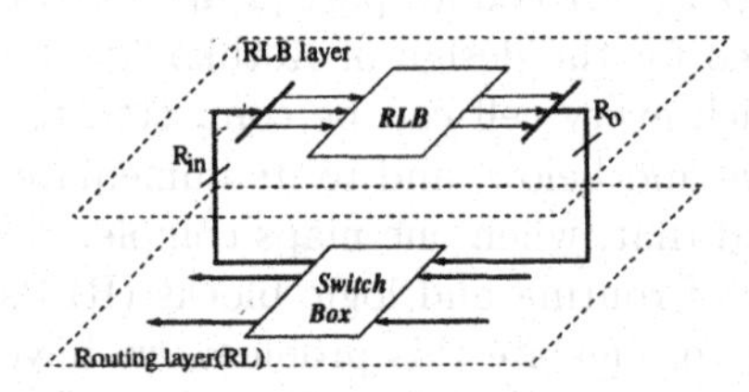

Fig. 2. Connections between an RLB and a switch-box

2.2 Routing Layer (RL)

The RL consists of switch-boxes that are organized in a mesh structure (without wraparound). The structure of the switch-box is presented in Fig. 3. The switch-box accepts nR_o signals from the RLB layer (where $n = m^2$ and it represents the number of RLBs in a cluster) and outputs nR_{in} signals to the RLB layer. The switch-box also accepts 4W inter-cluster signals and outputs another 4W inter-cluster signals, where W represents the number of connections, in each direction, between 2 adjacent switch-boxes. The switch-box is used to provide both intra- and inter-cluster connections. It is important to emphasize that the switch-box provides a limited connectivity between its incoming and outgoing signals. This is the result of two important architectural decisions. First, local routing is performed in the RLB layer, therefore it is not necessary to provide

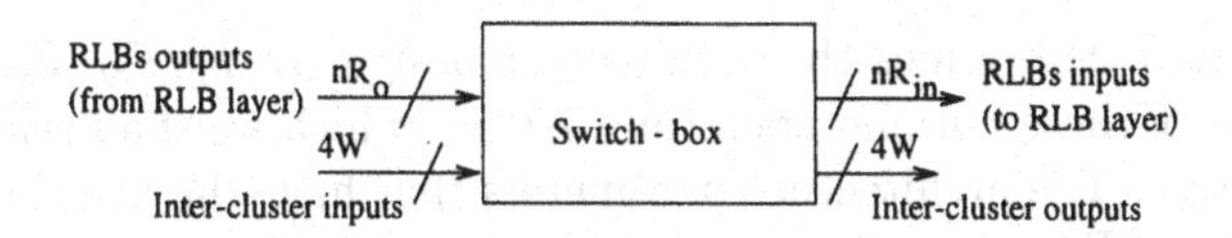

Fig. 3. General structure of the switch-box

100% intra-cluster connectivity. Second, a complete connectivity between connections will increase the switch-box area and the balance between the RLB and RL layer areas will be adversely affected.

The following notations are used in the discussion below. The total number of input signals which are connectable to an output signal o is defined as the cardinality of o, C_o. The domain of an output signal o, D_o, is defined as the set of input signals from which the connectable input signals are chosen. Fig. 4 presents the domain and the cardinality of an output signal o that has $Do = \{\alpha_i | i = [1 \ldots 5]\}$ and C_o=3. Each of the switch-box outputs going into the RLB layer has C_k=F_c (where k in $[1 \ldots nR_{in}]$). The domain of each of these output signals is composed of all the switch-box inputs (RLBs outputs plus the inter-cluster inputs, i.e., nR_o+4W). Each of the 4W inter-cluster output signals has two domains, D_1 and D_2. The first domain is composed of the inter-cluster input signals (4W in Fig. 3), while the second domain is composed of signals coming from the output of the RLBs in the given cluster (nR_o in Fig. 3). The cardinalities of the inter-cluster output signals in D_1 and D_2 are F_s and F_{c1}, respectively.

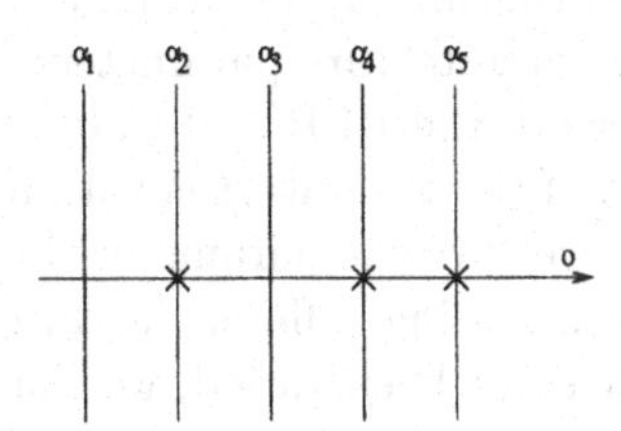

Fig. 4. The domain and cardinality of an output signal o

2.3 Memory Layer

The memory layer is used to store extra configuration bits for both the RLB and the routing layers. The incorporation of a memory layer enables the FPGA to perform a context switch without having to fetch the new configuration from an external memory. This feature helps to decrease the reconfiguration time. The memory can be updated through an external bus during the normal operation of the FPGA. This feature further decreases the reconfiguration time of our FPGA.

3 Experimental Methodology

This section describes the experimental methodology that was used for determining the architecture parameters. The following is a list of parameters (described in Section 2) that determines the architecture design: $W, F_c, F_{c1}, F_s, n, R_i$, and R_o. The values of these parameters form a 7-tuple which we shall refer to as the architecture state. The architecture parameters that have the most influence on the area of the RLB and routing layers form a set of primary parameters. The methodology steps are presented in Fig. 5.

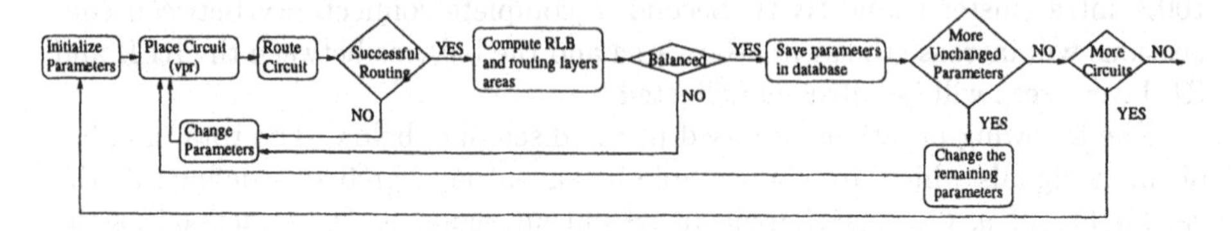

Fig. 5. Experimental methodology

3.1 Experimental Methodology Steps

The first step is to initialize the architecture parameters. The second step is to place a test circuit on the RLB layer. The placement is done using the vpr tool [5]. Circuits from the LGSynth93 suite [6] will be used as test circuits. All the circuits are optimized using SIS [7] and mapped to 4-input look-up tables. The third step in Fig. 5 is to route the test circuit. For this step we have created a customized routing tool. If the test circuit is successfully routed (at least a certain percentage of connections are routed), we proceed with the calculation of the areas of the RLB and routing layers. If the areas of these two layers are not balanced, we modify the architecture parameters (one at a time) and route the circuit again. If the areas of the RLB and routing layers are balanced (the difference between the areas of the two layers is within a tolerable percentage of the larger of the two areas), we save the parameters in a database. Note that we can successfully route a circuit using different architecture states. If there are primary parameters that have not been varied, we change these parameters and go back to step two. We repeat the above steps for different circuits. The final architecture parameters are computed by taking an average of the best obtained solutions over all the circuits in the benchmark.

4 Area Analysis

This section presents the area estimation models that were used to calculate the areas of the RLB and routing layers. We also rank the architecture parameters based on their influence on the areas of the two layers.

4.1 Area Estimation for the Routing Layer

In Section 2 we have presented the organization of the routing layer. Recall that the basic constituent block of the routing layer is the switch-box. The switch-box can be built using three basic primitives: partially populated crossbar switches, multiplexers, or subset selectors.

A subset selector uses the fact that some of the switch-box outputs are interchangeable since they represent inputs of look-up tables.

The structure that will be used to implement the switch-box has to satisfy the following requirements: 1)the area has to be small; 2)the number of transistors that have to be switched on in order to connect two paths has to be minimized; 3)the interconnection pattern can be described with a minimum number of configuration bits.

It is important to note that the above conditions cannot be simultaneously satisfied by any of the above structures. After a careful analysis including, among others, area estimation and reconfiguration speed, we eliminate the sub-set selector from the list of candidates, mainly due to its design irregularity and large area. Throughout this section, we will consider that the switch-box to be implemented has n inputs and k outputs.

The areas of the crossbar implementation and the multiplexer implementation are found to be:

$$A_{crossbar} = kn(f(A_m + A_t) + 2L_w(L_w + \sqrt{A_m})) \quad (1)$$

$$A_{mux} = k \log_2 n A_m + n(2n-2)A_t + 2L_w n\sqrt{A_t}(n+6)\lceil k/6 \rceil, \quad (2)$$

where A_m represents the area of a memory cell ($\approx 1k\lambda^2$)[1], A_t represents the area of a pass transistor($\approx .2k\lambda^2$), L_w is the wire pitch ($\approx 8\lambda$), and f represents the occupancy percentage of the crossbar.

Comparison of Crossbar and Multiplexer Implementations In order to evaluate these implementations, we have to consider two factors: the areas and the reconfiguration times of these two implementations. The merits of the two implementations will be evaluated by using a cost function that allows us to explore the area-configuration speed trade-off. The cost function is the product between the area of the implementation and the configuration delay. The configuration delay is a function of both the number of configuration bits and the settling time.

Fig. 6 plots the difference between the areas occupied by the multiplexer implementation and the crossbar implementation for 3 different values for k and 2 different values for f. After analyzing the areas of the two implementations, it turns out that if $n > 20$ or if the crossbar is partially populated, it is more economical, silicon wise, to implement the switch-box using a crossbar like structure (see Fig. 6).

After carefully considering the configuration times of the two implementations, and computing the cost function for the two implementations, we have decided to use partially populated crossbar-switches as switch-boxes.

4.2 Area Estimation for the RLB Layer

The area of a cluster is the product between the area of an RLB cell and the number of the cells that form the cluster. The area of an RLB cell is given

[1] λ =1/2 of the minimum feature size for a VLSI technology.

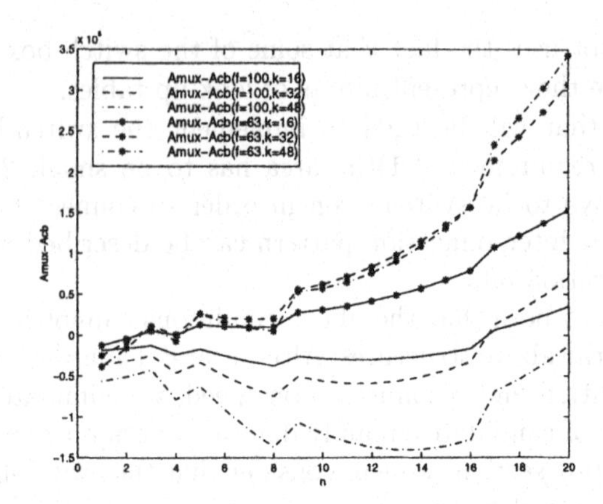

Fig. 6. $A_{mux} - A_{cb}$ for different values of k and f

by the sum of the areas of its constituent components, the area of the internal routing, and the area of the vertical 3-D vias. The configuration memory is also included in the RLB layer estimation. The internal structure of an RLB cell is presented in Fig. 7. Recall from section 2.1 that each RLB can send to the RL R_o signals. We have decided to set R_o to 1. In Fig. 7, R_1 and R_2 represent inputs from the routing layer. R_{out} is the RLB's output signal that is being sent to the routing layer. The N_{in}, E_{in}, W_{in}, and S_{in} are inputs coming from the RLB's corresponding neighbors, while $N_{out}, E_{out}, W_{out}$, and S_{out} are outputs going to the RLB's corresponding neighbors. The RLB requires 24 configuration bits. The internal routing area was taken to be ≈20% of the area of the RLB constituent components. The area of a vertical 3-D via, for a .25μm process is 9.77μm^2. However, the 3-D technology requires an additional spacing (used for the vertical alignment of the layers) in one of the layers. We have decided to include the additional spacing in the RLB layer, therefore the area of a via in the RLB layer is 13.14μm^2 and is larger than the area of a via in the RL.

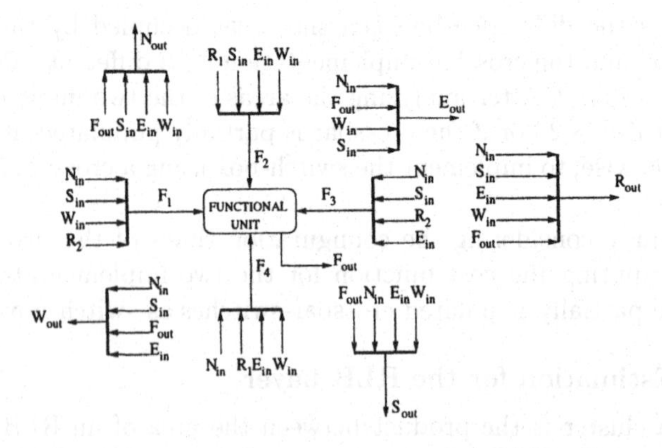

Fig. 7. The internal structure of an RLB cell

4.3 Parameter Ranking

The areas of the RLB and routing layers are most sensitive to the size of the cluster. We have investigated multiple sizes for the cluster. Recall that our primary goal is to balance the areas of the RLB and routing layers. The cluster size that accomplishes our goal is sixteen[2]. Therefore, the cluster will be organized as an array of 4×4 RLBs.

W and R_{in} have a great impact on the area of the routing and RLB layers. Fig. 8 presents the area of the switch-box versus R_{in} and W. The plots in Fig. 8 assume that F_c and F_{c1} were kept constant at 60% of their maximum values. It can be easily seen that the switch-box is more sensitive to R_{in} than to W. The next logical step is to determine how F_c and F_{c1} affect the area of the switch-box. We kept R_{in} and W constant (to some values that allowed the router to achieve a high routability percentage). We let F_c and F_{c1} to vary between 70 and 90 percent of their maximum allowable values and we compute the area of the switch-box that is implemented as a crossbar. Fig. 9 shows a variation of the area of the switch-box with respect to F_c and F_{c1}. From Fig. 9 we can see that the switch-box is more sensitive to an increase in F_c than to an increase in F_{c1}. This variation is expected since the two parameters represent percentages from two different quantities. By analyzing Fig. 8 and 9 we can see that the area of the switch-box is more sensible to an increase in R_{in} than to an increase in F_c. This conclusion is very important and it will guide us in the process of choosing the architecture parameters.

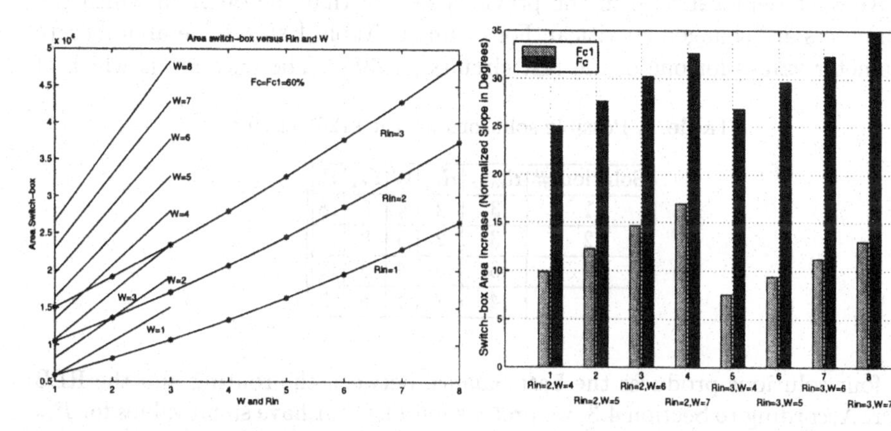

Fig. 8. Area switch-box versus R_{in} and W showing that the switch-box is more sensitive to R_{in} than to W

Fig. 9. Area switch-box (implemented as a crossbar) versus F_c and F_{c1}. The switch-box is more sensitive to F_c than to F_{c1}

[2] Due to the space limitation the analysis will not be shown. Interested readers can contact the authors for further information.

5 Experimental Results

We have applied the methodology described in section 3 to 10 circuits from the LGSynth93 benchmark suite. The FPGA architecture was created to fully support partial reconfiguration [8]. During a normal operation, a part of the FPGA will contain one circuit while another part can be reconfigured to carry out a different application. We decided that a circuit that has a size of 300 to 400 4-input LUTs is a reasonable candidate for being reconfigured. Therefore, we have optimize the architecture parameters to minimize the reconfiguration times of circuits of this size.

During our experiments we have fixed the values of two parameters: n (cluster size) and F_s. We have previously shown that a cluster formed by an array of 4×4 RLBs yields a balanced implementation. F_s was kept constant at 3.

We have demonstrated in the previous section that the order in which the parameters are changed is critical. For example, Table 1 shows the architecture parameter values for one of the test circuits (s1238). The question is which of

Table 1. Possible solutions for the s1238 circuit

Solution #	R_{max}	R_{in}	W	F_{c1}	F_c
1	3	2	4	8	18
2	3	2	5	8	12
3	3	3	4	8	12
4	4	2	5	8	4

the four solutions produces the best balance between the routing and the RLB layer. According to Section 4.3, we prefer solutions that have small values for R_{in} (since the area of the routing layer is more sensitive to an increase in R_{in} than to an increase in W). However, a smaller R_{in} yields a larger W. This behavior is not unexpected since the router tries to compensate for a loss in vertical flexibility by adding more inter-cluster wires.

Table 2 presents the best solutions for the 10 circuits. The best solutions have the primary parameters listed from left to right in the order they affect the balance between the areas of the routing and RLB layers. The last parameter, R_{max}, is not a physical parameter. However, this routing parameter affects the quality of the results. If R_{max} is too large, then a large number of medium size nets will be routed in the RLB layer, and the delays of these nets will become significant (affecting the critical path). If the value of R_{max} is too small, then a lot of resources will be needed in the routing layer, making its area unnecessarily larger. The values listed under the F_c and F_{c1} lines represent the flexibilities of the outputs of the switch-box (see Section 4.1). The average values for F_c and F_{c1} are 27.6% and 58.75 %, respectively, of their maximum possible values.

For the average values presented in Table 2, we have plotted in Fig. 10 the difference between the areas of the routing and RLB layers. We can see that the areas of the two layers are almost balanced. The "extra" area that is in the routing or RLB layers can be used to store additional configuration bits for both

Table 2. Best solutions for the 10 circuits

Circuit	s1	t481	mm9b	C880	s953	styr	s1196	duke2	s1238	e64	average
R_{in}	2	3	2	2	2	2	2	3	2	3	2.3
W	6	7	5	6	7	6	6	6	4	7	6
F_c	16	4	12	16	24	24	16	16	18	32	16.8
F_{c1}	12	8	8	10	8	10	8	12	8	10	9.4
R_{max}	3	3	3	3	3	3	3	3	3	3	3

the routing and RLB layers. The number of configurations that can be stored in the "extra" area is presented in Fig. 11.

The size of the memory layer will be determined by the largest of the routing and RLB layers. The memory layer will be used to store the configurations of both the routing and RLB layers. The number of configurations that can be stored in the memory layer, for a cluster of 4×4 RLBs and average architecture parameters, is presented in Fig. 12.

6 Future Research

This section presents some ongoing improvements and future direction of research. Since the methodology that was employed is experimental, the results are affected by the quality of the CAD tools used. We have mentioned in Section 3.1 that the vpr tool has been used to place the circuits. This tool does not take advantage of our new FPGA architecture. A tool customized for this FPGA can produce better placement results and thus routing results.

Another source of improvement is in the area estimation. Each basic element of the architecture was broken down into small constituent blocks (i.e., multiplexers, latches, memory elements, etc.). The area of each of the constituent blocks was estimated by measuring the actual layout. However, the exact area occupied by the FPGA layers cannot be found unless we evaluate the entire layout.

It is interesting to further investigate how the introduction of a limited number of long segmented wires to connect non-immediate switch-box neighbors will

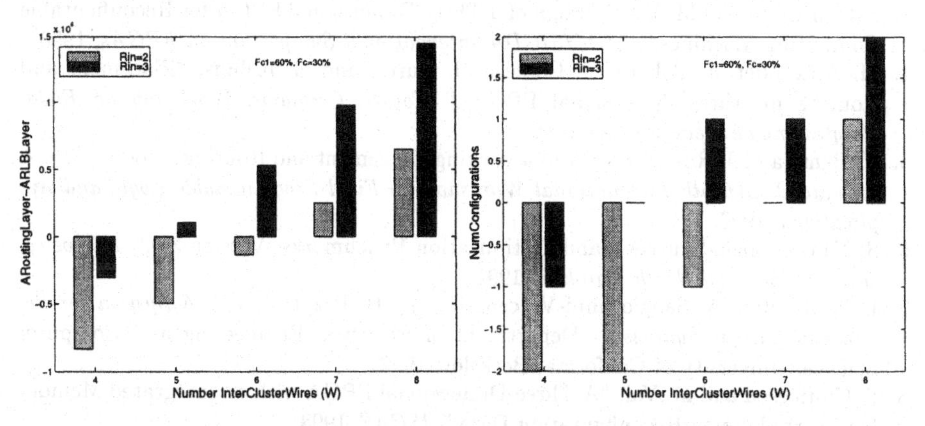

Fig. 10. The difference between the ares of the routing and RLB layers

Fig. 11. The number of configurations that can be stored in the "extra" area

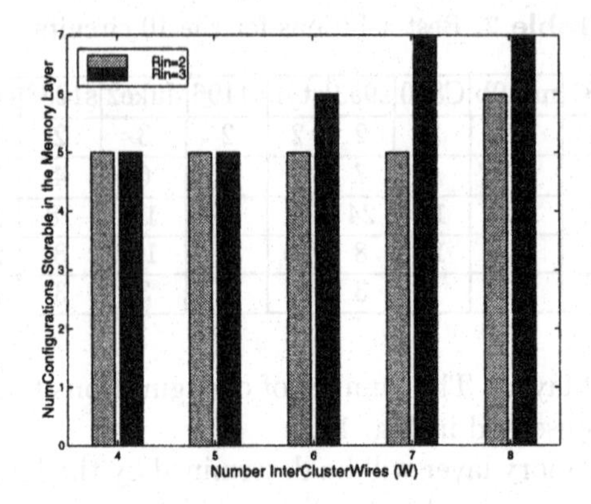

Fig. 12. The number of configurations that can be stored in the memory layer

affect the architecture. We are also in the process to create a software package that will manage the partial-reconfiguration process.

7 Summary

In this paper, we have analyzed some of the parameters involved in the design of a 3-D FPGA. We have ranked the parameters based on the impact that they have on the area of the FPGA's constituent layers. We have found out that n (the number of RLBs in a cluster), R_{in} (the number of RLB inputs coming from the routing layer), and W (the number of inter-cluster wires) have the largest impact on the area of the routing and RLB layers. We have successfully routed multiple designs on our architecture using different architecture states. These values were used to determine our final FPGA parameters.

References

1. P. Zavracky, M. Zavracky, D.-P. Vu, and B. Dingle
2. M. Leeser, M. Vai, W. Meleis, S. Chiricescu, W. Xu, and P. Zavracky, "Rothko: A Three Dimensional FPGA," *IEEE Design and Test*, 1998.
3. S. Chiricescu and M. Vai, "Design of a Three Dimensional FPGA for Reconfigurable Computing Machines," *ACM/SIGDA International Symposium on FPGAs*, 1998.
4. M. Alexander, J. Cohoon, J. Colflesh, J. Karro, and G. Robins, "Placement and Routing for three dimensional FPGAs," *Fourth Canadian Workshop on Field-Programmable Devices*, May 1996.
5. V. Bentz and J. Rose, "Vpr: A new packing, placement and Routing Tool for FPGA Research," *Seventh International Workshop on Field-Programmable Logic and Applications*, 1997.
6. S. Yang, "Logic Synthesis and Optimization Benchmarks, Version 3.0," *Microelectronics Centre of North Carolina*, 1991.
7. E. Sentovitch, A. Sangiovanni-Vincentelli, and R. Brayton, *SIS: A System for Sequential Circuit Synthesis.* Department of Electrical Engineering and Computer Science, University of California, Berkeley, 1992.
8. S. Chiricescu and M. Vai, "A Three-Dimensional FPGA with an Integrated Memory for In-Application Reconfiguration Data," *ISCAS*, 1998.

Tabu Search: Ultra-Fast Placement for FPGAs

John M. Emmert[1] and Dinesh K. Bhatia[2]

[1] Department of Electrical Engineering, University of Kentucky
Lexington, KY 40506, USA

[2] Design Automation Laboratory, Department of ECECS, University of Cincinnati
Cincinnati, OH 45221, USA

Abstract. Two-dimensional placement is an important problem for FPGAs. Current FPGA capacity allows one million gate equivalent designs. As FPGA capacity grows, new innovative approaches will be required for efficiently mapping circuits to FPGAs. We propose to use the tabu search optimization technique for the physical placement step of the circuit mapping process. Our goal is to reduce the execution time of the placement step while providing high quality placement solutions. In this paper we present a study of tabu search applied to the physical placement problem. First we describe the tabu search optimization technique. Then we outline the development of a tabu search based technique for minimizing the total wire length and minimizing the length of critical path edges for placed circuits on FPGAs. We demonstrate our methodology with several benchmark circuits available from MCNC, UCLA, and other universities. Our tabu search technique has shown dramatic improvement in placement time relative to commercially available CAE tools (20×), and it results in placements of quality similar to that of the commercial tools.

Key Words: Placement, FPGA, Tabu Search, CAD

1 Introduction

Placement is an extensively studied topic. However, the importance of FPGA placement cannot ever be ignored due to changing design complexities and requirements. Commercially available field programmable gate array (FPGA) devices can map up to one million gate equivalent designs[5], and some of the newly announced products like Altera's APEX series will map over two million gate equivalent designs[4]. Such complex design densities demand tools that can efficiently and quickly make use of available gates. In general, placement is an NP-hard problem. For FPGAs, it is more difficult due to fixed logic and interconnection resources. Improvements in CAD tools for FPGAs have not kept pace with hardware improvements. The available tools typically require from minutes to hours to map[1] designs (or circuits) with just a few thousand gates, and as design sizes increase, execution times also increase.

We describe a two-dimensional placement algorithm that greatly reduces the execution time of the circuit mapping process. Our two-step placement algorithm maps graph based netlists on a two-dimensional array of point locations. The first step of our placement algorithm, TS_TWL, minimizes the total wire length of the netlist, thereby enhancing circuit routability. The second step of our placement algorithm, TS_EDGE, minimizes the length of critical graph edges, thereby enhancing circuit performance.

[1] typical mapping steps include technology mapping, placement, and routing

2 Tabu Search Overview

Combinatorial optimization is the mathematical study of finding an optimal arrangement, grouping, ordering, or selection of discrete objects usually finite in number[8]. A combinatorial optimization problem can be specified by a set of instances where each instance is associated with a solution space Ω, a *feasible* space X ($X \subseteq \Omega$) that is defined by problem constraints, and an *objective* function C that evaluates each solution $S \in X$.

In general the combinatorial optimization problem is defined:

$$P: \text{ Optimize } C(S) \text{ subject to } S \in X \subseteq \Omega$$

where X and Ω are discrete. The goal of the combinatorial optimization is to find the *optimal* feasible solution $S^* \in X$ such that $C(S^*)$ is optimal $\forall\, S \in X$. Typical combinatorial optimization functions are easy to define, but difficult to solve.

Tabu search (TS) is a meta-heuristic[2] approach to solving optimization problems that (when used properly) approaches near optimal solutions in a relatively short amount of time compared to other *random move* based methods [3]. Unlike approaches that rely on a good random choice (like simulated annealing or genetic algorithms), TS exploits both good and bad strategic choices to guide the search process. In TS, a list of possible moves is created. In the short term, as moves in the list are executed, *tabu*, or restrictions, are placed on executed moves in order to avoid local optima. This *tabu* is typically in the form of a time limit, and unless certain conditions are met (e.g. *aspiration criteria*), the move will not be performed again until the time limit has expired. This short term phase is associated with intensification of the search strategy. Diversification moves the current solutions to unexplored areas of the search space. A general TS procedure is described in procedure GENERIC_TS.

Procedure GENERIC_TS()
  begin
    Initialize solution;
    repeat
      Perform IntenseSearch();
      Perform DiverseSearch();
    until stop criterion met;
    return BestSolution;
  end;

In procedure GENERIC_TS, two basic stages, IntenseSearch and DiverseSearch, are repeated until some stop criteria is satisfied. The range of choices for stop criteria is very broad, and there is no set method. One method is to perform the search until a specific time or number of iterations has elapsed. Alternatively, a reduced rate of improvement can be used to stop the search. When a certain time or number of iterations has elapsed without any improvement in the current best solution, the search terminates. Regardless of the method chosen to terminate the search, the TS methodology can be used to improve most problem solutions found by local heuristics.

A typical intense search is shown in the procedure IntenseSearch. Key to any TS is a move list. A move list consists of legal or valid moves which transform the solution

[2] guides local heuristic search procedures beyond local optima

space from one configuration to another. During the intense stage of TS, new solutions are typically chosen from a local neighborhood, $N(S) \subseteq X$, and when a solution is found that improves the current best solution, it is stored in long term memory. $N(S)$ consists of solutions that can be reached from the current solution space by a *move*. Associated with each move in the move list is an evaluation criteria or cost, $C(S)$, which provides a quantitative means for comparing the relative merits of moves in the list. Typically, a move is found by applying one local heuristic or another to the current solution space. In the procedure IntenseSearch, the function Pick_Move searches the move list of $N(S)$ for the current best move. By successively picking moves that improve the current solution a local optima is determined. In this way, the approach is not so different from typical greedy procedures. At this point we diverge from usual heuristics by applying a short term memory field called *tabu tenure* to each move. The tabu tenure prohibits recently executed moves from being reexecuted. Additionally, it means the current best move may not always be the one chosen. By using a tabu tenure and accepting non-improving moves, TS can climb out of local optima and find solutions more along the lines of a global optima. As stated earlier, updating move list and

```
Procedure IntenseSearch()
begin
    repeat
        Pick_Move();
        Execute_Move();
        Set_Tabu_Tenure();
        Update_Move_List();
        if current solution is an improvement then
            Store current solution;
        end if;
    until stop criterion met;
end;
```

objective function values after every move can be highly computationally intensive. Therefore candidate list strategies have been devised to reduce the computational effort of the TS[3].

The diversification stage moves the current solution out of the local neighborhood $N(S)$ and discourages cycling. For example, the frequency of certain move attributes is a good candidate for long term memory storage. Penalties can be placed on attributes with high frequencies in order to explore other attributes that may lead to better solutions. Another way to diversify the search is to use random methods during the search process. One way is to start over with a new random input. However, this does not take advantage of information from the previous search cycles. A better way to accomplish random diversification is to randomly pick moves from the move list and execute them regardless of effect on the current solution. This is the random method we used in our search.

3 Placement Problem

Given a set of blocks $B = \{b_1, b_2, ..., b_n\}$ and a set of signals $S = \{s_1, s_2, ..., s_q\}$, we associate with each block $b_i \in B$ a set of signals S_{b_i}, where $S_{b_i} \subseteq S$. Similarly, with each signal $s_i \in S$ we associate a set of blocks B_{s_i}, where $B_{s_i} = \{b_j \mid s_i \in S_{b_j}\}$. B_{s_i} is said to be a *signal net*. We are also given a target set $L = \{l_1, l_2, ..., l_p\}$ of locations, where $p \geq \mid B \mid$. The placement problem then becomes how to assign each block $b_i \in B$

to a unique location $l_j \in L$ such that an objective function is optimized. For the case of mapping $b_i \in B$ to a regular two-dimensional array, each $l_j \in L$ is represented by a unique (x_j, y_j) location on the surface of the two-dimensional array where x_j and y_j are integers. Figure 1 shows the 16 element set L for an example 4×4 two-dimensional array. It can be noted that given the sets L and B where $|L| \geq |B|$ there are ${}_{|L|}P_{|B|} = \frac{(|L|)!}{(|L|-|B|)!}$ placements or permutations for the set B mapped to the set L.

Many recent papers have addressed placement for regular arrays. Rose et. al. address ultra fast placement[9]. Mathur et. al. studied the placement problem and presented methods for re-engineering of regular architectures[7]. Callahan et. al. developed a module placement tool for mapping of data paths to FPGA devices[1]. Lim, Chee, and Wu have developed a placement with global routing strategy for placement of standard cells[6]. In most search based methods, there is a tradeoff between the execution time and the quality of the results.

4 TS Based Placement

The primary motivation behind development of our placement strategy was to decrease the execution time of our macro based floorplanner [2]. In this section we formally describe our two-step, TS based placement algorithm for mapping graphs to two-dimensional arrays.

4.1 Placement Model

For our two-dimensional placement solution we convert the hyper-graph input circuit model described above (section 3) to a graph $G = (V, E)$ where $V = \{v_1, v_2, ...v_n\}$, $|V| = n$, $E = \{e_1, e_2, ...e_m\}$, and $|E| = m$. Each vertex $v_i \in V$ corresponds to a circuit block $b_i \in B$; therefore, there is a one to one mapping of the elements of V and the elements of B and thus $|V| = |B|$. Each $v_i \in V$ has a corresponding set of edges E_{v_i}, where $E_{v_i} \subseteq E$. Each edge $e_i \in E$ connects a pair of vertices $(v_j, v_k) \mid v_j, v_k \in V$. The elements of E are created by considering each signal, $s_i \in S$. If we let $b_j \in B_{s_i}$ be the source block for signal s_i then an edge (v_j, v_k) is added to E for each $b_k \in B_{s_i} \mid j \neq k$. At any given time, each element of V is mapped to a unique element of L, and the minimum requirement for mapping is $|V| \leq |L|$. Therefore if we define X as the set of all solutions to the placement problem, then $|X| = \frac{(|L|)!}{(|L|-|V|)!}$.

For the first step of our TS based placement strategy, TS_TWL, we seek to enhance routability by minimizing total wire length (TWL). We conservatively estimate TWL as the sum of the Manhattan length of each edge $e_i \in E$, and use minimization of TWL as our optimization function.

$$TWL = \sum_{\forall e_i \in E} MLength(e_i)$$

For the second step of our TS based placement strategy, TS_EDGE, we seek to enhance circuit performance by minimizing the length of critical circuit edges. To accomplish this, we traverse G and apply a weight, w_i, to each edge in E. Edges in critical paths receive a higher weight. To determine the weight for each edge, we determine a path weight pw_i for each path $p_i \in P$ where P is the set of all paths for G. For simplicity we let pw_i be the maximum level for each $p_i \in P$. Figure 2 shows an example circuit with six paths. In figure 2 path p_1 is at level 1; paths p_2 and p_3 are at level 2; and paths p_4, p_5, and p_6 are at level 3. We associate with each edge $e_j \in E$ a set of paths P_{e_j}, where $P_{e_j} \subseteq P$. For example, in figure 2 we have $P_{e_1} = \{p_1\}$ for $e_1 \in E$, $P_{e_2} = \{p_2, p_3, p_4, p_5, p_6\}$ for $e_2 \in E$, $P_{e_3} = \{p_4, p_5, p_6\}$ for $e_3 \in E$, $P_{e_4} = \{p_2\}$ for $e_4 \in E$, $P_{e_5} = \{p_3\}$ for $e_5 \in E$, $P_{e_6} = \{p_4\}$ for $e_6 \in E$, $P_{e_7} = \{p_5\}$ for $e_7 \in E$, and $P_{e_8} = \{p_6\}$ for $e_8 \in E$. Then, we determine a weight w_j for each edge $e_j \in E$.

$$\forall\, e_j \in E, w_j = max(pw_i) \;\forall\, p_i \in P_{e_j}$$

For example, in figure 2 the weight for edge e_2 is the maximum path weight for each path in the set $\{p_2, p_3, p_4, p_5, p_6\}$ or $w_2 = 3$. Similarly for e_3 in figure 2, w_3 is the maximum path weight for all paths in the set $\{p_4, p_5, p_6\}$ or $w_3 = 3$. The determination of the edge weights is accomplished with a breadth first search. Then, we weight the Manhattan length of each edge $e_j \in E$ by multiplying the Manhattan length of edge e_j by its corresponding weight w_j. For TS_EDGE, we use a two part optimization function. First, we minimize the weighted length of the longest edge. Second, since many configurations may have the same weighted longest edge length, we add together N of the longest edges (NLE) and minimize NLE.

$$NLE = \sum_{i=1}^{N} MLength(e_i) \times w_i$$

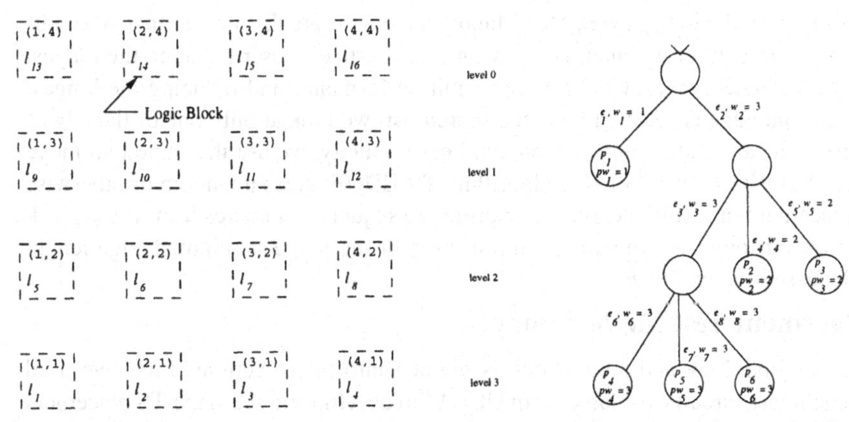

Fig. 1. *Example two-dimensional array $L = \{l_1, l_2, ..., l_{16}\}$ of physical logic block locations.*

Fig. 2. *Example circuit with six paths, $P = \{p_1, p_2, ..., p_6\}$.*

4.2 Total Wire Length Minimization

For TS_TWL, our search list U consists of all possible swaps of vertices occupying adjacent locations in L. This implies two basic swap moves: horizontal (swap of adjacent vertices with the same y coordinate) and vertical (swap of adjacent vertices with the same x coordinate). Valid swaps also include the exchange of a vertex from a position in L into an adjacent empty location in L. There are two reasons this move type was chosen: 1) to keep the move list short, and 2) to minimize the overhead of updating the move list after a move is executed. Given a two-dimensional array L of width W_L units and height H_L units, there are $|\, U \,| = ((H_L \times (W_L - 1)) + ((H_L - 1) \times W_L)) \approx 2(H_L \times W_L)$

possible swaps or moves in U. Therefore $U = \{u_1, u_2, ..., u_n\}$ where $n = \mid U \mid$. Figure 3 and4 show example horizontal and vertical swap moves respectively. For TS_TWL, given a random initial placement in L (by selecting an appropriate sequence of moves from U) we seek to optimize our objective function, minimization of TWL.

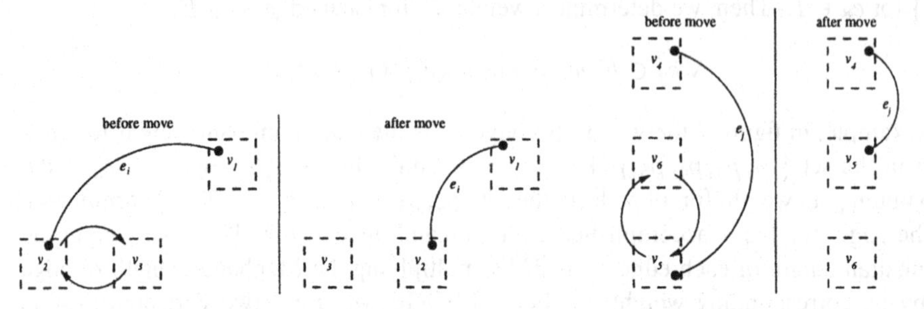

Fig. 3. *Horizontal move.* **Fig. 4.** *Vertical move.*

4.3 Timing Driven Placement

For our timing driven, TS based algorithm, TS_EDGE, we use the edge list E as our search list. We order our edge list E in descending order according to each edge's weighted Manhattan length. Then, we search E looking at each of the two vertices attached to each edge as possible candidates for a move. Therefore, in algorithm TS_EDGE, $E = \{e_1, e_2, ..., e_n\}$ where $n = \mid E \mid$ is our search or move list. The vertices attached to the edges with the longest weighted Manhattan lengths are the most attractive candidates for moving closer together. By moving these vertices closer together, the longest edges are shortened, thereby enhancing circuit performance and reducing the longest paths. Once an edge is selected from the search list, we look at only one of the edge's two vertices as a possible move candidate. For simplicity, we use the same two move types as TS_TWL. In our TS based algorithm, TS_EDGE, given a random or otherwise initial placement in L (by selecting an appropriate sequence of moves from E) we seek to optimize our objective function, minimization of the longest weighted edge length (or in the case of a tie, NLE).

5 Placement Test Methodology

Our test methodology and tabu search is divided into three sections. First, we used large benchmark circuits available from UCLA[3] to develop and setup the TS placement algorithm. We used this approach to setup and test the algorithm on large benchmark circuits (even though we cannot map these circuits directly to commercial FPGAs). Then, we used FPGA benchmark circuits available from MCNC[4] to compare the TS based approach to commercially available CAE tools. Finally, we compare the approach to the ultra fast placement work done by Rose et. al. at the University of Toronto[9] using their benchmark circuits[5].

Table 1 describes the circuit statistics for the UCLA benchmark circuits. We assume each of the vertices in the circuits can be mapped to one and only one location on the smallest square array L such that $\mid L \mid = \lceil \sqrt{\mid V \mid} \rceil^2$. In order to investigate tabu tenure

[3] http://vlsicad.cs.ucla.edu/~cheese/benchmarks.html

[4] email benchmarks@mcnc.org

[5] http://www.eecg.toronto.edu/~vaughn/challenge/challenge.html

and its effect on the short term search, we used one pass of algorithm IntenseSearch with two types of tabu tenure: *fixed* and *random dynamic*. For each tabu tenure type and each benchmark circuit, we executed IntenseSearch five times with five different random initial configurations. For each execution on a specific benchmark, we recorded average normalized TWL for the same set of iteration numbers. In order to investigate candidate list strategies, we used *aspiration plus* and the *elite* candidate list strategy[3]. For each circuit and candidate list strategy, we executed GENERIC_TS five times with five different random, initial configurations. We recorded the execution time using the `unix` time function. In order to investigate long term search strategies, we implemented a frequency based and a random move based diversification strategy. We tested the strategies on the benchmark circuits and recorded data for five different random initial configurations of each circuit.

For the benchmark circuits available at MCNC (see table 2), we placed the circuits using our TS_TWL, TS_EDGE, and TS_PLACE[6] algorithms as well as the `PPR` and `M1` tools available from Xilinx. All circuits were 100% routable after placement. For the MCNC circuits, the quality of circuit placement was based on performance speeds of the placed and routed circuits. The performance was measured using commercially available static timing analysis tools.

For the benchmark circuits from Toronto (see table 3), we placed the circuits using our TS_TWL algorithm and compared the execution time to the ultra fast placement [9] work done at Toronto. We assume each of the vertices in the circuits can be mapped to one and only one location on the smallest square array L such that $|L| = \lceil \sqrt{|V|} \rceil^2$.

Circuit Data		
Ckt	$\mid V \mid$	$\mid E \mid$
fract	149	147
baluP	801	735
structP	1952	1920
s9234P	5866	5844
biomedP	6514	5742
s13207P	8772	8651
s15850P	10470	10383
industry2	12637	13419
s35932	18148	17828
s38584	20995	20717
s38417	23849	23843

Table 1. *UCLA circuit statistics.*

Circuit Data		
Ckt	$\mid V \mid$	$\mid E \mid$
c432	88	237
c499	129	270
c880	181	417
c1355	288	581
c1908	185	563
c2670	398	758
c3540	361	1427
c6288	848	2992

Table 2. *MCNC circuit statistics.*

Circuit Data	
Ckt	**# Logic Blocks**
clma.net	8383
elliptic.net	3604
ex1010.net	4598
frisc.net	3556
pdc.net	4575
s38417.net	6406
s38584.net	6447
spla.net	3690

Table 3. *Toronto circuit statistics.*

6 Placement Results and Analysis

Figures 5 - 6 are representative of the test results for the *fixed* tabu tenure. The X-axis represents the iteration number for the execution of IntenseSearch with *fixed* tabu tenure, and the Y-axis represents the normalized value of the TWL. We use

$$\textbf{NormalizedTWL} = \frac{TWL}{TWL_{init}}$$

[6] two step process, first TS_TWL followed by TS_EDGE

(where TWL_{init} is the TWL of the initial random configuration) to represent the normalized value of TWL for each placement solution, S. In figures 5 - 6, there are five curves defined by $F \in \{0.025, 0.05, 0.1, 0.2, 0.4\}$. From the data, we determined $F = 0.2$ gives the best results relative to minimum value of **Normalized TWL**.

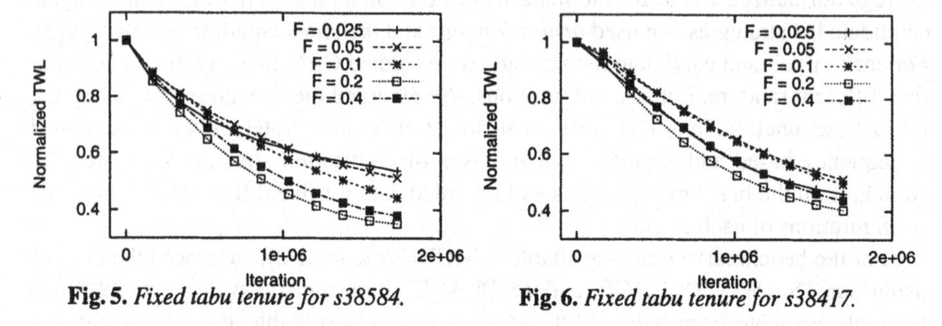

Fig. 5. *Fixed tabu tenure for s38584.* **Fig. 6.** *Fixed tabu tenure for s38417.*

Intuitively, the dynamic random tenure is less likely to cycle than fixed tabu tenure, but it is also less predictable than the fixed. This is supported by the data as described below. The standard deviation for the random tabu tenure and the fixed tabu tenure is shown in table 4 for each of the test circuits. Overall, the standard deviation for random tenure was much greater than for the fixed tenure. For the data in table 4, the standard deviation of the resulting **Normalized TWL** for dynamic tabu tenure was (on average) 11 times that of the fixed tabu tenure. Therefore, using fixed tabu tenure leads to more predictable solutions in the search for minimum TWL.

Tabu Tenure Data				
		Normalized TWL		
Ckt	**Type**	μ	**VAR**	σ
industry2	fixed	0.596	0.0010	0.03
industry2	random	0.597	0.0023	0.05
s35932	fixed	0.478	0.0010	0.03
s35932	random	0.487	0.0388	0.20
s38584	fixed	0.351	0.0004	0.02
s38584	random	0.388	0.0676	0.26
s38417	fixed	0.404	0.0001	0.01
s38417	random	0.466	0.0655	0.25

Table 4. *Standard Deviation for Fixed and Random Dynamic Tabu Tenure Types.*

Figures 7 - 8 are representative of the results for the candidate list strategy tests. From the graphs, we conclude that relative to solution quality (smallest **Normalized TWL**) and average execution time the *elite* candidate list strategy gives the best overall results.

Relative to the diversification strategies, the *frequency* based strategy shows consistently better results based on TWL minimization; however, we found the *random move* based approach executes 2.25× faster than the *frequency* based strategy for the same number of iterations. A tradeoff can be made between execution time and result quality or both methods can be used in conjunction.

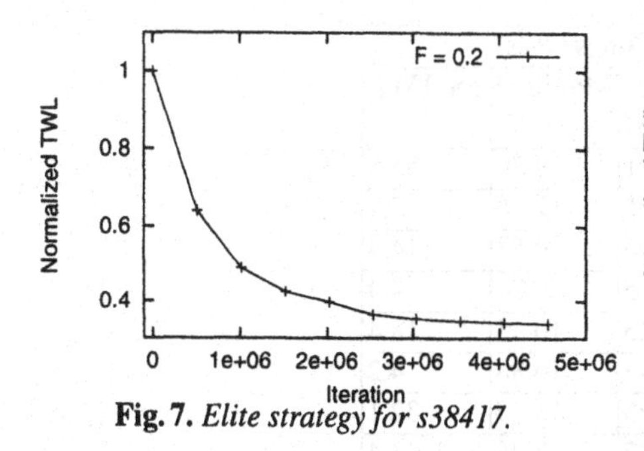

Fig. 7. *Elite strategy for s38417.*

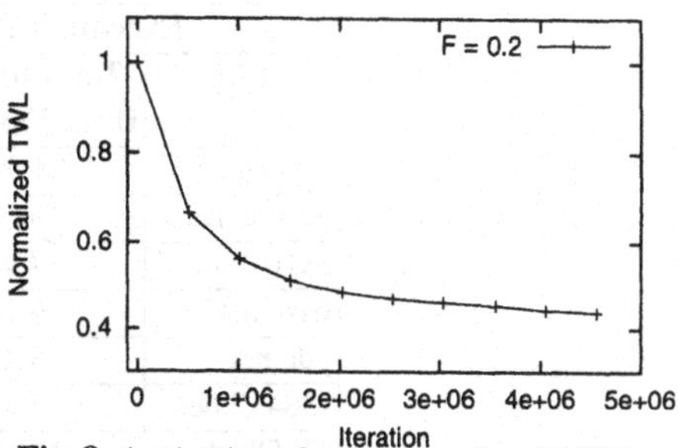

Fig. 8. *Aspiration plus strategy for s38417.*

For placement of the benchmark circuits available from MCNC, table 5 shows average execution times for placement by PPR, M1, TS_TWL, TS_EDGE, and combined (TS_TWL then TS_EDGE). All times were taken from an ULTRASPARC2 workstation. Times for PPR and M1 were taken for the default tool settings and just used for comparison. We found that a two step approach (where the first step sought to minimize TWL and a second step sought to minimize critical edge lengths) could greatly reduce the execution time of our TS based algorithms while achieving similar or better quality results when compared to the single step approaches. Obviously a true comparison cannot be made between TS and the commercial tools since the objective functions of the commercial tools are unknown; however, we provide data available from the tools for informational purposes. We see the two step TS is approximately 25 times faster than PPR and 20 times faster than M1. This is not surprising since the methods used by the commercial tools are probably more sophisticated than simple objective functions based on TWL and the minimum longest edge length. Table 6 shows the static delay calculations done on the postrouted circuits. The worst case static pad-to-pad delay for the two step TS is very similar to that of the circuits placed by PPR and M1.

Execution Times (secs)					
Ckt	PPR	M1	TS_TWL	TS_EDGE	TS Two Step
c432	16	27	0.6	1.9	0.9
c499	19	34	0.7	2.0	0.9
c880	34	35	0.8	3.8	1.3
c1355	37	40	1.4	2.7	1.8
c1908	51	75	1.7	3.0	2.3
c2670	90	75	2.1	4.8	3.8
c3540	166	159	4.5	7.2	6.2
c6288	355	169	8.7	15.9	13.1

Table 5. *Placement execution times (secs).*

Static Timing Analysis (ns)					
Ckt	PPR	M1	TS_TWL	TS_EDGE	TS Two Step
c432	63.4	67.6	65.9	62.8	63.8
c499	40.0	45.3	40.1	37.4	40.0
c880	55.7	49.0	54.5	56.4	54.8
c1355	58.3	61.7	63.9	65.6	64.3
c1908	56.8	67.1	60.2	62.5	59.7
c2670	100.6	118.5	112.0	110.0	93.6
c3540	82.9	79.6	115.5	98.4	91.5
c6288	378.0	380.8	429.0	414.9	398.0

Table 6. *Placement static timing analysis.*

Table 7 shows the execution time comparison of TS_TWL to the Ultra Fast placement tool and the modified VPR tool[7] from the University of Toronto[9]. A direct comparison of the algorithms cannot be made since they have different goals and cost functions; however, relative to execution time, the TS method is on the same order as that of the tools from Toronto.

[7] modified to improve execution time at the cost of placement quality[9]

Execution Times (secs)			
Ckt	Toronto Ultra Fast	Modified VPR	TS_TWL
clma.net	21.71	29.79	55.5
elliptic.net	6.05	7.16	2.1
ex1010.net	7.96	10.53	14.4
frisc.net	6.15	7.04	4.4
pdc.net	8.43	10.35	18.4
s38417.net	13.33	16.88	4.2
s38584.1.net	14.55	18.35	32.1
spla.net	6.37	7.26	13.4

Table 7. *Execution time comparison to Toronto's Ultra Fast Placement[9].*

7 Conclusions

We have described the development of a two-step, TS based algorithm for placement of circuits on two-dimensional arrays. We developed and demonstrated the approach with circuits from UCLA, MCNC, and the University of Toronto. Despite having a limited knowledge of gate level architectural details for the Xilinx architecture and CAE tools, our comparison shows significantly faster execution times (20×) as compared to the commercially available Xilinx tools with very similar results as determined by Xilinx static timing analyzers. Our timing also compares favorably to the Ultra Fast placement work at the University of Toronto; however, to truly compare the approach, the cost function of the TS should be modified so it is similar to that used by Toronto's Ultra Fast Placement tool. Additionally, our placement method is suitable for quickly initializing the inputs to other placement algorithms. Our work in progress includes the placement with routability estimation and techniques for controlled TS.

References

1. T. J. Callahan, P. Chong, A. DeHon, and J. Wawrzynek. Fast Module Mapping and Placement for Datapaths in FPGAs. In *ACM/SIGDA International Symposium on Field-Programmable Gate Arrays*, pages 123–132, Feburary 1998.
2. J. M. Emmert and D. K. Bhatia. A Methodology for Fast FPGA Floorplanning. In *ACM Seventh International Symposium on Field-Programmable Gate Arrays*, pages 47–56, Feburary 1999.
3. F. Glover and M. Laguna. *Tabu Search*. Kluwer Academic Publishers, 1997.
4. Altera Inc. *http://www.altera.com.*
5. Xilinx Inc. *http://www.xilinx.com.*
6. A. Lim. Performance Driven Placement Using Tabu Search. *Informatica*, 7(1), 1996.
7. A. Mathur, K. C. Chen, and C. L. Liu. Re-engineering of Timing Constrained Placements for Regular Architectures. In *IEEE/ACM International Conference on Computer Aided Design*, pages 485–490, November 1995.
8. I. H. Osman and J. P. Kelly. *Meta-heuristics: Theory and Applications*. Kluwer Academic Publishers, 1996.
9. Y. Sankar and J. Rose. Trading Quality for Compile Time: Ultra-Fast Placement for FPGAs. In *ACM Seventh International Symposium on Field-Programmable Gate Arrays*, pages 157–166, Feburary 1999.

Placement Optimization Based on Global Routing Updating for System Partitioning onto Multi-FPGA Mesh Topologies

Juan de Vicente[1], Juan Lanchares[2], Román Hermida[2]

[1] E.T.S.I.A.N., Arturo Soria, 287, Madrid (Spain)
email: juan.vicente@etsian.mde.es

[2] Dpto. Arquitectura de Computadores y Automática,
Facultad de Ciencias Físicas, Universidad Complutense de Madrid (Spain).
email: {julandan, rhermida}@dacya.ucm.es

Abstract. The drive of Multi-FPGA systems has been hindered due to the low gate utilization percentages they present. The scarce inter-FPGA communication resources turn the circuit partitioning phase into a complex one. The problem is becoming more acute for mesh routing topologies since the communication between distant FPGAs increases the IO-pin consumption of intermediate FPGAs. For these topologies the integration of placement and routing phases in the partitioning process may be the natural approach to take into account all communication constraints. In this research, we have modeled the Multi-FPGA mesh topology as a large single FPGA where the borders between neighboring FPGAs are represented by a superimposed template. Then, a placement optimization based on Simulated Annealing is performed to minimize the cut-sizes on this partition template. An initial routing is updated in each placement iteration to keep account of the actual number of nets crossing between FPGAs. Two different problems, mapping onto a fixed Multi-FPGA mesh or finding the Multi-FPGA mesh that fits best for a given circuit, can be treated. This second problem has been solved for a set of benchmark circuits. A large cut-size reduction has been achieved for all circuits being sufficient to get a high gate utilization percentage in some cases.

1. Introduction

Field Programmable gate arrays (FPGAs) provide flexibility and reconfiguration advantages making them an interesting alternative to implement many types of digital systems. When the capacity of a single FPGA is surpassed, the user should turn to Multi-FPGA systems. There are two basic Multi-FPGA systems that differ in the topology chosen to interconnect their components. Crossbar topologies interconnect the chips through *Field Programmable Interconnect Components* (FPICs) with the consequent area and delay increments that it involves. In mesh routing topologies the FPGAs are connected directly to their neighbors relieving additional inter-chip communication delays. Nevertheless, the potential of mesh routing topologies has been truncated due to low levels of *Configurable Logic Block* (CLB) utilization commonly achieved (less than 20%)[1]. The few IO-pins of an FPGA is the major drawback to obtaining higher gate utilization levels since it limits the inter-chip communication bandwidth. Therefore, new research efforts must be carried out to develop efficient tools that perform high quality layouts.

The usual physical design cycle for Multi-FPGA systems goes through various phases: Partitioning, Global Placement, Global Routing, FPGA Placement and FPGA Routing [16]. Partitioning breaks the netlist into subcircuits to fit them into individual FPGAs. Global placement allocates these subcircuits to specific FPGAs. Inter-FPGA signals are then routed during Global Routing. Finally FPGA placement and routing are performed for individual FPGAs. Partitioning has been the objective of multiple studies considering different goals. A low final cost led to the partitioning method proposed by Kuznar [2] for heterogeneous systems. In [3] and [4] the routability of component FPGAs was also considered by means of different models. Fang [5] carried out the partitioning by taking into account path delays and design structure information.

The studies described above assume that there is no restriction on which partitions can communicate, so they may be suitable for topologies using FPICs. In this paper, we focus on Multi-FPGA mesh topologies for their widespread interest (Fig. 1(a)). For these systems the connectivity among FPGAs is drawn for the mesh topology (FPGAs are connected only to their closest neighbors). Therefore partitioning tools must be aware of interconnection constraints of distant FPGAs. Thus, the additional IO-pins consumption of intermediate FPGAs must be taken into account by them. Moreover global I/O circuit signals can not be assigned to internal FPGAs since the I/O-pins of these are fully occupied constituting the mesh. Therefore a careful global input/output signal assignment should be made marking out the FPGAs according to their position on the mesh.

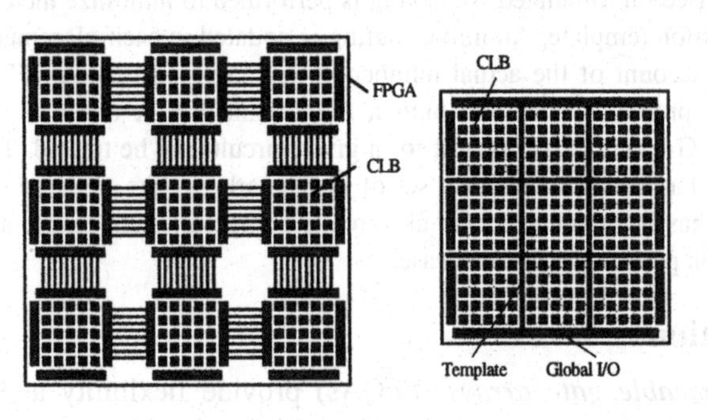

(a) Multi-FPGA mesh topology *(b) SFPGA model*

Fig. 1. *Multi-FPGA system and its simplified model*

The natural approach to meet all these constraints may be simultaneous Partitioning and Placement. In [6], Hauck recognizes that a high quality approach for mapping circuits onto Multi-FPGA systems may be a simultaneous partitioning and global placement guided by Simulated-Annealing (SA) [7], such as that developed in [8] [9]. So, he indicates that during the processing the tool knows which FPGA a given logic element will be assigned to, and it performs routing of inter-FPGA signals to determine how expensive a given connection pattern is, and looks for severe congestion points in the system. Nevertheless, he considers that it is a very slow process when it is also necessary to perform routing as part of the cost metric. In [10] was presented the Rectilinear Steiner Regions (RSR) algorithm, a new fast Rectilinear Steiner

Minimum Tree approximation for routing multiterminal nets. The speed of RSR algorithm allowed the SA process to relieve the FPGA routing congestion by simultaneous placement and routing. In this paper we describe a similar method for placement/partitioning onto Multi-FPGA mesh topologies. *Placement&Routing-based Partitioning* (PRP) performs the partitioning after a cut-size Placement&Routing optimization. Thus logic blocks are distributed on the FPGAs in such a way that the number of nets crossing inter-FPGA borders are diminished. Both approaches, mapping onto a predetermined Multi-FPGAs mesh or finding the Multi-FPGA mesh that fits best for a given circuit can be treated. The second approach has been applied to a set of MCNC benchmark circuits obtaining interesting CLB utilization percentages for some circuits.

In the next section of this paper, a model for the Multi-FPGA mesh topology as well as problem formulation for placement/partitioning within this model are introduced. Section 3 describes the strategy designed to obtain a good *CPU-time/min-cut-quality* trade-off in the SA optimization process. Section 4 presents the experimental results. The conclusions follow in section 5.

2. Multi-FPGA Mesh Topology Model and Problem Formulation

In Multi-FPGA mesh topology, the individual FPGAs are connected to their closest neighbors through their IO-pins (Fig.1(a)). For our purpose we can model the Multi-FPGA system with a large single FPGA (SFPGA), where the borders among neighboring FPGAs are represented by a superimposed template T constituted of a set of line segments $T=\{T_1,..,T_n\}$ (Fig.1(b)). Let C_i be the number of nets crossing the border T_i after placement and routing. We define the *cut set* through T as the set $C=\{C_i\}$ for i=1.. n and define C_m as the maximum of C. The objective of placement and routing tools must be to optimize the *cut set* C to satisfy inter-FPGA communication constraints.

There are two basic approaches for mapping circuits in Multi-FPGA systems. In the first one, Fixed Mesh (FM), the number and size of FPGAs as well as their arrangement is determined by the system architecture. In this case our Multi-FPGA mesh model is constructed joining the component FPGAs in a large SFPGA and storing their separation borders in a template (Fig.1(b)). After placement and routing optimization the circuit will fit into this Multi-FPGA system if the *cut set* (C_i values) meets the inter-FPGA communication bandwidth (Fig.2(b)). In the second approach, the Multi-FPGA mesh that fits better for a given circuit must be determined. Therefore, the necessary number of FPGAs, their size and arrangement to host the circuit netlist is explored. We refer to this approach as Adapted Mesh (AM) throughout the rest of this paper. In this approach the SFPGA model may be constructed with the minimum size that can house the circuit. Then a partition template is tried on the SFPGA to model the partition of the circuit (Fig.2(a)). After the *cut set* placement optimization, the value of C_m fixes the size of the necessary FPGAs (NFPGA) and the partition template their arrangement (Fig.2(b)). Assuming that the number of IO-pins per side for an FPGA of NxN CLBs is 2N, the size of the NFPGA will have to be at least $(C_m/2) \times (C_m/2)$ for even values of C_m, or $(C_m+1)/2 \times (C_m+1)/2$ for odd values of C_m. The partition will be valid if resultant NFPGAs are not larger than available FPGAs. When this constraint is not met additional FPGAs are necessary and a finer-grain partition scheme must be explored. Due to the low inter-FPGA communication bandwidth, the communication between FPGAs is much more expensive than intra-FPGA communication. Therefore large partitions meeting the library size-constraints lead to higher CLB utilization percentages.

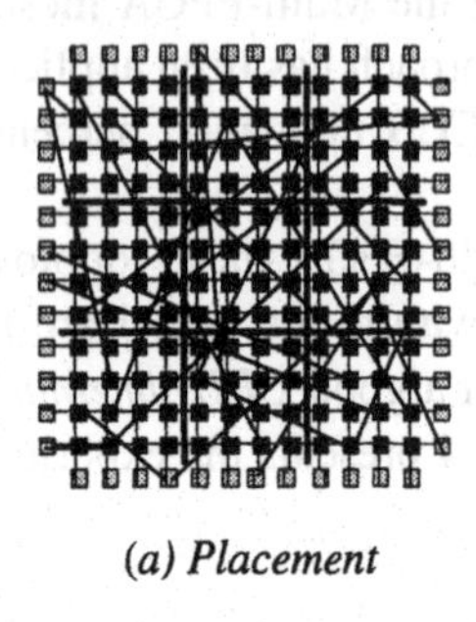

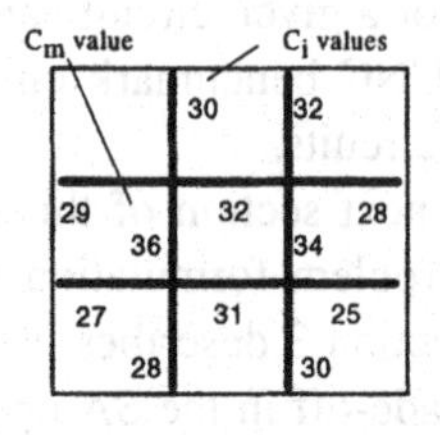

(a) Placement *(b) Cut set(C) after the routing*

Fig. 2. *Placement on the SFPGA model*

3. Placement Optimization Based on Global Routing Updating

Considering the model described in the previous section, common techniques can be applied for placement and routing the circuit on the SFPGA to meet the interconnection constraints. SA has been the most successful algorithm for placement circuits to date [11][12]. Our method performs two phases of SA placement optimization with a cooling schedule similar to that described in [12].

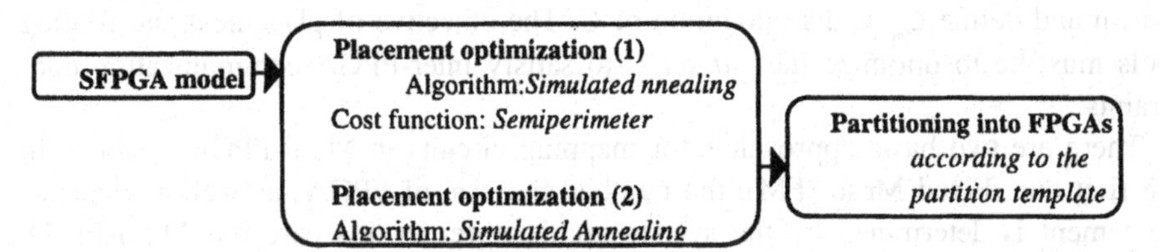

Fig. 3. *Metodology overview*

3.1. Cut-set Placement Cost Function for SA

The inter-FPGA communication bandwidth is the bottleneck in Multi-FPGA systems. The objective of the placement tool is to distribute the nets of the circuit on the SFPGA in such a way that the *cut set* produced in the partition template is adjusted to communication capacity. To achieve this *cut set* placement optimization a SA process is performed in two phases. The first one is driven by the semiperimeter cost function (ϕ_1) that approximates the length of a net by half of the perimeter (P_i) of the rectangle that includes it (Eq. 1). In this way related logic blocks are gathered in the same SFPGA area. Thus total wire length and hence the *cut set* through the partition template are reduced.

$$\phi_1 = \frac{1}{2} \sum_{i \in nets} P_i \tag{1}$$

Although this phase attains a large *cut set* reduction (Table 3), additional improvements can be achieved on this critical value. To be able to minimize the nets crossing through the partition template it is necessary to estimate or measure them. To estimate the *cut set* at the placement stage is a meaningless task since they depend strongly on the routing process. Fig. 4a shows an example of possible crosses on the partition template for a multiterminal net. It must be noticed that the high indetermination for the future path makes estimation of the *cut set* difficult. Fig.4b shows the same example once a multiterminal net routing has been performed. In this case the indetermination is resolved in favor of one alternative and hence the *cut set* can be minimized.

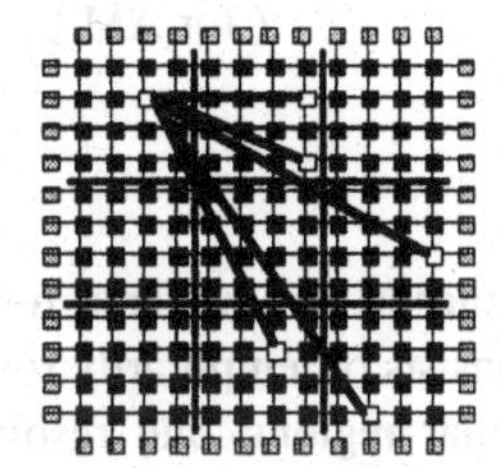

(a)Path indetermination (possible net-cuts for a multiterminal net)

b) Path determination (after RSR routing, the cut set can be measured and hence minimized)

Fig. 4. *Cut set for a multiterminal net*

Therefore, the second placement optimization phase performs an initial routing and updates affected paths due to the movement of logic blocks in each SA placement iteration. This method could be too slow using the common maze routing strategies [13]. Nevertheless, the development of fast Rectilinear Steiner Minimum Tree (RSMT) approximations [14][10], allows the integration of the routing as part of the placement optimization process. So, we use the fast RSR router ($o(n^2)$) [10] to update the routing in each placement iteration. This router described in section 3.2 performs high quality RSMT approximations while maintaining short source-sink path lengths and hence short signal delays. On the other hand, the intra-FPGA routing resource details may be neglected for the study of inter-FPGA circuit communication. Thus, a simple rectilinear grid of CLBs (Fig.4) is assumed instead of the intra-FPGA routing resources to speed up the routing process.

Let G(V,E) be a rectilinear grid where $V=\{v_1,..., v_{|V|}\}$ is the set of SFPGA's CLBs, and $E=\{e_1,.., e_{|E|}\}$ the set of edges connecting the CLBs with their neighbors. Let T be the partition template superimposed on G (Fig.2(a)). The *cut set* C (Fig.2(b)) can be measured in each placement iteration after an RSR routing update of the nets on G (Fig.4b). The cost function (ϕ_2) in this second optimization phase depends on the problem treated, Fixed Mesh or Adapted Mesh. In FM the objective is to meet the interconnection constraints. Thus, if *S* is the number of wires between two neighboring FPGAs, the values of the *cut set* above the *S* level must be minimized. This can be efficiently achieved by the first term of Eq. 2. In AM, the value of C_m determines the size of the FPGAs. For this approach the value C_m is minimized by the first term of Eq. 3.

Since the routing is updated in each placement iteration, the second term of ϕ_2 (Eq.2, Eq.3) minimizes the occupation O_j of the edges $e_j \in E$ that exceed the channel width (W) of the FPGAs, with an insignificant extra computational effort. This term is a measurement of the local congestion, and allows the placement process to uniformly distribute the nets in the Multi-FPGA mesh to facilitate the routability of individual FPGAs.

$$\phi_2 = \sum_{c_i \in C \,/\, c_i > S} (c_i - S)^2 + \sum_{e_j \in E \,/\, O_j > W} (O_j - W)^2 \qquad \text{(for FM)} \tag{2}$$

$$\phi_2 = \sum_{c_i \in C} c_i^2 + \sum_{e_j \in E \,/\, O_j > W} (O_j - W)^2 \qquad \text{(for AM)} \tag{3}$$

3.2. Rectilinear Steiner Regions (RSR) Algorithm

The *RSR* algorithm approximates the Rectilinear Steiner Minimum Tree for routing multiterminal nets. The basic idea of the algorithm consists of temporarily routing the sink vertices with indeterminate multi-path rectangular regions, and resolve these regions in favor of one alternative path when routing later sink vertices, based on their common paths. Let G (V,E) be the rectilinear grid and let N be the multiterminal net. Figure 5 introduces some new terminology to describe the algorithm.

Definitions

Distance *Manhattan or rectilinear distance*

Region *$R \equiv R(v_{SOURCE}, v_{SINK})$:Set of vertices of G that belong to the perimeter of the rectangle that delimits the source and sink vertices of a two terminal connection.*

Size of region *$s(R)$=distance(v_{SOURCE}, v_{SINK})*

Closest vertex between a vertex and a region
$v_{cl} \equiv$closest_vertex(v_k , R):$v_i \in R$ /distance(v_k,v_i) is minimum.

Distance vertex-region
Distance(v_k,R):distance(v_k,v_{cl})

List of regions *Set of regions: $\mathfrak{R} = \{R_1,\ldots, R_l\}$*

Common path region-vertex *Common path (R ,v_k):It is an operator that returns the list of regions as follows, $\mathfrak{R} = \{R(v_{SOURCE}, v_{cl}), R(v_{cl}, v_{SINK}), R(v_{cl}, v_k)\}$*

Closest vertex between a vertex and a list of regions
$v_{cl} \equiv$Closest vertex(v_k , $\mathfrak{R}$):$v_{clj} \in R_j \subseteq \mathfrak{R}$ /distance(v_k ,v_{clj}) is minimum.

Closest region between a vertex and a list of regions
Closest region(v_k , $\mathfrak{R}$):$R_i \subseteq \mathfrak{R}$ /$v_{cl} \in R_i$

Common path list of regions-vertex
Common path($\mathfrak{R}$,v_k):It is an operator that substitutes the region R=Closest region(v_k , $\mathfrak{R}$), for the sublist of regions provided from Common path(R ,v_k).

Fig. 5. *Definitions for RSR algorithm*

The RSR algorithm (Fig. 6) constructs a list of regions ($\mathfrak{R}$) through which the net can be routed. The algorithm begins ordering the sink vertices according to their distance from the source. Then, for each sink vertex (v_{di}) taken in order, determines their *Common path* with the path already routed ($\mathfrak{R}$), and creates new regions for their non-common paths. Although the first part of RSR algorithm produces some *regions* with more than one alternative path of the same length, the final path can be fixed in such a way that the congestion is avoided. Since the centre routing resources of the FPGA are the most requested, the farthest path from the centre of the FPGA will be chosen among the different alternatives. It is done with the *Fix_multi_alternative_paths* function. Figure 7 shows a simplified diagram of RSR construction from the source to sink vertices. In the figures 7(a) and 7(b) two sink vertices are routed with indeterminate rectangular regions. In the figure 7(c) the second region becomes resolved in favor of one alternative when routing the third sink vertex. In the figure 7(d) the fourth sink vertex is routed with a new rectangular region. This region is resolved (fig. 7(e)) in favor of one alternative when routing the fifth vertex. Finally, the rectangular regions that still remain are resolved in favor of the farthest path from the centre of the FPGA (fig. 7(f)).

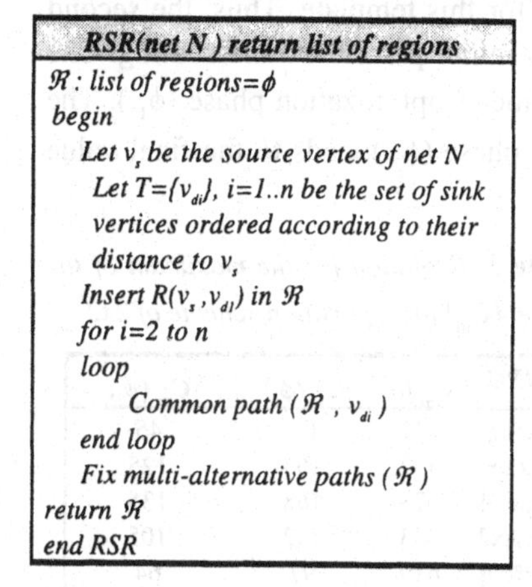

RSR(net N) return list of regions

```
ℜ : list of regions=φ
begin
  Let v_s be the source vertex of net N
    Let T={v_di}, i=1..n be the set of sink
    vertices ordered according to their
    distance to v_s
  Insert R(v_s ,v_d1) in ℜ
  for i=2 to n
  loop
      Common path (ℜ , v_di )
  end loop
  Fix multi-alternative paths (ℜ )
return ℜ
end RSR
```

Fig. 6. *RSR algorithm*

Example of RSR regions construction

(a) (b) (c)

(d) (e) (f)

Fig. 7. *Example of RSR construction*

If $|\Re|$ is the number of regions found for the net N, the time complexity of the RSR algorithm is $O(|\Re| n)$. The *Common path* operator produces one or two new regions every time it is applied, depending if the closest vertex is a terminal point or a Steiner point respectively. Then $|\Re| < 2n$ and the RSR algorithm time complexity is $O(n^2)$.

4. Experimental Results

A large number of studies relating to partitioning have been published, but few of them perform the global placement and routing phases to try out their results. Other approaches [15] perform the mapping for specific board architectures making direct comparisons difficult.

We have tried out the PRP optimization method with a set of MCNC benchmark circuits [17]. Their characteristics as well as the size of the SFPGA model are shown in Table 2. The demanded channel width (DCW), assuming the mapping is performed onto a large single FPGA (using their general purpose interconnection resources), was obtained with VPR (Versatile Place and Route) tools [17]. This value is also shown in Table 2 to highlight the degree of routability of these circuits. For our study we assumed the availability of a set of Xilinx's FPGAs [18] described in table 1.

Table 1. *Sizes of the FPGAs chosen for the study*

DEVICES	*CLB Matrix*	*CLBs*	*Max. IO*
XC4013XLA	24x24	576	192
XC4020XLA	28x28	784	224
XC4028XLA	32x32	1024	256

To facilitate direct comparison with future studies our tools look for the maximum percentage of CLB utilization achieved for the benchmark circuits. To get this coefficient, the Adapted Mesh approach described in section 2 is applied. Thus, a first 2x2-FPGAs partition scheme was tried on all circuits. Table 3 draws the evolution of C_m value during the placement optimization process for this template. Thus, the second column shows the maximum of the *cut set* for an initial placement and routing. The third column displays C_m value after the semiperimeter optimization phase (ϕ_1). The simultaneous placement and routing optimization phase (ϕ_2) leads to the final value of C_m shown in the fourth column.

Table 2. *Benchmark circuits: CLBs, IOBs, demanded FPGA sizes and channel widths*

CIRC	*#CLBs*	*#I/O*	*SFPGA size*	*DCW*
tseng	1047	174	33x33	7
ex5p	1064	71	33x33	11
apex4	1262	28	36x26	10
misex3	1397	28	38x28	9
diffeq	1497	103	39x29	7
alu4	1522	22	40x40	8
seq	1750	76	42x42	9
s298	1931	10	44x44	6
apex2	1878	41	44x44	9
dsip	1370	426	54x54	7
bigkey	1707	426	54x54	7

Table 3. *Evolution for the maximum of the cut set (C_m) for a partition scheme of 2x2*

CIRC.	C_m *(i)*	C_m *(ϕ_1)*	C_m *(ϕ_2)*
tseng	435	67	45
ex5p	439	153	125
apex4	473	168	135
misex3	533	142	105
diffeq	619	97	64
alu4	520	129	94
seq	679	151	113
s298	621	98	53
apex2	774	163	121
dsip	575	31	20
bigkey	730	40	27

Note that ϕ_1 phase achieves a large *cut set* reduction. The use of the *semiperimeter* cost function for the SA process groups related logic blocks together. This phase is fast since semiperimeter calculation has a low computational cost. The second optimization phase is slower, since the routing is updated in each placement iteration but it achieves a new reduction in the critical value of C_m.

The value of C_m fixes the minimum size of the NFPGAs (section 2). Also, the size is constrained to the available FPGAs described in table 1, so the maximum number of IO-pins that communicate two neighboring FPGAs is 64 (2x32). Thus only the circuits with C_m smaller than or equal to 64 accept the 2x2 partition scheme. Table 4 shows the size of the NFPGAs and the CLB utilization percentage (UP) (Eq.4) for these circuits.

$$UP = 100x\frac{UsefulCLBs}{TotalCLBs} \quad (4)$$

where

TotalCLBs=Patitions x NFPGA size and *UsefulCLBs=CLBs of the circuit*

Table 4. *Partition scheme of 2x2. The maximum value of the cut set (C_m), the size of FPGAs needed to support this cut-set, and the utilization percentage are shown for the circuits that accept this partition scheme*

CIRC.	$C_m(\phi_2)$	*#Partitions*	*Arragement NFPGA array*	*NFPGA size*	*UP*
tseng	45	4	2x2	24x24	45 %
diffeq	64	4	2x2	32x32	37 %
s298	53	4	2x2	28x28	62 %
dsip	20	4	2x2	28x28	44 %
bigkey	27	4	2x2	28x28	55 %

When the value of C_m is greater than 64 a finer-grain partition schemes are tried on and the second phase of the SA optimization is repeated. Table 5 shows the circuits that accept 3x3 and 4x4 partition schemes.

Table 5. *Partition schemes of 3x3 and 4x4. The maximum value of the cut set (C_m), the size of FPGAs needed to support these cut-sets, and the utilization percentage are shown*

CIRC.	$C_m(\phi_2)$	*#Partitions*	*Arragement NFPGA array*	*NFPGA size*	*UP*
misex3	59	9	3x3	32x32	15%
alu4	60	9	3x3	32x32	17%
ex5p	60	16	4x4	32x32	6 %
apex4	60	16	4x4	32x32	8 %
seq	56	16	4x4	28x28	14%
apex2	58	16	4x4	32x32	11%

The final UP obtained for all circuits ranges from 6% to 62%, with an average percentage of 28. As was expected, those circuits with the highest demanded channel width for a large single FPGA (Table1), achieve the lowest UP onto Multi-FPGA systems.

5. Conclusion

Many published Multi-FPGA partitioning studies actually skip over the interconnection topology which they are directed to. Nevertheless, it is an important issue since Multi-FPGA interconnection resources are strongly limited. In this paper, we have presented a SA-based method for mapping circuits onto Multi-FPGA mesh routing topology. Both problems, mapping onto fixed FPGA arrangements or finding the arrangement that fits better for a given circuit can be treated. By means of a two-phase placement optimization process the actual *cut set* is minimized to meet the inter-FPGA bandwidth constraints. Furthermore, global I/O signals are assigned in a natural way, to peripheral FPGAs that can support them.

The method has been applied to a set of MCNC benchmark circuits. An initial 2x2-partition scheme has been explored for all circuits. The placement optimization process has achieved a large decrease in the *cut set* for all circuits. For the circuits with medium DCW values (tseng, s298, dsip, bigkey), this high reduction has been enough to obtain interesting increases in the CLB utilization percentages. For circuits with higher DCW values (ex5p, apex4, seq, apex2), in spite of the high *cut set* decreases, the shortage of FPGA IO-pins still forces the application of finer-grain partition schemes. Nevertheless, the *cut set* achieved for these circuits may be an unbeatable starting point for the application of other techniques such as virtual wires [19].

References

[1] S. Hauck, "The Roles of FPGA's in Reprogrammable Systems", Proc. IEEE. Vol. 86. NO. 4. April 1998.

[2] R. Kuznar, F. Brglez, and K. Kozminski. "Cost minimization of partition into multiple devices". 30th ACM/IEEE DAC, pages 315-320, Dallas, Texas, June 1993.

[3] I. Hidalgo, M. Prieto, J. Lanchares, F. Tirado. "A parallel gentic algorithm for solving the partitioning problem in Multi-FPGA Systems". Proc. of 3rd International meeting on vector and parallel processing. VECPAR 98, Porto, 1998.

[4] P.K. Chan, M.D.F. Schlag, J.Y. Zien. "Spectral-Based Multi-Way FPGA Partitioning". IEEE Trans. on CAD of IC's and systems. Vol. 15 NO. 5 May 1996.

[5] W. Fang, C.-H. Wu. "Performance-Driven Multi-FPGA Partitioning Using Functional Clustering and Replication". 35th DAC Proceedings, San Francisco, June 1998.

[6] S. Hauck, "Multi-FPGA systems". Ph. D. Thesis. University of Washington, 1994.

[7] S. Kirkpatric, C. D. Gelatt, and M. P. Vecchi, "Optimization by simulated annealing," Science, vol. 220, no. 4598, pp. 671-680, May 1983.

[8] K. Roy, C. Sechen, "A Timing Driven N-Way Chip and Multi-Chip Partitioner", International Conference on Computer-Aided Design, pp. 240-247, 1993.

[9] G. Vijayan , "Partitioning Logic on Graph Structures to Minimize Routing Cost". IEEE Trans. on CAD. Vol. 9 NO. 12. Dec. 1990, pp. 1326-1334.

[10] J. De Vicente, J. Lanchares, R. Hermida, "RSR: A new Rectilinear Steiner Minimum Tree Approximation for FPGA Placement and Global Routing". EuroMicro 98, Västerås. IEEE Press, 1998, pp 192-195.

[11] C. Sechen, and A. Sangiovanni-Vincentelli, "The TimberWolf placement and routing package," in Proc. Custom Integrated Circuit Conf. (Rochester, NY), 1984, pp. 522-527.

[12] V. Betz and J. Rose, " VPR: A new Packing, Placement and Routing Tools for FPGA Research", International Workshop on Field Programmable Logic and Applications, 1997.

[13] C. Y. Lee, "An algorithm for path connections and its applications," IRE. Trans. Electron. comput., vol. EC-10,PP. 346-365, 1961.

[14] P.Berman, U. Föβmeier, M. Karpinski, M. Kaufmann, A. Zelikovsky. " Approaching the 5/4-Approximations for Rectilinear Steiner Trees". LNCS 855, 60-71, 1994.

[15] U. Ober, M. Glesner. "Multiway Netlist Partitioning onto FPGA-based Board Architectures". ICCD 95, pp 150-155.

[16] S. Hauck, G. Borriello, "Pin Assignment for Multi-FPGA Systems". IEEE Trans. on CAD of IC's and systems. Vol. 16 NO. 9 Sep. 1997.

[17] V. Betz and J. Rose, "Effect of the Prefabricated Routing Track Distribution on FPGA Area-Efficiency". IEEE Trans. on VLSI systems. Vol. 6 NO. 3 Sep. 1998, pp. 445-456.

[18] Xilinx,"The Programmable Gate Array Data Book",1994.

[19] J.Babb, R. Tessier, M. Dahl, S.Z.Hanono, D.M. Hoki, A.Agarwal, "Logic Emulation with Virtual Wires", IEEE Transactions on CAD of IC and Systems, no.6 Jun. 1997, pp 609-626.

Hierarchical Interactive Approach to Partition Large Designs into FPGAs

Helena Krupnova and Gabriele Saucier

Institut National Polytechnique de Grenoble/CSI,
46, Avenue Felix Viallet, 38031 Grenoble cedex, France
{bogushev,saucier}@imag.fr

Abstract. The paper addresses the design partitioning into multiple FPGA devices. The industrial experience shows that the designers were never satisfied by the full automatic partitioning: as the design size grows, it takes longer CPU times and produces poor results. The present paper proposes an algorithm which may be integrated into the mixed interactive manual/automatic partitioning framework. The hierarchy nodes of the design are selected one by one and assigned to a defined set of FPGA devices. The automatic partitioning algorithm is called to split a big node among the selected subset of devices taking into account previous assignments to these devices. Experimental results show that the proposed approach works well for big industrial circuits.

1 Introduction

The capacity of FPGA chips is significantly increased and they are able to implement circuits of up to 1 Mln gates. Partitioning is needed for huge circuits of several million gates. This requires the ability to handle the designs of big complexity. To cope with the complexity problems and to obtain better results, clustering of logic was traditionally applied. Bottom-up ([11], [2], [8]), top-down ([4]) and the combination of both clustering methods ([3]) are known. The hierarchical representation is a natural basis for the reduction of complexity. A number of recently appeared methods ([6], [7], [5], [1], [9]) address the hierarchy-based partitioning.

Automatic algorithms are usually "blind", they are not flexible and work in a pre-defined way. The user intervention may guide to more efficient solutions. From one side automatic partitioning needs more flexibility, and from the other side, manual partitioning needs more sophisticated assistance. User intervention may consist in guiding the partitioning process by controlling the clustering granularity, invoking automatic partitioning on selected nodes or tuning the parameters of the automatic algorithms for better results. A typical scenario of the partitioning session starts by assigning some blocks to devices manually, then invoking automatic partitioning on selected blocks, iterating this process several times, and finally assigning the remaining blocks to partially filled devices.

The given paper extends the hierarchical FPGA partitioning approach (1) to deal with pre-assigned blocks (2) to be applied on selected nodes of the design hierarchy. The paper is organized as follows. The next section presents the hierarchical circuit modeling. Next, node partitioning problem formulation is given. The next section describes the node partitioning algorithm handling pre-assigned blocks. Finally, the experimental results on big industrial designs are presented. Partitioning session examples are explained, then mixed automatic/manual partitioning is compared with full automatic partitioning, and finally results with different sets of pre-assigned blocks are presented.

2 Hierarchical Circuit Modeling

It is supposed that the netlist is already mapped on the target FPGA technology by one of the commercially available synthesis tools. The synthesized netlist is composed of primitive, or glue logic cells. Primitive cells are defined as "indivisible" physical entities with respect to partitioning (FPGA CLBs, flip-flops, etc.) and organized in hierarchy blocks. Each hierarchy block may contain primitive cells and other hierarchy blocks.

The circuit hierarchy may be represented by a hierarchy tree where each node corresponds to a hierarchy block (Figure 1). As a result of partitioning, the hierarchy tree will be modified and the direct successors of the top block will be blocks $\{d_1, d_2, ..., d_k\}$ corresponding to the filled FPGA devices $\{D_1, D_2, ..., D_k\}$. Blocks $\{d_1, d_2, ..., d_k\}$ are called *device nodes*. The successor nodes of each device node d_i are nodes of the initial hierarchy. Each device D is characterized by the size and I/O pin constraints, S_{MAX}, T_{MAX}.

An example is shown in Figure 1. Two devices (D_1 and D_2) are filled during partitioning. The first one contains the blocks $\{A, H, I\}$, and successor blocks $\{E, F\}$ of block A. The second device contains the blocks B and L, as well as the successor nodes $\{J, K\}$ of node L. The block C is ungrouped during partitioning and is no more visible in the hierarchy tree: its successor blocks $\{L, H, I\}$ are assigned to different devices.

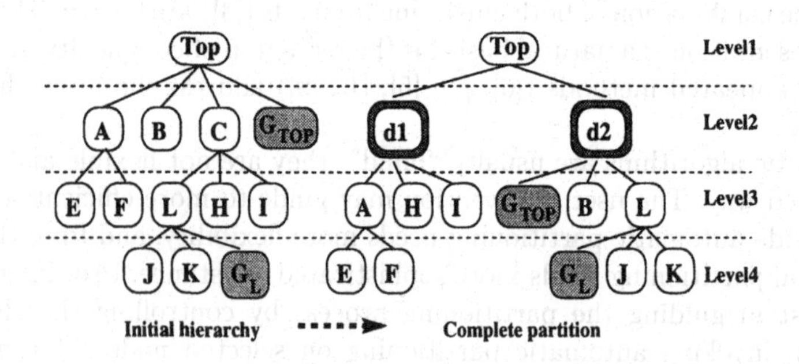

Fig. 1. Design hierarchy transformation.

At each step, current partitioning status may be described by a corresponding hierarchy tree containing zero or more device nodes.

<u>Definition</u> The hierarchy node is *assigned* to device if it is a successor (direct or not) of the device node in a hierarchy tree. Otherwise the hierarchy block is *not assigned*, or *free*.

<u>Definition</u> Partition is *complete* if all the hierarchy nodes are assigned to devices. The complete partition contains only the device nodes of the Level2 of the hierarchy tree. The hierarchy tree on the right side of Figure 1 corresponds to the complete partition.

<u>Definition</u> A partition is *incomplete* if there exist free nodes in the corresponding hierarchy tree. In other words, not all direct successors of the top node correspond to device nodes. An example is presented in Figure 2b. The node A is assigned to the device D_1, the nodes B and G_{TOP} are assigned to the device D_2, and the node C is free.

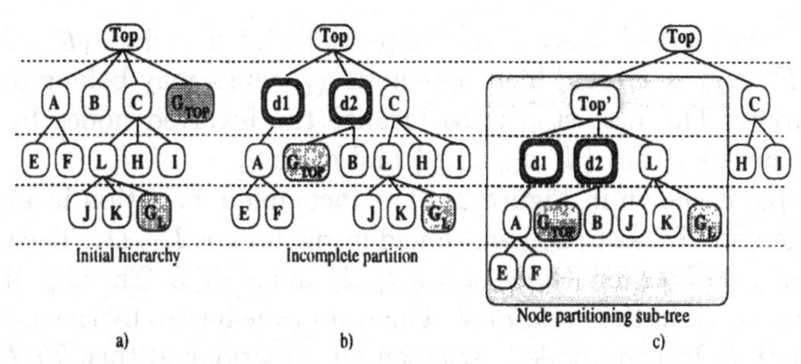

Fig. 2. Hierarchy transformations in node partitioning: a) initial hierarchy; b) incomplete partitioning solution; c) node partitioning sub-tree.

If some logic is assigned to the device, then the device constraints may be written in terms of *available free space*: $S_{AVL}(D_i) = S_{MAX}(D_i) - \sum S_j$, $T_{AVL}(D_i) = T_{MAX}(D_i) - \bigcup T_j$, where $\sum S_j$ is the summary size of j blocks assigned to the device D_i, and $\bigcup T_j$ is the summary I/O pin number occupied by all the assigned blocks. The available size shows the maximal size of additional block(s) which may be assigned to the given device. It may be considered as *firm* constraint. The available I/O pin constraint shows how many I/O pins are available in the device, but for each free hierarchy block the requirement of additional pins, if assigned to the selected device, depends on its connections with the blocks already assigned to the device. The available I/O pin constraint is defined as *soft* constraint.

3 Problem formulation

Each hierarchy node may be either assigned to one device or partitioned among several devices. If already existing devices are not sufficient to receive the selected block, one or more new devices D_i^* may be created.

We define the *node partitioning problem* for node χ_i as assignment of all the successors of this node to a set of existing devices $D = \{D_1, D_2, ..., D_l\}$ and new devices $D^* = \{D_1^*, D_2^*, ..., D_n^*\}$ while minimizing the number of created new devices and the total number of occupied I/O pins: $MIN(\sum_{i=1}^{l+n}(T_i))$ & $MIN(n)$. The necessary but not sufficient condition for node χ_i to be assigned in existing devices set is that the summary free size in these devices $\sum_{i=1}^{l} S_{AVL}(D_i)$ is bigger or equal than the node size S_i. If it is not the case, additional new devices D^* are required.

Design partitioning then consists in successive calls of the node partitioning algorithm for different nodes of the hierarchy. If no blocks were pre-assigned and the top node is selected for node partitioning, the node partitioning problem is equivalent to the partitioning problem solved by the *HPART* algorithm ([9]).

4 Node partitioning algorithm

The input for the algorithm is a subset of partially filled devices $\{D_1, D_2, ..., D_l\}$, new device D^* and a selected hierarchy node χ_i, which may belong to any level of the hierarchy. The first step is to localize the involved nodes by grouping together the device nodes $\{d_1, d_2, ..., d_l\}$ and node χ_i.

An example is shown in Figure 2c. The incomplete partition in Figure 2b is taken and the node L has to be partitioned using devices D_1, D_2. The temporary node Top' encapsulates the device nodes d_1, d_2 and node L. The algorithm works on a sub-tree rooted in the node Top'. The rest of the hierarchy tree is not visible for the algorithm. If a top node is selected for partitioning, then $Top' = Top$.

If the selected hierarchy node does not fit in any of the selected devices, it is recursively clustered by the *HPART* algorithm ([9]). Each node may fit in zero or more devices. The hierarchy nodes may be classified in three categories: (1) *"coarse"* nodes which fit in no one device; (2) *"fine"* nodes which fit in all devices; (3) *"medium"* nodes which fit in a non-empty subset of devices.

To perform the hierarchical clustering, S_{MAX} and T_{MAX} constraints have to be defined. One possibility is to take the smallest free space as a reference. All the hierarchy blocks which exceed these constraints will be non acceptable and all the clusters will be "fine". However, this may be unnecessarily restrictive if some devices are filled by 80%-90%: many glue logic may be introduced in the final partitioning.

The hierarchical clustering is performed in several iterations. Initially, the largest free space $MAX_{i=1}^{l}(S_{AVL}(D_i), T_{AVL}(D_i))$ is selected as a constraint. As a result, at least one device may receive each cluster. The number of free I/O pins is considered as firm constraint during the hierarchical clustering. The hierarchical clustering algorithm produces "medium" or "fine" clusters which fit in at least one device.

Greedy arrangement by bin-packing algorithm with first-fit heuristic is used to obtain the initial assignment of hierarchy nodes to devices. Devices play a role of "bins". Initially, all nodes are sorted in decreasing lexicographical order (S, T). All devices are sorted in increasing lexicographical order of the available free space. This favors in first turn filling devices which have the smallest free space. Each node is examined and assigned to a device which may accept it without any size or I/O violation. I/O pin constraint is considered as soft constraint during the bin-packing. The node partitioning algorithm (*PREPART*) is presented below.

Algorithm 1: PREPART
Input: χ_i, $D = \{D_1, D_2, ..., D_l\}$, D^*
Output: k, $\{P_1, P_2, ..., P_k\}$
$Top' =$ **Group**$(\chi_i, d_1, d_2, ..., d_l)$;
/* *create subtree rooted at* Top' */
while $S(\chi_i) > \sum_{i=1}^{l} S_{AVL}$
    $D = D \bigcup D^*$;
end while
$\chi = \{\chi_i\}$;
$Proceed = TRUE$;
$S_{MAX} = MAX(S_{AVL})$; $T_{MAX} = MAX(T_{AVL})$;
while $\chi \neq \emptyset$ & $Proceed = TRUE$
    for *all* $\chi_i \in \chi$
        if $S(\chi_i) > S_{MAX}$
            $\chi = \chi \bigcup$ **HPART**(χ_i);
        end if
    end for
    Update=**Bin-Packing**(χ, D);
    if *Update=1*
        update S_{MAX}, T_{MAX} ;
    else $Proceed = FALSE$;
    end if
end while
if $\chi \neq \emptyset$
    create initial partition;
end if
$\{P_1, P_2, ..., P_k\}$ =**FPART**(Top');
remove envelope Top';

If free blocks exist at the end of bin-packing, S_{MAX} and T_{MAX} are updated and a new iteration is performed. The iterations stop when no more assignments are possible during the bin-packing pass. If there still exist free blocks, an initial partition has to be created. Finally, the flat partitioning algorithm, FPART ([10]), is applied to obtain a feasible solution.

The initial solution is obtained in the following way. The hypergraph model is created for Top' node. Each device node d_i is assigned to partition block P_i. All non assigned logic is considered as an additional partition block P_{k+1}. Only the moves of non assigned clusters from block P_{k+1} to remaining blocks are allowed. The objective is to empty the block P_{k+1} while satisfying constraints for at least $(k-1)$ blocks. If it is impossible, the block P_{k+1} is merged with the block having the smallest ST-quality.

At the end of the *PREPART* algorithm, the Top' node, containing only device nodes as direct successors, is ungrouped.

5 Experimental results

In this section we (1) give examples of manual/automatic partitioning which invokes the node partitioning algorithm *PREPART*; (2) compare partitioning results of full automatic approach with mixed manual/automatic approaches; (3) show how the performance of the partitioning algorithm may depend on the amount of pre-assigned blocks.

We experimented with two big industrial circuits mapped on Altera FLEX 10K technology. Both circuits are synthesized from VHDL description by Synopsys FPGA Compiler tool. The first example (Circuit1) is a telecommunication circuit. The mapped circuit size is about 29,800 Altera CLBs, which corresponds to about 600K equivalent gates. The circuit contains 6 FIR filters of about 4,700 CLBs each and a set of smaller hierarchy blocks. The initial hierarchy of the design is shown in the left part of Figure 3. Each hierarchy block contains several hundreds of smaller blocks which are not presented in the hierarchy tree in Figure 3 for simplicity. The majority of these blocks are 32-bit Altera DesignWare adders and multipliers. Each block has more than 100 I/Os. If blocks *R1_I*, *R1_Q*, *R2_I*, *R2_Q*, *R3_I*, *R3_Q* are ungrouped, it will be a hard task for the automatic partitioning algorithm to merge adder blocks together to form partitions.

The second example (Circuit2) is a Reed-Solomon decoder with a size of about 22,400 Altera CLBs, which corresponds to about 450K equivalent gates. The initial hierarchy of this circuit is presented in the left part of Figure 4. The particular feature of this hierarchy is that several blocks of the hierarchy (X, V, I, Y) are big blocks without sub-hierarchy and contain 20-30 thousands of glue logic cells. Ungrouping such a block immediately increases the problem complexity. Another point is that several blocks (V, E, I) have the number of I/O pins which exceeds the available I/O pin number of all the FPGA devices from the Altera FLEX 10K family. In Figure 4, not all the successor blocks (*J1*,...,*J20*) of block *E* are shown for simplicity.

5.1 Partitioning session examples

First, we present two examples of partitioning sessions which call the described *PREPART* algorithm. The first example with Circuit1 is presented in Figure 3. The Altera Flex EPF10K70-503 device with $S_{MAX} = 3,744$ and $T_{MAX} = 358$ is chosen for partitioning. The minimal number of devices required for partitioning is 9, which will produce an average of 88% devices filling. We start the partitioning manually. The blocks $R1\_I, R1\_Q, R2\_I, R2\_Q, R3\_I, R3\_Q$ have the size bigger than the device constraint. It is impossible to preserve them. We have 6 blocks that are bigger than the device constraint and we target 9 devices. The most easy solution is to refine the granularity by invoking the automatic partitioning on the selected blocks with the constraint $S_{MAX} = 3,700$. Four of six blocks are split into two blocks which will form a "seed" for 8 devices. These blocks are assigned manually. At this step, the *PREPART* algorithm may be called to finish the partitioning.

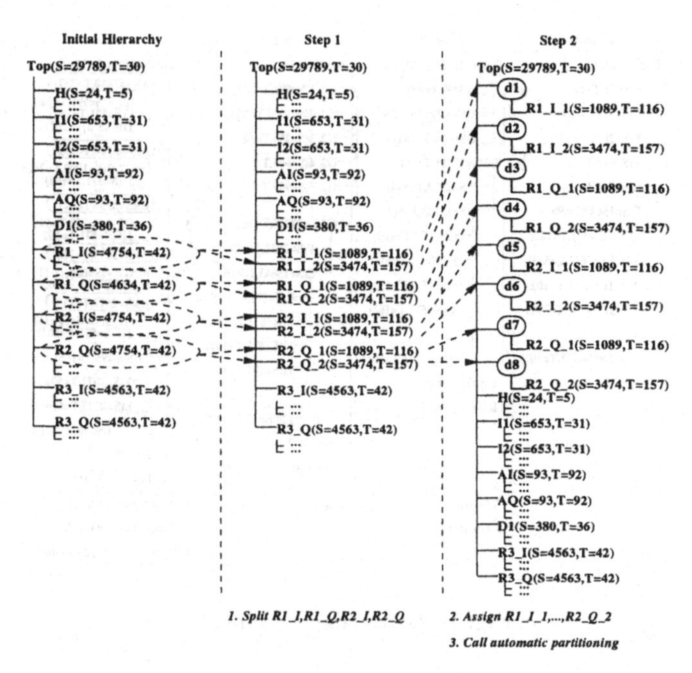

Fig. 3. Partitioning example 1.

The second example is the partitioning of Circuit2 using the Altera EPF10K100-503 device with $S_{MAX} = 4,992$ and $T_{MAX} = 406$. The minimal required number of devices is 5. The Circuit2 contains hierarchy blocks (V, E, I) with I/O pin number greater than T_{MAX}. Blocks V and I have no sub-hierarchy, each of them contains about 20,000 instances of glue logic. Performing the connectivity analysis on Level2, gives the possibility to re-organize the hierarchy by grouping the blocks (X, Y) and (V, E). New blocks $G1$ and $G2$ have small I/O numbers. The blocks which are most suitable to be assigned manually are the blocks $C, I, J1, ..., J20$. They may be assigned to devices $D_1, ..., D_5$ (about 4 blocks per device). To refine the granularity, the node $G2$ obtained by merging together the nodes X, Y, is split in 5 blocks which may fill the free space in the devices $D_1, ..., D_5$. At this step, we have non-assigned nodes F, R, which are "fine" nodes and are easy to assign, and nodes V, E which have big number of I/Os. At this point, it is difficult to continue manually and the *PREPART* algorithm finishes the partitioning. The amount of logic participating in the automatic partitioning is about 30% of the initial design size.

5.2 Full automatic versus mixed manual/automatic partitioning

Next, we performed the comparison of the quality of partitions obtained by full the automatic tool and using a mixture of manual and automatic approaches. The circuit1 was partitioned multiple times using different devices from the Altera FLEX 10K family. Each time, initially full automatic partitioning was performed. Then, the selected subset of hierarchy blocks was pre-assigned to

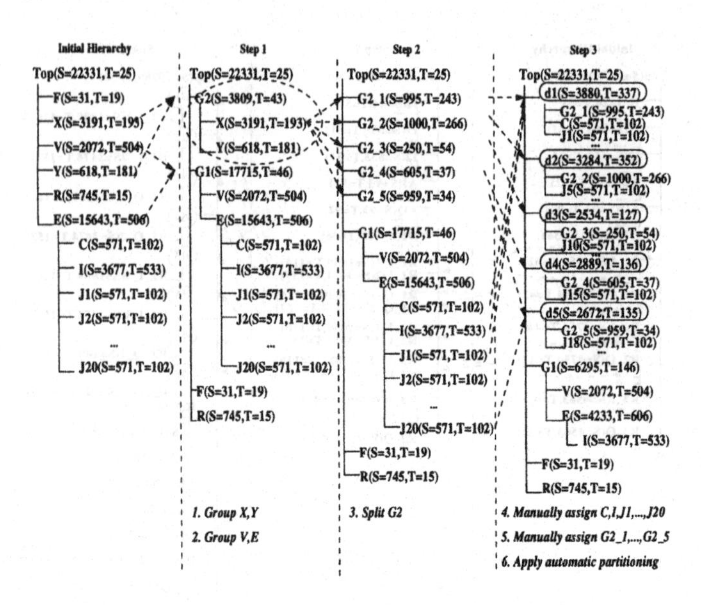

Fig. 4. Partitioning example 2.

devices, and the remaining logic was partitioned using the *PREPART* algorithm. The results are presented in Table 1.

	Device	MAX S	MAX T	Min #D	Automatic #D	Automatic F	PREPART #D	PREPART F
1	10K100-503	4992	406	6	7	85%	6	99%
2	10K70-503	3744	358	9	11	73%	10	80%
3	10K50-403	2880	310	11	16	64%	14	74%
4	10K50-356	2880	274	11	16	64%	13	80%
5	10K30-356	1728	246	18	23	71%	19	91%
	Average					64%		85%

Table 1. Comparison of full automatic and mixed automatic/manual partitioning.

For each experiment is presented the device type, its S_{MAX} and T_{MAX}, the minimal theoretical devices number M, the results for full automatic partitioning and the partitioning with pre-assigned blocks. Partitioning results are presented in terms of obtained FPGA devices ($\#D$) and average filling ratio for all the devices (F). The *HPART* algorithm ([9]) was used as automatic algorithm. In the testcase 1, six blocks (*R1_I*, *R2_I*, *R3_I*, *R1_Q*, *R2_Q*, *R3_Q*) were pre-assigned. The testcase 2 corresponds to the partitioning example presented in Figure 3. The blocks *R1_I*, *R1_Q*, *R2_I*, *R2_Q* are first split in two blocks and the obtained blocks are assigned to 8 devices. In the testcases 3 and 4, five blocks were initially split with the constraint $S_{MAX} = 3,700$ and ten obtained blocks were assigned to 10 devices. In the testcase 5, six blocks were split with the constraints $S_{MAX} = 1,600$. The 18 produced blocks were assigned to 18 devices.

Table 1 shows that partitioning with pre-assigned blocks gives better results than the full automatic partitioning. The node pre-assignment creates stable "seeds", smaller amount of logic participates in the automatic partitioning, solution space is smaller and it becomes easier to explore it.

5.3 Results with pre-assigned blocks

We performed also experiments which show gradually how the pre-assignment of blocks impacts the result quality. The testcases 2,3,4,5 from Table 1 were selected for detailed analysis. Initially, no blocks were pre-assigned. The partitioning result is the same as in Table 1. We performed tests with one block pre-assigned, two blocks pre-assigned, three blocks pre-assigned and finally four blocks pre-assigned. In the testcase 2, the blocks $R1\_I$, $R1\_Q$, $R2\_I$, $R2\_Q$ were initially split in two smaller blocks, as was shown in Figure 3, and then each of two sub-blocks was pre-assigned. The results are presented in Table 2. For each testcase is given a subset of hierarchy blocks which are pre-assigned, the percentage of the total design size which corresponds to the pre-assigned blocks, and the partitioning result in number of devices.

6 Conclusions

The present paper proposes an extension of a hierarchy-based partitioning algorithm to deal with pre-assigned blocks. The necessity in such an algorithm was justified by the observations that there always exist in the design a set of hierarchy blocks which can be easily managed manually. These blocks play a role of "seeds" for the blocks of partitions and the rest of the design is partitioned around these blocks. Experimental results show that mixed manual/automatic approach using the proposed PREPART algorithm allows to obtain average filling ratios of about 85% on big industrial netlists. In addition, big portions of the design are preserved after partitioning and may be easily located in the devices. This facilitates the engineering changes and allows to apply block-based debugging strategy in prototyping.

References

1. D. Behrens, K. Harbich, E. Barke, "Hierarchical Partitioning", *Proc. Int. Conf on Computed-Aided Design* (1996): 470-477.
2. T.N.Bui, C. Heigham, C. Jones, T. Leighton, "Improving the Performance of the Kernigan-Lin and Simulated Annealing graph bisection algorithms", *Proc. 26th Design Automation Conference* (1989): 775-778.
3. N.-C. Chou, L.-T. Liu, C.-K. Cheng, W.-L. Dai, R. Lindelof, "Circuit Partitioning for Huge Logic Emulation Systems", *Proc. 31st Design Automation Conference* (1994): 244-249.

Test Case (Table 1)		Blocks preassigned	% logic preassigned	#D
2	1.	∅	0%	11
	2.	R1_I	16%	11
	3.	R1_I,R1_Q	32%	11
	4.	R1_I,R1_Q,R2_I	47%	10
	5.	R1_I,R1_Q,R2_I,R2_Q	64%	10
3	1.	∅	0%	16
	2.	R1_I	16%	16
	3.	R1_I,R1_Q	32%	16
	4.	R1_I,R1_Q,R2_I	47%	15
	5.	R1_I,R1_Q,R2_I,R2_Q	64%	14
	6.	R1_I,R1_Q,R2_I,R2_Q,R3_I	80%	14
4	1.	∅	0%	16
	2.	R1_I	16%	16
	3.	R1_I,R1_Q	32%	16
	4.	R1_I,R1_Q,R2_I	47%	14
	5.	R1_I,R1_Q,R2_I,R2_Q	64%	14
	6.	R1_I,R1_Q,R2_I,R2_Q,R3_I	80%	13
5	1.	∅	0%	23
	2.	R1_I	16%	23
	3.	R1_I,R1_Q	32%	22
	4.	R1_I,R1_Q,R2_I	47%	19
	5.	R1_I,R1_Q,R2_I,R2_Q	64%	19
	6.	R1_I,R1_Q,R2_I,R2_Q,R3_I	80%	19
	7.	R1_I,R1_Q,R2_I, R2_Q,R3_I,R3_Q	96%	19

Table 2. Results of partitioning with different sets of pre-assigned blocks.

4. C-K. Cheng, Y-C. A. Wei, "An improved two-way partitioning algorithm with stable performance", *IEEE Trans. on Computer-Aided Design of ICs and Systems* 10/12 (1991): 1502-1511.
5. H-J. Eikerling, W. Rosenstiel, "Automatic Structuring and Optimization of Hierarchical Designs", *Proc. Euro-DAC with Euro-VHDL* (1996): 134-139.
6. W-J Fang, A. C-H Wu, "A Hierarchical Functional Structuring and Partitioning Approach for Multiple-FPGA Implementations", *Proc. Int. Conf on Computed-Aided Design* (1996): 638-643.
7. W-J Fang, A. C-H Wu, "Multi-Way FPGA Partitioning by Fully Exploiting Design Hierarchy", *Proc. 34-th Design Automation Conference* (1997): 518-521.
8. J. Garbers, H.J. Promel, A. Steger, "Finding Clusters in VLSI circuits", *Proc. Int. Conf. on Computed-Aided Design* (1990): 520-523.
9. H. Krupnova, A. Abbara, G. Saucier, "A Hierarchy-Driven FPGA Partitioning Method", *Proc. 34-th Design Automation Conference* (1997): 522-524.
10. H. Krupnova, G. Saucier, Iterative Improvement Based Multi-Way Netlist Partitioning for FPGAs *Proc. Int. Conf. Design, Automation and Test in Europe* (1999): 587-594.
11. T.K.Ng, J. Oldfield, V. Pitchumani, "Improvements of A Min-Cut Partition Algorithm", *Proc. Int. Conf. on Computed-Aided Design* (1987): 470-473.

Logical-to-Physical Memory Mapping for FPGAs with Dual-Port Embedded Arrays

William K.C. Ho and Steven J.E. Wilton

Department of Electrical and Computer Engineering
University of British Columbia,
Vancouver, B.C., Canada,
{williamh|stevew}@ece.ubc.ca
http://www.ece.ubc.ca/~ stevew

Abstract. On-chip storage has become critical in large FPGAs. This has led most FPGA vendors to include configurable embedded arrays in their devices. Because of the large number of ways in which the arrays can be combined, and because of the configurability of each array, there are often many ways to implement the memories required by a circuit. Implementing user memories using physical arrays is called *logical-to-physical mapping*, and has previously been studied for single-port FPGA memory arrays. Most current FPGAs, however, contain dual-port arrays. In this paper, we present a logical-to-physical algorithm that specifically targets dual-port FPGA arrays. We show that this algorithm results in 28% denser memory implementations than the only previously published algorithm.

1 Introduction

It has become clear that on-chip storage is critical in large FPGAs. As FPGAs grow, they are being used to implement entire systems, rather than the small logic subcircuits that have traditionally been targeted to FPGAs. One of the important differences between these large systems and smaller logic subcircuits is that the large systems often require storage. Although this storage could be implemented off-chip, on-chip storage has a number of advantages. Besides the obvious advantages of integration, on-chip storage will often lead to higher clock frequencies, since I/O pins need not be driven with each memory access. In addition, on-chip storage will relax I/O pin requirements, since pins need not be devoted to external memory connections. These advantages have led most FPGA vendors to produce architectures with significant amounts of on-chip storage.

Since the storage requirements of circuits vary widely, the FPGA memory architecture must be flexible enough to implement different numbers of independently addressable memories as well as different memory shapes and sizes. Many recent commercial devices, such as the Altera 10K and 10KE devices [1, 2], the Xilinx Virtex FPGAs [3], the Actel 42MX [4], and the Lattice ispLSI 6192 FPGAs [5], provide several large arrays embedded into the FPGA. Each array can typically be used in one of several modes, each with a different width and depth. As an example, the Altera 10KE devices contain between six and twenty 4-Kbit blocks, each of which can be used as a 2Kx2, a 1Kx4, a 512x8, or a 256x16 array.[1] These arrays can be combined to implement larger user memories.

The task of implementing the memories required by a user circuit using the FPGA embedded arrays is called *logical-to-physical mapping* [6]. Because of the large number of ways in which arrays can be combined, and because each array can be used in one of several modes (widths/depths), this problem is not trivial. Yet, it is vitally important – since each FPGA contains only a few memory arrays, a sub-optimal implementation that wastes even one memory array could very easily cause a circuit to not fit on a given FPGA. Even if the memory configuration does fit on the FPGA, minimizing the number of arrays needed to implement the storage part of the circuit is beneficial because unused memory arrays can be configured as ROM and used to implement logic [7,8].

In [6,9], logical-to-physical mapping for FPGAs with single-port embedded arrays (arrays in which only one access can be performed at a time) is discussed. Many recent FPGAs, however, contain dual-port arrays (so that two accesses can be performed by each array concurrently) [2–4]. Many applications require memories that can be accessed simultaneously by two separate subcircuits; these applications can most efficiently be implemented if the FPGA has dual-port arrays.

In this paper, a new logical-to-physical mapping algorithm that targets dual-port arrays is presented. We show that this new algorithm results in much more efficient implementations than if we simply extend the techniques targeting single-port arrays [6,9]. The user circuits are assumed to consist of both single and dual-port user memories; our improvement is obtained by intelligently packing the single-port user memories into the dual-port physical arrays. Under the right conditions, each dual-port array can implement two single-port memories (or parts of two single-port memories).

Besides [6] and [9], little work as been done in this area. Jha and Dutt describe an algorithm to map logical memories to physical library elements [10], but do not consider the optimizations that are possible when the physical elements are dual-port. Karchmer and Rose show how user memories can be implemented by larger physical memory chips, but only consider single-port physical arrays [11]. Their work is also different in that they consider discrete memory devices, which do not have the variety of modes that FPGA memory arrays have. There has also been considerable work mapping variables to both single and dual port memories during high-level synthesis in an attempt to minimize the execution time of an algorithm [12–14]. None of these papers consider physical memories with the configurability of FPGA arrays, however.

This paper is organized as follows. Section 2 presents our assumptions regarding the FPGA architecture and the application circuits, and then gives a precise definition of the problem solved in this paper. Section 3 then describes our algorithm. Finally, Section 4 compares the results from our algorithm with

[1] In this paper, a *axb* memory has *a* words of *b* bits each.

Parameter	Meaning
N	Number of Arrays
B	Bits per Array
P	Ports per Array
M	Number of Modes for each array
W_i	Data width of array in mode i
l	Number of Logical Memories
d_k	Depth of Logical Memory k
w_k	Width of Logical Memory k
p_k	Ports in Logical Memory k

Table 1. Architectural and Circuit Parameters

those obtained by simply extending a previous algorithm that was developed to target single-port arrays.

2 Problem Definition

In this section, we first describe our assumptions regarding the target FPGA architecture and the user circuit that is to be mapped, and then present a precise definition of the Logical-to-Physical Memory Mapping problem.

2.1 Architectural Framework

The top half of Table 1 summarizes the parameters that define the FPGA embedded memory array architecture. The number of embedded memory arrays is denoted by N, the number of bits in each array is denoted by B, the number of independent access ports in each array is denoted by P. Each array can be used in one of M different modes; each mode has a different width and depth. The width of each array in mode i is denoted W_i; the depth can be calculated as B/W_i. In the Altera FLEX10KE, $B = 4096$ bits, $P = 2$, $M = 4$, and $\{W_0, W_1, W_2, W_3\} = \{2, 4, 8, 16\}$, meaning each array is dual-port and can be configured to be one of 2048x2, 1024x4, 512x8, or 256x16.

In this paper, we will only consider dual port arrays, ie. $P = 2$. Note that some FPGA architectures, such as the Altera FLEX10KE, contain two independent ports, but one port is a dedicated read port and one port is a dedicated write port. This works well for many applications (such as a first-in first-out buffer that is used to temporarily hold data in a communication system), but there are many applications for which this is insufficient (a dual-port register file in a processor which must be read by two functional units simultaneously, for example). To implement these sorts of circuits, true dual-port memory arrays are required, in which the two accesses are independent, and either can be a read or write. With the increasing importance of embedded memory in FPGAs, and since true dual-port arrays appear to be a natural evolution from the restricted dual-port model used in many architectures today, we feel that true dual-port memories will be available in future devices. Thus, we focus our efforts on studying algorithms that target true dual-port memories.

2.2 User Circuit Assumptions

It is assumed that the user circuit to be implemented on the FPGA contains both logic and memory portions. In this paper, we are only concerned with the memory portion.

We assume that the memory portion of the circuit consists of l independent user memories. We refer to each of these user memories as a *logical memory.* The set of all logical memories required for a circuit will be referred to as that circuit's *logical memory configuration.* The depth of logical memory k ($0 \leq k \leq l-1$) will be denoted d_k, the width of logical memory k will be denoted w_k, and the number of ports required by logical memory k (maximum number of simultaneous accesses to memory k) will be denoted p_k. Unlike [6], we allow user memories that require either one or two ports. These parameters are summarized in the bottom half of Table 1.

2.3 Problem Statement

The problem studied in this paper can be stated as follows:

Given: 1. An FPGA architecture described by B, M and W_i ($0 \leq i < M$) as described in Subsection 2.1 (this paper only considers architectures with $P = 2$),

2. A Logical Memory Configuration described by l, d_k, w_k, and p_k ($0 \leq k < l$), as described in Subsection 2.2 (in this paper, $1 \leq p_k \leq 2$ for all k),

Find: An implementation of the logical memory configuration using n embedded memories.

Such that: n is as small as possible.

Note that the goal is to implement the logical memory configuration using as few physical arrays as possible. In an FPGA with N arrays, it may appear that minimizing the number of arrays required to implement the logical memory configuration is immaterial, as long as the implementation requires N or fewer arrays. Minimizing the number of arrays is important, however, since the remaining arrays can be configured as ROMs, and be very efficiently used to implement the logic part of the user circuit [7,8]. The fewer arrays that are used to implement memory, the more that will be available to implement logic.

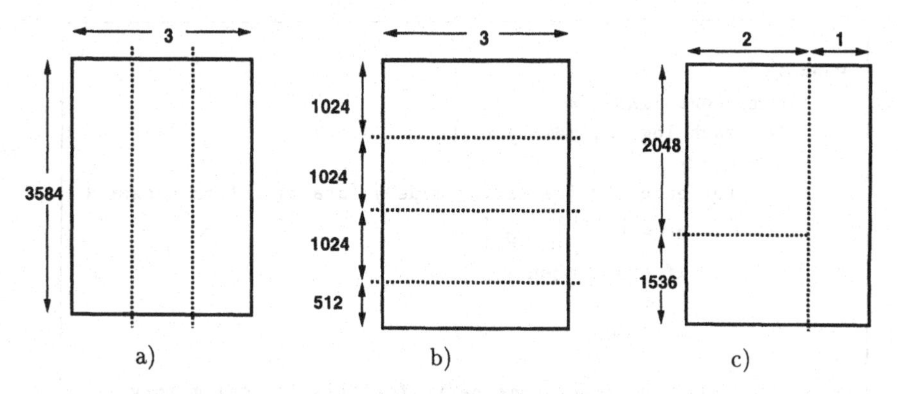

Fig. 1. Three ways to partition a 3584x3 logical memory

3 Logical-to-Physical Mapping Algorithm

In this section, our new logical-to-physical mapping algorithm is described. The algorithm consists of three phases: during the first phase, the logical memories are broken into *components*, in the second phase, these components are packed into the physical arrays, and in the third phase, these physical arrays are wired together to implement the original logical memory configuration.

3.1 Phase 1: Break Logical Memories into Components

The first phase of the algorithm partitions each logical memory into several *components*, each of which is small enough to fit into a single physical array. Each component represents a portion of the bits in the original logical memory, and can be described by its width, wc_i, its depth, dc_i, and the number of ports required, pc_i. In order for the component to fit in a single physical array, wc_i and dc_i must satisfy the following inequalities:

$$wc_i \leq W_j \tag{1}$$

$$dc_i \leq B/W_j \tag{2}$$

for some value of j between 0 and $M - 1$ (recall that a physical array can be used in one of M modes, each of which has a different width/depth). The number of ports required by each component, pc_i, is the same as the number of ports required by the original logical memory. We also define a quantity mc_i for each component which indicates the physical array mode(s) that can be used to implement this component.

As an example, Figure 1 shows three ways in which a 3584x3 logical memory could be broken into components. In each case, it is assumed that each physical array consists of 4096 bits ($B = 4096$) and can be used as a 4096x1, 2048x2, 1024x4, or a 512x8. In Figure 1(a), $wc_i = 1$ for each component, while in Figure 1(b), all components have $wc_i = 3$. In Figure 1, one of the components has $wc_i = 1$, while the others have $wc_i = 2$.

```
Phase1:
    component list = φ
    for each logical memory i {
        c = ∞
        for each physical array mode j (starting from widest){
            cj = ⌈wi/Wj⌉ ⌈di/(B/Wj)⌉
            if cj < c then {
                c = cj
                m = j
        }
        construct c new components for this logical memory
        calculate wck, dck, pck for each new component k
        mck = m for each new component k
        component list = component list ∪ new components
    }
```

Fig. 2. Summary of Phase 1 of the Algorithm

In general, there are many ways to partition each logical memory. To simplify the task, we only consider partitions in which each component has the same mc_i (that is, each component can be implemented by a physical array in the same mode). The partitions in Figures 1(a) and 1(b) would be considered, therefore, while the one in Figure 1(c) would not. Note that this only applies to components that make up a single logical memory. A logical memory configuration typically has several logical memories; components from different logical memories may correspond to different physical array modes.

Given this assumption, the number of components required to implement a logical memory i using physical array mode j is:

$$c = \left\lceil \frac{w_i}{W_j} \right\rceil \left\lceil \frac{d_i}{B/W_j} \right\rceil \tag{3}$$

To find the partition that results in the smallest c, we cycle through all possible array modes and choose the best result. This partitioning is done independently for each logical memory in the logical memory configuration. This is summarized in Figure 2.

3.2 Phase 2: Bin Packing

Given the list of components found in phase 1, it is possible to implement the logical memory configuration directly by using one physical array for each component. As will be shown in Section 4, this often results in very poor utilization of the memory arrays. As an example, consider implementing two single-port

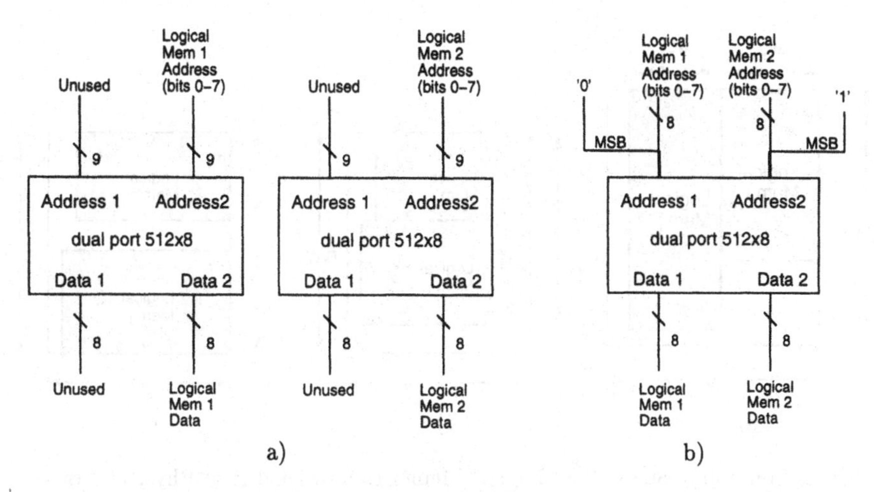

Fig. 3. Two ways of implementing two single-port 192x8 logical memories

192x8 logical memories using an architecture with $B = 4096$ and $P = 2$. Phase 1 would create two components, each 192x8. If these components were implemented directly, two memory arrays would be required, as shown in Figure 3(a). Even though the original logical memory configuration only consists of 3072 bits, a total of 8192 bits (two physical arrays) are used to implement it. An alternative implementation is shown in Figure 3(b); in this implementation, each logical memory is mapped to a portion of a single array, and one of the array's two ports is used for each logical memory. The upper order address bit of port 1 is tied to 0, while the upper order address bit of port 2 is tied to 1. This ensures that each port sees a different 256x8 portion of the physical array. Since both ports are independent, both logical memories can be accessed independently. This example illustrates the goal in phase 2: the components from phase 1 are packed into the available memory arrays such that the total number of required arrays is as small as possible.

Informally, two single-port components can be packed into a dual-port physical array "vertically" or "horizontally". Consider packing two arrays i and j "horizontally" as shown in Figure 4(a). In this case, the physical array is of width W, and the two components are of width wc_i and wc_j. Each word in the memory contains wc_i bits for component i and wc_j bits for component j. By supplying an address to the first port's address bus, and accessing data through bits 0 to $wc_i - 1$ of the first port's data bus, the first component can be accessed. The second component can be accessed in the same way using the second port's address bus and bits wc_i to $wc_i + wc_j - 1$ of the second port's data bus. Since the ports are independent, both components can be accessed independently.

In order to combine arrays in this way, it is sufficient that:

$$W \geq wc_i + wc_j \tag{4}$$

where W is the physical array width in the mode chosen to implement the components.

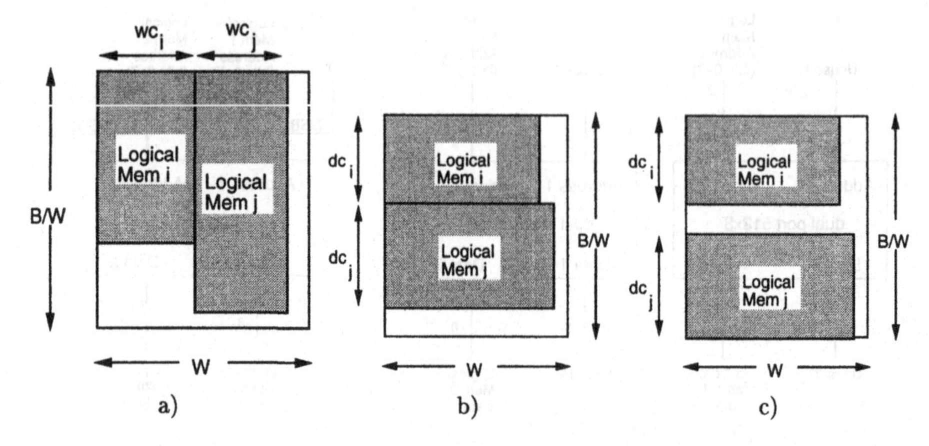

Fig. 4. Combining Single-Port Logical Memories into Dual-Port Physical Arrays

Figure 4(b) shows how two components i and j can be combined "vertically". In this case, dc_i of the words are used to implement component i and dc_j are used to implement component j. By accessing words 0 to $dc_i - 1$ through one port and words dc_i to $dc_i + dc_j - 1$ through the other port, we can access each component independently. A straightforward implementation of this, however, would require an adder on the path feeding the second address port, since an offset of dc_i must be added to the second component's address. This adder would be on the critical path of the memory access, which could slow down the circuit. We can eliminate the need for an adder as long as the following condition holds:

$$B/W \geq dc_i + \text{pow2}(dc_j) \quad (5)$$

where pow2(x) means x is "rounded-up" to the next highest power-of-two.

As long as condition 5 holds, we can pack the components as shown in Figure 4(c). Component i is implemented starting at word 0, so no adder is needed on the first address port. Component j is implemented in the top pow2(dc_j) words. Then, $log_2(\text{pow2}(dc_j))$ address lines in the second address port can be used to address component j, while the remaining lines in the second address port are set to '1'. In this way, both memories can be accessed independently, and no adder is needed on either address port. The example in Figure 3(b) was implemented in this way.

Note that condition 5 is sufficient but not necessary. By employing the addressing techniques in [12], the constraint could be relaxed somewhat. Experimentation has shown, however, that for our problem, these more complex addressing schemes rarely lead to a final implementation that uses fewer memory arrays.

The above discussion has assumed that both components are single-ported. If either component requires two access ports, then it can not be packed with any other component, and must be implemented in its own physical array. In general, the following condition must be true to combine arrays:

$$P \geq pc_i + pc_j \tag{6}$$

Most FPGAs with dual-port arrays require that each port in a physical array be used in the same mode (eg. one port can not be used as a 4Kx1 while the other is used as a 512x8). Thus, one final constraint is that two components i and j can only be packed together if:

$$mc_i = mc_j \tag{7}$$

With these constraints, we can formulate the packing problem as a multi-dimensional bin-packing problem. The physical arrays are bins, and the components are the objects to be packed. In order for two components to be packed in the same bin, constraints 6 and 7 as well as either 4 or 5 must be satisfied. Figure 5 summarizes this phase of the algorithm.

3.3 Phase 3: Wire together the Memories

After phase 2, the components implementing each logical memory may be scattered among several physical arrays. The final step in the algorithm is to combine the arrays and connect them to the rest of the circuit. If the "horizontal" partitioning was used in phase 1, the address ports can be simply wired together. If the "vertical" partitioning was in phase 1, a multiplexor and decoder are needed to connect the components. Both of these techniques are described in [6] and [9] and so will not be discussed further here.

4 Results and Discussion

Our algorithm was implemented in a program called SPACK. To evaluate SPACK, we compare it to results obtained from the algorithm presented in [6]. That algorithm maps single-port logical memories to single-port arrays. Each logical memory is broken into components and each component is implemented by a single physical array (this is the same as our algorithm, without Phase 2). Although all logical memories considered in [6] were single-port, the algorithm will support dual-port memories, as long as the physical arrays are dual-port. Thus, we can compare it directly with our algorithm using benchmark circuits containing both single and dual-port logical memories.

Table 2 shows the results from SPACK and the previous algorithm for 19 benchmark circuits. The circuits and their logical memory configurations are shown in the first two columns of Table 2. Each circuit was mapped to 4Kbit physical arrays, each of which can be used as a 4Kx1, 2Kx2, 1Kx4, or 512x8. The number of arrays required to implement each benchmark using each algorithm is shown in the third and fifth columns of the table. Averaged over all benchmark circuits, the previous algorithm required 9.89 arrays, while SPACK required only 7.1 arrays.

```
phase2:
     bin list = φ
     sort component list (largest component first)
     for each component i {
          for each bin j in the bin list {
               if fits(component i, bin j) {
                    add component i to bin j
                    go to next component
               }
          }
          b = new bin with component i as its only occupant
          bin list = bin list ∪ b
          }
     }

fits(component i, bin j) {
     k = component already in bin j
     if (P ≥ pc_i + pc_k) and
        (mc_i = mc_k) and
        [(W_mc_i ≥ wc_i + wc_k) or (B/W_mc_i ≥ dc_i + pow2(dc_k)) or
                                   (B/W_mc_k ≥ dc_k + pow2(dc_i))] {
          return(yes)
        } else {
          return(no)
        }
     }
}
```

Fig. 5. Summary of Phase 2 of the Algorithm

Note that the purpose of Table 2 is *not* to compare FPGAs with single-port arrays to those with dual-port arrays. If the FPGA had only single-port arrays, many of the circuits in the table could not have even been implemented (unless the dual-port user memories were implemented by time-multiplexing two ports onto a single physical port). Rather, the purpose of the table is to show that when targeting FPGAs with dual-port arrays, our algorithm performs considerably better than the previous algorithm. Since we expect most future FPGAs to contain dual-port arrays, this is an important result.

The fourth and sixth columns of Table 2 show the *utilization* of the arrays. The utilization is defined as:

$$\text{utilization} = 100 \frac{\text{number of bits in logical memory configuration}}{(\text{number of bits in each array})(\text{number of arrays used})}$$

A utilization of 100% means that every bit in the physical arrays was used, while a utilization lower than 100% means that some bits in the arrays were wasted (because each physical array could not be completely filled with logical memories). Using the previous algorithm, the utilization is only 35.4% averaged over all circuits. SPACK results in a significantly higher utilization of 51.7%.

5 Conclusions

In this paper, we have presented a new logical-to-physical mapping algorithm that targets FPGAs with dual-port embedded arrays. The purpose of the algorithm is to map the memories required by a circuit to the physical FPGA memory resources. This is an important problem, since an implementation of a user's memory that requires even one more physical array than necessary could very easily cause a circuit to not fit on a given FPGA.

Previous work has studied FPGAs with single-port embedded arrays. Most current FPGAs, however, contain dual-port arrays. We have shown that by explicitly taking advantage of the dual-port nature of these arrays, our algorithm produces considerably more efficient implementations of the memory parts of circuits. Specifically, we have shown that under the right conditions, we can pack two single-port user memories (or parts of two single-port user memories) into a dual-port array. Our algorithm results in memory implementations that use, on average, 28% fewer arrays than an algorithm that does not take advantage of the dual-port arrays in this way.

Acknowledgments

This work was supported by Cypress Semiconductor, the Natural Sciences a nd Engineering Research Council of Canada, and UBC's Centre for Integrated Computer Systems Research.

References

1. Altera Corporation, *Datasheet: FLEX 10K Embedded Programmable Logic Family*, May 1998.
2. Altera Corporation, *Datasheet: FLEX 10KE Embedded Programmable Logic Family*, August 1998.
3. Xilinx, Inc., "Virtex: Our new million-gate 100-MHz FPGA technology." XCell: The Quarterly Journal for Xilinx Programmable Logic Users, First Quarter 1998.
4. Actel Corporation, *Datasheet: Integrator Series FPGAs: 40MX and 42MX Families*, April 1998.
5. Lattice Semiconductor Corporation, *Datasheet: ispLSI and pLSI 6192 High Density Programmable Logic with Dedicated Memory and Register/Counter Modules*, July 1996.
6. S. J. E. Wilton, *Architectures and Algorithms for Field-Programmable Gate Arrays with Embedded Memory.* PhD thesis, University of Toronto, 1997.
7. S. J. E. Wilton, "SMAP: heterogeneous technology mapping for FPGAs with embedded memory arrays," in *ACM/SIGDA International Symposium on Field-Programmable Gate Arrays*, pp. 171–178, February 1998.

8. J. Cong and S. Xu, "Technology mapping for FPGAs with embedded memory blocks," in *Proceedings of the ACM/SIGDA International Symposium on Field-Programmable Gate Arrays*, pp. 179–187, February 1998.
9. Altera Corporation, "Implementing RAM functions in FLEX 10K devices." Technical Note, Nov. 1995.
10. P. K. Jha and N. D. Dutt, "Library mapping for memories," in *Proceedings of the 1997 European Design and Test Conference*, March 1997.
11. D. Karchmer and J. Rose, "Definition and solution of the memory packing problem for field-programmable systems," in *Proceedings of the IEEE International Conference on Computer-Aided Design*, pp. 20–26, 1994.
12. H. Schmit and D. Thomas Jr., "Address generation for memories containing multiple arrays," *IEEE Transactions on Computer-Aided Design of Integrated Circuits and Systems*, vol. 17, May 1998.
13. P. R. Panda and N. D. Dutt, "Behavioral array mapping into multiport memories targeting low power," in *Proceedings of the 10th International Conference on VLSI Design*, Jan 1997.
14. M. Balakrishnan, A. Majumdar, D. Banerji, J. Linders, and J. Majithia, "Allocation of multiport memories in datapath synthesis," *IEEE Transactions on Computer-Aided Design*, vol. 7, April 1988.

Circuit	Memory Requirements	Previous Algorithm [6]		SPACK Algorithm	
		Number of Arrays Req'd	Utilization	Number of Arrays Req'd	Utilization
Variable Length CODEC	one 768x16, two 32x7, one 512x1	7	46.2%	5	64.7%
Discrete Cosine Transform Chip	two 16x16	4	3.13%	2	6.25%
Video Compression	one 24x112 (dual port) one 16x96 (dual port)	26	3.97%	20	5.16 %
Encryption Circuit	one 256x16	2	50%	1	100%
Robot Controller	one 172x20	3	28%	2	42%
Filter	two 8x24 (dual port) one 320x24	9	21.9%	9	21.9%
Neural Network Chip 1	one 160x8, one 32x8	2	18.8%	1	37.5 %
Neural Network Chip 2	one 1310x24, one 1024x16	13	89.8%	13	89.8%
DMA Chip for LAN	one 15x24, one 16x4, one 256x32	8	26.3%	4	52.6%
Translation Look-aside Buffer	two 256x59, one 16x18 (dual port)	19	39.2%	11	67.7%
Proof-of-Concept Viterbi Decoder	three 128x8, one 28x3 (dual port)	4	19.3%	3	25.7%
Image Backprojector	two 128x22	6	22.9%	3	45.8%
DSP Control Unit	one 1024x32, one 128x16, three 64x16, two 24x16	20	47.2%	14	67.4%
Vector Processing Unit	two 256x9, two 256x8, three 128x9, three 128x16 (dual port)	18	24.8%	12	37.2%
Communications Circuit 1	six 88x8, one 64x24	9	15.6%	5	28.1%
Communications Circuit 2	three 736x16	12	71.9%	9	95.8%
Communications Circuit 3	four 368x16, one 736x16	12	71.9%	11	78.4%
Communications Circuit 4	two 1620x3, two 168x12, two 366x11	12	44.4%	9	59.1%
Communications Circuit 5	one 192x12	2	28.1%	1	56.3%
Average		9.89	35.4%	7.10	51.7%

Table 2. Experimental Results

DYNASTY: A Temporal Floorplanning Based CAD Framework for Dynamically Reconfigurable Logic Systems

Milan Vasilko

Microelectronic Systems Research Group
School of DEC, Bournemouth University, Talbot Campus
Fern Barrow, Poole, Dorset BH12 5BB, UK

E-mail: M.Vasilko@computer.org

Abstract. This paper presents DYNASTY—a new CAD framework aimed at supporting research of design techniques, algorithms and methodologies for dynamically reconfigurable logic (DRL) systems. Design flow implemented in the DYNASTY Framework is based around a *temporal floorplanning (TF)* DRL design abstraction, which allows simultaneous DRL design space exploration in spatial and temporal dimensions.
The paper introduces temporal floorplanning and its implementation in the DYNASTY Framework. Methodologies based on temporal floorplanning promise reduction of design time and elimination of costly design iterations present in traditional DRL design methodologies.

1 Introduction

Traditionally, designs for circuits with partial run-time reconfigured modules have been implemented manually with little support from available CAD tools. Most of the reported DRL methodologies were designed around a sequential design flow adopted from conventional design methodologies for *static* hardware. These methodologies require a large number of iterations due to their inability to provide accurate configuration overhead and layout metrics early in the design flow, and inability to backtrack design decisions from earlier design stages (Fig. 1(a)).

Recently several authors have advocated the use of FPGA layout metrics at earlier stages of a DRL design flow (e.g. [1, 2]) in order to remedy these problems. Accurate estimation of these metrics requires placement and detailed routing of the design modules for each of the investigated design architectures and configuration schedules, which is impractical for all but very simple designs.

This situation has motivated our work into integrated design techniques, which permit delaying design decisions depending on the physical design metrics until later stages of the design flow.

In this paper we present a new DRL design framework, which implements a Library server-based DRL design methodology, introduced in [1] (design flow is reprinted in Fig. 1(b)). The implementation is based on abstracting DRL design

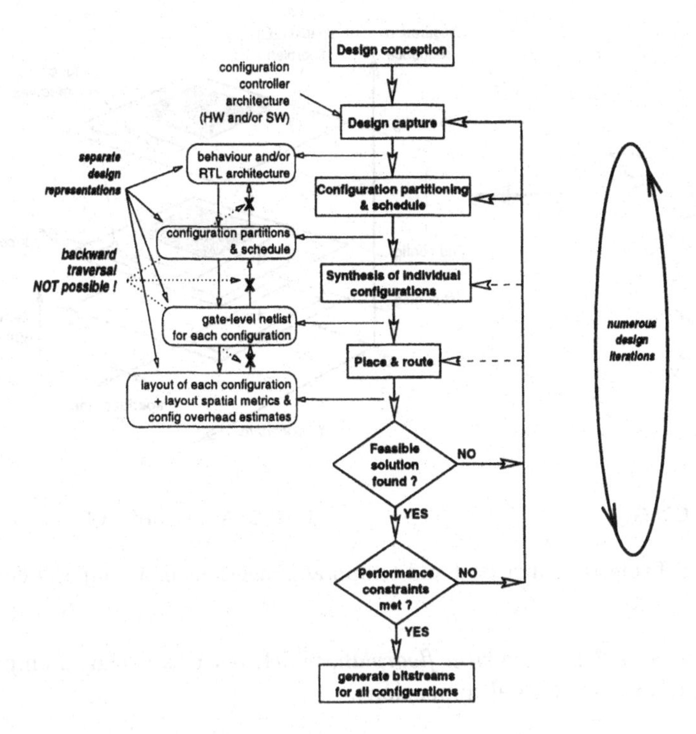

(a) Traditional (sequential) DRL design flow

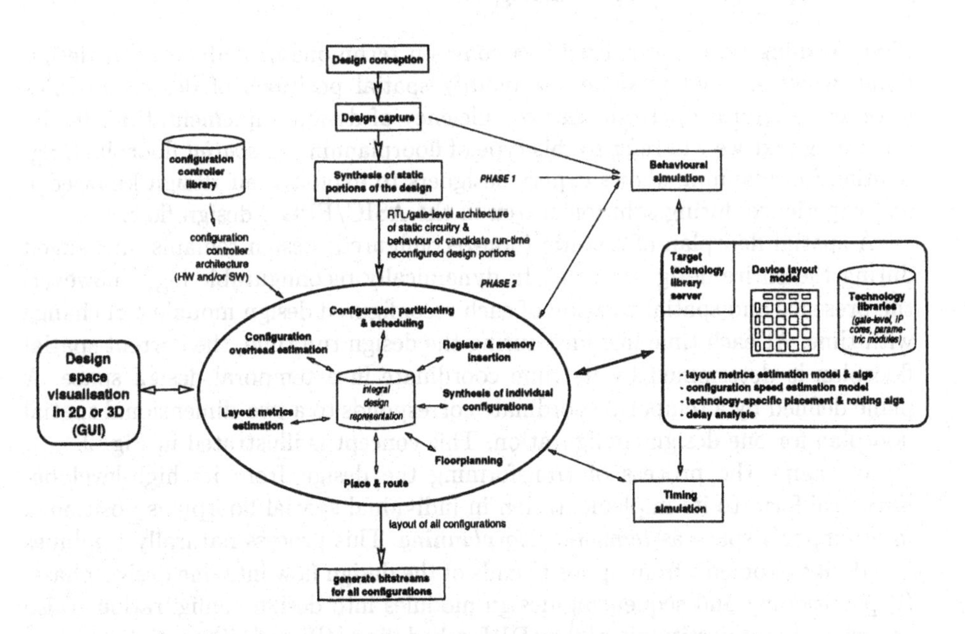

(b) Two-phase DRL design flow introduced in [1]

Fig. 1. Traditional vs integrated two-phase DRL design flow.

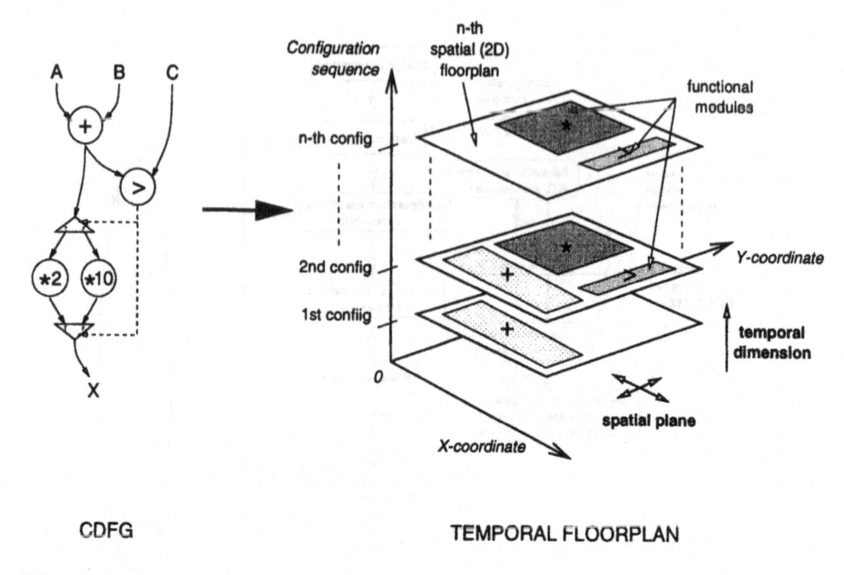

Fig. 2. Projection of spatial and temporal dimensions in a temporal floorplan.

search space using a *temporal floorplan*, which permits design manipulation in both spatial and temporal dimensions.

2 Temporal Floorplanning

Floorplanning at a layout level is a common technique in ASIC/FPGA design flows, which is used to define or modify spatial positions of design modules in order to improve performance or efficiency of design implementation. In the remaining text we will refer to this type of floorplanning as *spatial* floorplanning. Spatial floorplanning allows expert designers to exploit their design knowledge and experience during solution search in the ASIC/FPGA design flow.

A spatial floorplan of a static (non-reconfigured) design remains unchanged during the entire design lifetime. In dynamically reconfigurable logic, however, the presence and spatial position of each reconfigured design module can change with time. At each time instance during the design run time, the current spatial floorplan is determined by its time coordinate in a temporal design space. A plane defined by a temporal coordinate corresponds to a two-dimensional spatial floorplan for one design configuration. This concept is illustrated in Fig. 2.

We name the process of transforming the design from its high-level behavioural form to its implementation in individual spatial floorplans positioned in a temporal space as *temporal floorplanning*. This process naturally combines two design problems from opposite ends of the design flow into one design phase: (i) partitioning and sequencing design modules into design configurations (also called temporal partitioning [3] or DRL scheduling [4]), and (ii) spatial positioning of design modules & wiring within the reconfigurable area of each configuration.

In order to allow full design space exploration of run-time reconfigured portions of a design, the temporal floorplanning needs also to address other closely related design problems: (iii) control step partitioning (scheduling), (iv) functional unit allocation, and (v) register allocation.

Such combination of NP-complete design problems makes an automatic solution search in a temporal floorplan a difficult task. Although some relevant techniques have been proposed (e.g. [3–5]), these are not yet capable of solving practical DRL synthesis problems. These techniques in general use simplified models of dynamically reconfigurable systems, which for example ignore the impact of routing or reconfiguration resource sharing in order to reduce complexity of a DRL design search space.

We believe that temporal floorplanning offers better abstraction of the DRL design space to a *human* designer. The conventional DRL design flow (Fig. 1(a)) does not provide visualisation of a DRL search space, thus making its exploration a difficult task. Temporal floorplanning on the other hand is well suited for design visualisation of both the temporal and spatial dimensions in 2 or 3-D.

Temporal floorplaning may also prove a suitable abstraction for automatic synthesis and guidance techniques for DRL design methodologies. Temporal floorplan allows embedding of behavioural design constraints (data and control dependencies) and structural DRL design constraints (size of FPGA, module shapes, overlap, etc.) in the design flow. The architecture of a DRL design (module selection & binding) can be modified to reflect layout constraints imposed by the target technology.

3 DYNASTY Framework

DYNASTY is an extensible and portable design framework, designed for research of DRL design techniques and methodologies. The Framework implements a TF-based design flow discussed in the previous section.

3.1 Internal Design Representation

In order to support full exploration of the reconfigurable design space late in the design flow, the internal design representation provides design views at various abstraction levels: (i) behavioural, (ii) structural (RTL and gate-level), (iii) temporal floorplan, (iv) detailed layout. One robust design representation implemented as a collection of C++ objects is being used to capture design solutions throughout the entire design flow. Another set of C++ data structures is used to implement library management based on a common library server, including technology cell/module libraries.

The internal design representation allows combination of design views during DRL design exploration. For example during temporal floorplanning various portions of the design could be available in architectural and behavioural views. Incomplete design representations are also supported to facilitate late insertion

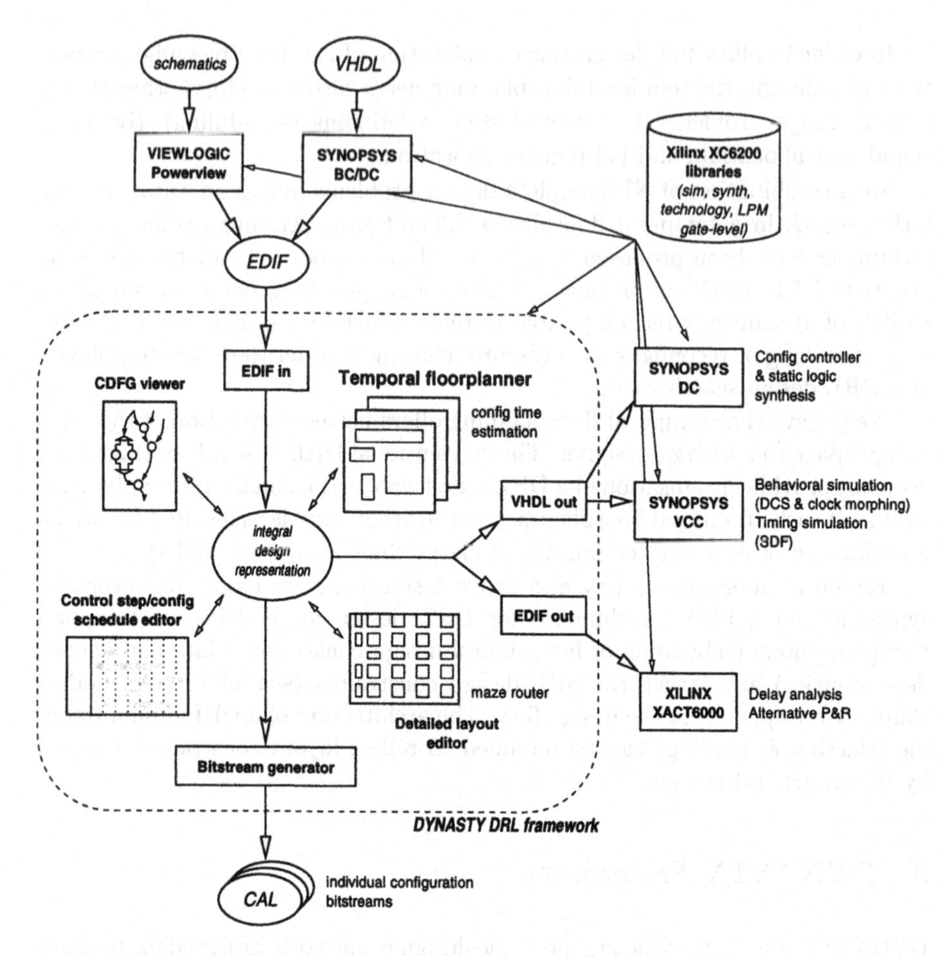

Fig. 3. Temporal floorplanning-based design flow in the DYNASTY Framework

of configuration controllers or other static circuits. The internal design representation further facilitates analysis of design metrics and storage of analysed results.

3.2 Temporal Floorplanning Implementation and Design Flow

Temporal floorplanning design flow implemented in the DYNASTY Framework is depicted in Fig. 3. The current implementation of the Framework supports only the Xilinx XC6200 [6] family of dynamically reconfigurable FPGAs as a target technology. However, the same considerations will apply to other DRL device technologies or systems.

Design manipulation and visualisation. The following DYNASTY Framework tools allow exploration of the DRL design search space:

- *Control/Data Flow Graph (CDFG) Viewer* provides information about system behaviour, which will be implemented during the design flow.
- *Design Hierarchy and Structure Browser* provides details of all design elements and their hierarchical relationship. It also allows allocation of technology library modules/elements to CDFG operators and control/data edges.
- *Control Step and Configuration Schedule Editor* is an analysis tool, which allows tracking and controlling design execution and configuration schedules during design space exploration.
- *3-D Floorplanner* visualises and allows manipulation of design module relationships within a spatial floorplan and between two or more configurations.
- *Detailed Layout Cell Editor* allows manipulation of detailed layout elements such as routing and logic switches.

The structure and parameters of a DRL design solution can be directly manipulated via the DYNASTY Framework graphical user interface. A change in any view will be propagated to relevant design representation in the other design views. For example, when a design module is placed in a configuration which results in violation of data dependencies in the design CDFG, the control step and configuration schedules are automatically recalculated to reflect resulting design latency and configuration overhead.

The core tool is a 3-D floorplanner. It uses 2-D design visualisation to display spatial floorplans for each configuration, which can be traversed in time (third dimension) using configuration navigation controls. The floorplanner uses a number of features which facilitate manipulation of multiple and partial design configurations:

- simultaneous display of multiple spatial floorplans at the same time can be used to check for resource sharing conflicts between configurations
- colour coding of design modules/logic blocks/configurations, functional module types and logic cells with identical configuration guides a designer to minimise partial configuration overheads via sharing of the entire functional modules or small sections of logic cells
- configuration time estimator calculates the number of configuration memory bits needed to change between the two configurations or for the entire design. In the current implementation, the XC6200 library server provides a simple configuration overhead estimation algorithm, based on calculation of bit correlation between successive configuration bitstreams.
- it is possible to mark the "don't use" regions, which are to be left unused after the routing. This is useful for late insertion of design modules (such as configuration controllers or static design modules) or controlling the locality of the design routing

DRL Simulation. The simulation of DRL designs is supported at two abstraction levels:

- The temporal floorplanner can generate a VHDL simulation model of the DRL design. The user can select between two abstract simulation techniques supported by the Framework:

  - *Dynamic Circuit Switching (DCS)* [7]. DCS implementation in DYNASTY Framework uses virtual multiplexors, demultiplexors and switches, which were implemented as parametric VHDL procedures in a separate VHDL package.
  - *Clock Morphing (CM)* [8]. CM implementation is based on a modified IEEE `std_logic_1164` package which implements a new virtual signal value and supports its propagation in target technology libraries (see [8] for details).
- The finalised design can be imported into the Xilinx XACT6000, where a detailed timing model can be generated and simulated by third party tools.

DYNASTY interfaces to third party tools. The DYNASTY Framework provides interfaces to third party tools via EDIF 2 0 0 [9] and VHDL interfaces.

Input designs can be read in EDIF format, which can be exported from most design entry tools. Design behaviour is stored in EDIF as a structural representation of a design CDFG. DYNASTY Framework EDIF interface supports Xilinx XACT6000 placement attributes, which can be used to restrict the placement of a DRL design from the VHDL or schematic design entry.

At the time of writing the DYNASTY framework has been tested with two commercially available design entry tools: Synopsys Design Compiler and Viewlogic ViewDraw, which are also supported by the standard XC6200 design flow [6].

Design can be exported from the DYNASTY Framework either in an EDIF or VHDL format. This can be used for simulation or synthesis using third party tools. A finalised DRL design can be generated using a bitstream generator, which produces design configuration files (CAL format for XC6200).

The DYNASTY framework provides a simple maze router, which can be used to route the XC6200 design. This router will respect the "don't use" attributes set by the user in the temporal floorplanner and will also allow only partial net routing.

Xilinx XACT6000 (XC6200 P&R tool) is used for the detailed timing analysis of design configurations produced by the temporal floorplanner. This tool can be also optionally used to place and/or route the DRL design as the design file formats are interchangeable between the DYNASTY framework and XACT6000.

Design layouts for each spatial floorplan can be also exported for manipulation with the VPR generic FPGA research P&R tool from University of Toronto [10] (not shown in Fig. 3).

Synthesis of a configuration controller and static design modules. Automatic synthesis of a configuration controller is not directly supported by the DYNASTY framework. However, the temporal floorplanner can generate a configuration control schedule in a text file (both pseudocode and VHDL formats), from which such controller can be constructed using standard design or compilation tools.

In the present implementation of the DYNASTY Framework, the reconfiguration controller contribution to the overall configuration latency has to be stated explicitly.

Our Framework could provide a library of reconfiguration controllers suitable for the selected target DRL technology (shown in Fig. 1(b)). Such a library could be used by the library server configuration overhead estimation algorithms in order to provide estimates on non-deterministic metrics such as overheads due to random interrupts or memory contention, etc.

Design modules which are not reconfigured during run-time can be synthesised outside the DYNASTY framework and then imported back into the reserved floorplan positions and connected with the remaining parts of the design through unique net labels.

3.3 Designing with the DYNASTY Framework

From the user's perspective the DYNASTY Framework provides a collection of tools allowing the designer to construct and analyse various DRL design solutions in an interactive environment. In the current implementation of the Framework this is mostly a manual process, which is supported by automatic tracking of design changes and estimation of layout metrics. A typical design sequence in the Framework includes the following steps:

1. *Design conception & capture* using either schematic or HDL design entry.
2. *Selection of the static parts of the design* which should not be subjected to DRL design exploration. These are marked as static throughout the design flow in order to identify conflicts with reconfigured modules.
3. *Design exploration of a DRL design search space* using temporal floorplanning. This involves the use of the tools described in Sect. 3.2. Typically, a good candidate solution is created first and various implementation and scheduling options are then explored in order to meet the design criteria.
   An initial solution can be created by allocating modules from technology libraries to nodes in a design CDFG (using Hierarchy Browser) and placing these modules in a 3-D floorplan (using 3-D floorplanner). The design performance can then be estimated (using Schedule Editor). Design exploration is performed by gradual modification of design parameters (module allocation & binding, execution and reconfiguration schedule, spatial and temporal partitioning, etc.). Execution and configuration schedules are analysed throughout the design exploration in order to (i) verify design performance and (ii) preserve data, control and configuration dependencies in constructed solutions.
   Once a satisfactory design solution has been created, it can be exported from the Framework for a detailed timing analysis. Any violations of timing constraints are used to adjust design solution parameters until all design constraints are met.
4. *Generation of final solution.* The configuration bitstreams are generated for the final design.

4 Conclusions

We have presented the concept of temporal floorplanning and its implementation, which enables simultaneous DRL design space exploration in both temporal and spatial dimensions.

At the time of writing only a small set of circuits has been implemented using the TF-based methodology in the DYNASTY framework. These include a text pattern matcher [11], and dynamically reconfigured versions of an elliptic wave filter [12] and differential equation solver [13]. Even on these small examples we were able to achieve a dramatic reduction of design time, when compared to designing the same DRL circuits using standard VHDL XC6200 design flow.

The DYNASTY Framework is the first tool providing a DRL search space abstraction, which allows easier manipulation and verification of a design in time and space. Many difficulties of current DRL design methodologies result from their inability to explore the entire design search space in one intuitive design environment. With tools based on the concept of temporal floorplanning, system designers will be able to explore the reconfigurable search space more efficiently, while solving multiple DRL design problems in one transparent environment.

Development of our framework is only the first step towards a successful design methodology for dynamically reconfigurable logic systems. The framework reflects our belief that an expert human designer should have firm control over the entire DRL design process, however, automatic design techniques should provide guidance and acceleration of computationally intensive tasks. Push-button DRL methodologies are also needed for scenarios where design or compilation time is a primary objective, whilst possible implementation inefficiencies can be tolerated.

Our future work will concentrate on development of suitable automatic synthesis and estimation algorithms for the DYNASTY Framework as well as support for other DRL technologies and systems. Visualisation of a temporal floorplan in 3-D could provide a more intuitive interface for the manipulation of DRL design objects in time and space. More investigation is needed into the possibilities of temporal floorplanning compilation strategies in reconfigurable computing tools and general HW/SW co-design environments, offering DRL technology as one of many implementation options.

Acknowledgements

This work was supported in part by donations from Xilinx, Inc., California and Xilinx Development Corporation, Scotland.

References

1. M. Vasilko, D. Gibson, D. Long, and S. Holloway, "Towards a consistent design methodology for run-time reconfigurable systems," in *IEE Colloquium on Reconfigurable Systems, Digest No.99/061*, Glasgow, Scotland, pp. 5/1–4, Mar. 10 1999.

2. P. Lysaght, "Towards an expert system for a priori estimation of reconfiguration latency in dynamically reconfigurable logic," in Luk *et al.* [14], pp. 183–192.
3. M. Kaul and R. Vemuri, "Optimal temporal partitioning and synthesis for reconfigurable architectures," in *Design, Automation and Test in Europe Conference*, Paris, France, Feb. 23–26, 1998.
4. M. Vasilko and D. Ait-Boudaoud, "Architectural synthesis techniques for dynamically reconfigurable logic," in *Field-Programmable Logic: Smart Applications, New Paradigms and Compilers, (Proc. FPL'96)* (R. W. Hartenstein and M. Glesner, eds.), LNCS 1142, pp. 290–296, Springer-Verlag, 1996.
5. K. Bazargan, R. Kaster, and M. Sarrafzadeh, "3-D floorplanning: Simulated annealing and greedy placement methods for reconfigurable computing systems," in *Proc. IEEE Workshop on Rapid System Prototyping (RSP'99)*, Clearwater, FL, USA, June 16–18, 1999.
6. Xilinx, *XC6200 Field Programmable Gate Arrays*, Advanced Product Information Xilinx, Inc., Apr. 1997. Version 1.10.
7. P. Lysaght and J. Stockwood, "A simulation tool for dynamically reconfigurable field programmable gate arrays," *IEEE Transactions on VLSI Systems*, vol. 4, no. 3, pp. 381–390, Sept. 1996.
8. M. Vasilko and D. Cabanis, "Improving simulation accuracy in design methodologies for dynamically reconfigurable logic systems," in *Proc. IEEE Symposium on Field-Programmable Custom Computing Machines (FCCM'99)*, Napa, CA, USA, Apr. 21–23, 1999.
9. P. Stanford and P. Mancuso, eds., *EDIF Electronic Design Interchange Format Version 2 0 0*. Electronic Industries Association, 2nd ed., 1990.
10. V. Betz and J. Rose, "VPR: A new packing, placement and routing tool for FPGA research," in Luk *et al.* [14], pp. 213–222.
11. P. Foulk and I. Hodson, "Data folding in SRAM configurable FPGAs," in *Proc. IEEE Workshop on FPGAs for Custom Computing Machines* (D. A. Buell and K. L. Pocek, eds.), Napa, CA, USA, Apr. 5–7, 1993.
12. P. Dewilde, E. Deprettere, and R. Nouta, "Parallel and pipelined VLSI implementation of signal processing algorithms," in *VLSI and Modern Signal Processing* (S. Kung, H. Whitehouse, and T. Kailath, eds.), pp. 257–264, Prentice Hall, 1985.
13. P. G. Paulin and J. P. Knight, "Algorithms for high-level synthesis," *IEEE Design and Test of Computers*, vol. 6, no. 3, pp. 18–31, Dec. 1989.
14. W. Luk, P. Y. K. Cheung, and M. Glesner, eds., *Field Programmable Logic and Applications (Proc. FPL'97)*, LNCS 1304, Springer-Verlag, 1997.

A Bipartitioning Algorithm for Dynamic Reconfigurable Programmable Logic

E. Cantó, J.M. Moreno, J. Cabestany, J. Faura[+], J.M. Insenser[+]

Department of Electronic Engineering, Technical University of Catalunya (UPC)
c/ Gran Capitá s/n, 08034 Barcelona, Spain
e-mail: canto@eel.upc.es
[+]SIDSA, PTM, Torres Quevedo 1, 28760 - Tres Cantos (Madrid), Spain
e-mail: faura@sidsa.es

Abstract. Most partitioning algorithms have been developed for conventional programmable logic (especially FPGAs), being their main goal the minimisation of the signals constituting the interface (cutsize) between partitions, while balancing partition sizes. New families of dynamic reconfigurable programmable logic (DRPL) offer new possibilities to improve functional density of circuits, but traditional partitioning techniques are not able to exploit the novel features offered by these devices. A new family of partitioning techniques for DRPL should be developed, being its main goal the maximisation of the functional density on balanced partition sizes. This paper presents a new partitioning algorithm based on a temporal separation of the system functionality. As our experimental results will show, the algorithm is able to benefit from the dynamic reconfiguration properties of FPGA devices.

1. Introduction

Partitioning techniques focus on large electronic designs, where the entire circuit functionality cannot be physically mapped in just one device. Most traditional partitioning techniques physically divide the design in two (bipartition) or more (multiway partition) circuits, being the goal to minimise the communication signals between partitions (cutsize) while balancing partition sizes with a given tolerance, and ensuring these sizes are not larger than the capacity of the target devices.

Custom computing machines (CCM) can be used for many computationally intensive problems. Architectural specialisation is achieved at the expense of flexibility, because they are designed for one application and are inefficient for other application areas. Reconfigurable logic devices achieve a high level of performance on a wide variety of applications. The hardware resources of the device can be reconfigured once a function is completed for another different one, achieving a high level of performance on a wide variety of computations. A step forward is run-time reconfiguration or dynamic reconfiguration, where hardware resources can be reconfigured while the system is in normal execution, increasing the efficiency of CCM systems.

One way of improving efficiency is to replace idle hardware with more useful circuitry. Another way is to partition a large computing architecture on a limited set of hardware resources, so that a partition is being executed while another partition is being configured. The efficiency of the hardware resources can be measured with some functional density parameters. Improvements on functional density of dynamic reconfigurable programmable logic (DRPL) is achieved by minimising reconfiguration time. This reconfiguration time has been minimised on previous FPGAs with partially reconfiguration capabilities and increasing memory bandwidth (i.e.: Xilinx XC6200 FASTMAP interface [1]) to transfer configuration bits from off-chip memory into device configuration bits. Another FPGA family, the Field Programmable System on Chip (FIPSOC [2], [3]) device, contains a partially dynamic reconfigurable FPGA, where configuration bits can be loaded from internal memory in just two write memory cycles as explained later. As new logic functions are required they can be loaded in configuration memory from internal memory replacing or complementing the active logic when required, without loss of data (stored on flip-flops or latches). This is a very important property for efficient DRPL devices, since it makes them suitable for reconfigurable computing. DRPL devices are able to cover a wide range of applications, such as adaptive signal processing, reconfigurable processors, self-repairing hardware [4], parallel cellular machines [5], and others.

This paper will describe the FIPSOC architecture in section 2, and how it fits for dynamic reconfiguration. In section 3 we will describe the concept of functional density and the proposed algorithm. Section 4 will show experimental results on ISCAS benchmark circuits and others. Section 5 will describe the conclusions and future work to improve the performance provided by the algorithm.

2. FIPSOC Architecture and Dynamic Reconfiguration Principle

The FIPSOC is a new family of FPGA devices that will be introduced on market very soon. This device includes a 8051 microcontroller core, a FPGA, a configurable analog block (CAB) which includes a programmable A/D, D/A acquisition section, RAM memory for the configuration bits and/or for general purpose user programs and some additional peripherals. All configuration bits are mapped on the microprocessor address space. This provides a large flexibility for the chip, because the internal microprocessor is able to run user program and can dynamically read or write the configuration of the FPGA or CAB blocks in terms of memory access. Furthermore all the outputs of the cells constituting the programmable digital section are mapped into memory and any point of the CAB can be dynamically routed to the internal ADC. Therefore the microcontroller can be used to probe any analog signal or any output of the digital blocks in real time.

This device configuration makes it a good tool for prototyping or developing mixed-signal processing, hardware/software co-design or reconfigurable computing applications, or a combination of above three characteristics in just one programmable device. The global architecture of the FIPSOC device is depicted in figure 1(a).

The programmable digital section consists of an FPGA, whose building blocks have been termed DMCs (digital macro cells). The DMC, whose internal structure is depicted in figure 1(b), is a large granularity, 4-bit wide programmable cell which contains a combinational block and a sequential block interconnected through an internal router. Both combinational and sequential blocks have 4 bit outputs, plus some extra outputs for macro modes. The DMC output block has four bit outputs individually multiplexing combinational or sequential outputs, plus two auxiliary outputs.

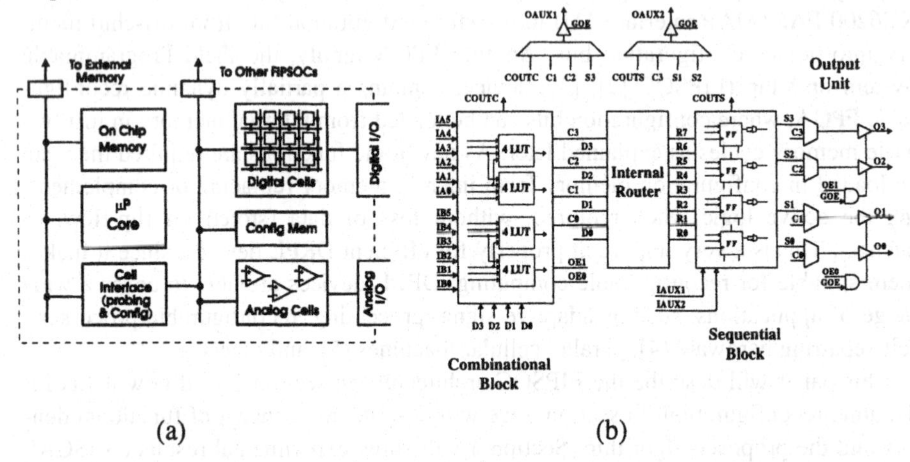

Fig. 1. (a) FIPSOC global organisation and (b) DMC internal architecture

The combinational block is implemented with four 16-bit Look Up Tables (LUTs), than can be programmed to provide any combinational logic function. Each two LUTs constitute a tile, which has six inputs and two outputs. The combinational block can be used either as RAM, as a 4 to 1 multiplexer, or as a 4-bit adder/subtractor macro mode with carry signals.

The sequential block is composed of four registers, which can be configured independently as different types of D flip-flop or D latches, with different clock or enable polarities and synchronous or asynchronous set/reset signals. It also contains macro modes for a 4-bit up/down counter with load and enable, and a 4-bit shift register with load and enable.

The combinational and sequential blocks can be configured in static or dynamic modes independently. Dynamic mode provides two time-independent contexts that can be swapped with a single signal. The contents of the registers are duplicated so that they can be stored when active context is changed and restored back when the context becomes active again. Also each configuration bit is duplicated for dynamic reconfiguration capability, exception made for the LUTs. The combinational block functionality differs on static or dynamic modes. In static mode for the combinational block each tile includes two 16-bit LUTs sharing two inputs that can be programmed to implement two 4-bit boolean functions, or they can grouped to implement one 5-bit boolean function, or can be used as a 16x2 bits RAM. In dynamic mode each tile can implement two 3-bit boolean functions without shared signals, a 4-bit boolean function, or a 8x2 bits RAM. Static and dynamic modes for sequential block have the same functionality.

As depicted in figure 2, there is a physical isolation between active configuration bits and the mapped memory, which can be used for storing the configuration of the two contexts, or as a general purpose RAM once its contents have been transferred to the active context.

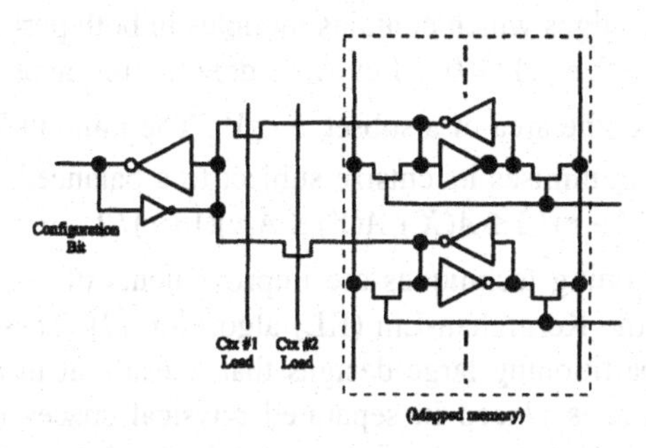

Fig. 2. Mapped contexts memory and configuration bits for the active context.

FIPSOC features a partially reconfiguration scheme (represented in figure 3), so that the microcontroller can select rectangular sections of the FPGA for context swapping while the rest is in normal operation without stopping the system. It is a very fast strategy because it only requires two memory write cycles to perform the context swap, one to select logical rectangles using row and column masks, and another to establish the new configuration bits for selected DMCs.

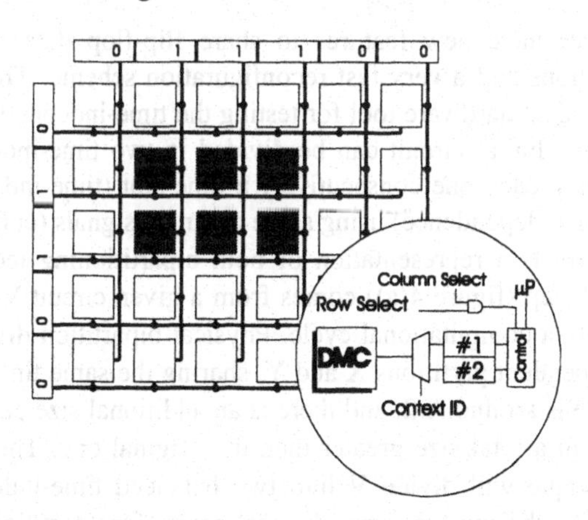

Fig. 3. Dynamic reconfiguration on a group of DMCs

A new scheme for reconfiguring DMCs has been recently added to the FIPSOC circuitry enabling a faster reconfiguration time. Each DMC have a extra bit accessible for any of their outputs able to trigger the context swap, reducing context swap time to just one clock cycle.

3. Proposed Bipartitioning Algorithm

Most partitioning algorithms focus on bipartition a graph G={V,E} or hypergraph H={V,E} (containing n modules $V=\{v_1,v_2,...v_n\}$ and m nets $E=\{e_1, e_2,...e_m\}$) in a pair of disjoint clusters (subset s of V) X, Y. This bipartition P={X,Y} must accomplish $X \cup Y = V$ and $X \cap Y = 0$. The cutsize of the bipartition P={X,Y} is the number of edges or hyperedges which contains modules in both partitions X and Y, that is $cut(P) = \{e \mid e \cap X \neq 0, e \cap Y \neq 0\}$. Let A(v) denote the area of a module and let $A(S)=\Sigma_{v \in S}A(v)$ denote the area of a subset $S \subseteq V$. The min-cut bipartition algorithms seek a solution that minimises its cutsize subject to a balanced solution with a given tolerance r [6]: $A(v)(1-r)/2 \leq A(X), A(Y) \leq A(v)(1+r)/2$.

Traditional partitioning techniques are improvements of the Fiduccia-Mattheyses (KLFM) variant of the Kernighan-Lin (KL) algorithm [7]. These bipartitioning techniques focuses on partitioning large designs that doesn't fit in a constrained area or device. Each partition is placed in separated physical spaces (space-independence) sharing the same time slot (time-dependence) and joined using their common signals (cutsize).

DRPLs enables the device to have different behaviours in time, using the reconfigurable ability of their hardware resources. This feature is normally used in order to have different digital circuits mapped on the same physical space but in different time slots, running the same active logic for several clock cycles. Some newer DRPLs (ATMEL AT40K, FIPSOC) can store their flip-flop or latch data without loss when dynamic reconfiguration is done, so new active logic can easily share signals with the old active logic.

FIPSOC devices have new features to share flip-flop data between two time-independent partitions and a very fast reconfiguration scheme. These new properties make FIPSOC an ideal hardware tool for testing the time-independence of a hardware bipartition. A large digital circuit can be divided in two time-independent partitions (contexts), assigning each one consecutively a time slot (time-independence) sharing the same area (space-dependence) using some common signals (cutsize).

In figure 4 there is a representation of both bipartitioning techniques explained above. The initial step (figure 4(a)) comes from a given circuit V of size A taking a time T to complete a computational cycle. Physical bipartition divides V in two balanced space-independent partitions X and Y, sharing the same time slot (figure 4(b)). Each partition size is around A/2, and there is an additional size cost (Ac) due to their cutsize, resulting in a total size greater then the original one. The new proposed bipartitioning technique will divide V into two balanced time-independent partitions (contexts) X and Y, sharing the same physical-space. One partition will be the active context in the programmable logic while the other one will be a virtual context (their configuration bits are in memory, but not yet used) during one computational cycle. In the next computational cycle virtual context will be the new active one and the old active context will be virtual, and so on. Each partition will need a computational cycle and there is an additional time cost (Tc) due to cutsize signals and reconfiguration, but total size will be about a half of the original size. FIPSOC reconfiguration scheme enables signals to be shared between different time-windows using their flip-flops/latches without taking extra time. The only extra time penalty will be given by the need to change the active context, minimising extra time in the cutsize. Additionally, in reducing the complexity of the partitions it is possible to increase the clock frequency, thus reducing the overall computing time.

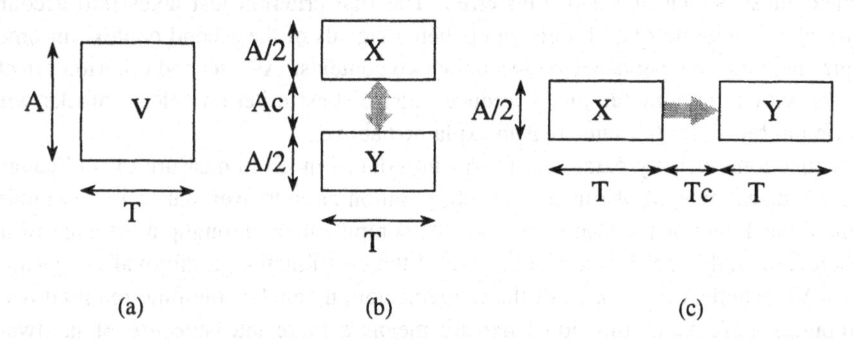

Fig. 4. (a) Circuit to be partitioned (b) in two space-independent partitions and (c) in two time-independent partitions.

A combinational circuit can be represented in a graph as depicted in figure 5(a), where arrows represent time-dependence between combinational blocks. A space-independent bipartition graph is shown in figure 5(b), whose cutsize signals point in both directions, so the final solution is reached after two context swaps, and more context swaps could be necessary when circuits are larger. A time-independent partition is show in figure 5(c) whose cutsize signals point in just one fixed direction, so final solution is always reached after just one context swap.

Bearing in mind the previous considerations, the proposed algorithm for combinational logic has been developed as follows. An initial solution is generated by placing in the first context all combinational blocks (one output, several inputs) that are time-independent from the rest. These are the blocks whose inputs are exclusively system inputs, placing the rest of the blocks in the second context. An iterative process is executed, which consist on selecting and transferring blocks from the larger partition to the smaller one until partition sizes are balanced or there was no block selected.

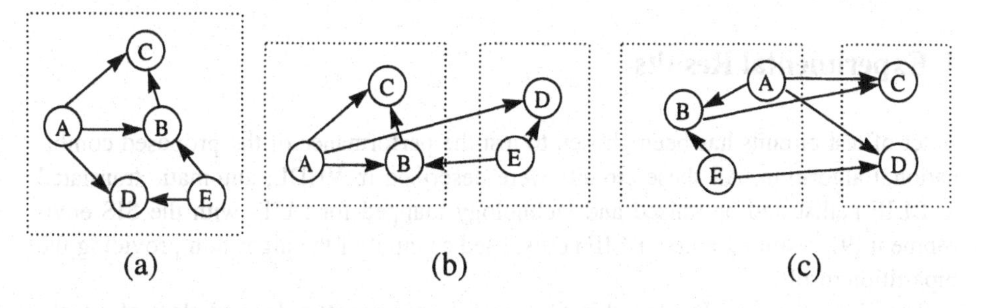

Fig. 5. (a) Graph of circuit (b) Graph of space-independent partition (c) Graph of time-independent partition.

There are two different cases to consider, depending on the relative size of both initial partitions. If the size of the first partition is greater then the second one, blocks must be selected from the first partition, using as a criterion the maximum fan-in block in order to minimise cutsize. The other possibility is given when the initial solution produces a second context larger than the first one, so blocks of the second context must be selected and transferred. The first criterion just takes into account those blocks with no one of their inputs being signals of the second context, in order to provide time-independence between the two contexts. As a second criterion selects blocks with minimum fan-out to reduce cutsize. Last criterion selects blocks with maximum fan-in for the same reason explained above.

A functional density parameter D was introduced in [8] to measure overall advantage (or disadvantage) of run-time reconfiguration circuits over static circuits implementations. Functional density measures the computational throughput for a hardware circuit area, and is defined as the inverse of the cost function, traditionally measured as C=AT, where A is the area of the implementation and T is the time required for a computation. A better functional density means a more intensive use of hardware resources for a fixed time. Computational time for a static implementation T_S is lower than time for a reconfigurable implementation T_R (including reconfigurable time), but the area of the static implementation A_S is greater than the area of reconfigurable implementation A_R. Due this fact the functional density for a reconfigurable implementation D_R can be greater than the functional density for the static implementation D_S, meaning that the circuit should be implemented in reconfigurable scheme, otherwise static implementation is preferred.

Additionally, a parameter able to indicate the improvement in the functional density is defined in [8] as the relationship between the incremented functional density of the reconfigurable implementation over the functional density of the static implementation, in relative terms with respect the static one. This improvement can be expressed as:

$$I = \frac{D_R - D_S}{D_S} = \frac{A_S \cdot T_S}{A_R \cdot T_R} - 1 \tag{1}$$

4. Experimental Results

A set of test circuits has been chosen to test the performance of the proposed combinational algorithm. All these circuits were described in VHDL, automatic translated to BLIF netlist and optimised and technology mapped for LUTs with the SIS environment [9], resulting a new BLIF netlist used as input of the algorithm providing the bipartition result.

The algorithm was developed in C++ and it implements a derived class of a technology base class to implement the behaviour of the FIPSOC device, so it can be easily ported to any other technology with a new derived class for a specific device.

There were development four functional versions of the algorithm, differing between them in internal data actualisation. The first version recalculates entire graph when a block move was done, resulting a large complexity algorithm. The second version calculates only partial portions of the graph affected by a move. The third version uses a cache evaluation data structure for the implementation of the algorithm criterion described above, reducing complexity of the algorithm. The fourth version of algorithm includes a faster FIPSOC technology class for the calculation of partition sizes. Last version outperforms the algorithm run-time for larger circuits.

The circuits tested are a carry lookahead adder (cla16), a 4-bit slice ALU (alu4), a 16-bit, 16 operations ALU (alu16), a 15x15-bits Wallace-tree multiplier (mult15) and several ISCAS circuits. Experimental results are presented in table 1 in terms of 2-, 3- and 4-input LUT functions (first column), number of inputs, outputs and internal nets (second column), cutsize (third column) and overall execution time of the algorithm (fourth column - data obtained in a Pentium PC, 133 MHz, 32 MB main memory).

	#LUTs	#In/Out/Internal	cut(P)	Time(s)
cla16	0/32/0	33/17/15	1	1
alu4	5/26/18	12/5/44	11	1
alu16	54/61/115	36/21/238	51	5
mult15	25/173/486	30/30/654	162	28
C432	4/9/57	36/7/63	28	1
C499	8/0/66	41/32/42	13	1
C880	4/34/112	60/26/124	33	2
C1355	0/0/114	41/32/82	32	2
C1908	13/66/87	33/25/241	34	5
C3540	29/165/486	50/22/658	110	25
C6288	240/2/717	32/32/927	175	50

Table 1. Experimental results corresponding to cutsize and execution time.

Table 2 shows the functional density (D_S for the static implementation, D_R for the dynamically reconfigurable one) and the improvement ratio (I) given by equation (1). The area figures (A_S for the static implementation, A_R for the dynamically reconfigurable one) are computed based on the number of DMCs required to map the LUTs constituting the circuits. The delay time (T_S for the static implementation, T_R for the dynamically reconfigurable one) is computed by the maximum logic depth, as a approach to the critical path delay. The area for the reconfigurable implementation is the maximum area of both partitions, while delay time is the sum of the delay time for both partitions.

	A_S	T_S	$1/D_S$	A_R	T_R	$1/D_R$	I
cla16	8	16	128	4	16	64	100%
alu4	13	9	117	7	10	70	67.1%
alu16	72	25	1800	36	27	972	85.2%
mult15	188	30	5640	95	36	3420	64.9%
C432	23	20	460	11	21	231	99.1%
C499	23	5	115	12	7	84	36.9%
C880	46	13	598	16	7	368	62.5%
C1355	31	6	186	16	7	112	66.1%
C1908	72	15	1080	36	19	684	57.9%
C3540	182	21	3822	91	30	2730	40%
C6288	296	31	9176	148	33	4884	87.9%

Table 2. Results corresponding to area retard time and functional density for static and reconfigurable implementations.

As it can be deduced from these tables, a rather large improvement is to be expected from dynamically reconfigurable implementations of a system. It has to be noted that the ideal 100% improvement factor corresponds to an exact bipartition in size with no penalty in execution time. Therefore, our proposed algorithm shows a promising behaviour for considering reconfigurable alternatives in the first stage of the design process.

5. Conclusions and Future Work

This paper has presented a new approach for a new family of time-independence bipartitioning algorithms for DRPLs than can be developed further to improve functional density of fast DRPL devices such as FIPSOC or future FPGAs families. The main goal to improve functional density of a large circuit is achieved through a balanced time-independent partitions (or contexts) with a very short reconfigurable delay due FIPSOC reconfiguration scheme. This algorithm can be used as a start point to improve functional density and other parameters for this new family of bipartitioning techniques.

Additional features that could be included for future advanced algorithms, based on the one proposed in this paper, are clustering and multilevel clustering, multiway partition, generalisation for asynchronous circuits and for partitions covering several clock cycles, or the use of genetic algorithms.

Our current work deals with the physical characterisation of the dynamically reconfigurable features explained previously on the first prototypes of the FIPSOC devices.

References

[1] S. Churcher, T. Kean, B. Wilkie, "The XC6200 FastMap Processor Interface", Field Programmable Logic and Applications, Proceedings of FPL'95, pp. 36-43, Springer-Verlag, 1995.

[2] J. Faura, C. Horton, P. van Doung, J. Madrenas, J.M. Inserser , "A Novel Mixed Signal Programmable Device with On-Chip Microprocessor". Proceedings of the IEEE 1997 Custom Integrated Circuits Conference (1997) 103-106.

[3] J. Faura, J.M . Moreno, M.A. Aguirre, P. van Duong, J.M. Inserser, "Multicontext dynamic reconfiguration and real time probing on a novel mixed signal programmable device", Field Programmable Logic and Applications, 7th international workshop, FPL'97, pp. 1-10, Springer-Verlag, 1997.

[4] J. Madrenas, J.M. Moreno, J. Cabestany, J. Faura, J.M. Insenser, "Radiation-Tolerant On-Line Monitored MAC Unit for Neural Models Using Reconfigurable-Logic FIPSOC Devices", 4th IEEE International On-Line Test Workshop, pp. 114-118, Capri, Italy, July 1998.

[5] J.M. Moreno, J. Madrenas, J. Faura, E. Cantó, J. Cabestany, J.M. Insenser, "Feasible Evolutionary and Self-Repairing Hardware by Means of the Dynamic Reconfiguration Capabilities of the FIPSOC Devices", Evolvable Systems: From Biology to Hardware, M. Sipper, D. Mange, A. Pérez-Uribe (eds.), págs. 345-355, Springer-Verlag, 1998.

[6] Charles J. Alpert, Hen-Hsin Huang, Andrew B. Kahng, "Multilevel Circuit Partitioning", IEEE on Computer Aided Design of Integrated Circuits and Systems, Vol. 17, No. 8, pp. 655-667, August 1998.

[7] S. Hauck, G. Borriello, "An evaluation of bipartioning techniques", IEEE Transactions on Computer-Aided Design of Integrated Circuits and Systems, Vol. 16, No. 8, pp. 849-866, August 1997.

[8] Michael J. Wirthlin, Brad. L. Hutchings, "Improving Functional Density Using Run-Time Circuit Reconfiguration". IEEE Transactions on VLSI Systems, Vol. 6, No. 2, pp. 247-256, June 1998.

[9] E. M. Sentovich, K.J. Singh, L. Lavagno, C. Moon, R. Murgai, A. Saldanha, H. Savoj, P.R. Stephan, R.K. Brayton, A. Sangiovanni-Vincentelli, "SIS : A System for sequential circuits synthesis", Technical Report UCB/ERL M92, U.C. Berkley, May 1992.

Self Controlling Dynamic Reconfiguration: A Case Study

Gordon McGregor[1] and Patrick Lysaght[2]

[1] Motorola, Ltd., Sherwood House, Gadbrook Business Centre,
Rudheath, Cheshire, CW9 7TN
Gordon.Mcgregor@motorola.com
[2] University of Strathclyde, George St., Glasgow, G1 1XW
p.lysaght@eee.strath.ac.uk

Abstract. The design and physical implementation of a self-controlling dynamically reconfigurable system is described in detail. The reconfiguration control logic and target application execute in parallel within the same FPGA. In addition, the data required for each reconfiguration is generated on demand. A pattern-matching algorithm is used to investigate the viability of systems that exhibit self-control of reconfiguration management. The system was engineered on a custom development platform, based around a Xilinx XC6216 FPGA. The methodology and development tools used in the design are also evaluated. Initial measures of performance and recommendations for future device design are presented.

1 Introduction

Dynamic reconfiguration of Field-Programmable Gate Arrays (FPGAs) is recognised as an advanced application area within reconfigurable logic, with an emerging commercial future [1]. Currently only a small subset of available FPGAs are capable of being reconfigured in this way. There is, however, a growing trend in the industry to provide dynamically reconfigurable devices with varying degrees of configuration flexibility.

Contemporary FPGAs such as the Xilinx Virtex family and the Atmel At40k family are dynamically reconfigurable. The degree of support for the technique varies across the device families, depending mainly on the flexibility of the configuration mechanisms. To date, the Xilinx XC6200 devices provided the most flexible and complete configuration access methods on a dynamically reconfigurable array. Although these devices are now obsolete, some of the features have been adopted in the Virtex family. The case study considered in this paper is based on a custom platform, which integrates a XC6216 device with supporting hardware and software development tools [2].

All dynamically reconfigurable devices are characterised by their ability to continue to operate *without interruption* while sub-sections of the array logic are being reconfigured. The ability of the array to continue to function while being modified presents designers with intriguing new design possibilities. In particular it allows the development of applications that maintain their own internal configuration state, without the need for an external controller [3].

To investigate the practicality of such applications, a proof-of-concept design has been developed. Several novel tools were employed to accelerate the specification and design of the exemplar system. The design is described in detail in this paper, along with the associated methodology.

2 Types of Dynamic Reconfiguration and Controllers

Two types of DRL tasks can be identified [4]. Some tasks intrinsically require dynamic reconfiguration and cannot execute on devices that are not dynamically reconfigurable. These we will call examples of *intra-task* dynamic reconfiguration. Other tasks are functionally complete and would successfully operate if placed on devices that were not dynamically reconfigurable. Systems comprised of such tasks can also benefit from dynamic reconfiguration, particularly as FPGAs become larger. This statement is based on the observations that as an FPGA accommodates more tasks, it becomes increasingly unlikely that all of these tasks need to execute simultaneously. Inactive circuits can be dynamically reconfigured to allow more functions to be performed with smaller devices. This modification of operation at the system level is termed *inter-task* dynamic reconfiguration. Support for inter-task reconfiguration requires fewer modifications to current design methodologies than the design of intra-task dynamically reconfigurable circuits. The use of inter-task reconfiguration is especially applicable to modular system design methodologies using standard communication interfaces and protocols. These methodologies are typical of those being developed to enable IP reuse and System-On-a-Chip (SoC) design. Increasing convergence of these design methodologies is likely in the near future [4].

Both intra-task and inter-task reconfiguration require a controller to schedule and implement rcconfiguration. All FPGAs have a simple, configuration controller integrated into the array, which loads the initial configuration from an external source. Currently, no FPGAs integrate controllers complex enough to support and schedule reconfigurations. Hence, external devices have always been required to manage and load each bitstream, after the initial configuration has been made.

The reconfiguration controller can exist in several forms. Software controllers executing on a standard microprocessor are feasible, as are dedicated hardware controllers. The optimal selection of controller is intimately tied to the nature of the reconfiguration required by the application. Inter-task reconfigurations are characterised by large functional changes. Each task typically executes for a significant proportion of time, before the next reconfiguration is required. In contrast, intra-task reconfigurations typically require more frequent, and smaller reconfigurations. The scheduling complexity of intra-task systems is also comparatively simpler, due to the reduced logical dependencies between the functional changes. Often, pre-determined sequences of reconfiguration can be implemented to provide the required operation.

Based on these characteristics, one can generalise that software control is better suited for inter-task reconfigurable systems. Interrupt latencies in the processor when handling reconfiguration requests can be amortised with the longer execution times of the tasks. Intra-task systems often require dedicated controllers to implement simple scheduling at the fastest possible speed. Large reconfiguration latencies cannot be tolerated, due to the frequent reconfigurations.

A novel way to obtain the flexibility of a processor and the dedicated speed of hardware is to implement the controller on an FPGA. With the support of new CAD tools, a fully customised implementation of a controller can be developed for each application. The scheduling and sequencing operation of the application is captured early in the design process as part of the methodology, and the aim is then to transfer this information directly into the controller logic that is used. At the present time, the controller design must be manually developed based on simulations of the reconfigurable operation of the application.

3 Self-Controlling Reconfigurable Systems

With a dynamically reconfigurable device, there is no compelling reason to use an external FPGA to implement the reconfiguration controller. The controller can be hosted within the array that is being dynamically reconfigured. There are several desirable features of such an arrangement. Firstly the control logic is as close to the array as possible, thus minimising the latencies associated with accessing the configuration port. Secondly, fewer discrete devices are required, reducing the overall system complexity.

To implement a self-controlling dynamically reconfigurable application, certain practical problems have to be overcome. The first issue is gaining access to the configuration control ports of the FPGA from within the array. In the XC6216, this feature is very well supported. All configuration signals can be driven from internal sources, via the I/O ring. Status information describing the state of the array is available to the application logic, allowing the reconfiguration controller to monitor the current configuration environment.

The second main potential pitfall occurs when the device is initially configured. Contention over control of the configuration interface can occur, as two controllers co-exist in the system. A more subtle problem can also arise, as the standard controller can potentially configure the array so that it no longer controls the configuration port, although still actively using the port. If this situation occurs, it is impossible to complete the instantiation of the reconfiguration controller. Multiplexing can be used to isolate the two controllers but this is an overly complex solution, requiring external routing and additional logical components.

Note that hosting the reconfiguration controller within the array does not necessarily imply a self-modifying system. The controller does not actively reconfigure its own internal logic. While reconfiguring the controller is technically possible with this arrangement, it is not seen as a desirable feature and should be avoided in most designs. The controller task should be treated as a static task on the array, throughout the execution of the application.

4 System Overview

A proof-of-concept application has been developed to investigate internally managed control of dynamically reconfigurable FPGAs. The system is the first working example of a self-configuring, dynamically reconfigurable application. It is also the first to exhibit internal generation of bitstream data. A custom, in-house prototyping platform [5] was used during the initial development and also to host the final application. The platform comprises two FPGAs and associated support logic. A Xilinx 4013E device provides debugging facilities and manages communications with the host system for interactive development. The second device is a Xilinx XC6216 FPGA and is the target for the dynamically reconfigurable application.

The application implements a pattern-matching algorithm, which requires intra-task reconfiguration. The algorithm is based on a concept known as *data folding* [6] or more recently known as run-time constant propagation [7]. The principle of data folding is quite simple: fixed-coefficient implementations of circuits such as pattern matchers or multiplication units execute more quickly, in a smaller area, than the equivalent variable coefficient implementations. Data folding exploits this speed and area advantage of fixed implementations but extends it to support operations with slowly varying parameters or operands. In this context, slowly varying means that the coefficients change sufficiently infrequently that the logic can be reconfigured without any appreciable impact on the execution time.

Fig. 1 illustrates a standard one-bit pattern matching circuit, and the equivalent dynamically reconfigurable implementation, using data folding. To understand the operation, consider the case where MatchData is fixed as a logical 0. When TestData is equal to 1, the output of the standard implementation is not asserted. When TestData is 0, the Match signal is asserted. For the equivalent dynamically reconfigurable implementation, if the coefficient to match is fixed at 0, the inverter/buffer section is configured to implement a buffer. This gives the equivalent logical operation of the standard implementation. Conversely to match a coefficient value of 1, the gate is configured to be an inverter.

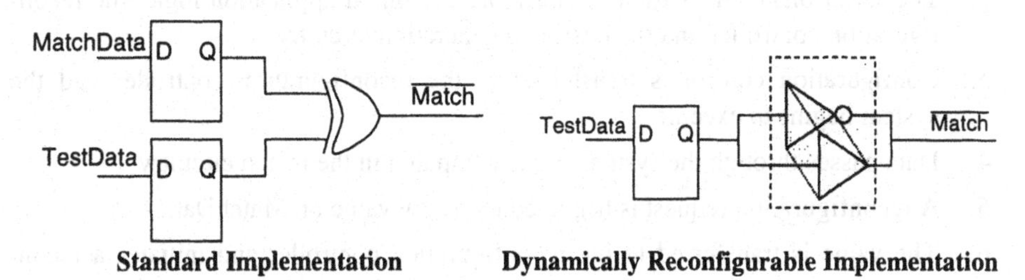

Fig. 1. Implementations of a one-bit pattern matching circuit

The data folded implementation executes at a faster rate due to the reduced propagation delay. The circuit footprint is also substantially reduced. For the XC6216 implementation, the variable coefficient implementations of the single bit pattern matcher logic are 50% slower than the equivalent data folded implementations.

Fig. 2 highlights the main components of the self controlling DRL system. The region within the dashed rectangle delimits the user-configurable region on the array. The grey section indicates the actual pattern matching logic. The system tasks manage the flow of data through the application and initiate requests for reconfiguration. The reconfiguration controller is responsible for processing these requests and converting them into appropriate signals to select the correct partial configuration to load. The bitstream generation circuitry produces the required configuration data on demand, in response to these requests.

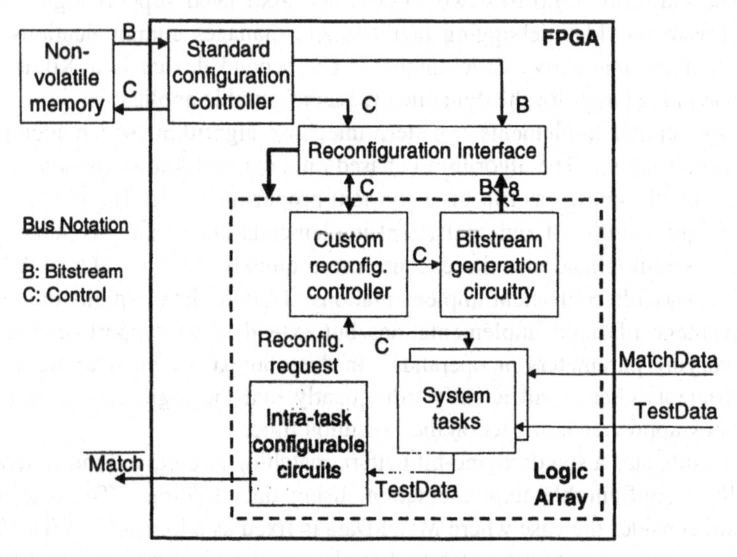

Fig. 2. System block diagram for pattern matching application

The profile of system execution proceeds as follows:

1. At power-on, the standard configuration controller transfers bitstreams from the non-volatile memory to the array
2. The serial bitstream is used to configure the initial application logic, the reconfiguration controller and the bitstream generation circuitry
3. Configuration control is transferred to the reconfiguration controller and the system begins to execute
4. Data passes through the system and is compared in the match circuitry
5. A reconfiguration request is triggered by a new value on MatchData
6. The value is transferred to the reconfiguration controller and initiates a reconfiguration request
7. The next reconfiguration is selected and an associated ID is passed to the bitstream generation circuitry
8. The reconfiguration controller manages the dynamic loading of the internally generated bitstream onto the array, which reconfigures the pattern matching logic

9. Indication of successful completion is sent to the system tasks and the pattern matching continues
10. Repeat to 4

Note that once each reconfiguration is complete, the pattern matching continues immediately. There is no requirement to reinitialise or reset existing tasks. Throughout the reconfiguration interval the system continues to execute, although the pattern logic must wait until successful completion of reconfiguration to avoid producing erroneous data. The overall system is required to suppress the use of these incorrect results. One method is to stall the stream of data through the circuit. The alternative is to just ignore the results. In the example design, the Match output is suppressed during reconfiguration, essentially ignoring any erroneous false matches. Once the system has been initially configured it requires no further external support, other than the application data. The internal generation of bitstreams enables this unique standalone operation of the dynamic system.

5 Design Methodology

Three distinct stages existed in the development of the self-controlling system.

1. Modelling and simulation
2. Prototyping and interactive debugging
3. Stand-alone implementation

The entire design was captured in structural VHDL, with the same files used for simulation and final implementation. The continuity that was provided by reusing the source files increased the likelihood of a correct design. Each of the main stages in the design is now considered in turn.

Modelling and Simulation

Dynamic Circuit Switching (DCS) [8] is a technique developed at the University of Strathclyde to simulate dynamically reconfigurable logic, using extensions to standard simulation tools. Initially developed for schematic capture, it has been extended to work with VHDL simulation packages. The first stage of the design required the capture of the operation of the dynamically reconfigurable tasks. In this example, DCS was used to describe the operation of the pattern matching logic.

Initially, all the possible permutations of logic to be swapped are described in the VHDL source files. For this design, the first step was to describe a single bit, pattern-matching circuit. Fig. 3 shows the code used to describe the buffer/ inverter pair for DCS. The structural VHDL does not represent a valid circuit at this stage. DCS tools are used to modify the design to produce a simulatable model of the circuit. Virtual components are inserted into the design to model the dynamic reconfiguration. These virtual components implement switching logic, to correctly isolate or integrate logic that is configured on the array. The *voff* and *von* tags in the source code allow the Xilinx netlist generation tools used later in the design to extract the correct initial configuration of the system. In this case the first configuration will implement a buffer in the pattern matcher. Careful use of these control comments allowed both the DCS compatible and the implementable[1] versions of the design to be produced from the same source files.

```
-- This is the dynamically reconfiguring section
   swap_buf     : buf port map( I=> supply, O=>tobuf);
-- voff
   swap_inv     : inv port map( I=> supply, O=>tobuf);
-- von
```

Fig. 3. pattern-matching logic : structural VHDL

Additional information is required to elaborate the conditions under which the logic is active in the circuit. A schedule file is generated for the design, which provides information to DCS for the simulation model. Fig. 4 shows part of this schedule file, for one buffer element in the design. The full path to the entity is given, along with estimates of the reconfiguration time. In the case of the pattern matcher, estimates based on the size of the logic and clock speed of the configuration interface can be used to give a very accurate initial estimate, particularly as the logic being modified is small. For more complex designs, estimation tools are being developed to aid the designer in accurately predicting the performance. Support for hierarchical designs allowed larger pattern matching circuits to be investigated, once the single bit matching circuit was captured.

```
{ RE_BLOCK
// Remove time is 0 NS as blocks are written over
   HIERARCHYPATH =     TopLevel(Arch)\g1:Bit[0],PatternBit(arch)\swap_buf
   FILEPATH      =     patternmatcher.vhd
   LOADTIME      =     40 NS
   REMOVETIME    =     0 NS
   ACTIVATE      =     work.control_pkg.ConfigBit(0)
   DEACTIVATE    =     NOT work.control_pkg.ConfigBit(0)
}
```

Fig. 4. DCS Schedule file example

In addition to the location and timing information, activation and deactivation conditions for the block are specified. These describe the conditions under which the logic is reconfigured in the system. The reconfiguration controller's operation is essentially specified by these conditions, for all of the elements in the design. By using simulation the designer can gain detailed understanding of reconfiguration requirements and develop the control logic. In this design the controller logic was manually developed, based on the simulation results. The potential does exist to generate the controller automatically from the DCS design information [10].

---

[1] Note that the most recent version of DCS [9] will create implementable designs without the requirement for these control flags.

Implementation and Interactive Debugging

The remainder of the system was designed using standard simulation tools. After simulating the operation of the main functional blocks, the design proceeded to the implementation phase. The VHDL source files were converted to XC6216 netlists using a Xilinx utility, called VELAB. These netlists were then placed and routed using the Xact6000 APR tools.

At this point some of the unique characteristics of the XC6216 were used to accelerate the debugging of the design. The values of any combinatorial or sequential signal in the system can be monitored due to the very flexible interface to the FPGA. Interactive tools have been created [2] that allow a design to be placed on the array and then executed under software control. Concepts borrowed from software engineering, such as watches and breakpoints, are used to debug the design. The interactive tools are integrated with the APR tools, accessing symbolic information about the location of structures in the design (e.g., registers). Using this information, portions of the array can be monitored and referred to symbolically, again greatly simplifying the debugging process. This very flexible debug environment was used to prove the operation of each task in the system in turn. Data from the original VHDL testbenches provided the stimulus to the task being tested on the array. The data is passed into the system using the debugging tools and the output was captured and compared with expected results.

Once all the static tasks had been proven and the reconfiguration controller logic had been exercised, the reconfigurable operation of the design was closely investigated. The developed tools allow partial reconfigurations to be loaded onto the array, under interactive or programmatic control. As the system clock can also be driven from software, the state of the array is accessible during the reconfiguration intervals, allowing inspection during reconfiguration. The bitstreams required to implement the pattern matcher were tested using partial reconfiguration data generated by the APR tool.

The partial reconfiguration bitstreams produced by the vendor's design tools are not optimal. To achieve the smallest and fastest possible reconfiguration, only the required configuration data should be downloaded. Information about the current and future configuration of the array at each stage is required to generate the optimal reconfiguration. With this temporal information available, the required differences between the logic can be evaluated and an optimal bitstream generated. The bitstream generation tools to process a design in this manner are currently being developed [11]. For this example, the optimal configurations for each reconfiguration were manually generated by inspection. Once the required design changes were known, the bitstreams were derived based on information available in the XC6200 databook. These hand-generated bitstreams were tested using the interactive tools to confirm that they performed the equivalent operations to those produced by the layout software. The manually generated bitstreams were two orders of magnitude smaller than the automatically generated equivalents.

Using the configuration data, a bitstream generation circuit was designed and tested in-situ. This circuit responds to signals from the reconfiguration controller which identify the required set of reconfiguration data. The data is then produced sequentially on each clock edge. In the final implementation, thc reconfiguration controller manages the transfer of this data to the reconfiguration interface.

Stand-Alone Implementation

Two versions of the final application were developed. The first implemented a byte wide pattern matching circuit. This was used to search ASCII-format text files and find selected strings. The reconfiguration controller for this design existed partly in software and partly in hardware. The understanding gained from completing this design accelerated the implementation of the final stand-alone version. Due to area limitations on the device, the final circuit implements a single bit pattern matcher, suitable for operating on streams of serial data. As this is a completely hardware controlled solution, the single bit version actually executes faster than the software controlled byte widc implementation.

The XC6200 provides a unique answer to the previously noted problem associated with boot-strapping a dynamically reconfigurable system. It has multiple, mutually exclusive, access points to the configuration interface. The exclusivity of the ports means that there is no contention over access during the initial boot-strapping of the reconfiguration controller. The two ports are a slow serial port and a high-speed parallel port, both of which are accessible from the user logic. The parallel interface is uncommon on FPGAs, providing a high-bandwidth configuration access point. This solution to the problem of contention at initialisation has not been available with any previous field programmable device. The application is initially configured via the serial interface. The custom reconfiguration controller manages subsequent re-configurations of the array via the fast, parallel interface. The ConfigOK signal, generated by the configuration interface, is used to indicate to the reconfiguration controller that it should take over control of the system. This signal also switched the reconfiguration interface from serial to parallel operation.

Fig. 5 shows actual timing data taken from the operation of the final stand-alone design. The RECONF signal is the request for reconfiguration sent to the controller which initiates the change in pattern matching logic. The two TASKI signals represent the identifier signals for the reconfigurable elements that initiate the bitstream generation. The BUFOUT signal changes when the pattern matcher has completed the reconfiguration.

The fastest possible reconfiguration of the array takes 40ns. This is the optimum time to configure one 32-bit wide pattern matcher. The practical reconfiguration time is actually larger than this value, and in the final system was measured as 80ns for a single reconfiguration, due to the response time of the reconfiguration controller to the request. The implication is that on average, only two patterns need to be compared between each reconfiguration to begin to be comparable to the static implementation.

The execution of the system is limited by the clock period that the configuration interface can support. At present this is 40ns, and is the limiting factor on the pattern matching application. The array logic can potentially execute at a faster rate, depending on the actual design. The main observation to be made is that the reconfiguration speed of the device is slower than the potential execution speed of logic on the array. To improve the overall system performance of dynamically reconfigurable systems, the impact of the reconfiguration latency has to be reduced. The most obvious method to achieve this would be configuration mechanisms that execute at a higher speed. Currently, careful controller design can effectively reduce the impact of reconfiguration by reconfiguring the device just prior to the logic being required on the array. One advantage of the comparatively slow reconfiguration rate is that the execution speed of the custom reconfiguration controller is not critical, presenting the opportunity for synthesis from a behavioural description [10].

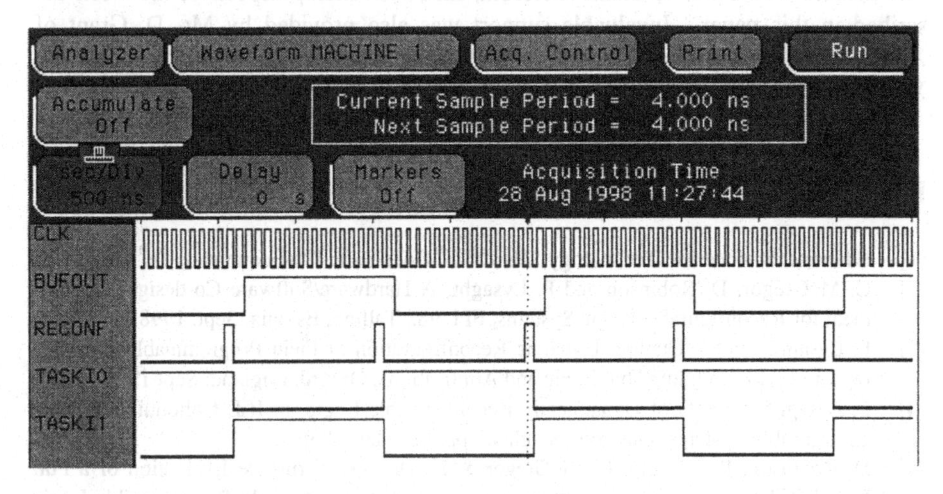

Fig. 5. Stand-alone implementation: actual operation

6 Conclusions

The design presented in this paper demonstrates the potential for entirely self-contained FPGA systems. The use of novel development tools and advances in FPGA technology provides new approaches to system development that have not previously been possible. The custom tools that have been developed to support the design flow have all been designed to inter-operate and integrate with standard commercial CAD tools. The advantages that these tools provide to the designer were very evident in this application.

The example was limited by the relatively small array size of the XC6216, with the reconfiguration controller using approximately quarter of the available logic. The rapid increases in FPGA densities will make systems similar to the one described here increasingly practical. It is unfortunate that there are currently no commercially available devices that can fully support the techniques described in this paper. Larger devices such as the Virtex FPGAs, while attempting to move dynamic reconfiguration techniques into the mainstream, are not adaptable enough to implement such novel applications. Features such as fully addressable configuration logic and support for dynamic configuration are vital, but careful consideration of the configuration access requirements must be made for future devices. Greater flexibility and accessibility in the configuration control is required for the true potential of dynamic reconfiguration to be achieved.

7 Acknowledgements

The authors would like to thank Motorola, Inc., for funding aspects of the work described in this paper. Invaluable support was also provided by Mr. D. Grant of Xilinx, Inc., in support of the VELAB VHDL elaboration tools and details of the XC6216 bitstream formats.

References

[1] Editorial, Science and Technology : Hardware Goes Soft, The Economist, 22nd May 1999

[2] G. McGregor, D. Robinson and P. Lysaght, A Hardware/Software Co-design Environment for Reconfigurable Logic Systems, FPL'98. Tallinn, Estonia. Sept. 1998

[3] P. Lysaght and J. Dunlop, Dynamic Reconfiguration of Field Programmable Gate Arrays, in Field Programmable Logic and Applications, Oxford, England, Sept 1993

[4] P. Lysaght, Aspects of Dynamically Reconfigurable Logic, in IEE Colloquium on Reconfigurable Systems, Glasgow, Scotland, pp 1-5, Mar 1999

[5] D. Robinson, P. Lysaght, G. McGregor & H. Dick, Performance Evaluation of a Full Speed PCI Initiator and Target Subsystem using FPGAs, In Field Programmable Logic and Applications, Proceedings of FPL'97, pp. 41-50, W. Luk & P. Cheung Eds., Springer-Verlag, 1997

[6] P. W. Foulk & L.D. Hodson, Data Folding in SRAM Configurable FPGAs, IEEE Workshop on FPGAs for Custom Computing Machines, pp. 163-171, Napa, CA, Apr. 1993

[7] M. J. Wirthlin and B. Hutchings, Improving Functional Density Through Run-Time Constant Propagation, In 1997 ACM/SIGDA International Symposium on Field Programmable Gate Arrays, pp. 86-92, 1997

[8] P. Lysaght and J. Stockwood, A Simulation Tool for Dynamically Reconfigurable Field Programmable Gate Arrays, IEEE Transactions on Very Large Scale Integration (VLSI) Systems, Vol. 4, No. 3, pp. 381-390, Sept. 96

[9] D. Robinson, P. Lysaght and G. McGregor, New CAD Framework Extends Simulation of Dynamically Reconfigurable Logic, in Field Programmable Logic and Applications, pp 1-8, Hartenstein, R. and Keevallik, A. (Eds), Tallinn, Estonia, Sept 1998

[10] D. Robinson and P. Lysaght, Modelling and Synthesis of Configuration Controllers for Dynamically Reconfigurable Logic Systems using the DCS CAD Framework, in *Field-Programmable Logic*, Glasgow, Scotland, Aug 1999

[11] N. Shirazi, W. Luk and P.Y.K. Cheung, Automating production of run-time reconfigurable designs, in Proc. IEEE Symposium on Field-Programmable Custom Computing Machines, K.L. Pocek and J. Arnold (Eds), IEEE Computer Society Press, 1998

An Internet Based Development Framework for Reconfigurable Computing

Reiner W. Hartenstein, Michael Herz, Ulrich Nageldinger, Thomas Hoffmann
University of Kaiserslautern
Erwin-Schroedinger-Strasse, D-67663 Kaiserslautern, Germany
Fax: ++49 631 205 2640, email: abakus@informatik.uni-kl.de
www: http://xputers.informatik.uni-kl.de/

Abstract. The paper presents a development framework for the Xputer prototype Map-oriented Machine with Parallel Data Access (MoM-*PDA*). The MoM-*PDA* operates as a reconfigurable accelerator to a host computer. It utilizes the KressArray as a coarse-grained reconfigurable architecture and implements concurrent data access to parallel memory banks. An Internet-based development framework allows remote prototyping of accelerator applications and enables worldwide access to the prototype via the Internet.

1 Introduction

The increasing interest in the Internet of the recent years results in a growing number of users as well as improved network infrastructure, quality and performance. This is the basis for a new kind of software, which is running via the Internet [5][6]. That means the software is installed on a server connected to the Internet and accessed and executed by a remote client. Such Internet based software provides several benefits. Once distributed by a server, Internet based software is available world-wide and depending on its implementation it can be executed without previous installation on the client machine. Further software updates and patches are not distributed to the users any more. Only the server has to be updated and the new version is available to all users immediately. For commercial use it is not necessary that customers buy the software any more, they download and pay only the modules they need. This pay by use method makes very expensive design software also available to small companies. It is also possible to place a high-performance or application specific computer at the server side and provide computation time for special tasks to the users.

In the area of reconfigurable computing the described techniques allow to provide public access to accelerator prototype hardware and related development software. Since experimental prototypes mostly consist of especially designed components, which are also very expensive, Internet based access increases the number of users efficiently. One of the first approaches, where remote testing was implemented, is Web-Scope [8][16]. A different approach, which is more going in the direction of operating systems, has been published in [4]. Here reconfigurable hardware is regarded more abstract and its computational resources can be used in a client server based approach.

In this paper a remote prototyping framework is presented, whereas the complete design software is executed via the Internet. For testing of applications a prototype hardware can be accessed remotely. The software part of the framework forms the Xputer Multimedia Development System (XMDS, [14][17][21]). It incorporates the advantages of Internet based software mentioned above. Further the XMDS enables access to the Map-oriented Machine with Parallel Data Access (MoM-*PDA*, [13]) for remote testing.

In the following section the MoM-*PDA* is presented first. After that the XMDS is introduced and it is explained how the prototype hardware is accessed via the Internet.

2 The Map-Oriented Machine with Parallel Data Access

In this section the MoM-*PDA* is introduced. It is used as an accelerator for a host computer. The machine architecture is based on the Xputer paradigm [9] and uses the KressArray [13] as a reconfigurable ALU (rALU). To achieve high data throughput possible through parallelism on hardware level, the MoM-*PDA* has parallel access to computation data [13]. Therefore a dedicated memory organization scheme has been implemented.

For the readers convenience the Xputer principles are briefly summarized. The basic operation principles are shown and the prototype MoM-*PDA* is presented.

2.1 The Xputer Machine Paradigm

The basic Xputer architecture consists of a data sequencer, which can be seen as the control part, a 2-dimensionally organized data memory and a rALU (figure 1). All parts of the Xputer are programmable. While the datapath is typically built of reconfigurable devices, the data sequencer is fixed and programmed by only a few parameters for generic address generation [10]. During operations the data sequencer generates an address stream for the data memory and sequences data to the rALU and results back to the memory. The accessed data is passed from the data memory through a smart interface which optimizes and reduces memory accesses. It stores interim results and holds data needed several times. All data manipulations are performed by the rALU.

To clarify how computations are performed an execution model is shown in figure 1. A large amount of input data is typically organized in arrays (e.g. matrix, picture) where the array elements are referenced as operands of a computation in a current iteration of a loop. These arrays can be mapped onto a 2-dimensionally organized memory without any reordering. The part of the data memory which holds the data for the current iteration is determined by a so called scan window, which can be of any size and shape. The scan window can be seen as a model for the smart interface which holds a copy of the data memory. Each position of the scan window is labeled as read, write or read and write. The labels indicate the performed operation to the specified memory location. The position of the scan window is determined by the lower left corner, called handle (see figure 1). Operations are performed by moving the scan window over the data map

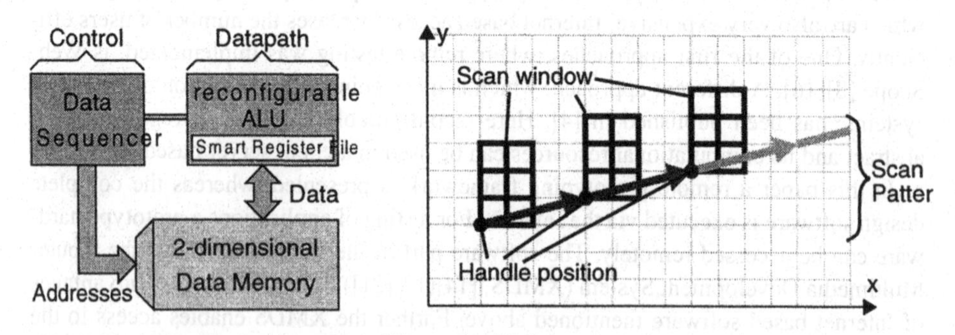

Fig. 1. Basic Xputer Architecture and Execution Model

and applying the configured operator of the rALU on the data in each step. Thus this movement of the scan window called scan pattern is the main control mechanism.

2.2 The Map-Oriented Machine with Parallel Data Access (MoM-*PDA*)

The MoM-*PDA* is an accelerator to be connected to a host computer via a PCI interface. It is based on the Xputer paradigm and utilizes the KressArray [13] for reconfigurable computing. This section gives a short overview on the overall hardware structure of the MoM-*PDA*.

The most important new feature of the MoM-*PDA* prototype is parallel high speed access to the data [13]. Therefore it has 2 parallel banks of Multibank DRAM (MDRAM, [18]). Addresses for the MDRAMs are computed by the Data Sequencer and extended with burst information by the Burst Control Unit (BCU, [3]). In the current prototype implementation the Data Sequencer of the MoM-*PDA* is mapped to an Altera FLEX10K100 device [1].

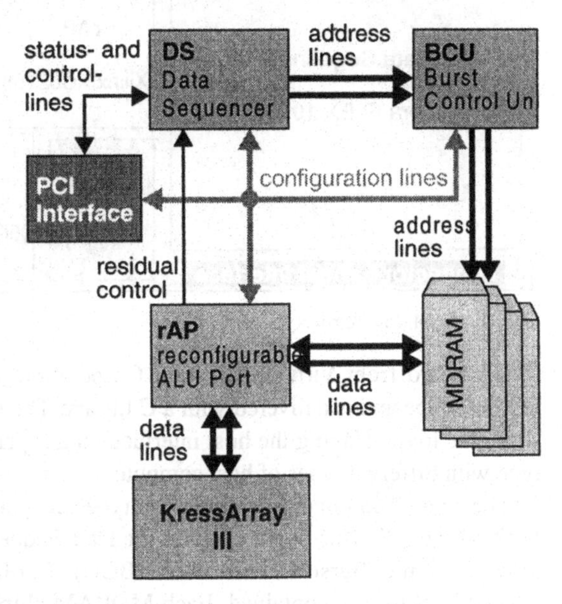

Fig. 2. MoM-*PDA* machine overview

The Data Sequencer handles up to 16 parallel tasks each consisting of a complex scan pattern [10]. Therefore up to 16 scan windows can operate on the data memory concurrently. The computation of the parallel tasks is done like known from multitasking systems. To access several locations in the data memory concurrently the Data Sequencer generates two parallel address streams. Based on these addresses the BCU performs MDRAM accesses. Usually computations are performed by the KressArray. Therefore data is routed to a reconfigurable ALU port (rAP, figure 2). It is implemented with a Xilinx XC6216 [20]. The rAP holds two functional units. One unit is the parallel memory interface for the MDRAMs and the smart interface. This unit is the same for every application and is configured once at power up. The remaining programmable space can be used in two different ways:

- as rALU: Computations are performed directly in the rAP. This allows to build a small version of the MoM-*PDA* without the KressArray. For this an operator library has been implemented, mainly containing image processing applications [7][12].
- as interface to the KressArray: In that case the rAP optimizes data exchange between the KressArray and the data memory.

2.3 MoM-*PDA* Prototype Implementation

Except the KressArray the MoM-*PDA* prototype implementation is based on FPGAs. The complete prototype consists of 3 printed circuit boards.

The interface to the host computer is implemented with a commercially available

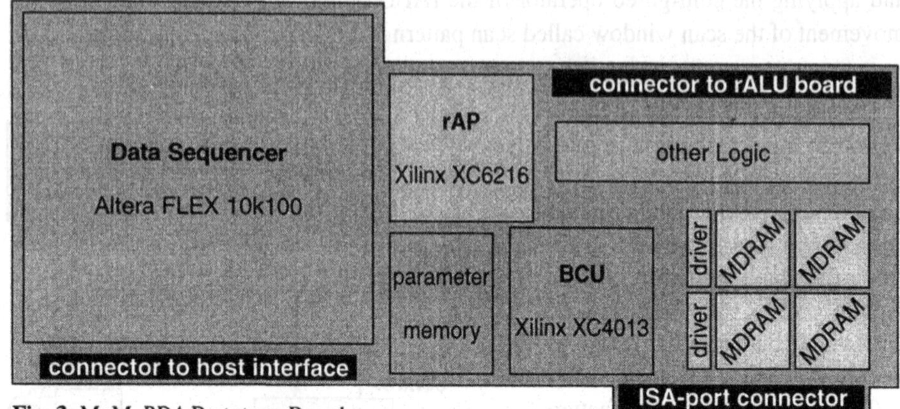

Fig. 3. MoM-*PDA* Prototype Board

FPGA board from Virtual Computer Corporation (VCC, [19]). It contains already a PCI interface and is delivered with a C library. The FPGA is used to implement necessary glue logic. Having the host interface on a separate board allows to use the prototype with different types of host computers.

The main components of the prototype are placed on the MoM-*PDA* prototype board (figure 3). This board contains the Data Sequencer, the reconfigurable ALU Port (rAP) and the Burst Control Unit (BCU). Further 2 parallel banks of Siemens MDRAM chips are contained. Each MDRAM chip has a capacity of 1MB and operates at a clock frequency of 120 MHz.

The MoM-*PDA* prototype board has 3 connectors:

- An ISA-port connector is used for power supply.
- A connector to the host interface.
- The connector to the rALU board allows to connect an external board with a KressArray. Since the rAP is able to perform computations, a KressArray is not always needed.

To set up the development framework all components are plugged into a Pentium processor PC running also the Xputer Multimedia Development System (XMDS). The XMDS integrates a special runtime system which enables users to test applications on the MoM-*PDA* via the Internet. This Internet based software is explained in the next section.

3 The Xputer Multimedia Development System (XMDS)

The most important part of the introduced Internet based development framework is the XMDS [14][17][21]. It is an Internet based CAD system which provides a comprehensive toolset for application development. Besides this remote access to conventional Xputer design software [2] written in C programming language is made possible. Further the XMDS enables remote testing of applications by executing them on the MoM-*PDA* prototype.

3.1 XMDS Concepts

The main part of the XMDS system is the XMDS server which holds the complete XMDS system including an user data library, a runtime system for the Xputer hardware and the MoM-*PDA* prototype hardware (figure 4). Parts of the XMDS system are executed on the server and other parts are downloaded on the fly to the users computer and executed there. For communication between these computers a special protocol (XMDS messaging, [17]) has been developed. Further an Unix host is connected to the XMDS server holding conventional design software written in C programming language.

Several concepts had been focussed during the XMDS implementation. In fact, the XMDS is a project that combines the features of a powerful Internet based CAD system with the specific requirements to access the experimental Xputer hardware and the traditional (C-programmed) software. This led to the following specification items. The XMDS should provide:

- a user front-end that is accessible via WWW for several platforms,
- a central user administration,
- a modularity of its components,
- a built-in support for further extensions,
- powerful network capabilities, and
- an embedding of multimedia features.

In the following, the realization of these concepts is described in detail.

WWW front-end and the Client /Server Model

Because the World Wide Web (WWW) is commonly explored with a WWW browser, the user front-end of the XMDS must have a WWW address, like normal WWW documents. By pointing his browser to that particular URL (Uniform Resource Location), a user of the XMDS may download the front-end to his client machine and display it on the screen. Therefore, the part of the XMDS that should display on the user's machine must be executable by the most common WWW browsers. Therefore a Java implementation has been chosen. The fact that a WWW browser forms the environment of the front-end ensures a certain independence against the underlaying operation system (e.g. Windows, MacOS, Unix). Portations of a particular WWW browser to different platforms ensure therefore the availability of the XMDS.

The different components that form the XMDS are realized following the client/server paradigm. Components are distributed in a logical manner on the client and on the server part. Every user front-end which is downloaded by a user forms a client of the XMDS. The user can reach it by requesting a specific URL targeting the computer where the XMDS files are installed. To allow the publishing of URLs, this computer must run a WWW server. The HTML document loaded in the users browser

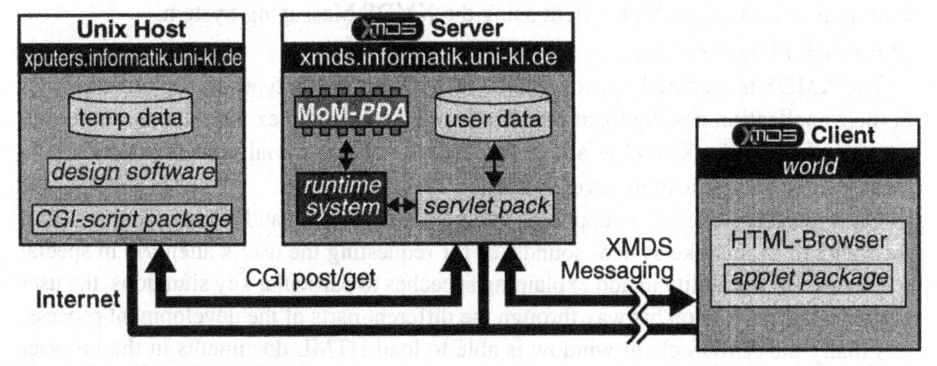

Fig. 4. Overview of the XMDS

window contains a reference to the XMDS client program.

The complete XMDS system is programmed in Java. While the Java applet technology realizes a dynamic download of the XMDS client to a users machine via Internet, the XMDS server makes use of the Java servlet technology to accept the connections from the client and perform the server-side actions.

The advantages of such a dynamic client/server model compared to a static installation of client software are obvious. With this concept, the XMDS

- is available world-wide,
- may be used without previous installation,
- is always available in its latest version,
- downloads only the components that are actually needed by the user,
- enables the users to evaluate the MoM-*PDA* without acquiring the hardware.

This concept realizes an easy way for distribution and provides easy access for interested persons.

Central User Administration

The XMDS server also contains the central user administration: the user database and the user management system.

The users of the XMDS need to store the data, the state and the configuration of their XMDS sessions in order to continue their work in a next session. Because the XMDS-Client has no permission to store data on the client's machine due to the applet limitations [17], all the user-specific data must be stored on the XMDS server. Therefore the users of the XMDS must be registered. To secure intellectual property (IP) all users receive a personal password to protect their data.

Network capabilities

A very close binding between the client and the server is essential for a powerful distributed system like the XMDS. This is realized by the XMDS Messaging system which will not be discussed here. Refer to [17] for a detailed presentation.

An important task involving a network communication is to invoke the design software contained on an Unix host. Because this host runs a Web server, a medium-level communication is possible involving the Common Gateway Interface (CGI) of the Unix host. Because Java applets (like the XMDS client) are not allowed to communicate with an Internet host different from their own home host, it must be the XMDS server that establishes this communication. Therefore the XMDS server provides a module for a comfortable use of the CGI-request-response interaction and extends the communication to the XMDS client using the XMDS Messaging System.

Multimedia User Interface

The XMDS is prepared to cooperate with different media types to support the users in their application development process. It integrates complex animations within the user interface of the XMDS to allow visualization of non-trivial processes which efficiently supports application development.

On a more basic level, a support is integrated for playing audio files. This is useful for standard events like system sounds, or for requesting the user's attention in special situations. By the ability to add explaining speeches to particular key situations, the user is supported actively on his way through the different parts of the development process.

Finally the XMDS client window is able to load HTML documents in the browser window. This allows the construction of a context-sensual help system integrating text, image and hyperlink elements. The on-line-help is in fact expanded to a complete on-line-tutorial providing the most actual help documents available at the XMDS Server.

3.2 The XMDS Runtime Support

The XMDS provides a runtime system to test applications on the real hardware (figure 4). As a distributed system it accesses the Xputer prototype on the server side. The MoM-*PDA* is directly connected to the server which hosts the XMDS. While the front-end of the XMDS (client) realizes the interface to the user, the server part is designed to access the Xputer hardware through different communication layers. The purpose of the runtime system is testing of applications only. It has no operating system features and does not support host/accelerator co-processing. Further also the multitasking feature of the MoM-*PDA* is not used.

On a low level the runtime system consists of a set of functions accessing the MoM-*PDA* via the PCI bus of a PC. These functions are implemented in C because they have to operate on a sub-operating-system-level, which can not be reached by a conventional Java program. They form the lowest communication layer connecting directly the hardware ports of the PC. On a higher an enveloping program realizes the control flow and calls the C functions. In approach to the XMDS modularization concepts [17], the program is implemented as an XMDS server module in Java, and calls the C functions using the Java Native Interface (figure 5). The Java Native Interface allows to utilize special capabilities of a computer, operating system or non-Java software that the Java Class library does not already provide.

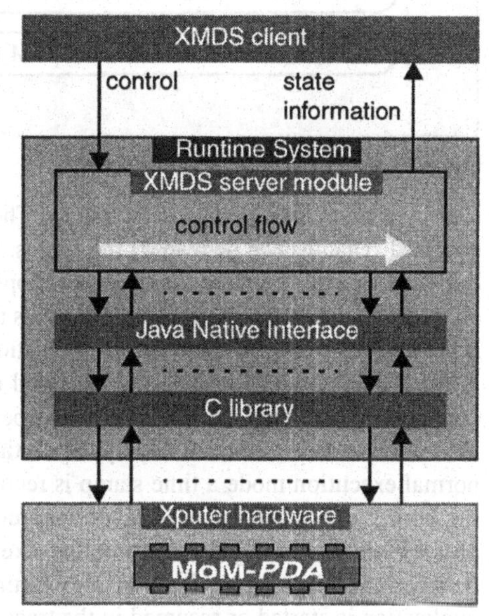

Fig. 5. Overview of the Runtime System

The status visualization of a running application and the information about the application queue on the server is displayed for the user by a specialized XMDS client module. All process control comes from this client module. The user specifies the application to be executed and provides the test data. Since the test data is generated on the users machine it has to be transferred to the XMDS server. This is done using the file transfer protocol (ftp). After this preparations a start request is sent to the XMDS server. Besides normal execution the application can also be processed in a trace mode. In that mode the application is executed step by step and for evaluation the internal state of the MoM-*PDA* is sent to the client and displayed each time.

Incoming applications for the MoM-*PDA* are organized in a sequential way. Only if one application is executed at a time precise measurement of execution time is possible. Therefore the XMDS integrates a time-management for multiple requests. This is real-

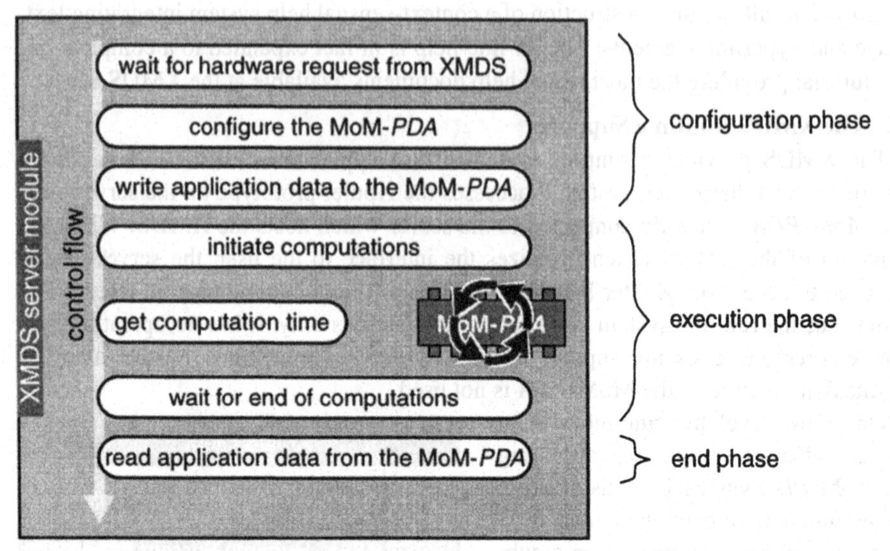

Fig. 6. Control Flow of the Runtime System

ized by a first-come-first-served strategy. The server holds a list of waiting jobs to be executed. This list can be read by the clients. In case that the hardware is currently used by another client, a started application is appended to the list of waiting jobs. Then the client has to wait until all earlier applications are processed or he may cancel his request. If the hardware is not used when an application is started, it is immediately executed.

Before an application is really executed on the hardware, the hardware has to be configured first. Also the test data has to be stored in the data memory of the MoM-*PDA*. This is done automatically by the runtime system. If the application is invoked in normal execution mode a time stamp is recorded to measure the execution time. During normal execution the user has to wait until computations have finished. The only possible interaction is to interrupt the execution. In that case the user can select between canceling the application or restarting it in trace mode. If the execution of the application is started or resumed in the trace mode the user can watch all internal processes of the hardware but no measurement of overall execution time is possible. At the end of computation the state of the hardware, the results of computation and depending of the execution mode the execution time is transferred to the client.

Inside the XMDS server module, the control flow can be divided on an abstract level into 3 phases (figure 6). In a first phase, the necessary data and programs are loaded into the MoM-*PDA*; this is the configuration period. Next, the computation of the hardware is started. In normal execution mode a time stamp is recorded and the XMDS server module waits until the MoM-*PDA* reports an "end-of-computation"-signal. In the last phase, the execution time is determined based on the recorded time period, the results are read, and the XMDS client is informed.

If the hardware is invoked in the trace mode client and server are working interactively. The hardware stops each time the data sequencer generated a handle position (figure 1). The internal state of the hardware is transmitted to the client and displayed to the user. The user may manipulate internal registers and send the contents back to the server to reconfigure the hardware. Further the user triggers the hardware to perform computation steps. The user may operate the hardware in single steps, a specific number of steps or by setting a breakpoint on an internal state.

3.3 XMDS Design Flow

Depending on the users knowledge of the hardware principles the XMDS provides different design tools which results also in different degrees of performance. Unexperienced users may use automatic compilation of applications. For this the CoDe-X (Co-Design for Xputers, [2]) framework has been implemented. Because CoDe-X is programmed in C language, it is installed on an Unix host and accessible via the XMDS. There are also several tools implemented for application development on expert level. Figure 7 shows the XMDS client window with a clickable design flow on expert level. On this level there are 3 different possibilities to edit an application. Besides a specialized programming language (MoPL, [2]) a graphical task editor and a hardware level parameter editor are available. The design validation phase is supported by simulation tools animating the computations on a hardware model as well as on the execution model (figure 1). The Application Analyser calculates the exact runtime of an application. After design validation it can be downloaded to the users computer or executed with test data on the MoM-*PDA* prototype.

4 Summary

A development framework for reconfigurable accelerator applications has been introduced. The framework consists of an Xputer prototype (Map-oriented Machine with Parallel Data Access, MoM-*PDA*) and an Internet based development software (Xputer Multimedia Development System, XMDS). The XMDS provides a set of development tools for the MoM-*PDA* and supports also remote execution of other design software. Further the XMDS enables testing of applications on the MoM-*PDA* prototype hardware without acquiring the hardware.

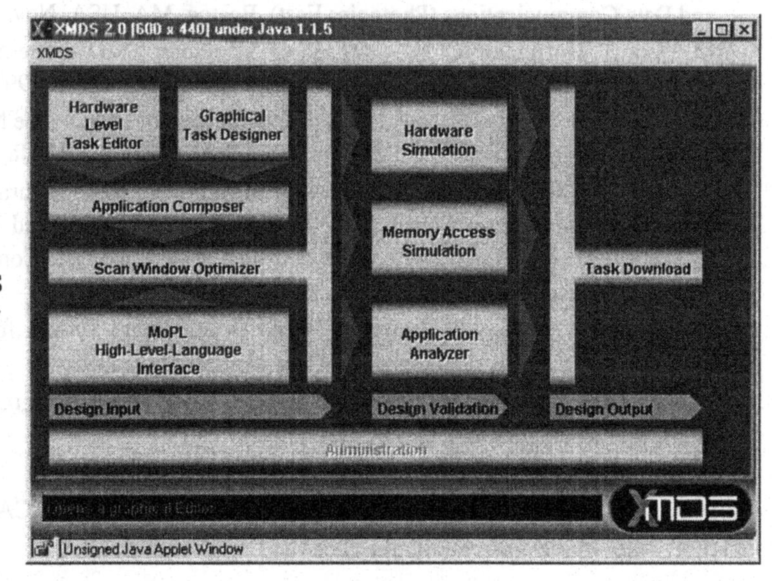

Fig. 7. XMDS client window showing a selection of tools as a design flow for expert level application development

References

[1] N.N.: Altera 1998 Data Book; Altera Corporation, San Jose, CA, USA, 1998.

[2] J. Becker: A Partitioning Compiler for Computers with Xputer-based Accelerators; Ph. D. Thesis, University of Kaiserslautern, 1997.

[3] M. Bednara: A Burst Control Unit to Perform Optimized Access to Multibank DRAMs; Diploma Thesis, University of Kaiserslautern, May, 1998

[4] G. Brebner: Circlets: Circuits as Applets; FCCM'98, Napa Valley, CA, USA, 1998.

[5] P. Denyer, J. Brouwers: Java in Electronic Design Automation; Proc. of ASP-DAC'97, Chiba, Japan, Jan. 28-31, 1997.

[6] L. Geppert: IC Design on the World Wide Web; IEEE Spectrum, June 1998.

[7] F. Gilbert: Development of a Design Flow and Implementation of Example Designs for the Xilinx XC6200 FPGA Series; Diploma Thesis, University of Kaiserslautern, May, 1998.

[8] S. Guccione: WebScope: A Circuit Debug Tool; FPL'98, Tallinn Technical University, Estonia, Aug. 31 - Sept. 31, 1998

[9] R. Hartenstein, A. Hirschbiel, M. Weber: A Novel Paradigm of Parallel Computation and its Use to Implement Simple High Performance Hardware; InfoJapan'90, Int'l. Conf. memorizing the 30th Anniv. of Computer Society Japan, Tokyo, Japan, 1990.

[10]R. Hartenstein, J. Becker, M. Herz, U. Nageldinger: A Novel Universal Sequencer Hardware; Proceedings of Fachtagung Architekturen von Rechensystemen ARCS'97, Rostock, Germany, September 8-11, 1997

[12]R. Hartenstein, M. Herz, F. Gilbert: Designing for the Xilinx XC6200 FPGAs; FPL'98, Tallinn Technical University, Estonia, Aug. 31 - Sept. 31, 1998

[13]R. Hartenstein, M. Herz, T. Hoffmann, U. Nageldinger: Using the KressArray for Reconfigurable Computing; Proc. of Conference on Configurable Computing: Technology and Applications, part of SPIE International Symposium on Voice, Video, and Data Communications (Photonics East), Boston, MA, USA, Nov. 1-5. 1998.

[14]M. Herz, T. Hoffmann, U. Nageldinger, C. Schreiber: XMDS: The Xputer Multimedia Development System; HICSS-32, Hawaii, USA, Jan. 5-8, 1999.

[15]M. Herz, T. Hoffmann, U. Nageldinger, C. Schreiber: Interfacing the MoM-*PDA* to an Internet-based Development System; HICSS-32, Hawaii, USA, Jan. 5-8, 1999.

[16]D. Levi, S. Guccione: BoardScope: A Debug Tool for Reconfigurable Systems; Proc. of Conference on Configurable Computing: Technology and Applications, part of SPIE International Symposium on Voice, Video, and Data Communications (Photonics East), Boston, MA, USA, Nov. 1-5. 1998.

[17]C. Schreiber: Design of an Internet Based Development System for Xputers in Java, Diploma Thesis, University of Kaiserslautern, July 1998.

[18]N.N.: Siemens Multibank DRAM, Ultra-high performance for graphic applications; Siemens Semicond. Group, Oct., 1996.

[19]URL: http://www.vcc.com/

[20]N.N.: The Programmable Logic Data Book; Xilinx Inc., San Jose, CA, USA, 1996.

[21]URL: http://xmds.informatik.uni-kl.de/

On Tool Integration in High-Performance FPGA Design Flows

Andreas Koch

Tech. Univ. Braunschweig (E.I.S.), Gaußstr. 11, D-38106 Braunschweig, Germany
koch@eis.cs.tu-bs.de

Abstract. High-performance design flows for FPGAs often rely on module generators to counter coarse logic-block granularity and limited routing resources, However, the very flexibility of current generator systems complicates their integration and automatic use in the entire tool flow. As a solution to these problems, we have introduced FLAME, a common model to express generator capabilities and module characteristics to clients such as synthesis and floorplanning tools. By offering a unified view of heterogeneous generator libraries, FLAME allows the seamless and efficient use of flexible generation techniques in automatic compilation flows targeting configurable hardware.

1 Introduction

While well known for decades, the use of algorithmic module generation in VLSI design flows has recently been exploited with renewed interest. Especially for FPGAs with their limited interconnect resources and coarse-grain logic blocks, module generators have traditionally been the tool of choice to quickly provide fast and dense circuits [1–4]. The need for structured circuit generation becomes even more pronounced when FPGAs are not just used to implement glue logic, but as compute elements in configurable computing machines (FCCM). Many research efforts on automatic compilation to FCCM targets include module generation as a crucial step [5–7].

However, the very flexibility of parametrized generators makes their integration with the main design flow (synthesis, floorplanning, place and route) difficult. The solution quality of any optimization performed by these tools will be limited by the information offered *about* the available module alternatives. For example, modern generators can completely restructure a general circuit description to optimally process constant inputs [3]. In an FCCM, this restructuring could even be postponed to take place at run-time (once the actual data values are available) instead of at compilation time. When one of the main flow tools requires information about a module this flexible, the sheer volume of the design space covered by each generator (such as behavior, time, area, power, and layout) precludes a simple enumeration of all variations.

Furthermore, common formats for characterizing library elements [8–10] are tailored for ASICs and do not cover the higher-level concepts required for efficient module embedding (e.g., function, control interface, sequential timing, etc.).

In contrast to the practical tutorial presented previously [14], this work concentrates on a high-level overview of the concepts underlying our proposed solution.

2 FLAME

FLAME, the Flexible API for Module-based Environments aims to resolve these difficulties. It is not a module generation system itself, but a wrapper around heterogeneous module libraries.

FLAME offers client tools (synthesis, floorplanning) a single unified means to access generators (termed servers) and determine module characteristics without regard for their original vendor, target technology, or implementation.

To enable this functionality, FLAME specifies a common model independent of an implementation language or platform to express generator capabilities and module design characteristics to client tools. Instances of this model can be represented in a portable manner and are exchanged between clients and servers via a simple, yet flexible active interface (API). Note that since FLAME aims at inter-tool communication, a GUI is not required.

3 Active Interface

To avoid the problems associated with static module description files, FLAME relies on a dynamic exchange of queries and replies between clients and servers instead. Figure 1.a shows such a sample dialog, using synthesis and floorplanning as clients in the main design flow. The retrieval of information (which could also include complete netlists or layouts) proceeds as a stepwise refinement by incrementally tightening constraints to narrow down the range of possible solutions. In addition to parameters such as operand widths and data types, queries are also hierarchically scoped (see Section 6) to further constrain solutions.

Figure 1. (a) Sample query/reply scenario, (b) System architecture

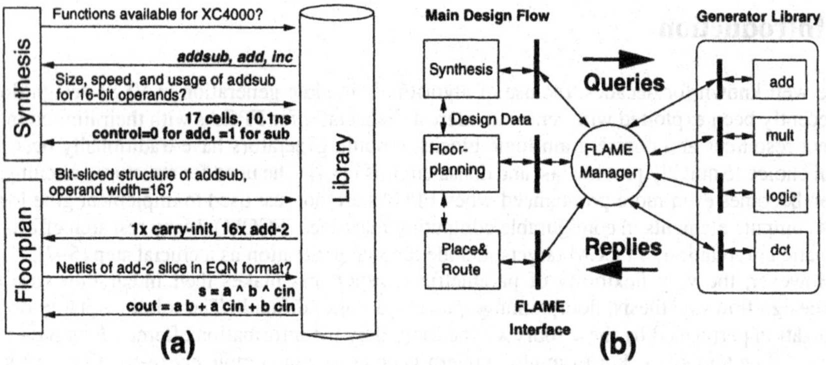

This approach reduces computation times by allowing results to be calculated only to the detail actually required by the current query abstraction level. For example, during the initial synthesis phases, only timing and area estimations are required. The generation of a complete layout at this step would be wasteful and is not necessary. On the other hand, once synthesis has decided on a certain implementation, the floorplanning phase requires precise shape information for optimal operation. This more detailed data, which might also be more time consuming to compute (especially for data-dependent units generated at FCCM run-time), can then be provided only for the single selected implementation. In a similar manner, local logic optimization [11] can retrieve netlists from a per-bit-slice scope instead of trying to find regularity in the flat, unstructured netlists of the entire module.

The internal FLAME architecture is shown in Figure 1.b. The communication between clients and servers is mediated by the FLAME Manager, which accepts queries from clients and forwards them to the appropriate servers. Next, the individual replies from the servers are assembled into a composite reply and passed back to the clients. Further capabilities can include the automatic

Table 1. FLAME views

Name	Contains	Required?
behavior	functionality, logical interface	yes
synthesis	timing, area, power, control, and physical interface	yes
structure	regularity (bit-slices), hierarchy	no
topology	layout shape, interconnect densities	no
netlist	device-independent netlist	yes
mapped	device-dependent netlist	no
placed	device-dependent placement	no
routed	device-dependent layout	no
simulation	simulation model	yes

translation between different FLAME representations and gatewaying between various communication mechanisms (direct function call, program invocation, network). In addition, a memoization mechanism in the FLAME Manager can serve to further increase efficiency by answering previously encountered requests out of a cache without forwarding them again to the generators.

4 Data Representations

FLAME data can be represented in multiple formats, which generally trade-off ease-of-use and portability vs. efficiency. A human readable text representation is easiest to process even for simple tools (such as Perl scripts), while a tokenized representation (keywords replaced by integer tokens) can be processed and transferred more efficiently between tools implemented in different languages and environments. For maximum performance, an implementation-language specific binary representation (generally built using pointers or references, e.g., [12]) can be passed between integrated tools using the same language binding. This approach results in much lower communications overhead than the traditional file-based approach and is one of the unique characteristics of FLAME.

5 Views

Views serve to group related items of data. Thus, a client only has to ask for a single view, instead of querying for each individual datum. For the server, views reduce the workload by allowing it to compute only the information in the specified view, instead of all information fulfilling the query constraints. As mentioned earlier, a synthesis tool is generally not interested in a placed and routed layout during data flow-graph (DFG) covering. It uses the "synthesis" view to only consider appropriate aspects such as module function, timing, and area. Table 1 shows some of the supported views.

FLAME defines required and optional views. The latter are used for advanced optimizations [15] [11], or to allow a client to pre-compute information for fast processing. This is most applicable to the "placed" view, in which the generator can offer a pre-placed layout which takes maximum advantage of a regular circuit structure.

6 Design Hierarchy

As another measure to limit the amount of data exchanged between clients and servers, FLAME operates on a hierarchy of design entities (Figure 2), where lower levels (more detail) are only accessed when required.

Figure 2. FLAME design entities

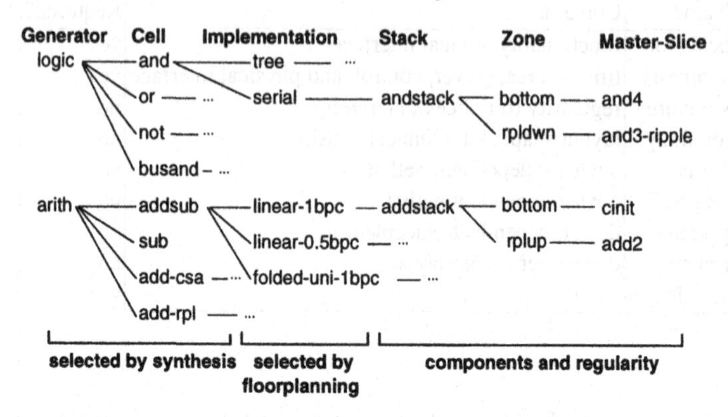

A *generator* is a concrete piece of code that creates different views of a circuit according to parametrized descriptions. A *cell* is a functional unit that can be generated by the specific generator. Different cells may supply identical or different functions and interfaces. An *implementation* is an actual circuit conforming to the behavior and interface of the enclosing cell. All implementations of a cell must have the same function and interface. In general, they differ in more physical aspects such as layout footprint, logical pitch, topology, low-level timing, and power consumption. The cell is composed from *sub-modules* (instances of other cells) and *stacks*. Stacks contain one or more *zones* of replicated logic (Figure 3.a). The smallest design unit is the *master-slice*, which will be iterated to form the zone.

To illustrate the hierarchy, consider the following example: A generator *arith* might provide the cells *addsub* (switchable adder-subtractor), *sub* (subtractor), *add-csa* (adder), and *add-rpl* (adder). The adder-subtractor is available in three implementations (*linear-1bpc*, *linear-0.5bpc*, and *folded-uni-1bpc*) that realize it in different physical layout styles. In the implementation *linear-1bpc*, the circuit consists of a single stack *addstack* defining two zones, *bottom* and *rplup*. The zone *bottom* holds a single iteration of the master-slice *cinit* (carry initialization), while the zone *rplup* contains multiple (up to the desired operand width) iterations of the master-slice *add2* (full-adder bit-slice).

When the synthesis system is covering the data-flow graph of the current design with modules in the library, it selects suitable cells. Since all implementations of a cell are guaranteed to have the same external interface (which includes, e.g., control specifications and pipelining), the floorplanner can then perform lower-level optimizations, such as matching the physical layouts across all modules by selecting appropriate implementations within the cells [15], while keeping the synthesized global controller intact.

Stacks, zones, and slices describe the regularity aspects of a design (e.g., the composition of a 16-bit ripple adder by replicating a 1-bit adder). This information can be exploited in later design flow steps to perform further optimizations (e.g, regular merging of modules) [11] and reduce computation times (by solving small problems and replicating the results) [15].

The use of hierarchical instead of flat composition also allows sophisticated run-time optimizations, such as reaching into a module to substitute faster/larger or slower/smaller submodules depending on whether they are located on or off the critical path at the system level.

7 Target Technology

FLAME abstracts key device features in a portable manner. This enables the development of technology-specific generators that offer a technology-independent interface for instantiation and composition. Design tools are thus presented with a uniform view of the different underlying FPGA architectures, allowing both the easy retargeting of designs between architectures as well as the development of portable CAD tools supporting multiple technologies.

Area is measured as a vector containing an entry for each kind of active resource (logic block, memories, arithmetic unit etc.), giving both the number of units used and the total number available. Thus, the area cost in terms of "scarcity" of resources is easily computed.

Routing density, which indicates the amount and/or type of routing resources used within a layout, is represented similar to area. This information is required for optimal datapath composition to avoid placing a very dense module in the middle of a linear arrangement of modules, thus making it very difficult to route signals through this obstacle.

The precise and complete description of the capabilities of on-chip storage or tri-state elements is crucial for efficient circuit synthesis. Modeled are the trigger type (edge or level) and the presence of set, reset, and enable signals including their polarities.

8 Function

In many compilation flows for FCCMs, the synthesis tool/compiler builds a data-flow graph of the operations required by the user program, which must then be mapped into hardware units. A common solution to this problem performs a covering of the operations in the data-flow graph with the hardware operations available in the module library, e.g. [16]. For this process (which in itself is not covered by FLAME), the function of each cell offered by the module generators must be described in a standardized way.

In FLAME, combinational logic and arithmetic functions are best formulated as an expression in infix notation using standard operators. For example, the expression `Y=A+B` could describe the semantics of an adder, while `Y=A&B&C` characterizes a 3-input AND.

Primitive (e.g., muxes, register banks, etc.) or high-level (FIR, FFT, processors, etc.) cells, which cannot conveniently be described by the infix expression, are accessed using well-known names. E.g, `LPM_REG` could be used for a register bank when using LPM [17] as guideline for module names.

In addition to modules performing a single function, FLAME also allows the specification of units computing multiple functions in parallel and/or in sequence. E.g., a cell could compute the sum and difference of operands simultaneously, or be a controllable adder/subtractor which can perform either one of these operations in sequence.

9 Interfaces

Describing only the function(s) of a cell is of course insufficient to allow its actual use in a circuit. To this end, the cell interface, which consists of port (how to connect the cell) and control (how to use the cell, Section 10) information has to be specified.

FLAME distinguishes between logical and physical interfaces. E.g., a sample cell *MulDiv* might logically accept operands *A* and *B* to compute a product or quotient *PQ*. Physically, however, it could accept the operands on the rising edges of successive clock cycles at a single input port *D*, after an opcode choosing either multiplication or division has been loaded (also through *D*, but with an asserted control input *Op*). When a control output *Done* becomes asserted, the result can then be retrieved from the physical output *Y*. FLAME easily allows the description of such a logical-to-physical mapping. By hiding these details in the physical interface, the initial DFG covering pass of synthesis only deals with the logical interface. This should lead to simplified tools and quicker execution (since fewer details have to be considered).

Both kinds of interface allow the definition of input, output and bidirectional ports as well as the application of constant and late-bound (loaded at run-time into the FPGA) values. FLAME distinguishes between data and control ports (to guide a regular datapath layout [15]). All ports are also characterized in terms of data types and bit widths. Currently, declarations for signed and unsigned integers as well as for fixed precision numbers are defined. If available on the target technology, outputs can also be registered and/or tristated on request.

10 Control

While the behavioral view describes the functionality of a cell, it does not specify how the functions can actually be *used*. This might range from a simple addition/subtraction switch by changing the value on a control input from 0 to 1, to a multi-cycle sequence of loading operands and opcodes into a complex functional unit that signals the end of a variable-length execution by asserting a control output (e.g., the example in Section 9). In FLAME, such control sequences for combinational and sequential cells can be specified in terms of six primitive statements: *LEVEL* asserts a signal combinationally (no cycles pass). *POSEDGE* and *NEGEDGE* assert signals in time for the positive (negative) clock edge, they take one clock cycle of time. *CONTINUE* waits until the specified signals have the specified values, but takes no time in itself. *START* marks an entry point for a new thread of control, which is spawned using *RESTART*. With this information, synthesis can automatically generate an appropriate FSM to integrate the cell into the host circuit. As an example, the control interface of the sample cell *MulDiv* (Section 9) in multiplication mode could be formulated as:

```
(POSEDGE  (("Op")    1) (("D")  0x42))  ; load magic number for multiply mode
(LEVEL    (("Op")    0))                ; now switch to operand input mode
(START)                                 ; label for starting another multiply operation
(POSEDGE  (("D")   ("A")))              ; load first operand A through port D
(POSEDGE  (("D")   ("B")))              ; load second operand B through port D
(CONTINUE (("Done") 1))                 ; wait until port Done becomes asserted
(RESTART)                               ; load new operands (fork to START) ...
(POSEDGE  (("Y")   ("PQ")))             ; ...  and simultaneously unload result through PQ
```

11 Timing

Once it is clear what a cell does and how it can be used, the single most important characteristic for high-performance designs is the cell timing. While the path-based modeling approach is more precise, it often becomes unmanageable due to the huge number of timing paths in larger circuits. The complexity of the slack-based model (timing specified in terms of required and arrival times) grows only linearly in the number of ports, but becomes pessimistic for circuits with wildly differing internal path lengths. FLAME supports both specifications and leaves the choice to the generator implementor.

In contrast to the simple combinational or CLK→Q delay values often found in ASIC-specific cell descriptions [9] [10], the often highly pipelined cells on FPGAs have additional requirements. All time specifications in FLAME also contain the number of the sequential cycle when an input must be valid or an output arrives (latency). This allows the description of pipelined or multi-cycle operations to synthesis, which can then also insert an appropriate number of deskewing registers when paths with different latencies converge. Furthermore, FLAME also allows the specification of the longest combinational delay before the first cell-internal storage element is reached. Together with the combinational delay beyond the last cell-internal storage element, the longest Q→D delay can be calculated across cell boundaries, thus allowing the determination of the datapath-wide clock period. Additionally, cells also specify their throughput as the number of clock cycles per datum to describe the performance of partially pipelined units.

To allow the delay comparison between units with variable execution times (Section 10), their timing is specified separately as best case, average, and worst case timing. This expanded information enables clients to make timing-based trade-off decisions.

12 Structure

As shown in [15], the exploitation of regular structures during datapath synthesis can lead to considerable reductions both in compile time as well as in delay/clock period. While it is to some degree possible to extract regularity from an unstructured netlist, e.g., [19], it is far more efficient to let the generator actually provide this information to the client. Since a generator usually composes regular circuits in a regular manner (e.g., by iterating a sub-circuit in a for-loop), this structural knowledge is already available, and just needs to be made visible externally.

Figure 3. (a) Regular Structure, (b) Pitch Matching

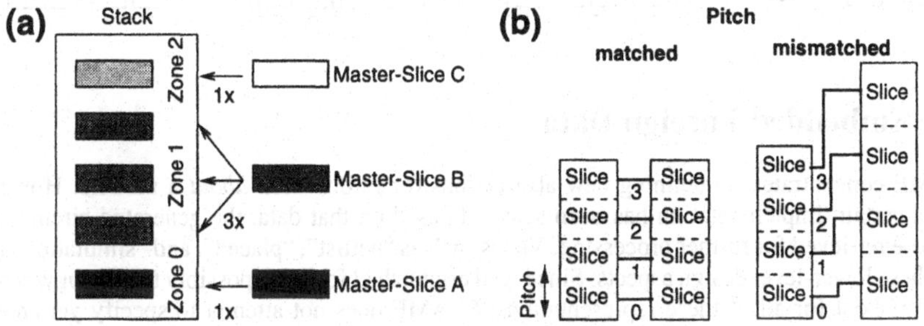

Following this approach, FLAME allows the description of regular structures based on the iteration (replication) of master-slices to compose zones of replicated logic (Figure 3.a). Note that this replication also includes regular connectivity (typically, but not limited to, next-neighbor connectivity as used in ripple-carry adders and shift registers). One or more zones make up a stack. E.g., a ripple-carry adder might consist of a zone for initializing the carry chain, a zone of replicated full-adders, and a zone for processing an overflow bit. An entire cell is then assembled from one or more stacks and/or sub-modules.

13 Topology

After synthesis has selected cells to cover the data-flow graph, (relying on the timing and area data retrieved from the generators), floorplanning [15] takes a more physical view to construct a high-performance datapath. To this end, FLAME models the topological characteristics of the layout. Included are the shape of the layout, the folding style for very wide modules, the pitch (spacing between bits of a regular bus-port), and the density (the availability of interconnect to route through the module). Furthermore, modules may be constrained to specific locations (e.g., to take advantage of specific function units such as memories).

Depending on the available chip area for the datapath, modules that are too tall might have to be folded to fit (Figure 4). In that case, the floorplanner needs to ensure that all of these modules use a consistent folding style (long routing delays would occur otherwise). Matching port pitch across all modules in the datapath also aims at reducing routing delays (Figure 3.b). As indicated in Section 7, the routing density inside a module is described to allow congestion management at the datapath level before performing detailed routing.

Combining regular structure exploitation with floorplanning as described in Sections 12 and 13 has improved design performance by as much as 33% and reduced tool runtimes by up to 80% [15].

Figure 4. Layout folding

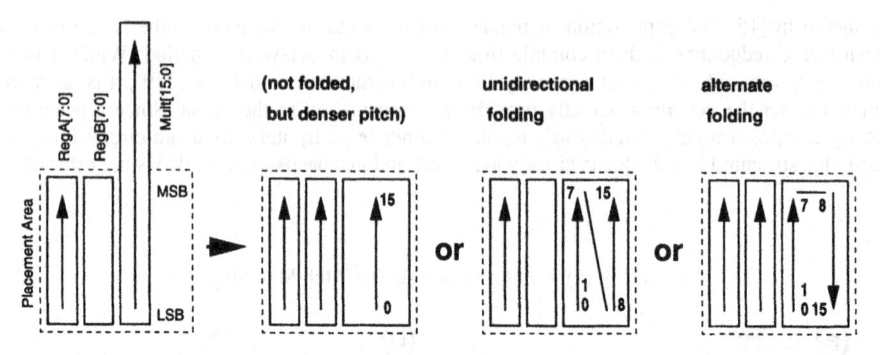

14 Embedded Foreign Data

FLAME concentrates on defining new abstractions for information *about* a module. However, once a certain implementation has been selected based on that data, the generated circuit itself has to be retrieved for further processing. Views such as "netlist", "placed", and "simulation" provide these lower-level design aspects. Since well established formats (possibly technology/vendor specific) exist for all of these representations, FLAME does not attempt to specify yet another "standard" in these areas. Instead, data in an existing format is wrapped in a FLAME shell for transfer. E.g., netlist data could be sent as EDIF, placement information on the XC4000 as XNF, and simulation models as VHDL or Verilog.

15 Results

The current FLAME technology demonstrator [12], containing the base library and a sample FLAME Manager, offers unified access to generators developed ad-hoc as well as to modules in the Xilinx CoreGen package [20]. In the near future, this prototype will be extended to also allow control of JHDL generators [21]. Since the system relies entirely on the high-performance FLAME binary representation (Sec. 4), and avoids cumbersome file operations, the overhead of the FLAME interface is negligible.

FLAME is documented in detail by a comprehensive manual [13]. Furthermore, a portable object-oriented model using Unified Modeling Language (UML) [22] for the binary representation has been developed. Since this model is exportable in multiple programming languages (e.g., C++, Java, Ada), it could be used as a starting point for individual implementation efforts. The prototype has been created in this manner by elaborating a model exported to Java.

The interface is currently used in academic as well as in industrial research projects on the next generation of EDA systems for reconfigurable computers,

16 Summary

We presented the motivations for and capabilities of FLAME, a new method for tool integration in generator-based EDA suites for FPGAs. FLAME encompasses all aspects of a design flow beginning with synthesis and ending at layout, modeling them by either introducing new abstractions, or encapsulating existing ones. By allowing the main flow tools easy and efficient access to a wide range of well-defined module parameters and representations, the full flexibility of a generator-based implementation method may be harnessed to create highly optimized circuits without human intervention.

Appendix: Examples

Loading the module library is generally the first step in any compilation flow run. In FLAME, this is formulated as a query for the "behavior" view (we are interested in all module functions) with the only constraint being the target technology. The corresponding FLAME expression in textual representation [13] is shown on the left, while the equivalent binary representation composed using the FLAME/Java interface [12] is shown in the right column.

```
(QUERY 1 1                              <--->   Query q = new Query(1, 1
 (TECHNOLOGY "Xilinx" "XC4000E"                   new Technology("Xilinx", "XC4000E",
             "XC4003EPG191" "-3"                                "XC4003EPG191", "-3",
  (VIEWS
   (VIEW "behavior"))))                            new VBehavior()));
```

For brevity, we assume that our library contains only a single cell, a bus-wide gate switchable between AND/OR operations (example code printed in two columns).

```
(REPLY 1 1
 (TECHNOLOGY "Xilinx" "XC4000E"
             "XC4003EPG191" "-3"
  (VIEWS
   (VIEW "behavior"
    (STATUS QUERYOK "view ok")
    (GENERATOR "andor" 1
     (CELL "andor" 1
```

```
(INTERFACE
 (LOGICAL
  (INPUT  (("A") ) (("B") ))
  (OUTPUT (("Y") ))))
(BEHAVIOR
 ("andmode"  (FUNCTION (INFIX "Y=A&B")))
 ("ormode"   (FUNCTION (INFIX "Y=A|B")))
)))))))
```

The reply describes the existence of a generator "andor", which can provide the single cell "andor", which in turn has the logical inputs "A" and "B" and the logical output "Y". It can perform two different operations, namely the logical AND of its inputs in "andmode", or the logical OR in "ormode". With this functional information, synthesis can now proceed to cover the data flow graph.

Further into the design flow, the synthesis tool needs the characteristics of a specific module instance to perform various trade-offs (area, time, ...). To this end, a "synthesis" view is requested by an appropriately constrained query.

```
(QUERY 1 2
 (TECHNOLOGY "Xilinx" "XC4000E"
             "XC4003EPG191" "-3"
  (VIEWS
   (VIEW "synthesis"
    (GENERATOR "andor" 0
```

```
(CELL "andor" 0
 (INTERFACE
  (LOGICAL
   (INPUT  (("A") (WIDTH 8) )
           (("B") (WIDTH 8) ))
   (OUTPUT (("Y") (WIDTH 8) ))))))))))
```

Here, we request information on an 8-bit wide instance with default data types (unsigned integer) and variable inputs. The resulting reply contains the control specification for the cell as well as the area and time point for an actual physical implementation.

```
(REPLY 1 2
 (TECHNOLOGY "Xilinx" "XC4000E"
             "XC4003EPG191" "-3"
  (VIEWS
   (VIEW "synthesis"
    (STATUS QUERYOK "view ok")
    (GENERATOR "andor" 1
     (STATUS QUERYOK "generator ok")
     (UNIT (TIMESCALE -10))
     (CELL "andor" 1
      (STATUS QUERYOK "cell ok")
      (INTERFACE
       (LOGICAL
        (INPUT  (("A") (WIDTH 8) (UNSIGNED))
                (("B") (WIDTH 8) (UNSIGNED)))
        (OUTPUT (("Y") (WIDTH 8) (UNSIGNED))))
      (PHYSICAL
        (INPUT
         (("A")      (WIDTH 8) (DATA) (UNSIGNED))
         (("B")      (WIDTH 8) (DATA) (UNSIGNED))
         (("mode")   (WIDTH 1)
                     (CONTROL) (UNSIGNED)))
        (OUTPUT
         (("Y")  (WIDTH 8)  (DATA)  (UNSIGNED)))))
```

```
(BEHAVIOR
 ("andmode"
  (FUNCTION (INFIX "Y=A&B"))
  (UCODE (LEVEL (("mode" 0 0) 0))))
 ("ormode"
  (FUNCTION (INFIX "Y=A|B"))
  (UCODE (LEVEL (("mode" 0 0) 1)))))
(IMPLEMENTATION "simple" 1
 (CATALOG
  ("structure" )
  ("topology" )
  ("netlist"     (FORMAT "verilog"))
  ("placed"      (FORMAT "xnf"))
  ("simulation"  (FORMAT "verilog")))
 (TIMING
  ( ("andmode" "ormode" )
   (FIXED
    (REQUIRED
     (("A" 7 0) ("B" 7 0) ("mode" 0 0) )
       0 0 0)
    (ARRIVAL (("Y" 7 0) ) 0 22)
    (CYCLETIME 22)
    (THROUGHPUT 1))))
(AREA ("CLBS" 4 4 100)))))))))
```

After declaring a time scale of 0.1ns per time unit, the generator confirms the constraints on the logical interface before revealing the physical interface. An additional input "mode" becomes visible, and the type and data/control nature of each port is declared. The behavior of the function is further refined by including control information: "andmode" is activated by applying a logical 0 on the "mode" input, "ormode" by a logical 1.

The cell is available in one concrete physical implementation named "simple", which can be retrieved in a number of views listed together with the formats for the non-FLAME views. The implementation timing for both operating modes is expressed in the slack-based model, stating that as long as all inputs arrive at time 0, the output will be valid no later than 22 time units (2.2ns) afterwards. Since the circuit is purely combinational, it could be clocked with a period of 2.2ns, and accept one new datum per clock cycle. It requires 4 of the chip resource "CLBS", of which 100 are available on the target architecture. See [12–14] for further examples.

References

1. Xilinx Inc., "X-BLOX Reference", *EDA tool documentation*, San Jose (CA) 1995
2. Dittmer, J., Sadewasser, H., "Parametrisierbare Modulgeneratoren für die FPGA-Familie Xilinx XC4000", *Diploma thesis*, Tech. Univ. Braunschweig (Germany), 1995
3. Chu, M., Weaver, N., Sulimma, K., DeHon, A., Wawrzynek, J., "Object Oriented Circuit Generators in Java", *Proc. IEEE Symp. on FCCM*, Napa Valley (CA) 1998
4. Mencer, O., Morf, M., Flynn, M.J., "PAM-Blox: High Performance FPGA Design for Adaptive Computing", *Proc. IEEE Symp. on FCCM*, Napa Valley (CA) 1998
5. Gokhale, M.B., Stone, J.M., "NAPA-C: Compiling for a Hybrid RISC/FPGA Architecture", *Proc. IEEE Symp. on FCCM*, Napa Valley (CA) 1998
6. Harr, R., "The Nimble Compiler Environment for Agile Hardware", *Proc. ACS PI Meeting, http://www.dyncorp-is.com/darpa/meeting/acs98apr/Synopsys%20for%20WWW.ppt*, Napa Valley (CA) 1998
7. Hall, M., "Design Environment for ACS (DEFACTO)", *Proc. ACS PI Meeting, http://www.dyncorp-is.com/darpa/meeting/acs98apr/defacto.ppt*, Napa Valley (CA), 1998
8. Electronics Industry Association, "EDIF Version 4 0 0", *ANSI/EIA 682-1996 Standard*, Washington (DC) 1996
9. Synopsys Inc., "Library Compiler User Guide Version 3.5", *EDA tool documentation*, Mountain View (CA) 1997
10. Open Verilog International, "Advanced Library Format for ASIC Cells & Blocks", *ALF Reference Manual Version 1.0*, Los Gatos (CA) 1997
11. Koch., A., "Module Compaction in FPGA-based Regular Datapaths", *Proc. 33rd Design Automation Conference (DAC)*, Las Vegas (NV) 1996
12. Koch, A., "FLAME/Java Release 0.1.1", *http://www.icsi.berkeley.edu/~akoch/research.html*, Berkeley (CA), 1998
13. Koch, A., "FLAME: A Flexible API for Module-based Environments – User's Guide and Manual", *http://www.icsi.berkeley.edu/~akoch/research.html*, Berkeley (CA), 1998
14. Koch, A., "Generator-based Design Flows for Reconfigurable Computing: A Tutorial on Tool Integration using FLAME", *Proc. PACT'98 Workshop on Configurable Computing*, Paris (France), 1998
15. Koch, A., "Regular Datapaths on Field-Programmable Gate Arrays", *Ph.D. thesis*, Tech. Univ. Braunschweig (Germany), 1997
16. Liao, S., Devadas, S., Keutzer, K., Tjiang, S., "Instruction Selection Using Binate Covering for Code Size Optimization", *Proc. ICCAD '95*, November 1995
17. EIA, "Library of Parametrized Modules", *EIA/IS-103 Standard*, 1993
18. Callahan, T., Chong, P., DeHon, A., Wawrzynek, J., "Fast Module Mapping and Placement for FPGAs", *Proc. ACM/SIGDA Symp. on FPGAs*, Monterey (CA), 1998
19. Nijssen, R.X.T., Jesse, J.A.G., "Datapath Regularity Extraction", in *Logic and Architecture Synthesis*, eds. Saucier/Mignotte, 1995
20. Xilinx Inc., "CORE Generator System User Guide", *EDA tool documentation*, San Jose (CA) 1998
21. Hutchings, B., et. al., "A CAD Suite for High-Performance FPGA Design", *Proc. FCCM '99*, April 1999
22. Fowler, M., Scott, K., "UML Distilled", *Addison-Wesley*, 1997

Hardware-Software Codesign for Dynamically Reconfigurable Architectures *

Karam S. Chatha and Ranga Vemuri

Department of ECECS,
University of Cincinnati, ML 30,
Cincinnati, OH 45220-0030.
Email: {kchatha,ranga}@ececs.uc.edu

Abstract. The paper addresses the problem of mapping an application specified as a task graph on a heterogeneous architecture which contains a software processor, a dynamically reconfigurable hardware coprocessor and memory elements. The problem comprises of three subproblems: partitioning of tasks between hardware and software, assigning tasks mapped on hardware to different temporal segments and scheduling task execution, reconfiguration of hardware, inter-processor and intra-processor communication. We present a heuristic based technique for solving the problem. The effectiveness of the technique is demonstrated by a case study of the JPEG image compression algorithm and experimentation with synthetic graphs.

1 Introduction

Embedded systems typically have heterogeneous architectures which contain both off the shelf software (SW) processors and custom application specific integrated circuits (ASICs) as hardware (HW) coprocessors. The SW processors provide flexibility and help in reducing the cost of the system. The HW coprocessors help in increasing the throughput of the system. The advent of dynamically reconfigurable field programmable gate arrays (FPGAs) offers a new and exciting opportunity to the embedded system designer. Since they are run time reconfigurable they are flexible and offer unlimited HW resources over time. Since they are available as off the shelf components, they are cheaper than custom ASICs and also have the advantage of reduced design time. The reconfiguration times for one generation old reconfigurable devices, for example Xilinx 4000 series FPGAs, is about 100 ms. Present day reconfigurable processors (Xilinx 6000 series FPGAs) have a reconfiguration time of about 0.5 ms. Time multiplexed FPGAs [1] which are probably the next generation of run time reconfigurable devices have a reconfiguration time overhead of only 5 ns. If the present trend continues, the dynamically reconfigurable devices will become even more attractive for

* This work was partially supported by the ARPA RASSP program and US-AF, Wright Lab, under contract numbers F33615-93-C-1316 and F33615-97-C-1043 respectively.

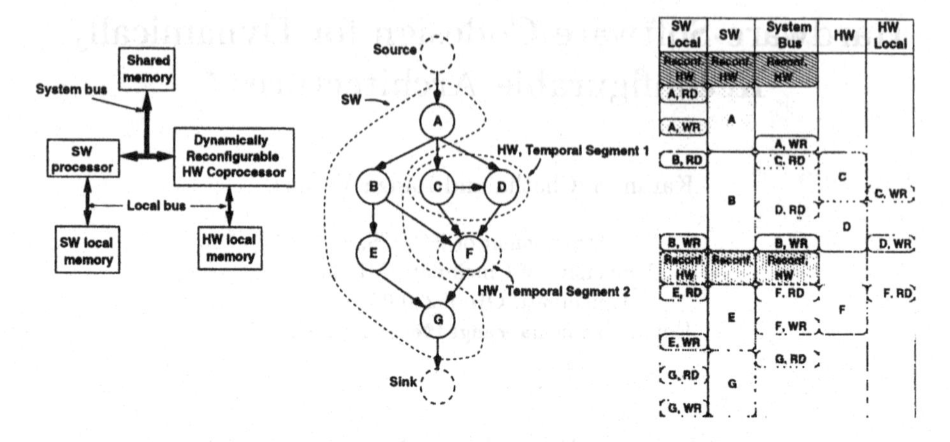

Fig. 1. Codesign Architecture, Task Graph and its HW-SW Implementation

embedded system design. This paper presents a technique for automatic mapping of an application described as a task graph on to a heterogeneous architecture which contains a SW processor, a dynamically reconfigurable HW coprocessor and memory elements.

Consider the example shown in Figure 1. The SW processor is a uniprocessing system and it can execute only one task at a time. The HW coprocessor can execute multiple tasks concurrently. In the task graph we assume that the source (sink) node writes (reads) from the SW local memory. The objective of the algorithm is to map the task graph on the dynamically reconfigurable codesign architecture such that the total execution time of the application is minimized. The schedule table for execution of the task graph is shown on the right hand side. In the schedule table SW (HW) local stands for the SW (HW) local bus. The schedule table specifies the processor mapping of the tasks and the order of execution of the tasks, inter-processor (HW-SW) communication, intra-processor (HW-HW, SW-SW) communication and HW reconfiguration. This is the final result produced by our algorithm. Among the HW tasks, C and D are assigned to temporal segment 1 and F is assigned to temporal segment 2. Communication between tasks C and D takes place through on-chip registers of the coprocessor and is not shown as a separate time in the schedule table. Communication from HW tasks C and D to HW task F takes place through the HW local memory since they belong to different temporal segments. All communication between HW and SW tasks takes place through the shared memory connected to the system bus. The reconfiguration of the HW coprocessor is done by the SW processor. We reconfigure the entire HW coprocessor at a time, we do not consider partial reconfiguration. The reconfiguration time is specified by the user, and we assume that it includes the time for reading the configuration from the SW local memory and then transferring it over the system bus to the HW coprocessor.

We present an integrated algorithm for HW-SW partitioning, temporal partitioning and scheduling. The objective of the algorithm is to: a) partition tasks between HW and SW, b) assign HW tasks to time exclusive temporal segments (henceforth called temporal partitioning), c) schedule task execution on HW and SW processors, d) schedule reconfiguration of HW coprocessor, e) schedule inter-processor and intra-processor communication through shared memory and local memories respectively, such that i) the time for one complete execution of the task graph is minimized, and ii) HW area constraint is satisfied. The algorithm is a modification of the approach proposed in [2] for traditional HW-SW systems.

Our technique concentrates on transformative or computation intensive applications. Typical examples are JPEG and MPEG image compression and decompression algorithms. Transformative applications can be easily modeled at a system level by a data dependency based task graph model. Towards the end of the paper we discuss a case study of the JPEG image compression algorithm which demonstrates the viability of HW-SW systems with dynamically reconfigurable HW. The paper is organized as follows: Section 2 discusses previous work, Section 3 presents the algorithm, in Section 4 we discuss experimental results and finally Section 5 concludes the paper.

2 Previous Work

There are a number of existing approaches to HW-SW codesign and they use different techniques for partitioning and scheduling. In [3], [4] and [5] nodes are bound at the partitioning stage and the scheduler is used to estimate the execution time of the partition. They all need scheduling to correctly evaluate the performance of the partitioned design. Our technique overcomes this limitation by using an integrated scheduler. On the other hand [6] and [7] map tasks at the scheduling stage. They both use a list scheduler which is a greedy algorithm. The algorithm maps tasks by trying to obtain a local minima, which does not necessarily result in a global minima. Our algorithm overcomes this limitation by mapping some of the tasks at the partitioning stage.

In this paragraph we discuss approaches for automatic temporal partitioning and scheduling for reconfigurable architectures. In [8], Gokhale et.al. have proposed a methodology for mapping "C" based high level description language on reconfigurable logic arrays. Vasilko et.al. [9] and Gajjalapurna et.al. [10] proposed a static list scheduling based approach for temporal partitioning and scheduling which is a greedy approach. Kaul et.al. in [11] have proposed an ILP based approach to temporal partitioning, HW design space exploration and scheduling. Although their approach generates optimal solutions it is limited by the high run times of the algorithm. Maestre et.al. in [12] have proposed a system level approach for temporal partitioning and scheduling for reconfigurable computers. They do clustering followed by scheduling. None of the approaches mentioned in this paragraph address the problem of HW-SW partitioning.

3 HW-SW Partitioning, Temporal Partitioning and Scheduling

The application is specified at a coarse level of granularity by a directed acyclic data dependency graph called the task graph $G(V, E)$. The vertices of the graph are tasks which may be bound to a HW or SW implementation. For each task $v \in V$, we know its execution time on the software processor (v_{sw}), its execution time on the HW coprocessor (v_{hw}) and its area when it is implemented on the HW coprocessor (v_{area}). For each edge $e \in E$ we know the number of data items (e_{data}) transferred across it. The SW time (v_{sw}) for a task v can be obtained by profiling on the SW processor. The HW time (v_{hw}) and area (v_{area}) can be obtained by using a high level synthesis tool. Although a task graph has only data dependencies, a task node may contain loops and control flow constructs inside it. The user specifies the area constraint on the HW coprocessor ($A_{constraint}$), the reconfiguration time for one complete reconfiguration of the HW coprocessor (T_R), the communication times per data item for inter-processor (*shrd*, *shwr*) and intra-processor (*swrd*, *swwr*, *hwrd*, *hwwr*) communication and the extra area overhead on the HW coprocessor for communication (a_{comm}). The communication overhead refers to the extra hardware needed for HW tasks to communicate to the shared memory and HW local memory.

We use an integrated partitioner and scheduler to obtain a mapping of the tasks on the target architecture. The partitioner maps some tasks to HW and SW respectively. The scheduler decides the mapping of the remaining tasks, maps the HW tasks to different temporal partitions and schedules the task graph. The feedback from the scheduler to the partitioner occurs in the form of execution time and mapping of the task graph.

3.1 Some Definitions

The mapping of a task to HW or SW depends on its processor timings (v_{sw} versus v_{hw}), area (v_{area}), communication time with neighboring tasks, mapping of the other tasks in the task graph, reconfiguration time of the HW coprocessor and area constraint. In this section we present some definitions which capture these relevant characteristics of the tasks.

Timing characteristics of a task : The *speed ratio* of a task v is given by :

$$v_{sr} = \begin{cases} \frac{v_{sw} - v_{hw}}{v_{sw}} & \text{if } (v_{sw} - v_{hw}) \geq 0 \\ \frac{v_{sw} - v_{hw}}{v_{hw}} & \text{if } (v_{sw} - v_{hw}) < 0 \end{cases}$$

It varies between -1 and 1. It is greater (smaller) than zero for a task whose run time on the SW processor is greater than (less than) its run time on the HW coprocessor. We scale the speed ratio of a task from 0 to 1 and define its *speed up* as $v_{speedup} = \frac{v_{sr}+1.0}{2.0}$. It is greater (smaller) than 0.5 for a task whose run time on the SW processor is greater than (less than) its run time on the HW coprocessor.

Area of a task : The area of a task does not influence its mapping if the area constraint ($A_{constraint}$) is not tight or if the reconfiguration time (T_R) is very low. We capture these architecture characteristics by defining C_{arch} as follows:

$$C_{arch} = \begin{cases} \frac{A_{constraint}}{A_{sum}} \times \frac{T_{sum}}{T_{sum}+T_R} & \text{if } A_{constraint} \leq A_{sum}, T_R > 0 \\ 1 & \text{if } A_{constraint} > A_{sum} \text{ or } T_R = 0 \end{cases}$$

where A_{sum} is the sum of areas of all tasks and T_{sum} is the sum of smaller processor (HW or SW) time of each task. C_{arch} lies between 0 and 1. A high value (closer to 1) of C_{arch} implies that the architecture does not impose tight area constraints. This might be due to a high value of $A_{constraint}$ or a low value of T_R. Let A_{max} (A_{min}) denote the maximum (minimum) area over all tasks. We define the *area ratio* of a task v_{ar} as $\frac{A_{max}-v_{area}}{A_{max}-A_{min}}$ if $v_{area} \leq A_{constraint}$ or 0 if $v_{area} > A_{constraint}$. The area ratio ranges from 0 to 1. The area ratio of a task is closer to 0 (1) if its area is nearer to $A_{max}(A_{min})$. We define the *area factor* (v_{area_factor}) of a task v by scaling its area ratio from C_{arch} to 1. The objective is to be able to distinguish between tasks in terms of their respective area depending upon the architecture characteristics. When the area constraint is very tight or reconfiguration time is high we see a larger difference in area ratios of two tasks as opposed to the case when the area constraint is very loose or reconfiguration time is low.

Suitability of a task to be mapped to HW : The suitability of a task v_{suit} to be assigned to HW is given by the product of its speed up and area factor. Since the area factor of a task depends on the architecture characteristics, the suitability of a task also shows a similar behavior. We scale the suitabilities of tasks from 0 to 1, and define suitability factor v_{sf} of a task as $v_{sf} = \frac{v_{suit}-minsuit}{maxsuit-minsuit}$ where *minsuit* (*maxsuit*) is the minimum (maximum) suitability over all tasks $v \in V$. Suitability factor gives the probability of a task to be assigned to HW.

Execution time and urgency of a task : We use a modified list scheduling algorithm. The list scheduling algorithm requires the execution times of the tasks. Initially the mappings of the tasks are unknown, therefore their execution times are estimated based on their suitability factors. We associate a read time and write time with every dependency based on the suitability factors of the predecessor and successor tasks. For an edge $e = (u, v) \in E$ the read time is given as follows: $e_{rdtime} = e_{data} \cdot (u_{sf} \cdot v_{sf} \cdot hwrd + (1-u_{sf}) \cdot (1-v_{sf}) \cdot swrd + (1-u_{sf}) \cdot v_{sf} \cdot shrd + u_{sf} \cdot (1-v_{sf}) \cdot shrd)$. The write time e_{wrtime} can be similarly defined. We assume that u and v communicate through HW local memory when they are both mapped to HW. They communicate through SW local memory when they are both mapped to SW. They communicate through shared memory when one of them is mapped to HW and the other is mapped to SW. The initial execution time v_{exec} of a task v is given by $v_{exec} = v_{rd} + v_{sf} \cdot v_{hwtime} + (1-v_{sf}) \cdot v_{swtime} + v_{wr}$ where v_{rd} (v_{wr}) is the read (write) time of the task. The read (write) time of task is calculated as the sum of the read (write) times of all its predecessor (successor) dependencies.

The list based scheduler is characterized by the heuristic function used to select a task from the ready list. We select a task from the ready list based on

its *urgency*. The urgency of a task v, $v_{urgency}$ is given by: $v_{urgency} = v_{exec} + max_{(v,w) \in E}(w_{urgency})$.
Threshold suitability of the task graph : We now define the *threshold suitability (Th)* of the task graph which is used to distinguish between tasks more suitable for a HW implementation as opposed to tasks more suitable for a SW implementation. Let Th' be the median suitability of all tasks $v \in V$. The median suitability is used to distinguish between tasks when $A_{constraint} = 0$. In the case of dynamically reconfigurable resource although it offers infinite HW area over time, there is a reconfiguration time penalty associated with it. We therefore increase the median suitability to obtain the threshold suitability as follows: $Th = (1 - C_{arch}) \cdot (maxsuit - Th') + Th'$.
Communication time: The scheduler decides the mapping of a task based on its suitability. The suitability of a task does not take the communication times in to account. Hence the suitability of a task is modified by multiplying it with the communication ratio v_{comm_ratio} of the task. It is given by ratio of communication time with task v in SW over communication time with task v in HW. During scheduling it is possible that the mapping of the successor tasks is not known. We calculate the communication to successor tasks based on their suitabilities.
Influence of mapping of other tasks: The effect of mapping of other tasks on the mapping of a task v is captured by its execution time ratio. Execution time ratio v_{exec_ratio} is the ratio of estimated execution time of the graph when v is mapped to SW over estimated execution time of the graph when v is mapped to HW. The estimated execution of the graph is given by the sum of the earliest schedule time of v on a resource plus the urgency of v.
Final suitability of the task graph: The scheduler decides the mapping of the task based on its final suitability given by $v_{final\_suit} = v_{suit} \cdot v_{comm\_ratio} \cdot v_{exec\_ratio}$.

3.2 HW-SW Partitioning

The partitioner tries to improve upon a solution generated by the scheduler. It works in two modes: SW move mode and HW move mode. In the SW (HW) move mode the partitioner selects a task mapped to the SW (HW) and moves it to HW (SW). A task during partitioning can be in either one of three states: *free, tagged or fixed*. Initially all tasks are in a *free* state. A task is in a *tagged* state when the partitioner has tentatively changed the mapping of the task and obtained the time for the new solution. A task is in a *fixed* state when the partitioner has permanently fixed the mapping of the task. During a move mode, say SW the partitioner picks a SW task which is in a *free* state. It then maps it to HW, temporally changes the task state to *fixed* and obtains the execution time from the scheduler. The task state is then changed to *tagged*. In the SW move mode the partitioner one by one maps all the free SW tasks to HW and selects a move which results in maximum improvement. The mapping of the selected task is changed to HW and its state set to *fixed*. The partitioner then sets the state of the *tagged* tasks to *free* and generates a new solution. The partitioner exits the SW move mode when it cannot find any more *free* SW tasks or no move results in an improvement. On exiting the SW move mode the partitioner enters the HW move mode. The partitioner alternates between the SW and HW move mode until neither of the two modes result in an improvement. The algorithm then returns the schedule table and execution time of the final solution.

3.3 Temporal Partitioning and Scheduling

The objective of the scheduler is to partition *free* tasks in to HW and SW, partition HW tasks in to time exclusive temporal segments, schedule reconfiguration of HW coprocessor, schedule task execution, schedule inter-processor and intra-processor task communication. The scheduler uses a modified list scheduling algorithm. As a first step we calculate the urgency values of different tasks. We then construct the ready list which is a list of all tasks which are ready to be scheduled. A task is ready to be scheduled if all its predecessor tasks have been scheduled. In the list based scheduler a *while* loop maps and schedules one task in each iteration. A task with maximum urgency is selected from the ready list. We then calculate its final suitability. If the final suitability is greater than the threshold or if the task's state is fixed and it is mapped to HW, the scheduler maps the task to HW coprocessor. Otherwise the task is mapped to SW processor. If the sum of the area of a task assigned to HW plus the area of the present temporal segment exceeds the area constraint, the algorithm assigns the task to a new temporal segment. It increases the number of temporal segments, it adds a reconfiguration task to the SW processor and sets the area of the present temporal segment as the sum of task area and communication area. If the task area does not violate the area constraint it is added to the present temporal segment. The task is then added to the schedule table. During the scheduling of a predecessor task u of v, the mapping of v is not known. The scheduler assumes that the communication between u and v occurs through shared memory. When the mapping of v is known, we update the communication and execution times of all predecessors of v. The ready list is updated by adding to it all successor tasks of v which are ready to be scheduled. The scheduler exits the *while* loop when all the tasks have been scheduled.

3.4 Time Complexity

In each of the move modes, the partitioner iterates over all the tasks and finds a move which results in maximum improvement. In the worst case to map all $N = |V|$ tasks the partitioner iterates $(\frac{N(N+1)}{2})$ times, hence the time complexity of the partitioner is $O(\frac{N(N+1)}{2})$. In one iteration of the *while* loop the scheduler maps one task, hence the loop iterates N times. Inside the *while* loop the function used to update the schedule table has a time complexity if $O(N + M)$ where $M = |E|$. Hence the time complexity of the scheduler is $O(N(N + M))$. Since the partitioner calls the scheduler for evaluating each move, the time complexity of the overall algorithm is $O(N^4 + N^3M + N^3 + N^2M)$.

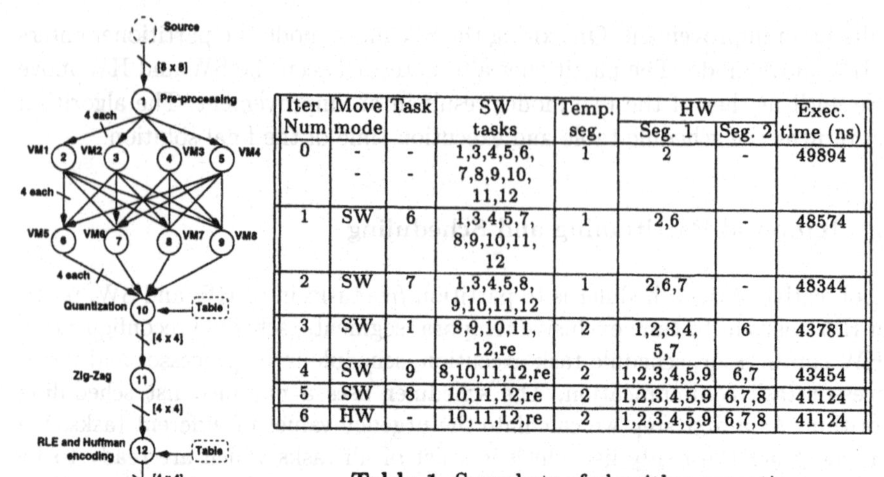

Fig. 2. JPEG

Iter. Num	Move mode	Task	SW tasks	Temp. seg.	HW Seg. 1	HW Seg. 2	Exec. time (ns)
0	-	-	1,3,4,5,6, 7,8,9,10, 11,12	1	2	-	49894
1	SW	6	1,3,4,5,7, 8,9,10,11, 12	1	2,6	-	48574
2	SW	7	1,3,4,5,8, 9,10,11,12	1	2,6,7	-	48344
3	SW	1	8,9,10,11, 12,re	2	1,2,3,4, 5,7	6	43781
4	SW	9	8,10,11,12,re	2	1,2,3,4,5,9	6,7	43454
5	SW	8	10,11,12,re	2	1,2,3,4,5,9	6,7,8	41124
6	HW	-	10,11,12,re	2	1,2,3,4,5,9	6,7,8	41124

Table 1. Snapshots of algorithm execution

4 Results

We demonstrate the effectiveness of our algorithm by a case study of the JPEG image compression algorithm. The JPEG algorithm was specified as a task graph with 12 nodes (see Figure 2). Our implementation of the JPEG algorithm differs from the standard JPEG since we operate on 4x4 matrix. The HW times and areas of the tasks for a Xilinx XC 4000 series FPGA were obtained by using a high level synthesis tool. The SW times of the nodes were obtained by profiling on a 100 Mhz pentium system. The shared memory access time was taken as 125 ns per data item, the SW memory access time was taken as 10 ns and the HW memory access time was 125 ns. We invoked the algorithm for different area constraints and HW reconfiguration times (see Figure 3). The algorithm took about 0.5 seconds to generate a solution for each of the design points. For the codesign architecture with 5000 ns reconfiguration time, none of the design points utilized the dynamic reconfiguration capability of the HW. This is because the high reconfiguration time over shadows any execution time improvement which might result due to partitioning of tasks in to multiple temporal partitions. For the codesign architectures with 500 ns and 50 ns reconfiguration times all but one design point utilized the dynamic reconfiguration capability of the HW. Since the design points for 5000 ns reconfiguration time architecture do not use the run time reconfiguration capability they are similar to traditional HW-SW systems. In comparison, the design points for 500 ns and 50 ns reconfiguration architecture result in much smaller execution times for the same area constraint. This result makes a strong case for use of dynamically reconfigurable HW for embedded system design. The case study also demonstrates that the algorithm adapts well to changing reconfiguration times for different codesign architectures.

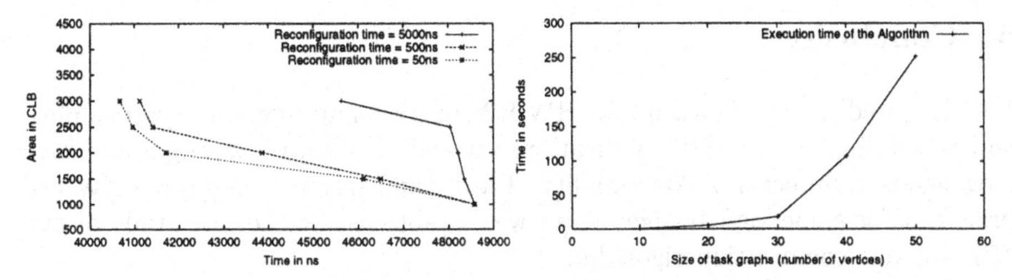

Fig. 3. Design space exploration

Fig. 4. Execution time of the algorithm

In Table 1 we give snapshots of algorithm execution on the JPEG task graph with reconfiguration time of 500 ns and an area constraint of 3000 CLBs. In the table, column 1 gives the iteration number of the partitioner, column 3 gives the task selected to be moved and column 8 gives the total execution time of the task graph. In the table "re" refers to the HW reconfiguration task which is scheduled on the SW processor. Iteration number 0 refers to the initial solution generated by the scheduler. A few important characteristics of the algorithm can be observed from the table. Every task which is selected to be moved to HW is a neighbor of some tasks already mapped to HW. This shows that the algorithm correctly maps neighborhood tasks to same processor elements since this minimizes inter-processor communication. In iteration 3 when task 1 is selected by the partitioner to be moved to HW, the scheduler maps tasks 3,4,5, and 7 to HW. This is a very important characteristic of the algorithm. The scheduler takes a partial mapping provided by the partitioner and maps the remaining tasks such that the total execution time is minimized. Our approach is characterized by the tight integration between the partitioning and scheduling operations. Other approaches [3] [4] [5] which bind all the tasks at the partitioning stage have the limitation that the scheduler does not have the freedom to map any tasks. Since the partitioner cannot correctly predict the overall execution time, they require more iterations to arrive at a good design. In iterations 4 and 5, tasks 8 and 9 are moved to HW respectively. Although they are in the neighborhood of tasks already mapped to HW, they are not mapped to HW by the scheduler. This highlights the limitations of a list scheduling based approach [6] [7]. Our algorithm overcomes the limitations of the list based scheduler by using an iterative partitioner.

We evaluated the run time of the algorithm by conducting a study with synthetic graphs. The graphs had 10, 20, 30, 40 and 50 task nodes respectively. The height, connectivity, task execution times and data items transfered across each dependence were varied to generate 10 graphs for each task number. We then obtained three solutions with different area constraints for each of the fifty task graphs on Sun SPARC 5 machine. The average execution times of the algorithm for each task number are plotted in Figure 4. For a graph with 50 nodes the average run time of the algorithm was 251.85 seconds (4.2 minutes). Since the application is modeled at a coarse level of granularity, 50 nodes are sufficient to specify a reasonably large application. The low run time of the algorithm makes it an attractive tool for designing HW-SW applications on dynamically reconfigurable architectures.

5 Conclusion

We presented a new algorithm for HW-SW partitioning, temporal partitioning and scheduling of HW-SW applications on a dynamically reconfigurable heterogeneous architecture. The viability of a dynamically reconfigurable HW coprocessor for embedded system design was established by the case study of the JPEG image compression algorithm.

References

1. S. Trimberger, D. Carberry, A. Johnson, J. Wong, " A Time-Multiplexed FPGA," *Proceedings of IEEE Symposium on FPGAs for Custom Computing Machines*, Napa Valley, CA, 1997.
2. K. S. Chatha and R. Vemuri, "An Iterative Algorithm for Partitioning and Scheduling of Area Constrained HW-SW systems", *Proceedings of* 10th *IEEE International Workshop on Rapid System Prototyping*, Clearwater, Florida, USA, June, 1999.
3. R. Gupta and G.D. Micheli "Hardware-software cosynthesis for digital systems," *IEEE Design and Test of Computers*, vol 10, no. 3, pp 29-41, 1993.
4. R. Ernst, J. Henkel and T. Benner, "Hardware-software cosynthesis for microcontrollers," *IEEE Design and Test of Computers*, pp 64-75, 1994.
5. K. S. Chatha and R. Vemuri, "Partitioning and Pipelined Scheduling of mixed HW-SW systems", *Proceedings of* 11th *International Symposium on System Synthesis*, Hsinchu, Taiwan, December 1998.
6. A. Kalavade and E.A. Lee, "The Extended Partitioning Problem: Hardware/Software Mapping, Scheduling and Implementation-Bin Selection", *Journal of Design Automation for Embedded Systems*, Kluwer, Vol. 2, No. 2, pp. 125-163, 1997.
7. S. Bakshi and D.D. Gajski, "A Scheduling and Pipelining Algorithm for Hardware/Software Systems," *Proceedings of* 10th *International Symposium on System Synthesis*, Antwerp, Belgium, September 1997.
8. M. Gokhale and A. Marks, "Automatic Synthesis of Parallel Programs Targeted to Dynamically Reconfigurable Logic Arrays," *Proceedings of* 5th *International Workshop on Field-Programmable Logic and Applications*, August/September 1995, Springer-Verlag publishers.
9. M. Vasilko and D. Ait-Boudaoud, "Scheduling for Dynamically Reconfigurable FPGAs," *Proceedings of International Workshop on Logic and Architecture Synthesis*, IFIP TC10 WG10.5, Grenoble, France, December 1995.
10. K. M. GajjalaPurna and D. Bhatia, "Temporal partitioning and scheduling for reconfigurable computing," *Proceedings of IEEE Symposium on FPGAs for Custom Computing Machines*, 1998.
11. M. Kaul and R. Vemuri, "Temporal Partitioning Combined with Design Space Exploration for Latency Minimization of Run-Time Reconfigured Designs," *Proceedings of Design, Automation and Test in Europe Conference*, Munich, Germany, March 1999.
12. R. Maestre, F.J. Kurdahi, N. Bagerzadeh, H. Singh, R. Hermida, M. Fernandez, "Kernel Scheduling in Reconfigurable Computing," *Proceedings of Design, Automation and Test in Europe Conference*, Munich, Germany, March 1999.

Serial Hardware Libraries for Reconfigurable Designs

W. Luk, A. Derbyshire, S. Guo and D. Siganos

Department of Computing, Imperial College
180 Queen's Gate, London SW7 2BZ, UK

Abstract. We describe serial hardware libraries which have been developed as building blocks for designs involving arithmetic, digital signal processing and video operations. These libraries are parametrised in various ways to provide the appropriate amount of serialisation and pipelining. They can be used for producing a wide range of implementations, including digit-serial circuits as well as LPGS and LSGP designs. The layout of such implementations can be controlled using device-specific placement constraints. Tools are being developed to assist production of correct, efficient and well-documented hardware libraries. Our experiments illustrate the conditions under which serial multipliers are more effective than parallel multipliers in Xilinx Virtex technology.

1 Introduction

Data parallel designs are fast but have a high demand on hardware resources. Serial designs involve reusing hardware resources during operation by feedback connections in the datapath. It is possible to achieve different trade-offs in throughput, size of computational blocks and number of interface connections by adjusting the degree of serialisation. Serial architectures are especially effective when the input is already in serial form, as is often the case in digital audio.

Many hardware designers recognise the importance of serial designs, particularly when reconfigurable devices, such as FPGAs, are involved. Since serial designs tend to be register-intensive while consuming little routing resources, they usually result in fast and efficient FPGA implementations. Moreover, reconfiguring the degree of serialisation enables designs to adapt to requirements on performance and resources for different applications, or for variations in operating conditions or in different stages of computations for a specific application.

Serial circuits have many advocates [1], [3], [10], [11], [12]. Previous work on serial designs, however, is often restricted to particular components such as multipliers [1], or to a particular serialisation style such as least-significant-digit first computations [3]. Some classes of serial circuits are not covered by existing development methods; for instance the unfolding method cannot be used to derive serial designs pipelined at the bit level [1], such as the serial adder shown in Figure 1. Moreover, array-based serialisation methods [2], [5] are often not supported.

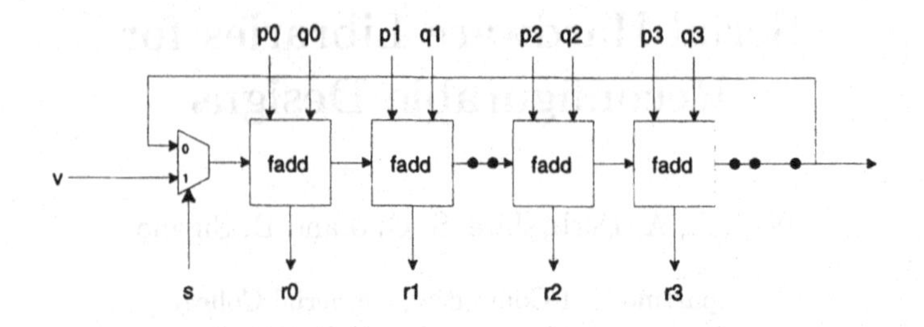

Fig. 1. A 4-bit partially-pipelined serial adder with 2 pipeline stages for computing 8-bit additions. The black discs represent latches. Delay compensation circuits are not included.

The aim of our research is to remove the restrictions mentioned above. We have devised serial hardware libraries, in addition to parallel ones [6], to be used as building blocks or synthesis targets for designs involving arithmetic, digital signal processing and video operations. The novel aspects of our approach include: (a) the library elements are parametrised to produce designs with an appropriate amount of serialisation and pipelining (Section 3); (b) many serialisation design styles are supported, including digit-serial schemes [3] and LSGP and LPGS methods [2] (Section 4); (c) technology-specific constraints, such as placement attributes, can be included (Section 5); (d) tools preserving design hierarchy have been developed to generate serial designs from parallel ones, and to optimise the composition of serial circuits; (e) the serial libraries can be reconfigured at run time to improve flexibility and efficiency. This paper will focus on (a)–(c); (d) and (e) will be covered separately.

2 Serial Designs

Our research aims to provide a framework which supports exploring and evaluating implementations in different design styles, so that the best can be adopted in the shortest time. Parallel designs often have high performance at the expense of size; serial designs provide an opportunity for achieving a range of trade-offs in size and performance.

Although the density of FPGAs continues to rise, applications involving a large volume of data or high numerical accuracy may still require a multi-device solution if parallel circuits are used. Appropriate serial circuits have the potential for a single-device solution while still meeting the performance requirements.

Although a serial design requires more cycles to produce a result than the corresponding non-pipelined parallel design, its smaller size often enables a shorter cycle time than the parallel version. There is a possibility for a serial circuit to operate at a higher clock speed than the parallel circuits that it interfaces to, such that the combination of the serial circuit and the associated serial-parallel data converters behaves like the corresponding parallel circuit but is smaller in size. This arrangement can be supported by recent FPGAs such as Xilinx Virtex devices, which include facilities for handling multiple clocks on-chip.

There are a number of serial design styles, each with its strength and weakness. The most common are probably LSD-first (Least-Significant-Digit first) serial architectures, in which all operands and results are in LSD-first serial form. For an n-bit operand and a k-bit digit, the serial hardware size is usually proportional to k while the number of cycles to complete each operation is proportional to (n/k). In contrast, the corresponding parallel circuit has size proportional to n and each operation takes a constant number of cycles.

Serial-parallel architectures are another popular serial design style [10]. In this case, some operands are in serial form while others are in parallel form. These architectures can be used to match data formats in minimising the need for converting between different data formats.

There are also serial designs adopting on-line arithmetic [11], with computations beginning from the most-significant-digit to achieve low latency in composite designs. These circuits often employ redundant data representation, which eliminates carry-propagation delay but requires more circuitry than the corresponding LSD-first designs. Many arithmetic and trigonometrical operators can be expressed in on-line arithmetic.

The above design styles cover the case when one or more operands are in serial form. For designs involving multiple-word operations, for instance, it is possible to serialise the operations rather than the operands. The serial version can be used to emulate different sections of the parallel circuit at different cycles. For regular circuits such as convolution filters, serialisation can be achieved in various ways, including the LSGP and LPGS methods [2]. Examples of these methods will be included in Section 4.

Given the variety of serial circuits, there are two main concerns. The first is correctness; the second is efficiency. Correctness issues can be dealt with by deriving the serial circuits from the parallel version in a rigorous manner. For instance, many of our circuits can be derived by provably-correct transformations using the Ruby [5] or Pebble [7] formalisms to reduce the chance of errors.

Efficiency concerns can be addressed by considering the overheads involved in serial designs. Given a circuit with an n-bit operand and a k-bit digit, ideally the serial version should be at least (k/n) times the size of the parallel version. In practice, the overheads of serial designs include: (a) feedback connections, (b) serial-parallel data converters, (c) control circuits, (d) multiple clocking, (e) all data having the same worst-case size to simplify control, regardless of local needs [10]. It is therefore important to evaluate which design has the best trade-offs in the given implementation technology. The next section shows how designs can be parametrised, and the effects of varying the parameters.

3 Design Parametrisation and Characterisation

Our libraries are parametrised in various ways to generate designs with different bit width, degree of pipelining, and degree of serialisation. For instance, a 32-bit

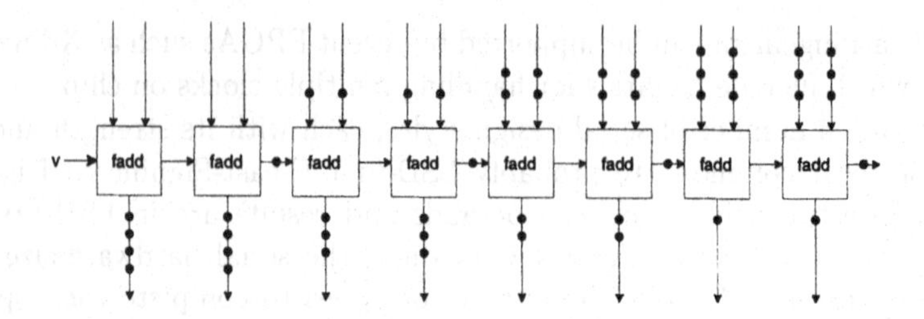

Fig. 2. An n-bit ripple-carry adder with m pipeline stages, where $n = 8$ and m=4. The black discs represent latches.

adder can be implemented with a digit size varying from 1 to 32. A fully parallel design is obtained when the digit size is 32; a bit-serial design is obtained when the digit size is 1. Space efficient designs, using only a single multiplexer (see Figure 1), can be achieved when the digit size is a factor of the word size; in this example these cases correspond to digit size being 1, 2, 4, 8 and 16. For other digit sizes, a scheme involving multiple multiplexers can be adopted [8].

Pipelined designs are supported in our libraries. Our serial designs, such as the one shown in Figure 1, can also be pipelined at bit level. An instance of a fully parallel implementation of n fulladders with m pipeline stages, where $n = 8$ and m=4, is shown in Figure 2. Delay compensation circuits in the form of triangular-shaped arrays of latches are required so that all the bits of an output word will be available at the same clock cycle.

Figure 1 shows the core elements of a 4-bit, 2-stage pipelined serial adder which can emulate the 8-bit, 4-stage pipelined adder in Figure 2. The s control input to the multiplexer selects the external v input and the feedback path in alternate clock cycles. After two clock cycles the least significant 4 bits will emerge at the outputs r_0 to r_3; after a further 5 cycles the most significant 4 bits will emerge. The input and output words will be 'skewed' unless the appropriate delay compensation circuits are included. This design can be automatically produced from the one in Figure 2; a generic version can be included as a serial building block once its size and performance have been characterised.

The libraries also include serial-parallel designs [10]. As an example, the serial-parallel version of an m-bit, n-tap Gaussian filter consists of an m-bit, k-tap Gaussian filter where k is usually a factor of n. Control circuits are used to select k of the n coefficients to compute a number of k-tap Gaussian convolutions sequentially, and the partial results are accumulated in internal or external memories to contribute to the final result.

Our libraries are written in a variant of the VHDL language known as Pebble [7]. Many designs include device-specific layout constraints to maximise hardware utilisation while minimising propagation delay. These constraints can be described using the facility for user-defined attributes in Pebble and VHDL. The support for user-directed layout is particularly useful for pipelined designs

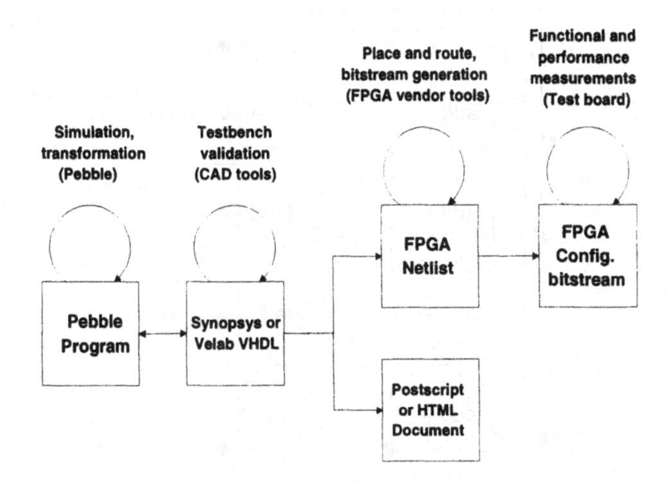

Fig. 3. Design flow for our Pebble-based system. Synopsys and Velab are VHDL tools.

which often involve triangular arrays of latches (Figure 2); they require careful mapping onto an FPGA device to optimise layout efficiency [7].

Figure 3 summaries the main elements in our design flow. Pebble designs are compiled into Synopsys or Velab VHDL, which can then be mapped into hardware by tools from FPGA vendors. The test boards that we use include ones for Xilinx Virtex FPGAs [4] and 6200 devices [9]. We have also developed a Library Documentation Tool (LDT), which automatically produces library documentation in various formats such as Postscript and HTML [6].

Our serial libraries provide a useful method to obtain a wide variety of implementations with different design trade-offs, between the extremes of parallel pipelined designs which are fast but require excessive hardware, and non-pipelined designs which are small but slow. Three methods have been used to characterise the libraries: (a) device-independent algebraic analysis showing, for instance, the variation of latency and critical path delay with digit size; (b) device-specific function and timing analysis using vendor tools; (c) function and timing analysis using FPGA-based platforms.

An example for method (a) is a circuit containing n fulladders with m pipeline stages, where m is a factor of n; an instance is shown in Figure 2. This design has a latency of m cycles, a critical path of n/m fulladder delays, n fulladders and $(n(3m-1)+2m)/2$ latches. Formulae like these enable users or synthesis tools to estimate the optimal amount of serialisation and pipelining to achieve, for instance, the smallest design for a given performance or the fastest design for a given size. Further examples will be described in the next section.

Our results for method (b) and method (c) are usually obtained by a toolkit for analysing and validating designs [9]. For instance, when dealing with Xilinx Virtex devices, the toolkit generates the appropriate hardware testbenches,

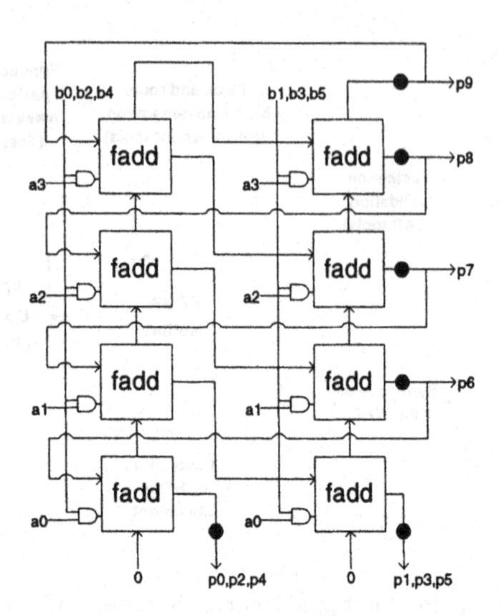

Fig. 4. An m by k serial-parallel multiplier for multiplying $a0 \ldots a(m-1)$ and $b0 \ldots b(n-1)$ to give $p0 \ldots p(m+n-1)$ (m=4, n=6, k=2).

invokes FPGA Express, and runs the Xilinx Foundation tools in order to automatically produce the size and performance statistics (Figure 8).

4 Examples

This section presents several serial designs to illustrate the effects of different parameters on size and performance. Both bit-level and word-level architectures will be covered. Their characterisation will be given in a technology-independent manner; technology-specific issues will be addressed in the next section. Also only the core datapath elements will be considered – interface circuits such as serial-parallel data converters will not be included in the resource calculations or in the diagrams for clarity.

Our first example is a serial-parallel multiplier based on the shift-and-add algorithm for multiplying an m-bit number a and an n-bit number b to give an $(m+n)$-bit number p (Figure 4). This design contains mk fulladders, mk and-gates, and $m+k$ latches. It takes n/k cycles to produce the product p, and its cycle time is around $(m+2k-2)$ fulladder delays when $m \geq k$. In contrast, a parallel version of this multiplier contains mn fulladders and mn and-gates, produces the product in a single cycle with a cycle time of around $(m+2n-2)$ fulladder delays, when $m \geq n$.

Next, we illustrate serial versions of the convolver in Figure 5, which computes an N-stage convolution on the input x to produce the output y. The coefficients are supplied to the unlabelled external inputs to the multipliers.

The first serial design of the parallel convolver can be obtained by applying the LPGS (Locally Parallel Globally Sequential) method [2]. In this method, the

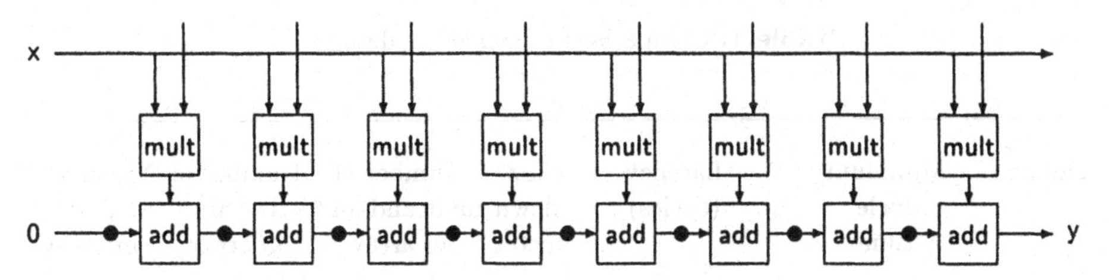

Fig. 5. Parallel convolver (N=8).

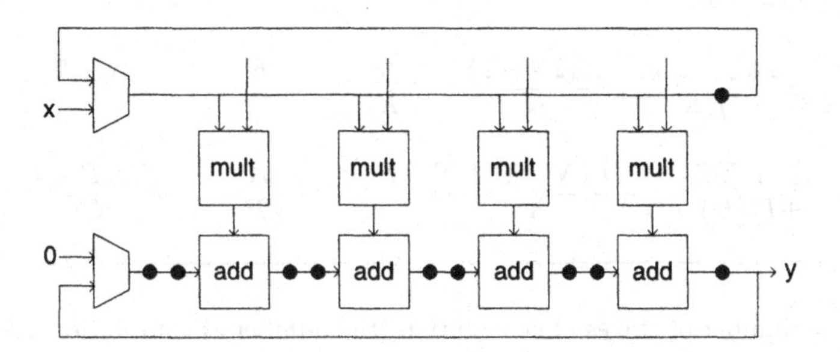

Fig. 6. An LPGS convolver ($N = 8$, $K = 4$).

components of a design are grouped into blocks, and serialisation is applied so that a single block is reused multiple times to produce each result. The LPGS convolver contains only K multipliers and K adders, in contrast to the N multipliers and N adders that the parallel version requires. However it has a pair of multiplexers selecting either the external or the feedback as input to the leftmost multiplier and adder (Figure 6). Other features of this design are given in Table 1. The slow-down factor, N/K, corresponds to the number of cycles between successive inputs, and similarly for outputs. In other words, the environment can supply data and receive results with a cycle time N/K times slower than the cycle time of the convolver array, because the hardware is used N/K times to produce each output. The formal derivation of this design can be found in [5].

The second serial convolver is obtained by applying the LSGP (Locally Sequential Globally Parallel) method. This method again involves grouping components into blocks, but this time serialisation is applied to each of the blocks. The LSGP convolver (Figure 7) contains (N/KP) blocks with feedback connections; each of these blocks contains K multipliers and K adders, and produces a result every P cycles. In contrast to the LPGS convolver, the LSGP convolver has shorter feedback connections but requires more latches and more multiplexers to control the inputs to each block.

Given the combinational delay of a multiplier and an adder, information such as that in Table 1 can be used to estimate performance and resource requirements, so that the optimal trade-offs can be achieved for a given application. Such information can also be useful for serial hardware compilers [12]. The serial-

Table 1. Comparison of convolver designs.

Design	Minimum cycle time	Latency (cycles)	Slow-down factor	Number of mult and add in array	Number of latches in array	Number of mux in array
parallel	$(N-1)T_w + T_m + T_a$	$2(N-1)$	1	N	N	0
LPGS	$(K-1)T_w + T_m + T_a + T_f(K)$	$\frac{N(2N-1)}{K}$	$\frac{N}{K}$	K	$N+2$	2
LSGP	$T_m + KT_a + T_f(K)$	$\frac{KP(N-1)+N}{K}$	P	$\frac{N}{P}$	$\frac{N(KP+2)}{KP}$	$\frac{2N}{KP}$

N: the number of stages of convolution (the number of convolution coefficients),
K, P, Q: serialisation parameters (see text),
T_m, T_a: the combinational delay of the multiplier and the adder,
T_w: the propagation delay of x across the wiring cell above the multiplier,
$T_f(n)$: the delay due to the feedback path (depending on n) and the multiplexer.

parallel multiplier presented earlier can be used in these convolvers, provided that additional synchronisation circuitry is included.

Other designs in our serial library include arithmetic circuits such as integer dividers, signal processing designs such as recursive and non-recursive filters, and non-numerical components such as insertion sorters.

5 Technology-Specific Mapping

Significant improvement in speed and resource usage can often be achieved by technology-specific optimisations. We shall illustrate such optimisations for the serial-parallel multiplier library in Section 4, targeting Xilinx Virtex devices.

Our mapping takes into account two Virtex-specific features: its cell structure and the fast-carry logic. Each Virtex Configurable Logic Block (CLB) contains sufficient resources to accommodate 4 fulladders and 4 and-gates. Hence an m by k multiplier requires $mk/4$ CLBs, where k is the digit size. To take advantage of the fast-carry logic, the multiplier array is oriented such that the carry path is vertical with the least-significant bit at the bottom (Figure 4).

The size and performance characteristics of this multiplier are shown in Figure 8. Note that the number of CLBs is proportional to the digit size. For a given word size, the leftmost point on the graph corresponds to the bit-serial design, while the rightmost point corresponds to the parallel design.

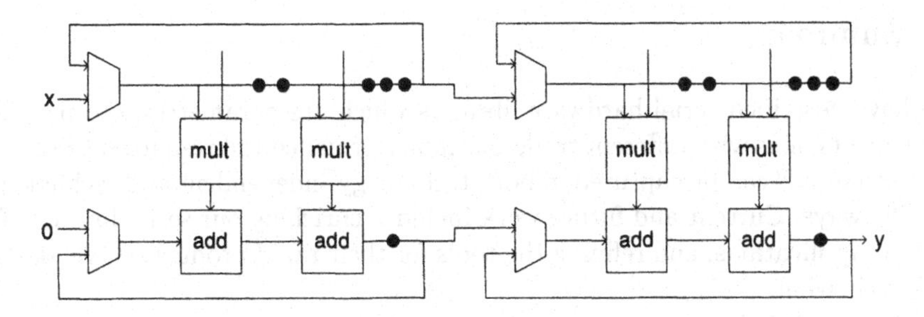

Fig. 7. An LSGP convolver ($N = 8$, $K = P = 2$).

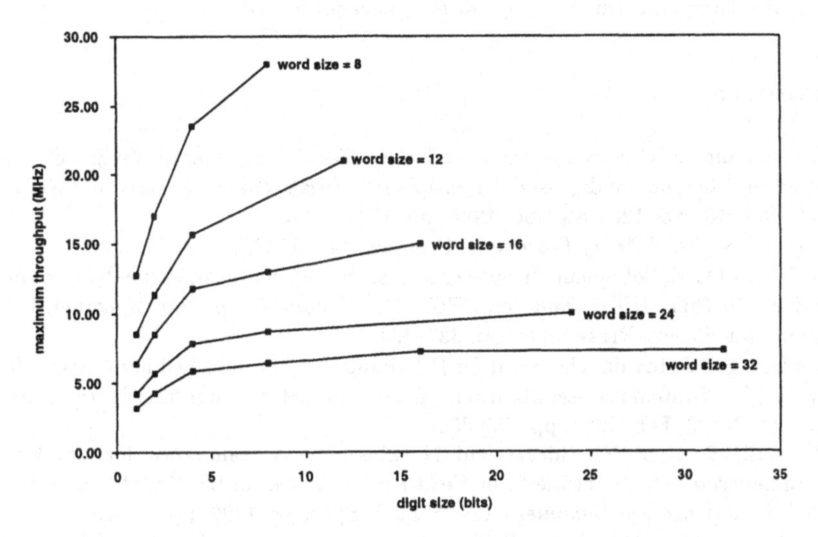

Fig. 8. Size and performance of Xilinx Virtex multipliers.

As expected, the parallel designs deliver the highest performance at the expense of size. For small word size, the serial circuits, while smaller than the parallel version, suffer from an appreciable speed reduction. For instance, when the word size is 8, the design with a digit size of 4, which is around 50% smaller than the parallel version, suffers a 15% speed reduction; however, the area advantage becomes insignificant when the serial-parallel data converters are included.

The performance reduction for serial designs becomes increasingly less pronounced as the word size increases. For instance, when the word size is 32, there is no appreciable reduction in speed with a digit size of 16, although the size is halved. Serial designs with a digit size of 8 and 4, which respectively are 1/4 and 1/8 of the size of the parallel version, suffer a 7% and a 15% speed reduction. Serial-parallel data converters contribute to an additional 10% size overhead and a 10% speed reduction. The results indicate that the serial-parallel multiplier is a good candidate for designs with large word size.

6 Summary

We have described serial hardware libraries which are parametrised to produce implementations with different trade-offs in performance and resource utilisation. Such trade-offs can be captured in both technology-independent and technology-specific ways. Current and future work includes enriching our serial libraries for various applications, and refining the tools for their rapid production, validation and evaluation.

Acknowledgements. Thanks to P. Kane, S. McKeever, S. Ludwig, R. Sandiford and S.P. Seng for help with this paper. The support of the UK Engineering and Physical Sciences Research Council (Grant number GR/24366, GR/54356 and GR/59658), Compaq Research and Xilinx Inc. is gratefully acknowledged.

References

1. Y.N. Chang, J.H. Satyanarayana and K.K. Parhi, "Systematic design of high-speed and low-power digit-serial multipliers", *IEEE Trans. Circuits and Systems - II*, Vol. 45, No. 12, December 1998, pp. 1585–1598.
2. S.Y. Kung, *VLSI Array Processors*, Prentice Hall, 1988.
3. H. Lee and G.E. Sobelman, "Digit-serial DSP library for optimized FPGA configuration", in *Proc. IEEE Symp. on FPGAs for Custom Computing Machines*, IEEE Computer Society Press, 1998, pp. 322–323.
4. S. Ludwig, "Virtex daughtercard for PCI-Pamette", unpublished note, April 1999.
5. W. Luk, "Systematic serialisation of array-based architectures," *Integration*, Vol. 14, No. 3, Feb. 1993, pp. 333-360.
6. W. Luk, S. Guo, N. Shirazi and N. Zhuang, "A framework for developing parametrised FPGA libraries", in *Field-Programmable Logic, Smart Applications, New Paradigms and Compilers*, LNCS 1142, Springer, 1996, pp. 24–33.
7. W. Luk and S. McKeever, "Pebble: a language for parametrised and reconfigurable hardware design", in *Field-Programmable Logic and Applications*, LNCS 1482, Springer, 1998, pp. 9–18.
8. W. Luk, C.S. Ng and F. Mang, "Serialising heterogeneous and non-factorisable processor arrays", in *Designing Correct Circuits*, Springer Electronic Workshop in Computing Series, http://ewic.springer.co.uk/workshops/DCC96/, 1996.
9. W. Luk, D. Siganos and T. Fowler, "Automating qualification of reconfigurable cores", in *Reconfigurable Systems*, IEE Digest, 99/061, 1999.
10. S.G. Smith and P. Denyer, *Serial-Data Computations*, Kluwer, 1988.
11. A.F. Tenca and M.D. Ercegovac, "A variable long-precision arithmetic unit design for reconfigurable coprocessor architectures", in *Proc. IEEE Symp. on FPGAs for Custom Computing Machines*, IEEE Computer Society Press, 1998, pp. 216–225.
12. L.E. Turner and P. Graumann, "Rapid hardware prototyping of digital signal processing systems using field programmable gate arrays", in *Field Programmable Logic and Applications*, LNCS 975, Springer, 1995, pp. 129–138.

Reconfigurable Computing in Remote and Harsh Environments

Gordon Brebner

Division of Informatics
University of Edinburgh
Mayfield Road
Edinburgh EH9 3JZ
Scotland

Neil Bergmann

Cooperative Research Centre for Satellite Systems
Queensland University of Technology
GPO Box 2434
Brisbane 4001
Australia

Abstract. Reconfigurable computing based on field-programmable logic offers exciting new options to the traditional system design space, yet has still to prove itself as a means of delivering performance improvement on the desktop. It seems that apparent speed-ups require much design effort, and can then be challenged by fast microprocessors or eclipsed by current computer system architectures. In this paper, a challenging application domain is investigated — computing on a spacecraft. Here, the environment is somewhat different from the typical desktop, in that real-time constraints and power consumption constraints suggest the use of circuitry rather than programs, and the lack of physical access demands configurability of the circuitry. This application domain is of genuine contemporary interest, and the paper explains how the authors propose to apply some of their earlier reconfigurable computing research, aiming at real system improvements.

1 Introduction

Like many promising high-speed computing technologies, reconfigurable computing is struggling to gain widespread acceptance in the general desktop computing market. While many researchers have reported significant algorithm speedups with the use of FPGA technology in fairly specific applications (e.g. image processing) generally results from reconfigurable computing have been disappointing. We explore some possible reasons for this, and identify a new application domain where reconfigurable computing shows promise: computing in space[1].

As a background to the nature of this application domain, it is useful to make some initial observations about computing systems in general, and the possible roles of reconfigurable computing. Computing system components can be categorised as follows:

- transformational: these components input data, and then output results computed using the data or translated from the data, at speeds determined by the computer and at times chosen by its user;

[1] That is, computing outside the Earth's atmosphere, rather than 'computing in space' using circuits as an alternative to 'computing in time' using programs!

- interactive: these components continually interact with a human user and the environment, at a speed and in a sequence determined by the computer and its user; and
- reactive: these components continually interact with the environment, at a speed and in a sequence determined by the environment.

Performance constraints for each of these types are different, and this greatly affects the usefulness of techniques which speed up algorithm execution, such as reconfigurable computing.

A large percentage of reported reconfigurable computing applications fall into the category of transformational system components. Such components typically have a vague performance constraint, namely that results are desired 'as soon as possible', and so any performance improvement derived from reconfigurability is useful, but has to be tempered by the effort required to achieve that improvement. Experience has shown that a very large development effort is required for fairly modest performance improvement, so the use of reconfigurable computing is not compelling at present.

For interactive components, especially GUI-based interfaces, there appears to be little benefit in using reconfigurable computing technology. The operational time-beat of such components is based on human response times. In terms of computation time, much of it is spent in the uninspiring task of shuffling data to the graphics controller; other compute-intensive sub-tasks are best regarded as transformational sub-components, and the comments above apply to these.

In contrast, for reactive components, the real-time constraints imposed by the environment often involve hard deadlines, and failure to meet these can be catastrophic. In practical terms, if a purely software approach cannot meet timing deadlines, then additional hardware must be added. In such cases, the designer may be willing to expend considerable extra design effort to gain a modest performance improvement in just one small component of a system, if this means that system deadlines are met. Given this compelling need in the traditional world of software/hardware systems, it is worth exploring the domain of reactive components in general, as a fruitful area for application of reconfigurable computing techniques, in which a more flexible view of 'hardware' and 'software' is possible.

In this paper, one particular application area within the reactive domain is considered: computing in space. Consider the characteristics of an unmanned spacecraft in operation. Such a system must operate in a remote and harsh environment where no physical repairs, modifications or upgrades can be made. The only contact with the spacecraft is via a radio link, often unavailable for long periods of time, often of low bandwidth and often with high error rates.

There are some features of this application that immediately suggest the use of configurability and programmability on a fairly coarse time grain, never mind the use of subtle reconfigurable computing. This is because it is often desirable to change the functional characteristics of a spacecraft in flight, perhaps to participate in a new mission, perhaps to take advantage of improved knowledge about some phenomenon, perhaps to overcome the effect of a faulty piece of equipment.

Where functions are necessarily implemented using circuitry, such changes can only be enabled by the use of programmable logic technology, because there is no possiblity of physically swapping chips, as in a terrestrial system. For a spacecraft, all management of the reconfiguration must be handled remotely through appropriate communications protocols.

In addition to investigating the problem of 'straightforward' reconfiguration, but at great distance, there are also opportunities for exploiting dynamic reconfigurability within the spacecraft system components during normal operation, in order to cope with the unusual operating circumstances of the remote and harsh environment that is space. Such methods may prove necessary not only because of the reactive time constraints but also because of constraints on system resources. Overall, this application domain exhibits very different characteristics to most others, and so requires and justifies rather different approaches to system design. In particular, it offers an opportunity to demonstrate the practical relevance of some of the more esoteric previous reconfigurable computing research carried out by the authors and others. Lest the reader may think otherwise, it should be stressed that computing in space is no artificial application area being pushed forward as the latest potential 'killer app' for reconfigurability; this fact is evidenced by the current interest being shown by organisations such as NASA and others, e.g. [9, 14].

The rest of the paper is organised as follows. Section 2 considers the environmental characteristics in more detail, in order to motivate the novel approaches required. Then Section 3 examines reconfigurable computing techniques that are apt for the spacecraft system itself, and Section 4 examines techniques for the required communication between ground control and the spacecraft, or between spacecraft. Section 5 briefly describes the practical details involved in constructing appropriate systems, and a first experiment involving a spacecraft that is under way. Finally, Section 6 contains some conclusions about this work.

2 Environmental characteristics

First, it is worth noting that space electronics must operate in very hostile conditions: for example, high mechanical and acoustic vibration during launch, microgravity, near vacuum, and a wide temperature range. The biggest problem is the high level of ionising radiation that the electronics is subjected to, both because this radiation causes soft errors, such as a few bit-flips in memory storage each day, and also because accumulated crystal damage eventually causes permanent failure. These two characteristics mean that special techniques are required when using reconfigurable logic technology in space. Such techniques are not considered further in this paper, but are discussed in detail in [8], as are the sorts of specialised FPGA chips required. Here, the interest is in how the technology is used.

One key influence on computing in space is that there may be very small power budgets, for example, as little as 20W in total. This makes the use of lower power consumption 8-bit or 16-bit microcontrollers more attractive than higher power consumption 32-bit microprocessors, for computing using sequential programs. Until recently, the typical processing capability on spacecraft was limited to little more than sensors, solid-state data recorders and communication devices to relay the data back to ground control. The use of programmable logic opens up the possibility of adding reactive functionality, achieved through the use of programmed circuitry, that cannot be achieved with a low-grade programmed processor. Examples include the use of smart sensing instruments, with reactive processing resources attached to the sensor, and the use of intelligent filtering on sensor data. Thus, there can be benefits that would not be possible under the more relaxed constraints of the terrestrial environment, where Shand, for example, has shown that it is hard for configurable logic to compete with fast microprocessors [15]. In addition to power constraints, there are constraints on mass and volume, which limits the physical size of the technology used, something which impacts on programmable logic, microcontrollers, microprocessors and memories alike. Advances in silicon technology, and the advent of system-level integrated chips, are of assistance in this matter.

The fact that this circuitry should be programmable, rather than supplied as custom hardware, follows from the remoteness feature of the environment, as discussed in Section 1. It means that both circuit and program errors can be repaired while the spacecraft is in flight, which allows reduced mission costs by minimising risks, especially for missions with very short lead times, but long lifetimes. However, the configurability is not just of use when correcting errors, but also allows ground control to adapt spacecraft functions in response to data observed as a mission progresses. A further justification is to allow the possibility of component re-use, or partial re-use, over a number of missions. Given that only a small number of spacecraft of a particular design (possibly only one) are ever built, there is a saving on bespoke hardware costs, and also an incentive for evolutionary innovation on each mission.

As far as communication between a ground station and a spacecraft is concerned, there could hardly be a more hostile environment. The wireless radio links used are low bandwidth, will have high latencies (seconds, minutes, even hours) for distant environments, may be unavailable for extended periods depending on orbits, and may have high error rates. Therefore, from a reconfigurability point of view, any communication must be modest but include protocols that ensure some guarantee of reliability. Ideally, this communication itself must embody some degree of configurability, to allow adaptation to unexpected circumstances during a space mission. Although the quality of communication may be low, there are still real-time demands that require circuitry rather than programs, e.g. signal modulation and demodulation for soft-radio technology.

Typically, unmanned exploration spacecraft are used to sense some aspects of the physical environment, either locally (as with a magnetometer) or remotely (as with a space telescope). In the future, it is expected that space missions will involve cooperating constellations of spacecraft, combining their observations to generate a coordinated view of particular space phenomena (for example, a near-Earth comet approach). It will not be possible to control such distributed computing systems from a ground station, so it will be necessary for the spacecraft to communicate, and so cooperate, amongst themselves. This poses additional computing and communication problems.

In summary, the remote and harsh environment that spacecraft operate in offers new challenges and opportunities for reconfigurable computing: on one hand, a rationale for the use of soft circuitry; on the other hand, a rationale for the use of lightweight, but flexible and powerful, communication protocols.

3 System techniques

In order to examine the potential uses of reconfigurable computing within components of spacecraft systems, it is useful to follow the three-level view of programmability for data flow models (i.e., circuits), suggested in [6]:

1. different circuits can be executed;
2. executing circuits can be modified;
3. dynamic behaviour through choice in data flows.

In general, there seems a much stronger case for programmability at the second level for circuits than there is for programs, largely because of the relative weakness of the third level for circuits.

3.1 Execution of different circuits

The first level of programmability encompasses most of the more routine uses of configurable logic, for example, prototyping ASICs and infrequent field upgrades. This sort of coarse-grain programmability is encouraged by the majority of FPGA technologies that do not allow partial reconfiguration. In the case of spacecraft, there is a need for such programmability, but it comes at a cost. For normal operation, there may be a range of different functions required over time, and complete circuit swapping is an efficient use of the programmable logic resource, given constraints on its size. For repairing severe faults, or performing upgrades, circuit swapping will be necessary.

The penalty for complete circuit swapping arises from the need to load new configuration data into the programmable logic device. In essence, there are two possible sources of the data: storage within the spacecraft system; and storage at the ground station. In the case of data stored locally, the general interfacing and control for reconfiguration is of the same type as used in normal terrestial systems. Added constraints are that the size of memory devices may be severely limited, and that a microprocessor or microcontroller supervising the task may be less powerful. Also, either the stored data is fixed from the time of the spacecraft's launch, or has been updated from a source at the ground station. These constraints suggest that the use of circuit swapping based on locally-stored data has to be used modestly and carefully.

The situation for circuit swapping based on remotely-stored data is equally challenging, but for different reasons. Clearly, there can be an effectively infinite storage facility and relatively unlimited processing capability at the ground station. However, the problem is that configuration data then has to be sent to the spacecraft via an unhelpful oommunication medium, and some degree of processing will be needed at the spacecraft end to process the received data. Some of the communication issues are addressed in the next section. However, a fairly obvious conclusion is that one does not want to make circuit swapping based on remotely-stored data a frequent operation. In practice, the best solution to enable circuit swapping is to use local memory on the spacecraft system as a cache between the configurable logic device and storage at the ground station. The cache would be relatively stable, in the sense that updates would occur at frequencies measured in hours at the fastest, more likely in days or weeks.

3.2 Modification of executing circuits

Given that there are problems associated with the volume of data required to swap complete circuits, here there is a particular motivation to consider modification, rather than replacement, of circuits. This is in addition to any time-related benefits that may result from such approaches. An essential underpinning for circuit modification is the use of a programmable logic technology that allows partial configuration, ideally while the circuit is still in operation. Two general approaches to circuit modification have been studied by the authors, and others, and both look to offer real benefits in this application domain, although the physical constraints mean that attention to parsimony is crucial.

The first approach is one where there is automatic system support for circuit modification, in a manner intended to be invisible to circuit designers. This encompasses the idea of virtual circuitry (also, and rather more commonly, known as 'virtual hardware'). Here, the general idea is to allow circuits to be larger than the actual area of configurable logic that is available, in a way akin to the use of virtual memory. This is particularly attractive in an environment of limited physical resource, both in terms of circuit area but also in terms of limiting power consumption only to parts of a circuit that are currently active; clearly it is neceassary to amortise the reconfiguration power overhead over time. Many of the issues reduce to the need for some form of inexpensive placement and/or routing of circuitry dynamically at run time, and an approach to dealing with this has been suggested by one of the authors in a series of papers [1, 3, 4]. The central concern is to ensure that the overhead incurred by using a microkernel to manage the virtual circuitry is kept to the minimum. In the author's model, this means ensuring that 'Swappable Logic Units' have very regular sizes, are placed in a very regular arrangement, and are supplied with very regular routing resouces. In time, it is to be hoped that some of the crucial functionality might be incorporated into the programmable logic technology itself.

The second approach is one where there is system support for circuit modification using run-time facilities anticipated by the circuit designer, and designed harmoniously. For example, modification of executing circuits can be used to achieve an effect similar similar to parameter-passing, by 'folding' parameter values into the functionality of the circuitry. This has been studied by various authors, using various terminologies, including 'partial evaluation' for example. It is also possible to change data flows by modifying interconnections between circuit components. The particular interest of the authors revolves around means of enabling dynamic circuit modification in a relatively dignified manner, distinct from writing binary configuration data as patches to circuitry generated by automatic plance and route tools with histories in ASIC design. Essentially, the approach revolves around circuit design and circuit modification at a uniform level of abstraction that is below the typical high level used by designers and above the typical low level used by reconfigurers. The CHASTE system of one of the authors [2] is an example of this approach in practice, as is the JBits (formerly known as JERC) system of Guccione [12]. This approach appears very promising for the sort of minimalist, precision circuit modification needs in the spacecraft application domain.

3.3 Summary

The authors strongly believe that computing in space offers a challenging, but ultimately fruitful, opportunity to demonstrate that their earlier work on soft circuitry, including virtual circuitry, has practical benefits. The third level of programmability listed above — dynamic behaviour through choice in data flows — is not considered further here, because it seems inappropriate apart from in very limited situations. Circuit size is very constrained, so it is not desirable to have 'dead' circuitry that is consuming power but not contributing to the computational data flow.

In [6], a very general view of design flows is presented, of which the three-level programmability is but one feature. Key components are control flow/data flow co-design (i.e., program/circuit co-design), algorithm/architecture co-design and softness/hardness co-design, embedded in a multiple-layer system architecture. This generality is a little grand for the necessarily parsimonious application here, but its central points have relevance. That is, the apt choice of what is best computed by programs and by circuits, the use of supportive architectures, and the commitment to hardness only when absoutely necessary.

4 Communication techniques

The essential problems involved in communication between a ground station and a satellite are taking account of high latencies, irregular availability and low bandwidth as operational constraints, together with unreliable transmission as a problem that has to be solved. The first two constraints rule out any use of communication that assumes either semi-real time or semi-continuous behaviour; the third constraint means that the amount of data transmitted has to be minimised, in particular the amount of configuration data in this case. Application-specific data compression has a significant role to play in ensuring that communication is minimised.

The unreliability problem for conventional communication channels is usually solved by acknowledgement-retransmission protocols. However, the high latencies involved here mean that this is not a good solution because of very high round-trip times for messages. A better solution is the use of error correction codes adding redundancy to the transmitted information — here, there is a tension with compression, since the redundancy uses up precious bandwidth. A feature of both error correction codes and compression is that appropriate computational power is needed at the spacecraft end of communication. Reconfigurable computing may be able to assist here, for example, circuitry for evaluating error correction codes is relatively straightforward.

In terms of application-specific compression, there is some existing work that can be built upon, although this tends to be FPGA-specific. Hauck et al [11] demonstate a method for compressing configuration data for the Xilinx XC6200 that exploits the built-in wildcarding facility allowing parallel configuration of multiple FPGA cells. Luk et al [13] discuss a method for computing differences between array configurations that exploits the partial reconfiguration capability of the XC6200. Both of these methods require non-trivial computation at the transmitter end, but no extra computation at the receiver end, which is ideal for this application.

A longer-term goal is to represent configuration information in a more compact and higher-level form, and there is initial discussion of such an approach in [5], which points to the desirability of standard 'instruction set architecture' equivalents to emerge for configurable logic. The aim is to introduce the concept of 'circlets' — a circuit analogue of program applets. It is intended that circlets will be investigated by the authors for communication in space, the basic trade-off being between compactness and portability of representation on one hand, and the computational needs of circlet interpretation on the other hand.

As well as communication protocols between ground stations and spacecraft, there is also a need for protocols for inter-spacecraft communication in multiple-craft missions. The characteristics of the communication medium are distinctly different in the latter case, and in some ways are more akin to terrestial communications. For implementing either type of communication protocol, to allow flexibility and evolution over time, the authors plan to investigate an 'active protocol' [10] approach, whereby new protocols can be dynamically implemented by assembling standard blocks that are in circuit and/or program form. This will add a 'configurable protocol' element to the overall picture of achieving adaptability at a distance.

5 Practicalities and experiments

The authors' first experiment to put these ideas into practice will involve a small low earth orbit satellite, FedSat [7], to be launched in 2000. This will include a board containing space-hardened SRAM-based FPGA chips together with control circuitry, forming the 'Adaptive Instrument Module' (AIM). The technical details of the AIM are discussed in [7]. Part of the experimentation will concern the investigation of techniques for dealing with radiation-induced errors. As far as reconfigurability is concerned, experiments will include:

- ensuring that basic communication mechanisms work;
- optimising the communication needed for reconfiguration;
- testing spacecraft-initiated dynamic reconfiguration; and
- investigating the practical usefulness of soft circuitry for achieving evolving mission requirements.

The entire reconfigurable computing set-up will be experimental so, although it will be tested as exhaustively as possible in a terrestial setting prior to launch, an essential feature will be crude hard-wired mechanisms for resetting the AIM to a known initialised state with a guaranteed capability for being configured from the ground station.

6 Conclusions

Clearly, the practical investigations associated with this work are some time off. However, this application area is not one in which researchers can do quick and early experiments in the real harsh and remote environment. What this paper has shown is that the unusual features of this environment make reconfigurable computing seem an attractive solution, in ways that are not applicable to a normal desktop setting. In particular, various reconfigurable computing ideas presented by the authors as the fruits of earlier research look like being able to make a genuine practical contribution.

Restrictions on power, mass and volume remove the microprocessor competition that typically threatens reconfigurable logic solutions which are proposed in order to meet real-time constraints. This gives a case for using soft circuitry rather than soft programs or, more likely, for using a combined soft circuitry/program approach that can be supported within the physical constraints. Limitations on circuit size makes the use of virtual circuitry an attractive possibility.

There are also challenges arising from the very limited communication interface between ground control and a spacecraft. This creates a heightened version of the 'system bus' bottleneck that threatens performance gains when reconfigurable logic chips are used alongside microprocessor chips. The solutions devised for appropriate communication will point to improved ways of within-system interfacing, in addition to coping with the difficulties of operating over very long distances in very harsh surroundings. In the future, communication between distant spacecraft themselves will also present challenges.

Acknowledgement

The support of the Australian Commonwealth Government and the Queensland State Government, through the Cooperative Research Centre for Satellite Systems, is gratefully acknowledged.

References

1. Brebner G, "A Virtual Hardware Operating System for the Xilinx XC6200", Proc. 6th International Workshop on Field-Programmable Logic and Applications, Springer LNCS 1142, 1996, pp.327–336.
2. Brebner G, "CHASTE: a Hardware-Software Co-design Testbed for the Xilinx XC6200", Proc. 4th Reconfigurable Architecture Workshop, ITpress Verlag, 1997, pp.16–23.
3. Brebner G, "The Swappable Logic Unit: a Paradigm for Virtual Hardware", Proc. 5th Annual IEEE Symposium on Custom Computing Machines, IEEE Computer Society Press 1997, pp.77–86.
4. Brebner G and A Donlin, "Runtime Reconfigurable Routing", Proc. 5th Reconfigurable Architecture Workshop, Springer LNCS 1388, 1998, pp.25–30.
5. Brebner G, "Circlets: Circuits as Applets", Proc. 6th Annual IEEE Symposium on Custom Computing Machines, IEEE Computer Society Press, 1998, pp.300–301.
6. Brebner G, "Field-Programmable Logic: Catalyst for New Computing Paradigms", Proc. 8th International Workshop on Field Programmable Logic and Applications, Springer LNCS 1482, 1998, pp.49–58.
7. Brown J, S Gardner, A Wicks, L Boland and E Graham, "The FedSat Spacecraft Design", 49th International Astronautical Congress, Melbourne, Australia, October 1998.
8. Dawood A and N Bergmann, "Enabling Technologies for the Use of Reconfigurable Computing in Space", submitted to FCCM'99.
9. Figueiredo K, K Winiecki, T Graessle and U Patel, "Study and Utilisation of Adaptive Computing in Space Applications", NASA GSFC, MAPLD Conference, September 1998.
10. Haeck S and G Brebner, "Active Protocols for Active Networks", submitted to 1st International Working Conference on Active Networks, Berlin, June/July 1999.
11. Hauck S, Z Li and E Schwabe, "Configuration Compression for the Xilinx XC6200 FPGA", Proc. 6th Annual IEEE Symposium on Custom Computing Machines, IEEE Computer Society Press, 1998, pp.138–146.
12. Lechner E and S Guccione, "The Java Environment for Reconfigurable Computing", Proc. 7th International Workshop on Field-Programmable Logic and Applications, Springer LNCS 1304, 1997, pp.284–293.
13. Luk W, N Shirazi and P Cheung, "Compilation Tools for Run-Time Reconfigurable Design", Proc. 6th Annual IEEE Symposium on Custom Computing Machines, IEEE Computer Society Press, 1997, pp.56–65.
14. Robinson S, M Caffrey and M Durham, "Reconfigurable Computer Array: the Bridge between High Speed Sensors and Low Speed Computing", Proc. 8th International Workshop on Field Programmable Logic and Applications, Springer LNCS 1482, 1998, pp.159–168.
15. Shand M, "A Case Study of Algorithm Implementation in Reconfigurable Hardware and Software", Proc. 8th International Workshop on Field Programmable Logic and Applications, Springer LNCS 1482, 1998, pp.333–343.

Communication Synthesis for Reconfigurable Embedded Systems

Michael Eisenring, Marco Platzner, and Lothar Thiele

Computer Engineering and Communication Networks Lab
Swiss Federal Institute of Technology (ETH)
email: {eisenring, platzner, thiele}@tik.ee.ethz.ch

Abstract. In this paper, we present a methodology and a design tool for communication synthesis in reconfigurable embedded systems. Using our design tool, the designer can focus on the functional behavior of the design and on the evaluation of different mappings. All the low-level details of the reconfigurable resource, e.g., a multi-FPGA architecture, are hidden. After the application's tasks have been bound to FPGAs and a multi-configuration schedule has been generated, the communication synthesis step provides the necessary interfaces between tasks. Interface circuitry is automatically inserted to connect communicating tasks, whether they are mapped onto the same or onto different FPGAs. This includes multiplexing several logical communication channels over one physical channel, inserting routing tasks, and providing dedicated interfaces that solve the problem of interconfiguration communication.

1 Introduction

During the last years, reconfigurable computing has gained interest not only as a potentially new paradigm for general-purpose computing but also for embedded systems. Reconfigurable systems are often classified according to their reconfiguration model into *compile-time reconfiguration* (CTR) and *run-time reconfiguration* (RTR) [7]. In CTR, hardware compilation and reconfiguration, i.e., downloading the design onto a reconfigurable device, are done at compile time. In embedded systems, this reconfiguration model is mainly used for rapid prototyping. However, reconfigurable hardware can be a viable alternative to ASICs for the final system implementation when the expected volume is rather low and performance constraints are met. In RTR, the configuration of the reconfigurable device is changed while the application is running. There are two application scenarios for RTR in embedded systems: The first scenario are embedded systems where frequent hardware updates are expected, e.g., network switches that may be reconfigured to support different topologies [1]. Usually these systems operate in a time-critical execution mode and a not time-critical update mode. The second scenario are embedded applications that are split into time-exclusive parts. A reconfigurable resource is used to execute these parts sequentially. By this, cost can be significantly reduced compared to an all-in-ASIC or an all-in-FPGA solution [6] [13] [14].

To develop system design tools for embedded reconfigurable systems, some of the synthesis tasks have to be revisited: scheduling/binding and communication synthesis. For RTR systems, the scheduling/binding step must respect that communicating hardware tasks can be grouped into several configurations that are executed sequentially on the same resource. A reconfiguration schedule has to be found and a controller that initiates and controls reconfiguration has to be synthesized. This problem has already received some attention [2] [11]. The second major issue is communication synthesis. Communication synthesis has been attacked in several projects [8] [12]. However, for embedded reconfigurable systems the automatic generation of communication and interface circuitry leads to new problems and is of utmost importance for the following reasons:

First, the manual interface construction for *multi-FPGA* systems is extremely tedious. The designer is considered with all the low-level details, including the number and positions of available I/O pins, I/O pins reserved for system signals, and I/O pins connecting to other FPGAs and memories. If communicating tasks are mapped onto different FPGAs, the available number of wires connecting the FPGAs may be insufficient to implement all the required communication channels. In this case, some form of multiplexing logical channels over physical wires must be used. Communicating tasks mapped onto FPGAs that are not directly connected require the insertion of routing nodes. Second, a problem unique to RTR systems is interconfiguration communication [7]. If two communicating tasks are executed sequentially, a means of transferring the data from the sender task to the receiver task must be provided. Given a partially reconfigurable FPGA device, such as the Xilinx XC62xx, part of the FPGA area can be reserved to implement a communication buffer. In the case that the resources can only be reconfigured globally, data has to be stored temporarily in an external memory.

In this paper we present CORES/HASIS, a communication synthesis tool set for embedded reconfigurable systems. Using this tool, the designer can focus on the functional behavior of the design and on the evaluation of different mappings. Interface circuitry is automatically inserted to connect communicating tasks, whether they are mapped onto the same or onto different FPGAs. This includes multiplexing, routing, and dedicated interfaces that solve the problem of interconfiguration communication by temporarily buffering data in the system's memories.

2 Synthesis of Reconfigurable Systems

Generally, system synthesis starts from a *specification model* and refines this model iteratively until the *implementation model* is reached. The three tasks that usually comprise system synthesis are *allocation*, *binding*, and *scheduling*.

The specification model consists of three parts, a *problem graph*, an *architecture graph*, and implementation *constraints*. The problem graph represents the functional objects of the application. In our methodology, the problem graph is a directed, acyclic graph with two kinds of nodes: tasks and queues. Tasks are functional objects with an associated executable code/circuit that is stored in an implementation library. Tasks read data from and write data to ports. The ports are characterized by their bit width and use handshake signals to implement an asynchronous communication protocol [4]. On termination, a task raises its *done* flag. Queues are buffers with FIFO semantics that connect two tasks. Directly connected tasks denote unbuffered communication, queues allow for buffering. A queue has a size (width × depth) and an access mode (blocking, non-blocking). Edges in the problem graph denote *data flow*. The architecture graph has three kind of nodes: *FPGAs*, *memories*, and *buses*. The edges of the architecture nodes indicate the connections between FPGAs, memories, and buses. An *allocation*

	FPGA	memory	bus
task	√	—	—
queue	√	√	—
routing node	√	—	—
interface node	√	—	—
data flow edge	√	√	√

Table 1. Possible bindings

is a subset of the architecture graph, i.e., a set of FPGAs, memories and buses that is sufficient to implement the problem graph. A *binding* maps each object of the problem graph (tasks, queues) onto some resource of the architecture graph. The possible bindings are shown in Table 1. A binding induces also a mapping of data flow edges to paths in the architecture graph. A *schedule* assigns a starting time to each task. In pure CTR systems, there exists only one configuration for each FPGA. All the specified tasks and queues are loaded (configured) onto the FPGA before the application starts. Tasks are active from beginning and can be blocked by a read from a channel, if the required data is not yet available. In RTR systems, each FPGA may undergo a sequence of configurations. A new configuration can be loaded, if all the tasks in the current configuration have completed. The individual done flags from the tasks are combined to form the configuration's done flag. All the configurations' done flags are collected by the *reconfiguration controller*. This function, usually located at the host, executes the static reconfiguration schedule.

Communication synthesis is a synthesis step that is applied after allocation, binding, and scheduling. Our communication synthesis tool suite CORES/HASIS expects as input a refined specification graph, i.e., allocation, binding, and scheduling have already been done by either a system synthesis tool or manually by the designer. The CORES/HASIS tool suite splits the refined problem graph into several smaller problem graphs, one for each resource and configuration. Then *interface nodes* and *routing nodes* are inserted. Interfaces nodes establish FPGA-to-FPGA and FPGA-to-memory communication and multiplex several logical channels over a limited number of physically available wires. Routing nodes are required when two communicating objects are mapped onto non-adjacent FPGAs. Interface- and routing nodes are bound to FPGAs, as shown in Table 1.

3 Communication Synthesis

Communication synthesis for reconfigurable systems deals with the connection of tasks and queues bound to arbitrary FPGAs and memories in one or several configurations. Fig. 1 presents a taxonomy of the different communication types for the basic task-queue-task system. In CTR systems, there is only *on/off-chip communication*. On-chip communication is the simplest communication type and occurs between tasks or tasks and queues. Off-chip communication involves two FPGAs or an FPGA and a memory. This communication type can induce routing and multiplexing. In RTR systems, *interconfiguration communication* is required.

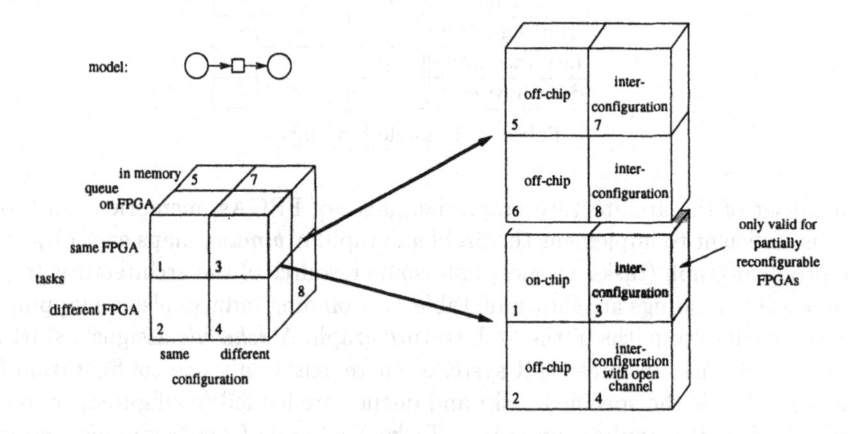

Fig. 1. Taxonomy of communication types for the basic task-queue-task system

3.1 On/off-chip communication

For on- and off-chip communication, we leverage on our object-oriented hardware/software codesign tool for communication synthesis, HASIS [3] [4]. HASIS copes with communication types 1, 2, 5 and 6 in Fig. 1. The key components of HASIS are configurable interface objects that generate dedicated VHDL interface code. HASIS maintains a set of object templates which can be easily extended by templates generated from scratch.

As an example, Fig. 2a) shows the implementation of a task. The core describes the node's functionality as defined in the problem graph. The wrapper around the core consists of FSMs that establish data transmission [5]. A queue implementation has a similar structure and consists of a container that is wrapped by FSMs for communication and access control. By using these interface objects, different protocols for synchronization and data transfer can

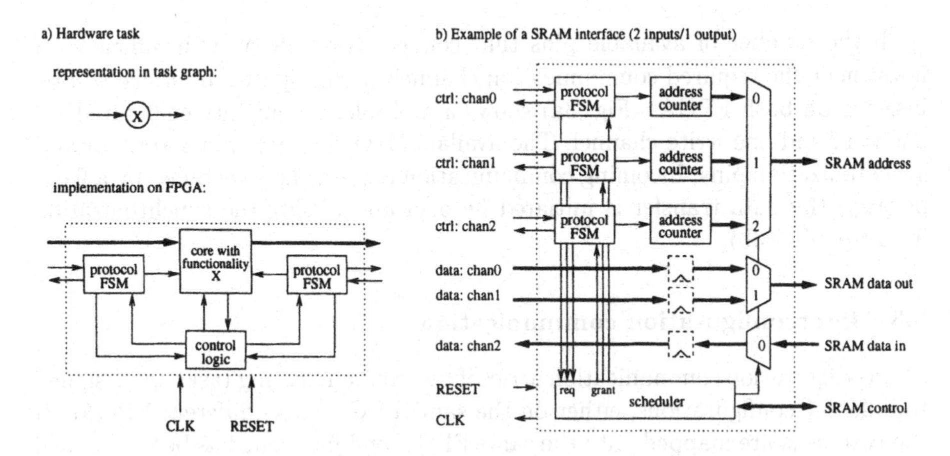

Fig. 2. a) Hardware task b) Example of an SRAM interface

be supported. Operations on queues may be blocking or non-blocking. Off-chip communication will be asynchronous, driven by handshake signals. This allows communication between FPGAs running at different clock speeds. Fig. 2b) shows a configured template of an SRAM interface with two write channels and one read channel. A scheduler manages the read/write requests. The access protocols are handled by FSMs.

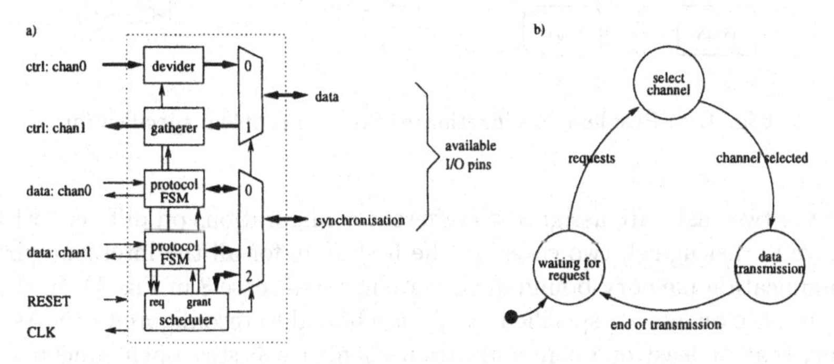

Fig. 3. a) Multiplexer template b) Main states of the scheduler

3.2 Routing and multiplexing

If two communicating tasks are bound to non-adjacent FPGAs, routing tasks with appropriate interfaces are inserted on intermediate FPGAs. A routing task acts like a blocking queue with depth one and is totally transparent to the communicating tasks. In larger systems, there may exist many alternative paths. In the current version of our tool, we expect that the synthesis tool or the designer chooses the path.

If the number of available pins that connect two FPGAs is insufficient to implement the required communication channels, a multiplexer interface will be inserted on both FPGAs. Fig. 3a) shows a multiplexer template configured for one read and one write channel. The available I/O pins are split into data and synchronization pins. Incoming communication requests are scheduled by a fixed priority; the data transfer is initiated by a protocol using the synchronization lines (see Fig. 3b).

3.3 Interconfiguration communication

Interconfiguration communication arises if two communicating tasks are assigned to different configurations, either on the same FPGA or on different FPGAs. If the two tasks are mapped onto the same FPGA and no queue has been specified between them, a partially reconfigurable FPGA is required (case 3 in Fig. 1). In the current version of our tool, this is not supported. Hence, we assume that either the designer or the system synthesis tool has inserted a queue of appropriate depth and mapped this queue to memory (case 7 in Fig. 1).

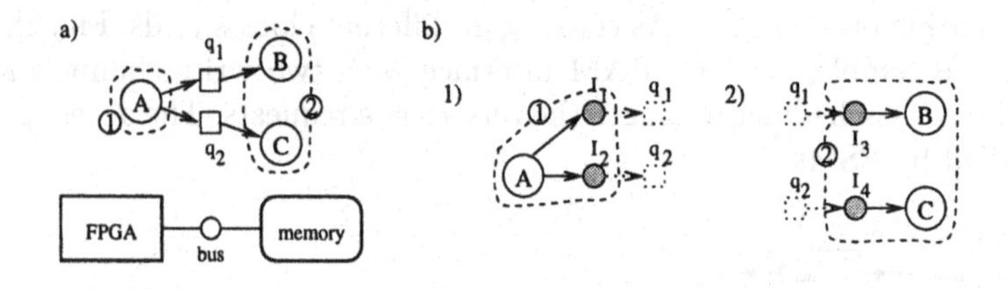

Fig. 4. a) Problem specification b) Splitted problem specification

If the two tasks are assigned to different configurations on different FPGAs, we have to distinguish two cases. In the first case, for all the interconfiguration communication memory-bound queues are inserted (case 8 in Fig. 1). In the second case, the queues - if specified at all - are bound to reconfigured FPGAs. This means, that at least one communication channel will stay open (unconnected) for a while (case 4 in Fig. 1). Using this technique requires FPGA technology that allows the physical implementation of open channels, e.g., tri-state I/O pins with pull-ups.

The example in Fig. 4a) shows a problem specification with three tasks A, B and C bound to one FPGA. Task A forms configuration 1, tasks B and C are assigned to configuration 2, and the queues are mapped to memory. Each edge of the task graph crosses exactly one configuration border. CORES allocates memory for each of the two queues q_1 and q_2 and refines the specification graph by splitting it up into one graph per configuration and by inserting interface nodes. The interface nodes I_1 and I_2 are inserted into configuration one and implement write channels to the memory. In configuration 2, the interface nodes I_3 and I_4 are inserted, which implement read channels. The write and read

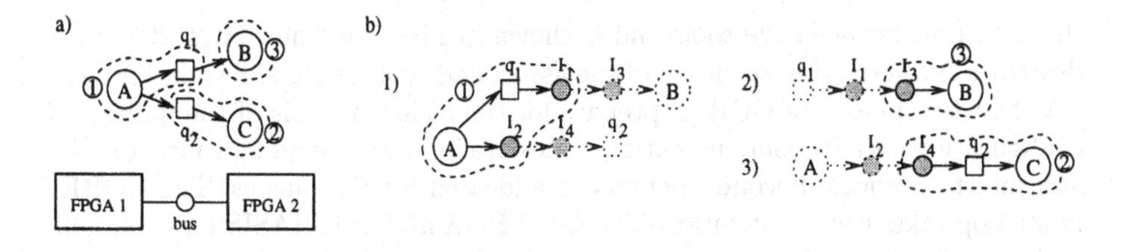

Fig. 5. a) Problem specification b) Splitted problem specification

interfaces are parameterized with the addresses of the queues in the memory and static priorities to resolve simultaneous requests.

The example in Fig. 5a) shows a problem specification with task A assigned to configuration 1 of FPGA 1 and tasks B and C assigned to configurations 2 and 3 of FPGA 2, respectively. The queue q_1 is assigned to configuration 1 of FPGA1, queue q_2 to configuration 2 of FPGA2 (FPGA2's first configuration). The edges q_1-B and A-q_2 cross two configuration borders, which denotes that the underlying communication type will have to use an open channel. Again, CORES allocates memory areas for the queues q_1 and q_2 and splits the problem specification into three graphs containing the necessary interface nodes.

4 Design Tool Flow

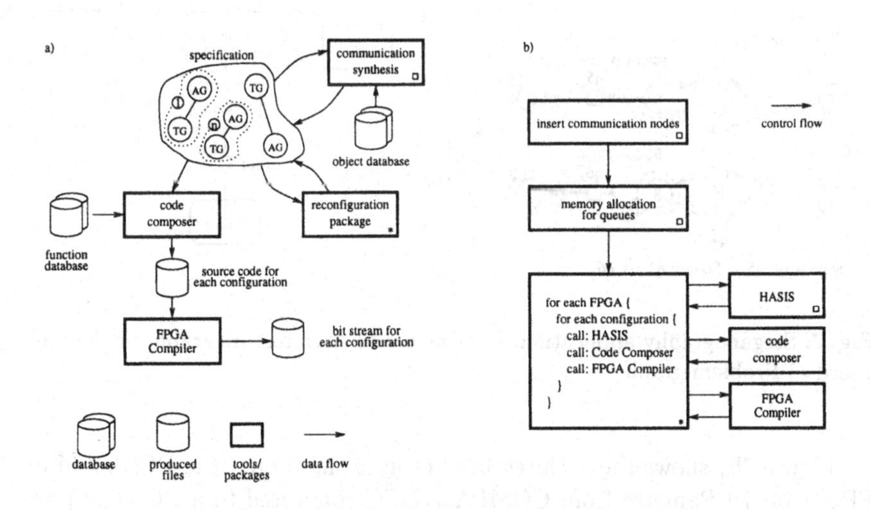

Fig. 6. a) Data flow view b) Control flow view

The input to CORES/HASIS is a problem graph, where the objects have been mapped to different resources and configurations. The design tool flow comprises a *data flow view* and a *control flow view*. The data flow view describes the data flow between the tools and is shown in Fig. 6a). The control flow view describes the activation sequence of the associated tools and is shown in Fig. 6b). The first two phases of CORES prepare for the code synthesis of the individual configurations. First, communication nodes are inserted including routing nodes and interface nodes. Second, memory is allocated for the queues. The CORES main loop takes each configuration of each FPGA and calls HASIS for communication synthesis, the code composer to assemble the top entity from all the tasks functionalities and interfaces, and finally Synopsys FPGA compiler to generate the bit stream.

5 Case Study

As proof of concept, we have implemented a steganography [10] [9] application. A message in form of an ASCII text is hidden within a stream of image data (see Fig. 7a)) by the Least Significant Bit (LSB) insertion method. The resulting data stream is then compressed by run-length coding.

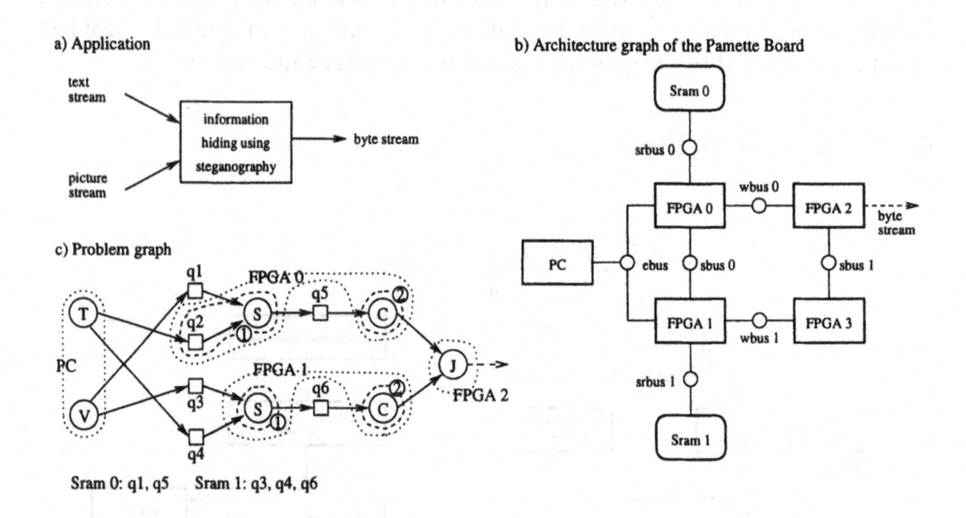

Fig. 7. Steganography application: a) principle b) Architecture graph for the Pamette board c) Problem graph

Figure 7b) shows the architecture graph of the target, the PCI-based multi-FPGA board Pamette from COMPAQ/DEC connected to a PC. The Pamette contains four FPGAs (FPGA0 ... FPGA3) of type Xilinx XC4028, two SRAM banks (SRAM0, SRAM1), and a number of buses. Fig. 7c) presents the problem graph for the steganography example. The tasks S (steganography), C (compress), and the queues q1, q2, and q5 are sufficient to implement the application. To increase the throughput we have doubled this processing line. The input data comes from two tasks running on the PC, V (image stream) and T (text), which write images and text to the queues q1, q2 and q3, q4 respectively. The output streams of the two processing lines are merged by task J (join).

We present results for two different mappings. In the first case, a CTR example, all tasks – except V and T – and queues are bound to FPGA0; FPGA2 performs routing. In the second case, an RTR example, FPGA0 and FPGA1 have two configurations, FPGA2 contains task J, and FPGA3 performs routing. This can be seen in Fig. 7b). The queue q2 is bound to FPGA1, all other queues are mapped to memory. The queues q1 and q3 are of size 16 ×128, queues q2 and q4 of size 8 × 32, and queues q5 and q6 of size 8 × 256. Table 2 summarizes the number of required CLBs for each FPGA for the RTR and CTR example. For each FPGA configuration the number of required CLBs is given, divided into interfaces (I), queues (Q), and tasks (T). Different configurations are separated by a comma, the task names are given in parentheses. R8(R16) denotes a 8(16)bit routing task.

		FPGA 0	FPGA 1	FPGA 2	FPGA 3
RTR	I	180, 29	210,29	0	0
	Q	38,0	0, 0	0	0
	T	37 (R16,S),38 (C)	42 (R8,R16,S),38 (C)	35 (J)	5 (R8)
CTR	I	65	-	0	-
	Q	514	-	0	-
	T	167 (S,S,C,C,J)	-	5 (R8)	-

Table 2. Amount of required CLBs for each FPGA configuration

The purpose of this case study was to proof the concepts of communication synthesis. Hence, the examples were chosen to include a variety of communication types; by no means these mappings are optimal with respect to performance. The results in Table 2 show that for CTR a considerable number of CLBs has been allocated for the queues. This is necessary, because all queues of the problem graph are mapped to FPGA0. In the RTR example, the number of interface CLBs is quite remarkable. This stems from the fact, that for queues mapped to memory, the interface implements the signaling protocol to the SRAM as well as the queue management, i.e., write and read access control. For example, FPGA1 in configuration 1 implements an interface to SRAM1 with 5 channels. Two of these channels write data from the tasks V and P (via routing tasks) into SRAM1, two channels read the data and feed it into task S, and the last channel writes the result of tasks S into SRAM1.

6 Future Work

Future work will include the investigation of system synthesis tasks for RTR, especially binding and scheduling. Communication synthesis plays an important role for system synthesis by providing good and reliable estimates for the final implementation. Based on the combined system and communication synthesis, a more accurate design space exploration for RTR systems will be possible.

Acknowledgment

We would like to thank Corneliu Tobescu for his contributions to this project and the fruitful discussions about interface implementations.

References

1. Saad AlKasabi and Salim Hariri. A Dynamically Reconfigurable Switch for High-speed Networks. In *IEEE 14th Annual International Phoenix Conference on Computers and Communications*, pages 508–514, 1995.
2. Robert P. Dick and Niray K. Jha. CORDS: Hardware-Software Co-Synthesis of Reconfigurable Real-Time Distributed Embedded Systems. In *IEEE/ACM International Conference on Computer-Aided Design (ICCAD)*, pages 62–68, 1998.
3. M. Eisenring and J. Teich. Domain-specific interface generation from dataflow specifications. In *Proceedings of Sixth International Workshop on Hardware/Software Codesign, CODES 98*, pages 43–47, Seattle, Washington, March 15-18 1998.
4. M. Eisenring and J. Teich. Interfacing hardware and software. In *8th International Workshop on Field-Programmable Logic and Applications, FPL'98, Lecture Notes in Computer Science, 1482*, pages 520 – 524, Tallinn, Estonia, August 31 - September 3 1998.
5. M. Eisenring, J. Teich, and L. Thiele. Rapid prototyping of dataflow programs on hardware/software architectures. In *Proc. of HICSS-31, Proc. of the Hawai'i Int. Conf. on Syst. Sci.*, volume VII, pages 187–196, Kona, Hawaii, January 1998.
6. Bernard Gunther, George Milne, and Lakshmi Narasimhan. Assessing Document Relevance with Run-Time Reconfigurable Machines. In *IEEE Symposium on FPGAs for Custom Computing Machines*, pages 10–17, 1996.
7. Brad L. Hutchings and Michael J. Wirthlin. Implementation Approaches for Reconfigurable Logic Applications. In *International Workshop on Field-Programmable Logic and Applications*, pages 419–428, 1995.
8. T. Ismail J. Daveau and A Jerraya. Synthesis of system-level communiction by an allocation-based approach. In *8th Int. Symp. on System Synthesis*, pages 150–155, September 13-15 1995.
9. Neil F. Johnson and Sushil Jajodia. Staganalysis of images created using current steganography software. In *Workshop on Information Hiding, Lecture Notes in Computer Science, 1482*, volume 1525, pages 273–289, Portland, Oregon, USA, 15 - 17 April 1998.
10. Neil F. Johnson and Sushil Jajodia. Steganography: Seeing the unseen. *IEEE Computer*, pages 26–34, February 1998.
11. M. Kaul and R. Vemuri. Optimal temporal partitioning and synthesis for reconfigurable architectures. In *Proceedings of Design, Automation and Test in Europe*, pages 389–396, 1998.
12. A. Kirschbaum and M. Glesner. Rapid prototyping of communication architectures. In *8th IEEE Int. Workshop on Rapid System Prototyping*, pages 136–141, June 24-26 1997, Chapel Hill, North Carolina.
13. Eric Lemoine and David Merceron. Run Time Reconfiguration of FPGA for Scanning Genomic DataBases. In *IEEE Symposium on FPGAs for Custom Computing Machines*, pages 90–98, 1995.
14. Brian Schoner, Chris Jones, and John Villasenor. Issues in Wireless Video Coding using Run-time-reconfigurable FPGAs. In *IEEE Symposium on FPGAs for Custom Computing Machines*, pages 85–89, 1995.

Run-Time Parameterizable Cores

Steven A. Guccione and Delon Levi

Xilinx Inc.
2100 Logic Drive
San Jose, CA 95124 (USA)
Steven.Guccione@xilinx.com
Delon.Levi@xilinx.com

Abstract. As FPGAs have increased in density, the demand for predefined intellectual property has risen. Rather than re-invent commonly used circuitry, libraries of standard parts have become available from a variety of sources. Currently, all of these offerings are based on the standard ASIC design flow and are used to produce fixed designs. This paper discusses *Run-Time Parameterizable* or *RTP* Cores which are an extension of the traditional static core model. Written in the Java (tm) programming language, RTP Cores are created at run-time and may be used to dynamically modify existing circuitry. In addition to providing support for run-time reconfigurable computing, RTP Cores permit run-time parameterization of designs. This adds flexibility and portablilty unavailable in existing design environments.

1 Introduction

In most FPGA designs, engineers make extensive use of preconstructed libraries. Rather than implement an entire design in low-level detail, these library elements or *Cores* are commonly used to build complex designs. These cores supply high level building blocks which can greatly simplify the design task for users. While a variety of core libraries have existed from both commercial vendors and independent sources, the increase in density in FPGAs has led to an increasing demand for cores. The complexity of these cores has also increased correspondingly.

The most basic type of core is a *fixed core*, which has a pre-defined size and cannot be modified by the designer. This type of core has been the basis of most FPGA design tools for the past decade. These cores typically supply standard functions familiar to board level designers, including components such as TTL 7400 series.

As larger cores have been added to libraries, the trend has been away from fixed cores to *parameterizable* cores. Parameterizable cores permit the user to enter information about the desired core, and a customized circuit conforming to the information supplied by the user is constructed.

One example of a parameterizable core would be an adder circuit. When requesting an adder core, a user could be asked to specify the bit width of the adder. This permits any size adder to be generated by a user. With fixed cores, the library would supply standard size adders, typically 4 bit, 8 bit and 16 bit adders. The user would select the pre-constructed circuit which best fit the needs of the circuit being designed.

Parameterization can extend beyond simple bit widths. Adders, for example, can be parameterized to provide various speed versus area tradeoffs. The ability to customize these cores is limited only by the skill of the core designer. With this ability to select from a larger number of cores, circuits can be better tailored to the particular design at hand, with less wasted circuitry and higher performance.

Recently, researchers and commercial vendors have focused attention on these parameterizable core libraries [3] [4] [6]. While these core libraries increase flexibility and provide better solutions to designers, their use is limited to what information can be provided at the time the circuit is designed.

This paper discusses a substantially new type of parameterizable core circuit, the *Run-Time Parameterizable* or *RTP* Core. This type of core is used exclusively in a *Run-Time Reconfigurable* or *RTR* system, where FPGA logic and routing are dynamically modified at run-time. These RTP Cores permit circuits to be constructed and instantiated at run-time, while the system is executing. This permits a new degree of flexibility in design, particularly in RTR systems. Designs may now react to information supplied in real-time, either by system software, user input or by real-time sensor data. This provides a high-level mechanism for performing true RTR using FPGAs. With this capability, systems can be designed which can interact with real-time data and host software.

2 Run Time Parameterization

Unlike compile-time parameterization, designs which are run-time parameterizable have an added degree of flexibility. The ability to create or modify circuits at run-time creates new design options that are unavailable in designs parameterized at compile-time. In general, RTP Cores provide a simple mechanism for producing designs which can vary in functionality based on various types of input. While this list is not exhaustive, some types of input and their use to parameterize circuits is discussed below:

Command line: Using *Command line* parameters, a command-line flag, or other forms of user input can be used to select a particular configuration. This is useful in applications where very similar circuits providing different functionality are supplied in a single RTR design. An example is an encode / decode circuit. A command line flag may be passed to the RTR software to permit either an encoder or a decoder circuit to be configured, depending on the desired mode of use.

Device type: In addition to reacting to command line parameters, it is also possible to parameterize circuits at run-time based on the actual *Device type* being used. For instance, if a small device is being used, a smaller, perhaps less accurate circut may be configured. If a larger device (or portion of a device) is available, a larger, perhaps more accurate circuit may be configured. For instance, in a DSP application, an 8-bit filter may be configured in a small device, where a 16 or even 32 bit instantiation may be used in a larger device. This not only allows one single compiled design to be used on many different sized devices across a family, but allows functionality such as accuracy or speed to change appropriately, depending on the hardware being used.

User input: *Command line* selection typically permits a simple configuration choice to be made at the beginning of execution. It is also possible with RTR to provide a user interface which permits circuit modification at the user's command. An example of *User input* configuration would be an image processing application which uses an FPGA to do coprocessing. A GUI which takes user inputs for processing parameters such as gain, offset, and coloring could be used to directly generate circuit parameters. These parameters may then be used to construct the RTP Cores used in the image processing circuit.

Real-time input: Where *User input* configuration requires a human to manually control RTR, it is also possible to query real-time data, typically from sensors, to drive RTR without human intervention. This permits capabilities such as adaptive digital filtering, where filtering is modified depending on various real-time system conditions.

Circuit state: Where *User and Real-time input* configuration provide for external control of RTR, it is also possible to query data within the currently configured circuit, typically by reading registers from software. The system could then use these values to perform RTR. While similar to using real-time inputs, the ability to probe the internal state of the device permits the possibility of simpler system integration. One example would be an internally configured counter circuit which keeps track of an external event count. Probing the state of such a counter permits software to react to internal values, and to use these internal values to perform circuit parameterization in real-time.

These represent broad classes of input data which can be used to drive RTR using RTP Cores. Using such data to construct circuit configuration parameters, RTP Cores can be instantiated and used to configure FPGAs at run-time. In addition to producing smaller, faster circuits, this approach permits the construction of circuits which would not be feasible using compile-time parameterization of circuits. Reacting to command line parameters, device type, user input, real-time input and circuit state all provide capabilities for circuit designers far beyond what is available using static circuit design tools such as schematic capture and hardware description languages.

3 The *JBits* System

Because RTP Cores are parameterizable at run-time, it is not practical to implement them using standard design tools. These tools, including schematic capture and hardware description languages, were originally designed to produce fixed, static circuits. Using these existing tools to support RTR and RTP Cores is currently not possible.

For this reason, RTP Cores are implemented using the Xilinx *JBits* interface [5]. This design environment is implemented completely in the Java (tm) programming language and currently provides RTR support for the Xilinx *XC4000EX* (tm) and *XC4000XL* (tm) series of FPGA devices. *JBits* provides an *Application Program Interface (API)* into the device configuration bitstream, permitting logic and routing to be modified at run-time.

It should be noted that *JBits* is based on earlier work on the Xilinx *XC6200* (tm) reconfigurable device. This work was know as the *Java Environemnt for Reconfigurable Computing* or *JERC6K* [7]. While a very small number of RTP Cores were supplied with *JERC6K*, the emphasis was on other aspects of RTR. Following the experience gained in development and use of *JERC6K*, *JBits* was implemented and focus quickly moved from low-level to high-level design details. This has led to more of an emphasis on cores and other support for high-level design activities.

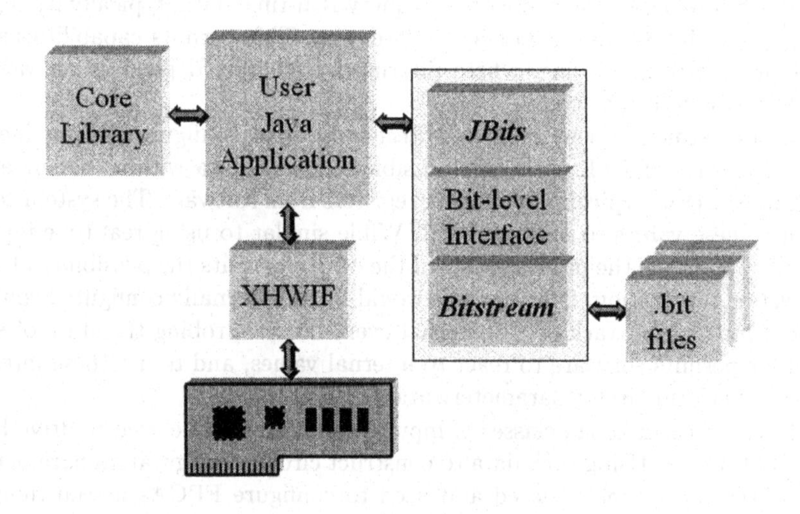

Fig. 1. The *JBits* system.

The *JBits* system views all devices in a given FPGA family as tiled arrays of *Configurable Logic Blocks* or *CLBs*. Associated with each CLB tile is some amount of routing. Partitioning the device in this way permits small sections of the device to be programmed, and the programming to be replicated in loops or implemented conditionally. This programming language support for repitition and conditionality is what provides the basic and necessary support for all paramtereizable cores, but particularly for RTP Cores.

Using this device model and programming interface, RTP Cores are defined as Java objects which can be constructed and then written as device configuration data. Because the device model is a simple two dimensional array of CLBs, the only difference between various devices in a family is the size of the array. This permits cores to be relocatable and device-independent. A *JBits* core used on the smallest XC4000EX family part will also work, unmodified, on the largest

XC4000EX family part. Of course, the size of a core may limit the devices with which it can be used. A core whose smallest instantiation is 20 x 20 CLBs clearly will never fit in a device which contains only 10 x 10 CLBs.

Finally, note that the details of the *JBits* interface are not discussed in detail in this paper. Suffice it to say that all device resources may be configured via *JBits*. This capability is used to build Java objects which construct higher level circuits which remove the need for the user to understand the low-level details of *JBits* and the underlying device architecture. For more a more detailed description of *JBits*, see [5]. Note that this paper refers to the *JBits* software as *"XBI"*. This was the internal development name of *JBits*.

4 Stitcher Cores

While RTP Cores implemented using *JBits* permit fast, run-time parameterization of circuits, one crucial piece is still missing. It is easy to construct and instantiate RTP Cores, but the issues of interconnecting these cores has until now been avoided, even in previous RTR efforts. Clearly, one possibility is using a general purpose router such as the one used in traditional placement and routing tools. Such a router would have to be more intelligent than current routers to keep track of the dynamically changing circuit within the device. In addition, all RTR code, including cores, would have to keep the router informed in some way of the resources being used. This system would be analogous to dynamic memory allocation in software. Because the usage of memory is varying at run-time, some mechanism, be it either an explicit relinquishing of resource, or some form of run-time garbage collection, must be performed to keep track of used and free resources.

While it is possible to implement a general-purpose router which works at run-time, there are two immediately obvious problems with this approach. First, indeterminate circuit generation times could be a problem for many systems, particularly those with real time constraints. And since it is assumed that the FPGA is being used as a coprocessor of some sort to off-load work from the main processor, it may be difficult to justify re-loading the processor with the complex task of performing routing. It is likely that any processor capable of performing real-time FPGA routing would have little need for a high performance FPGA coprocessor.

Finally, while router performance is an issue, completion is an even more severe problem. What should a system do if the router fails to find a solution within the given constraints? It is not likely that many systems would be able to function with even the possibility of such a failure.

One possible solution is to define a special type of RTP core called a *Stitcher*. This core is the same as any other RTP core, except that it has no logic and only modifies routing resources.

The operation of a Stitcher Core is very simple. The Stitcher Core abstracts away the underlying routing architecture in the same way that standard RTP Cores abstract away the underlying logic architecture. Inputs of an RTP core

```
/* Set up the JBits interface */
jbits = new JBits(deviceType);

/* Instanitate a 16-bit counter core at CLB(10,10) */
Counter counter = new Counter(16, 5);
counter.set(jbits, 10, 10);

/* Instanitate a 16-bit +7 constant adder core at CLB(10,11) */
ConstAdder constAdder = new ConstAdder(16, 7);
constAdder.set(jbits, 10, 11);

/* Stitch the counter and adder together */
Stitcher stitcher = new Stitcher(Stitcher.F1_IN, Stitcher.YQ_OUT, 16);
stitcher.set(jbits, 10, 11);
```

Fig. 2. A *JBits* RTP Core and Stitcher example.

are connected in some structured fashion to the outputs of another RTP core. The Stitcher must be aware of the geometry of the core inputs and outputs in question, but this has not been a problem. All cores in the library have so far fallen into a small number of input and output styles, the largest variation being the *stride* relative to the CLB array. In any case, if other more unusual RTP Cores are built in the future, it should be no problem to write new Stitcher Cores which work with these cores.

Using an RTP Stitcher Core is very simple. First, the two cores to be stitched are instantiated using their Java constructors. Then they are written to particular locations in the device using the *JBits set()* function. Once the RTP Cores are in place, the Stitcher may be used to connect them. The Stitcher is also instantiated, then written at the juncture of the two cores, again using the *JBits set()* function. This connects the cores.

Figure 2 shows actual *JBits* code to which contains a 16-bit counter (which is initialized to a value of "5") and a 16-bit constant adder which adds "7" to its input. The Stitcher core in this code connects the *YQ* CLB outputs of the counter to the *F1* LUT inputs of the adder. Note that since the stitcher has no width, it is *set()* to the same location as the constant adder. Also note that the Stitcher Core is treated just like any other RTP core.

While Stitcher Cores typically contain only routing resources, they are often represented as one-dimensional cores with no width. This indicates the absence of logic resources, and permits automatic placement and relative placement software to function in the presence of Stitchers. Figure 3 shows a diagram of such a Stitcher Core. However, it should be mentioned that it is not necessary that cores be abutted and Stitchers have zero width. Because *JBits* is a software solution, it is possible to define stitchers which operate in any mode and connect arbitrary cores at arbitrary locations. In this sense, Stitchers may be viewed as small, special purpose *auto-routers*. This permits them to execute very quickly,

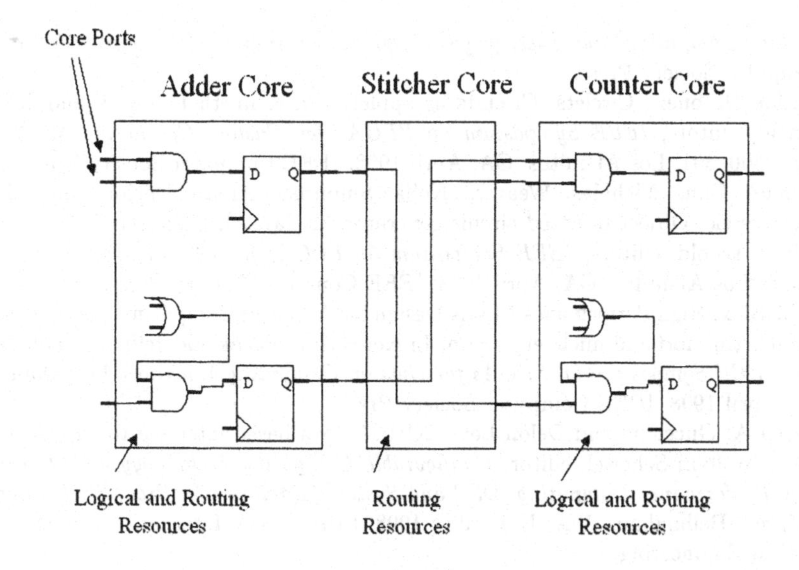

Fig. 3. A Stitcher Core.

with approximately one interconnect performed per line of Java code. Perhaps more importantly, RTP Stitcher Cores guarantee completion of routing.

5 Conclusions

Run-Time Parameterizable Cores or *RTP Cores* are high-level circuit components which can be configured at run-time. Unlike other existing systems which supply fixed circuit parameters at compile time, new levels of flexibility are possible using RTP Cores. Circuits which adapt based on user input, real-time input, the device being used or even internal FPGA state are possible with RTP Cores.

RTP Cores are currently implemented using the in the Xilinx *JBits*. This system currently supports run-time reconfiguration on the Xilinx XC4000EX and XC4000XL families of devices. Using this RTP core based design approach, fairly complex circuits using *Run-Time Reconfiguration* or *RTR* can be built.

In addition, the introduction of *Stitcher* cores to perform special-purpose routing tasks supplies the final piece of the design environment. While *JBits* and *RTP* cores have been in use by our group for a little over a year, the results have been very positive. A small handful of demonstration applications have been constructed and run on actual hardware. In most cases, the applications have been run on different hardware platforms with different FPGA devices, without re-compilation.

References

1. Gordon Brebner. The swappable logic unit: A paradigm for virtual hardware. In Kenneth L. Pocek and Jeffrey Arnold, editors, *IEEE Symposium on FPGAs for*

Custom Computing Machines, pages 77–86, Los Alamitos, CA, April 1997. IEEE Computer Society Press.
2. Gordon Brebner. Circlets: Circuits as applets. In Kenneth L. Pocek and Jeffrey Arnold, editors, *IEEE Symposium on FPGAs for Custom Computing Machines*, pages 300–301, Los Alamitos, CA, April 1998. IEEE Computer Society Press.
3. Michael Chu, Nicholas Weaver, Kolja Sulimma, Andre DeHon, and John Wawrzynek. Object oriented circuit generators in Java. In Kenneth L. Pocek and Jeffrey Arnold, editors, *IEEE Symposium on FPGAs for Custom Computing Machines*, Los Alamitos, CA, April 1998. IEEE Computer Society Press.
4. PAM-Blox: High Performancs FPGA Design for Adaptive Computing. Oskar mencer and martin morf and michael j. flynn. In Kenneth L. Pocek and Jeffrey Arnold, editors, *IEEE Symposium on FPGAs for Custom Computing Machines*, Los Alamitos, CA, April 1998. IEEE Computer Society Press.
5. Steven A. Guccione and Delon Levi. XBI: A Java-based interface to FPGA hardware. In John Schewel, editor, *Configurable Computing Technology and its use in High Performance Computing, DSP and Systems Engineering, Proc. SPIE Photonics East*, Bellingham, WA, November 1998. SPIE – The International Society for Optical Engineering.
6. James Hwang, Cameron Patterson, S. Mohan, Eric Dellinger, Sujoy Mitra, and Ralph Wittig. Generating layouts for self-implementing modules. In John Schewel, editor, *Configurable Computing Technology and its use in High Performance Computing, DSP and Systems Engineering, Proc. SPIE Photonics East*, Bellingham, WA, November 1998. SPIE – The International Society for Optical Engineering.
7. Eric Lechner and Steven A. Guccione. The Java environment for reconfigurable computing. In Wayne Luk and Peter Y. K. Cheung, editors, *Proceedings of the 7th International Workshop on Field-Programmable Logic and Applications, FPL 1997. Lecture Notes in Computer Science 1304*, pages 284–293. Springer-Verlag, Berlin, September 1997.

Rendering PostScript™ Fonts on FPGAs

Donald MacVicar[1], John W Patterson[1], Satnam Singh[2]

[1]Dept. Computing Science, University of Glasgow, U.K.
{donald, jwp}@dcs.gla.ac.uk
[2]Xilinx Inc., San Jose, California, U.S.A.
Satnam.Singh@xilinx.com

Abstract. This paper describes how custom computing machines can be used to implement a simple outline font processor. An FPGA based co-processor is used to accelerate the compute intensive portions of font rendering. The font processor builds on several PostScript components previously presented by the authors to produce a system that can rapidly render fonts. A prototype implementation is described followed by an explanation of how this could be extended to build a complete system.

1 Introduction

The way in which electronic documents are utilised has changed as computers have become more powerful. Standard page description languages, such as PostScript or PDF, along with powerful desktop publishing applications have increased the popularity of electronic documents. These documents often contain many different type faces at different sizes.

Fig. 1. Example Character.

Page description languages and computer graphics systems allow characters to be processed in the same manner as any other graphical object. Characters with similar stylistic properties are grouped together to form fonts. Each character within a font is defined in terms of its outline by a number of arbitrary shaped polygons. These outlines can then be scaled to the required size and rasterised when the document is rendered to either a computer screen or a printer. We concentrate on the processing of fonts for printed output by rendering fonts used within PostScript documents.

Some printers do have a small set of fonts built into ROM as pre-rendered bitmaps but other fonts require rendering from outline descriptions. The rendering of fonts adds more computation to the print time for a document. Desktop printers often contain a single processor which is used both to render fonts and interpret PostScript documents. This processor is often an off the shelf RISC processor which may not be able to supply data to the print engine at as high a rate to keep it fully utilised. Commercial high resolution printers often require a separate system to perform the PostScript interpretation. These external systems tend to be very powerful multi-processor workstations with large amounts of both memory and disk space. Even these powerful systems can require hours to render a document. Font processing becomes especially important when handling Eastern or Arabic languages which contain complex characters such as the one shown in Fig. 1.

Modern printers often have resolutions of around 5000dpi (dots per inch) and impose a severe computational load on the PCs and workstations that prepare images for them. Bitmap fonts at this resolution are large and can be much more efficiently rendered on the fly if the implementation is quick enough. We use FPGA technology to reduce the time spent rendering fonts. Our system does not have enough on-board memory to represent an entire A4 or US letter page for high quality printing. However font rendering makes localised and predictable memory accesses. We exploit this behaviour to transfer and format image memory in a way which reduces over-bus memory transfers.

Careful use of the memory hierarchy continues to the on-chip level where some of the 4K BlockRAMs on the Xilinx Virtex FPGAs are used to cache image data. Stack space needed for a recursive curve drawing algorithm is also implemented on chip.

2 Font Rendering

There are many different formats which can be used to define fonts we concentrate on the simplest which is PostScript Type3. Type3 PostScript fonts are described using the standard set of PostScript path operators with some additional font operators for precise layout control. Other formats are based on similar operators but employ more elaborate encoding schemes and modified operators which help to minimize the data. The FPGA is utilised for the most compute intensive part of the font rendering process which is the actual drawing of the characters.

There is a small basic set of operators required to allow the definition of any character. Only operators which allow the definition of moving to a point, straight lines, curved segments, ending a sub-path and filling the outline are required.

- `x1 y1 moveto`: Sets the current position to `(x1,y1)` used at the start of each subpath.
- `x1 y1 lineto`: Adds a straight line from the current point to `(x1,y1)` to the current subpath.
- `x1 y1 x2 y2 x3 y3 curveto`: Adds a curved segment to the current subpath

from the current point to `(x3,y3)` using the points `(x1,y1)` and `(x2,y2)` as control points for a cubic Bézier curve.

- `closepath` : Ends a subpath and joins the current point to the first point of the subpath if necessary.
- `fill` : Ends a characters description and fills the outline with the current colour.

2.1 A Simple Example

The character shown in Fig. 2 is defined using two subpaths: one defines the exterior outline and the other defines the interior outline. The interior outline uses the `closepath` operator to draw the straight line from the end of the last curve to the start of the first which reduces the data. A `lineto` operator could have been included to perform this. The internal and external outlines are also deliberately defined so that one winds clockwise and the other anti-clockwise. This allows the winding number rules to calculate which areas are interior and which are exterior achieving the same result no matter which rule is used.

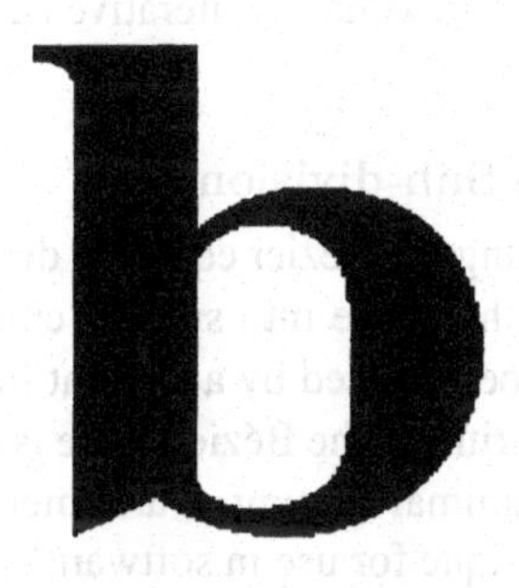

Fig. 2. Simple character extracted from a standard font

3 Processing Elements

The five operators only require implementations for line rendering, curve rendering and filling. Straight lines within a character are defined by the start and end points of the line. During rendering every point that lies on the straight line between the end points has to be calculated and coloured.

Cubic Bézier curves are described by a parametric (cubic) equation. These curves are used extensively in Postscript and other graphics systems, but are computationally expensive to calculate making them suitable for hardware rendering. As 3D imaging moves to curve rendering in preference to polygon rendering, the need for fast spline rendering will increase [3].

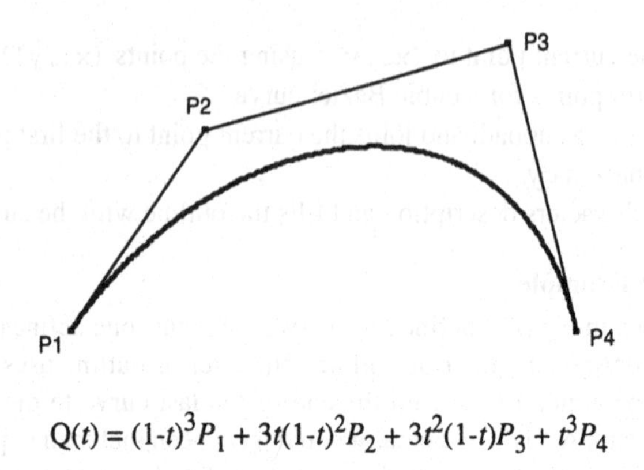

$$Q(t) = (1-t)^3 P_1 + 3t(1-t)^2 P_2 + 3t^2(1-t) P_3 + t^3 P_4$$

Fig. 3. Bézier Curve.

Whether rasterising Bézier curves for screen or printed output the general technique is to use a piecewise linear approximation of the true curve. There are then two main algorithms for obtaining this approximation, iterative calculation of points on the curve and recursive sub-division.

3.1 Bézier Recursive Sub-division

Rather than iteratively evaluating the Bézier equation directly to obtain the points on the curve it is possible to divide the curve into smaller curves until these curves become approximately flat enough to be replaced by a straight line. This is the principle behind the recursive sub-division algorithm. The Bézier curve is divided in half, division by two is used since it involves the minimal amount of arithmetic.

This is the preferred technique for use in software systems as it reduces the amount of computation required while still producing accurate results. The results from this method can be an improvement over the iterative calculation since the frequency of points around a sharp bend in the curve is increased.

3.2 Outline Fill

When performing a fill on arbitrary shaped polygons one of two rules is used to determine which areas are interior and which are exterior. The two rules are even-odd and non-zero winding number, which can produce different results given the same input data. The guide lines for creating fonts say that they should be defined in such a way that the correct result is achieved using either rule. When implementing the non-zero winding number rule the direction of each edge is important and has to be kept and is thus not suited to FPGA implementation because of the large data this would require.

We implement the even-odd rule which allows the outline to be rendered first. The fill is performed by working across each scanline from left to right. Whenever an edge is detected the fill value is inverted. In the bitmap black is represented by 1 and white by

0. As the current position on the scanline moves across from left to right (0 to F) the fill value changes at position 1 from white to black and then back again at position E.

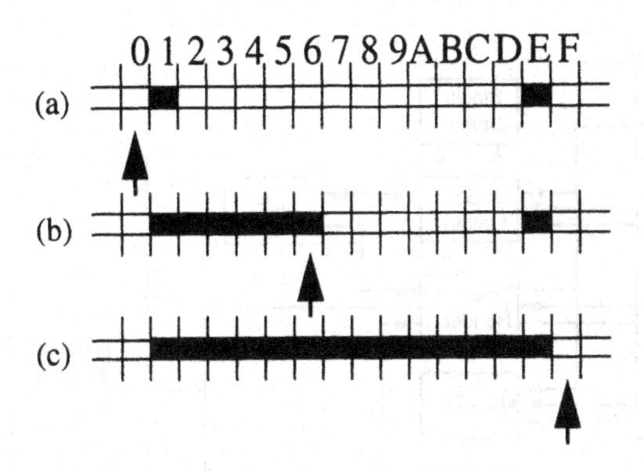

Fig. 4. Scanline Fill. (a) shows a scanline prior to filling, (b) shows the scanline partly filled and (c) shows the completed scanline

4 Target Architecture

The implementations are targeted to the Xilinx Virtex™ [11] series of FPGAs. We assume that the FPGA is situated on a PCI card which provides close coupling with the main system processor. We also assume that an adequate amount of fast SRAM is located on the same card as the FPGA. The host CPU passes the outline definition to the FPGA and then gets the result back when rendering is complete. There are several cards available which meet our requirements including the PCI Pamette with Virtex daughter card from Compaq Research Labs and the Wildstar™ range from Annapolis Micro Systems Inc. The following circuits are all implemented in VHDL and synthesised using either Synopsis Design Compiler or Synplicity for the Virtex XCV300 FPGA at speed grade 4.

5 Implementation

The input data describing the outline of a character is written into a number of blocks of the 4K BlockRAMs. Although the data is actually a number of 8-bit words and are read from the memory as 8-bit words they can be written as 16-bit words due to the dual ports available on the BlockRAMs. The wider data path reduces the time required to write the outline data to the FPGA. Each of the five operations is assigned an op-code which is 8-bits allowing extensions at a later date such as the more compact definitions used in Type 1 fonts. Due to the small number of operations the values 1, 2, 4, 8, and 16 are used as the op-codes since this makes the decoding simpler and thus faster.

The character in Fig. 2 requires 80 bytes of memory space. The more complex character in Fig. 1 contains 54 straight line segments, 15 curve segments and nine subpaths requiring 276 bytes. These memory requirements are using 8-bit coordinate values however a complete system would require larger coordinates of perhaps as much as 16-bits.

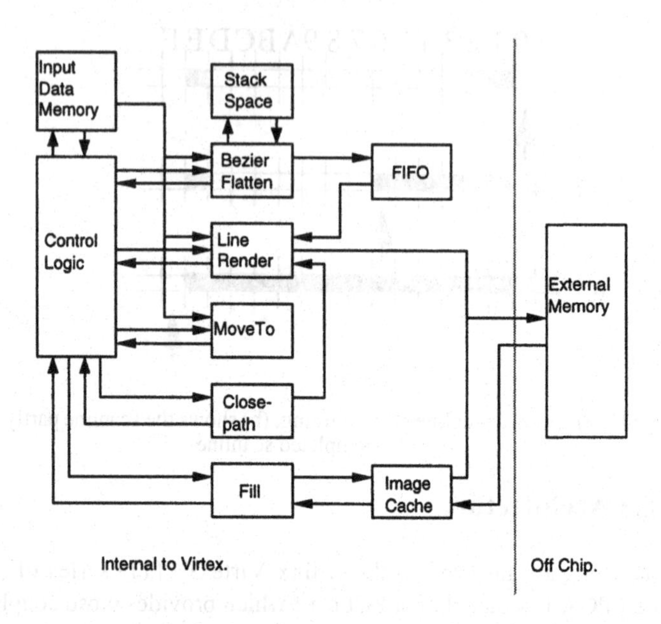

Fig. 5. Font Processor Block Diagram.

The processor is implemented using a fetch, decode and execute style architecture. A finite state machine is used to control the fetch, decode and execute stages. An op-code is read followed by the appropriate number of operands. The number of operands depends on the particular instruction that was previously read. Once an operation and its associated data have been retrieved from the BlockRAM it is passed onto the appropriate processing element where it is executed. Each of the five processing elements, which are Bézier flatten, line render, MoveTo, Closepath and Fill in Fig. 5 has a simple state machine to control the operation of that element. The fetch and execute stages can be pipelined as in modern processor architectures. While one operation is being executed the next can be loaded. This would improve performance but only marginally. The most common operations are the `lineto` and `curveto` which also have long execution times. These operations take many clock cycles each and thus the time taken to load data from memory is significantly less than that of the operation execution. The extra resources required to implement parallel fetch and execution may be better utilised elsewhere.

5.1 Straight Lines

A line rendering circuit is required for several of the operations, lineto, curveto and closepath. A single line rendering circuit is implemented on the FPGA. Depending on the current operation the input to the circuit is changed. When a lineto operation is being executed the current point is the start point and the point defined by the associated data is the end point of the line. When executing a curveto operation the input is taken from a FIFO buffer into which the Bézier curve circuit writes its results. When executing the closepath operation the start point is the current point and the end point is the first point in the current path. To produce the straight lines we use a simple implementation of Bresenhams Straight line algorithm. An optimised version of which runs at approximately 120Mhz. The output of the line circuit is written to external memory.

5.2 Recursive Bézier Implementation

We now return to implementing the Bézier calculation by using a recursive implementation. Dividing a Bézier curve in half requires very little computation. The only arithmetic involved is addition and division by two. The calculation is shown in Fig. 6

The main drawback with this calculation is the combinational delay through the calculation. The delay could be minimised by using a pipelined circuit, unfortunately this is used inside a loop where the next operation is dependant on the result of the previous operation thus making pipelining difficult.

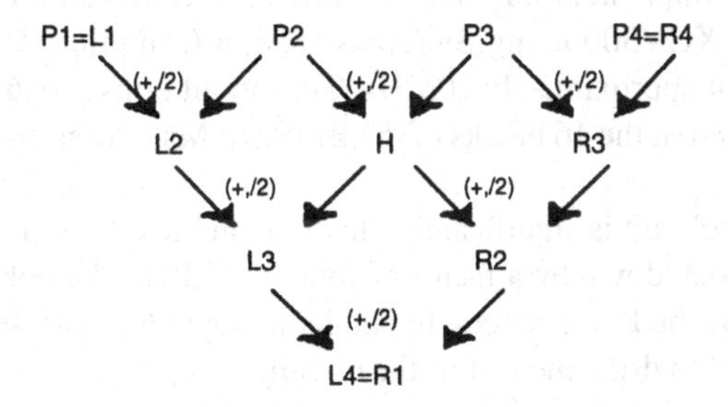

Fig. 6. Bézier division.

The other computational section of the recursive sub-divide method is the flatness test. This is achieve by comparing the gradients of the line connecting the control points. The gradient between P1, P2 and P1,P3 is compared to that of P1,P4 if these are suitably close then we have a straight line.When a curve is flat *delta1* and *delta2* are zero, since we are only calculating an approximation then we can assume the curve is flat when *delta1* and *delta2* are less than some tolerance value.

$$\begin{aligned} delta1 &= [(P2x - P1x) \times (P4y - P1y)] - [(P2y - P1y) \times (P4x - P1x)] \\ delta2 &= [(P3x - P1x) \times (P4y - P1y)] - [(P2y - P1y) \times (P4x - P1x)] \end{aligned} \quad (1)$$

The results from the flatness test are used to decide whether we have a point on the curve or if we need to do another division. When another division is required one curves is pushed onto the stack and the other is used for the next iteration. When a point on the curve is found the last control point is passed to the line rendering circuit and the next curve retrieved from the stack. The sub-division of the curve is complete when a point has just been output and there are no more curves left on the stack.

The internal BlockRAM is used for two purposes. Firstly to implement the stack which stores the waiting curves, and secondly as a FIFO buffer between the Bézier circuit and the line rendering circuit. The BlockRAM is ideal for both of these tasks due to the dual ports. The dual ports allow the Bézier circuit to produce the points at one speed and the line rendering circuit can obtain these values as it requires them.

The distributed nature of the BlockRAM is also an advantage for this application. A curve is stored on the stack by pushing four points. This requires eight values to be written to memory. We can use one BlockRAM per value and can thus write all eight values in parallel. Even when using 16-bit values the depth of 256 words is ample for the stack.

The control logic for this large circuit is simplified by the use of the clock Delay-Lock Loop (DLL) feature of the Virtex architecture. We use the four phases of the clock to control the various operations required to process each curve.

The hardware to implement this was described in behavioural VHDL and synthesised for the Virtex XCV300 using Synopsis Design Compiler. The recursive subdivision circuit runs at approximately 20MHz. The circuit utilises 636 slices (20% utilisation) as well as 10 out of the 16 blocks of the BlockRAMs. It uses one of the four clock DLLs.

Since the Bézier circuit is significantly slower than the others, a clock DLL is used to divide the input clock down by a factor of four. This divided clock is then passed to the clock DLL used by the Bézier circuit to obtain the four phases of the clock. The original clock is then used to drive the rest of the circuit.

5.3 Outline Fill

The process of filling the outline is relatively simple and thus fast but the data management techniques employed are vital to the performance. The fill process works on a scanline basis processing one scanline at a time. As an edge is found the position is written onto a small stack and once the end of the line is reached the space between pairs of edges is filled in the image. The outline fill process implemented in behavioural VHDL runs at approximately 90Mhz and uses 262 slices of an XCV300 when synthesised using Synplicity.

5.4 Image Cache

The only other circuit to directly alter the output image is the line rendering circuit. The line rendering circuit as previously described draws directly into the external memory. The arrangement of the image data in the external memory affects both the line rendering process and the filling process. We are only dealing with monochrome images and thus only require one bit per pixel. We assume that the external memory can be accessed using an 8-bit wide data bus and a 16-bit address. Thus each pixel is stored using 8-bits which requires 64Kbytes of memory.

The memory accesses performed by the fill process are highly predictable and thus the data can be cached easily. Two BlockRAMs are used as an image cache for the fill process. This allows for 32 scanlines to be held in the cache at one-time. This is adequate so long as the cache system can keep the fill process supplied with unprocessed scanlines. Assuming a read or a write to either BlockRAM or external memory takes one clock cycle. The load of a scanline takes 256 clock cycles and writeback takes 32 resulting in 73728 clock cycles to completely cache the image. When using 1-bit memory access for the fill process each scanline requires 256 reads and a number of write depending on the data resulting in a minimum of 65536 memory accesses. This requires 8192 writes to balance the access but is also equivalent to the number of cycles taken to load the cache.

6 Summary

The font processor circuit was implemented using the latest version of the Xilinx place and route software. Using the slowest XCV300 speed grade we achieved a primary clock input speed of 55MHz. The layout produced is shown in Fig. 7 Note that related chunks of logic get grouped together. The complete system uses only 1335 of the available 3072 slices on the XCV300 leaving room for a more pipelined approach and other additions.

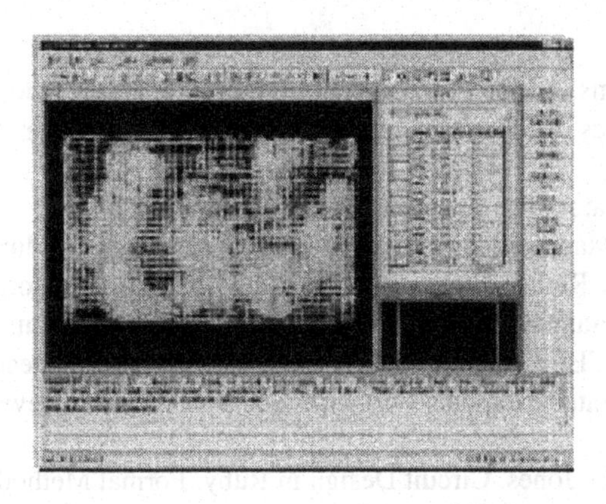

Fig. 7. The layout of the font processor

We have shown how a custom co-processor can be built using FPGA technology to solve the problem of font rendering. The increased size of current FPGAs allows us to implement in hardware algorithms which would previously only been possible in software. Specifically we show that a recursive algorithm can be implemented efficiently in hardware when fast access memory is available on chip.

There are several optimisations and extensions that can be integrated into the processor described above. The use of external memory which is cached using the BlockRAM could be improved. Since the majority of memory access comes from the fill process optimizing the caching behaviour for this would be optimal. When the memory access of the line rendering are compared which those of the fill process it can easily be seen which is more memory intensive. The line rendering process involves relatively few memory access and it may be worthwhile to pay the cost incurred when the image data is packed even when in external memory.

Many characters use circles or arcs as part of the outline for example the dot on the letter *i*. It is possible to implement circle rendering very efficiently using an FPGA [12]. An extra operator would be added which would allow circles and arcs to be drawn. Bézier curves cannot represent circles perfectly. Four Bézier curves are required to approximate a circle. If a circle operator were added this would not only produce improved visual results for circular objects but would also improve performance due to reduced data and more efficient implementation.

The definition language used is based on the set of PostScript path operators. This could be expanded to allow the definition of PostScript documents and not just PostScript fonts.

"Virtex" and "XCV300" are trademarks of Xilinx Inc.

References

[1] Adobe Systems. Adobe PostScript Extreme White Paper. Adobe System Inc. 1997

[2] Adobe Systems. Adobe PrintGear Technology Backgrounder. Adobe Systems Inc. 1997

[3] Smooth Operator. The Economist. Edition 6 March 1999.

[4] William H. Mangione-Smith, Brad Hutchings, David Andrews, André DeHon, Carl Ebeling, Reiner Hartenstein, Oskar Mencer, John Morris, Kirshna Palem, Viktor K. Prasanna, Henk A. E. Spaanenburg. Seeking Solutions in Configurable Computing. IEEE Computer, December, Vol. 30, No. 12. December 1997.

[5] Intel. Accelerated Graphics Port Interface Specification Revision 2.0. December 11, 1997.

[6] M. Sheeran, G. Jones. Circuit Design in Ruby. Formal Methods for VLSI Design, J. Stanstrup, North Holland, 1992.

[7] Satnam Singh and Pierre Bellec. Virtual Hardware for Graphics Applications using FPGAs. FCCM'94. IEEE Computer Society, 1994.

[8] Satnam Singh. Architectural Descriptions for FPGA Circuits. FCCM'95. IEEE Computer Society. 1995.

[9] J.D. Foley, A. Van Dam. Computer Graphics: Principles and Practice. Addison Wesley. 1997.

[10] Xilinx. XC6200 FPGA Family Data Sheet. Xilinx Inc. 1995.

[11] Xilinx Virtex™ 2.5V FPGA Product Specification. Xilinx Inc. 1998.

[12] D. MacVicar, S. Singh. Accelerating DTP with Reconfigurable Computing Engines. FPL'98. Springer-Verlag 1998.

[13] S. Singh, J. Patterson, J.Burns, M. Dales. PostScript™ Rendering with Virtual Hardware. FPL'97. Springer-Verlag 1997.

[14] http://www.xilinx.com/products/virtex.htm

Implementing PhotoShop™ Filters in Virtex™

Stefan Ludwig[1], Robert Slous[2] and Satnam Singh[2]

[1]Compaq Systems Research Center, Palo Alto, California, U.S.A.
Stefan.Ludwig@compaq.com
[2]Xilinx Inc., San Jose, California, U.S.A.
{Robert.Slous, Satnam.Singh}@xilinx.com

Abstract. This paper presents a complete system that utilises a FPGA-based co-processor to accelerate compute intensive image processing operations. Its main contributions are a methodology for incorporating hardware-based acceleration into a commercial image processing application by exploiting a plug-in architecture; a presentation of a new PCI-based FPGA accelerator system suited for image processing style applications; and theoretical calculations and empirical measurements of the system that was actually built.

1 Introduction

The design, implementation and performance analysis of a FPGA-based co-processor system for accelerating the image processing application Adobe Photoshop is presented. We describe a general purpose FPGA co-processor system using the Xilinx Virtex FPGA. We show how circuits performing various image processing applications are realised on this FPGA hardware. The software interface between the card and the Photoshop application is described as well as a description of how Adobe Photoshop was made to communicate with the FPGA hardware. We instrument the performance of software and hardware versions of two filters and compare against the theoretical performance of our hardware platform.

2 Image Processing with Adobe Photoshop

Adobe markets a series of applications for producing or processing drawings (Adobe Illustrator), photographic quality pictures (Adobe Photoshop) and video (Adobe Premier). In the paper we shall concentrate on the acceleration of Adobe Photoshop, but the principles and techniques are equally applicable to the hardware-based acceleration of the other applications. Indeed, the hardware and software we produce can be directly incorporated into Adobe Illustrator and Adobe Premier without change. Other third party tools also use Adobe-style plug-ins and these can also immediately benefit from our hardware-based filter accelerator.

Photoshop is a widely used image processing package which provides a modular architecture for extending its functionality based on plug-ins. Images to be processed are often in true-colour (24-bits) and may be sampled from a photograph or video camera at a high resolution. Photoshop provides filters that can manipulate an image in various ways including colour manipulation and filtering (e.g. Gaussian blur). For large images these filters can take a long time to run, and there is already a market for specialised DSP-based cards which can be used with plug-ins to accelerate Photoshop. The work presented in this paper has been carried out using Adobe Photoshop version 5.0.

By using a FPGA-based co-processor system, one can produce filters that are accelerated using specialised circuits that operate at hardware speeds. One can distribute image processing circuits as plug-ins, making them a commodity item that is conveniently packaged. If high speed filters can be produced then there may be a market for FPGA-based boards in the desktop publishing niche.

Adobe Photoshop provides a collection of 'filters' that perform various image processing operations. The filter menu of Photoshop is shown in Fig. 1 below. The available filters are not fixed, but instead are read as plug-ins to Photoshop. This means that a user can purchase or develop more filters and extend the functionality of Photoshop without access to the source code of the application.

We used the publicly available Photoshop Software Development Kit (SDK) to implement a variety of filters that use the Virtex FPGA to reduce image processing time. We have been concentrating on colour space conversion (RGB to greyscale conversion) and convolution style calculations (e.g. Gaussian Blur). Gaussian Blur is one of the slowest operations in Photoshop and is often used as a benchmark when assessing the performance of desktop publishing systems.

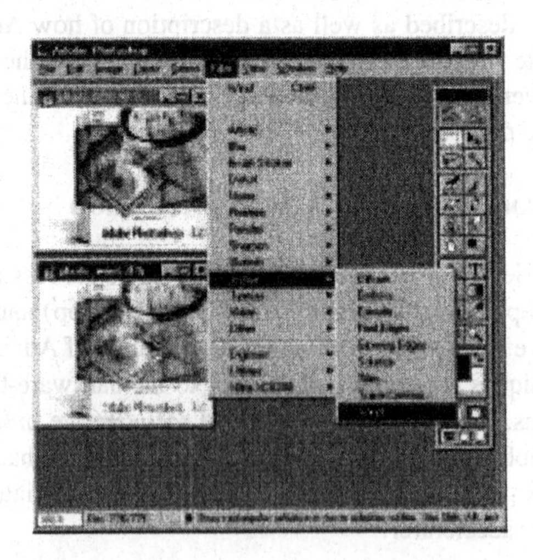

Fig. 1. The Filter Menu of Adobe Photoshop

The filter plug-ins were developed in C++ and compiled as Windows dynamic link libraries (DLLs). The binary programming information bitstream for the accelerators is compiled into the DLL allowing the hardware and software to be delivered in one convenient package.

3 Accelerating Photoshop Filters

Photoshop filters communicate with Adobe Photoshop using a series of messages that specify the nature of the image to be processed, as shown in Fig. 2.

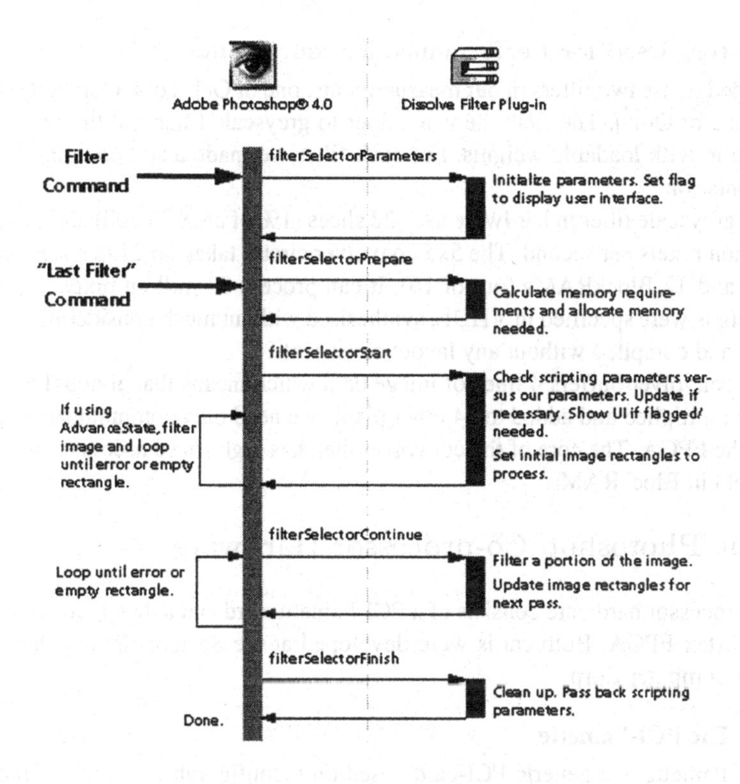

Fig. 2. Photoshop Plug-In Architecture

When Photoshop starts up it scans a series of directories containing plug-in DLLs and registers them (adding menu options to the Filter menu for each filter). Plug-ins respond to filter commands as shown in Fig. 2 which results in a series of messages begin passed, specifying the size and nature of the image to be processed.

The plug-in software that we produced asks for the image to be presented in red/green/blue/alpha format (32-bits). Every time one of our filters is selected Photoshop copies its internal working image into a buffer for our use. We then manipulate this buffer to calculate a new image which is placed into a destination buffer. When the filter completes, Photoshop copies the destination buffer back into its own internal buffer (which requires it to reformat the image to the internal representation). Consequently for both the software and hardware filters we incur an image movement cost that is not under our control. We do not instrument the cost of copying these buffers.

The filters that dispatch the image to the FPGA co-processor just pass the image data directly to the hardware and then read back the processed results into the destination buffer.

3.1 Filters Used for Performance Measurements

We decided to use two filters in our measurements: one of O(1) computational complexity and one of $O(n^2)$. The first one is a colour to greyscale filter and the second a 5x5 convolution with loadable weights. For each filter we made a software and hardware implementation.

The greyscale filter in hardware uses 22 slices (1% of an XCV300) and can process 100 million pixels per second. The 5x5 convolver circuit takes up 2790 slices (90% utilisation) and 12 BlockRAMs (out of 16). It can process 33 million pixels per second. Both designs were specified in VHDL, synthesised without much consideration for optimisation and compiled without any layout constraints.

The convolver buffers 4 lines of image data which means that although each pixel has to be multiplied and added to 24 other pixels we need only communicate each pixel once to the FPGA. The core of the convolver then has high speed access to the required pixels held in BlockRAMs.

4 The Photoshop Co-processor Hardware

The co-processor hardware consists of a PCI-Pamette card and a daughtercard using the Xilinx Virtex FPGA. Both cards were developed at the Systems Research Center of Compaq Computer Corp.

4.1 The PCI-Pamette

The PCI-Pamette is a generic PCI-card based on reconfigurable logic [3]. One Xilinx XC4010E FPGA implements a master and slave PCI interface supporting 32- and 64-bit transactions and contains a DMA engine capable of transferring data at full PCI-bus speed. The card features a PCI mezzanine card connector (PMC) for daughtercards.

4.2 The Virtex Daughtercard

The daughtercard is based on the new Xilinx Virtex FPGA series [8]. Fig. 3 shows a block diagram of the card and Fig. 4 a photograph. The daughtercard consists of the following components:

- 1 XCV300 Virtex FPGA in a BGA package
- 2 independent banks of synchronous ZBT SRAM (18-bits wide), 1 MB total
- 2 independent banks of synchronous DRAM (16-bits wide), 4 MB total
- programmable clock generator
- general-purpose 68-pin connector for input/output (34 signal pins)
- 64-bit PCI interface to PCI-Pamette
- clock buffer, 2.5 V switching power supply, temperature sensors

We use a Xilinx Virtex XCV300 FPGA in a Ball-Grid Array package. It is connected to a total of 5 MB of memory. We use new Zero-Bus Turnaround synchronous SRAMs [9] and more traditional synchronous DRAMs. The four independent banks of SRAM and DRAM allow for an aggregated memory bandwidth of over 1 GB/s.

The daughtercard has a flexible clocking scheme. Based on a 20 MHz oscillator or a signal from the PCI-Pamette [3] the clock generator is capable of generating any frequency between 0 and 90 MHz. This can be multiplied using the delay-locked loop circuits (DLL) of the Virtex FPGA to generate higher frequencies. The resulting signal is distributed to the RAMs and back to the FPGA itself by an external clock buffer. Using another DLL in the FPGA, we can generate a zero-skew copy of the board clock for the FPGA circuit. There are two additional clock sources available: a copy of the PCI clock, generated by a phase-locked loop on the PCI-Pamette and a clock signal coming from one of the FPGAs on the PCI-Pamette. The latter can be used, for instance, to implement a software clock.

A switching power supply is used to generate the FPGA core supply voltage of 2.5V. Two temperature sensors [2] are used to monitor the ambient temperature of the front and the back of the card. When mounted on a PCI-Pamette, most components of the daughtercard face the components on the PCI-card. The temperature sensors can be used to provide a shutdown function, should the boards get too hot.

4.3 Interface to PCI-Pamette and the Host-PC

A 64-bit PCI interface (98 wires) is provided through 3 of the 4 connectors (see Fig. 3). Nothing about the interface is PCI-specific except the pin-out as prescribed by the standard. PCI-Pamette is capable of transferring data to and from memory at a sustained 120 MB/s by using DMA transfers. This rate depends on the host-bridge and the system bus speed used in the PC. The figures we report here are for an Intel 440BX AGP chipset [5] running the system bus at 100 MHz.

The PCI-Pamette is configured with an interface, which passes data back and forth between the Virtex daughtercard and the PC's host memory. Currently, only programmed IO is supported, but a DMA mode is under development. The programmed IO mode allows for data transfer speeds of 50 MB/s to and 10 MB/s from the card. The interface is also used to connect to the configuration port of the Virtex FPGA.

4.4 Software Interface

To configure the Virtex FPGA with a circuit, we first configure the PCI-Pamette with the aforementioned interface and then download the accelerator to the Virtex FPGA. The software can move data to and from the daughtercard by writing to or reading from an address in PCI-space.

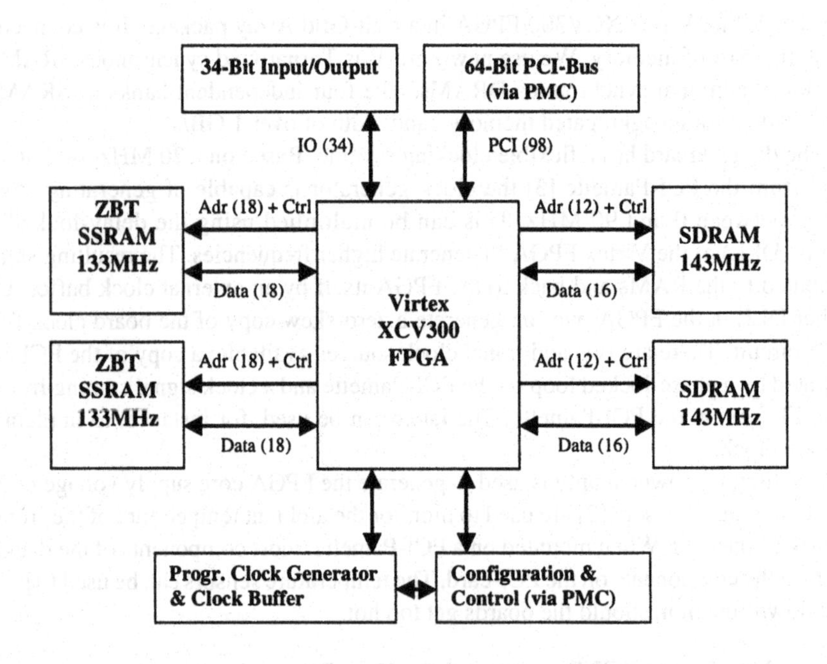

Fig. 3. The Virtex daughtercard architecture

5 Performance Analysis

5.1 Theoretical Performance

In the following, we calculate the maximal performance of a 5x5 convolution, implemented in software on a PC and as a circuit on the co-processor, respectively. For the analysis we consider a 6" x 4" photograph scanned at 600 dpi in 24-bit RGB colour (padded to 32 bits per pixel). This results in a 35 MB image of 8.6 million pixels, which has to be transferred from main memory to the computing device and back. Every colour value of every pixel is subjected to the filter, which results in 25 multiplications and 24 additions per colour.

On the PC, transferring the image from memory to the CPU and back takes 70 MB / 800 MB/s = 88 ms, or 1.25 ns/B. If we could fit all filter coefficients into the CPU's registers (this is not possible on a Pentium-III) we could do one multiply-accumulate step every 1.25 cycles. Ignoring all other overhead, calculating the filter takes 31.25 cycles per byte. At 500 MHz this is 187.5 ns per pixel or 1.62 seconds for the whole image. Thus, the time of filtering is dominated by the calculation time. This results in a theoretical performance of 5.3 million pixels per second.

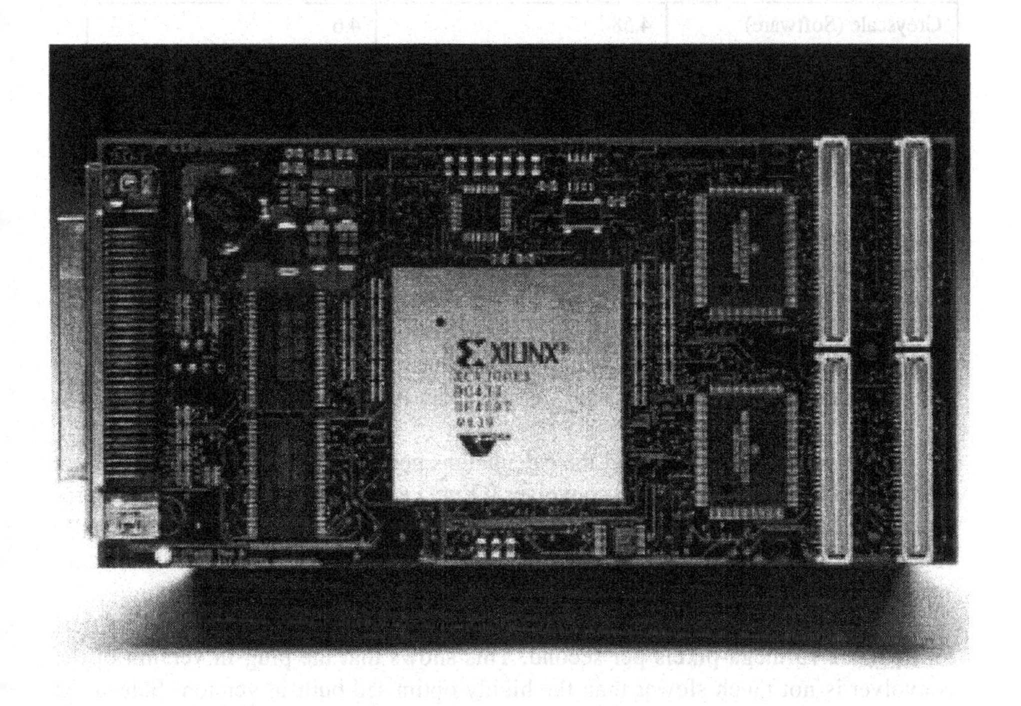

Fig. 4. A photograph of the Virtex daughtercard

On the daughtercard, transferring the image takes 70 MB / 120 MB/s = 583 ms, or 8.3 ns/B. We can fit the 5x5 filter completely in the Virtex FPGA and can therefore get 75 MACs (multiply-accumulates) per cycle. At 66 MHz this is 15.2 ns per pixel or 131 ms for the whole image. Since the performance of the hardware filter is dominated by memory transfer speed we achieve a throughput of 14.8 million pixels per second. Taking into account the transfer speed this gives us a factor of 3 improvement over the PC.

We have to emphasize that in the above calculations, very optimistic assumptions have been made for the performance of the software. In reality, we should be able to achieve higher speedups.

5.2 Measurements and Analysis

Approximate performance measurements are shown in Table 1. The measurements were

Table 1. Performance Measurements (in mega-pixels per second)

Filter	200K Pixels	3.4M Pixels
Greyscale (Software)	4.58	4.6
Greyscale (Hardware)	1.07 (speedup 0.23)	1.06 (speedup 0.23)
Convolver 5x5 (Software)	0.73	0.79
Convolver 5x5 (Hardware)	1.08 (speedup 1.48)	1.08 (speedup 1.37)

instrumented by calculating the total time taken to communicate the image data to and from the FPGA card, as well the time spent processing the image and performing the filter protocol commands. The software measurements were taken on a high-end personal computer with a 500MHz Pentium III processor and 192MB of memory. Each filter was executed five times for each image and an average of the processing time was calculated.

The measurements show that the calculations performed on the FPGA subsystem are memory bound since the 5x5 convolver [$O(n^2)$ i.e. 75 multiply-adds per pixel] proceeds roughly the same number of pixels per second as the far less computationally challenging greyscale operation [O(1)]. As expected, the software version of the convolver is compute bound, operating six times slower than the greyscale filter.

We also instrumented the built-in 5x5 convolver provided by Adobe and it performed at 1.13 mega-pixels per second. This shows that the plug-in version of the convolver is not much slower than the highly optimised built-in version. Indeed the built-in version only achieves a fifth of the theoretical throughput.

Although the results show very modest speedups, the relative results are encouraging. The memory transfer speed to the co-processor board is currently only 8 MB per second, but this can be improved by an order of magnitude using DMA. This would result in a speedup of 14 over our convolver plug-in or a speedup of 10 over the built-in version.

6 Related Work

FPGA-based Photoshop accelerators have been previously designed and built by the authors using a XC6200 [7] based system on a PCI-card [6] which contained 2MB of SRAM. This system demonstrated how Photoshop, plug-ins, device drivers, programming bit-streams and run-time control could be used to accelerate commercial applications with FPGAs. However the XC6200 system used a considerably smaller FPGA device which lacked on-chip memory blocks. Furthermore, the previous system required carefully hand-crafted filters with each cell being manually placed. Our current system is synthesised from a high level description and contains no location constraints (except for the IOs).

7 Summary and Conclusions

This project is still in progress and we have just presented preliminary results. The speedups at this stage of the project are very small, but with further enhancement of the communication infrastructure we expect speedups that make a FPGA-based Photoshop accelerator board competitive with high-end workstations or DSP-based co-processor boards. Not all types of filtering operations are suitable for acceleration with a FPGA-based co-processor and one has to carefully analyse the computational complexity of the filter, its memory requirements and its memory access behaviour.

The experiments we have performed show that the hardware solution scales linearly since the pixels per second performance for a large image is no worse than for a much smaller image. The experiments also show that the measured performance of both the hardware and the software is significantly less than the theoretical values that we have calculated. However, the hardware solution suffers from limited memory bandwidth, which we can improve in the future.

The filters that we have designed and implemented can be directly used to accelerate other Adobe Photoshop applications like Illustrator (for producing drawings) and Premier (for processing moving images). Although we have concentrated on the acceleration of a specific application using an FPGA-based co-processor we believe the general technique is applicable to a wide class of problems. Many modern computer applications are designed to use plug-ins to extend their capability after shipment, e.g. Netscape Communicator. Applications that require a significant amount of computation per datum after transfer are good candidates for acceleration using our technique. Examples include encryption/decryption, encoding/decoding and compression/decompression.

The measurements here have been performed without fully exploiting all the features of the Virtex daughtercard. For example, we could use the on-board SRAMs to cache large portions of the image to implement larger filters (e.g. to time-multiplex a 7x7 convolver which would not otherwise fit into the FPGA). Alternatively we could use the SRAMs to convolve images which are too wide to fit into BlockRAMs.

"Virtex "XCV300", "XC4010E" and "XC6200" are trademarks of Xilinx Inc.

References

[1] Compaq Professional Workstation AP200 Series. Compaq Computer Corp. 1998. http://www.compaq.com/products/workstations/ap200/index.html

[2] DS1820, 1-Wire™ Digital Thermometer. Dallas Semiconductor Corp. 1998. http://www.dallassemiconductor.com/Prod_info/Thermal/thermal.html#1820

[3] PCI Development Platform. Compaq Computer Corp. 1996. http://www.research.digital.com/SRC/pamette

[4] J.D. Foley, A. Van Dam. Computer Graphics: Principles and Practice. Addison Wesley. 1997.

[5] Intel 440BX APGset. Intel Corp. 1998. http://developer.intel.com/design/chipsets/440bx/index.htm

[6] Satnam Singh and Robert Slous. *Accelerating Adobe Photoshop with Reconfigurable Logic*. FCCM'98. Napa, California. IEEE Computer Society Press, 1998.

[7] Xilinx. XC6200 FPGA Family Data Sheet. Xilinx Inc. 1995.

[8] Xilinx Virtex™ 2.5V FPGA Product Specification. Xilinx Inc. 1998. http://www.xilinx.com/products/virtex.htm

[9] Pipelined ZBT™ Synchronous Fast Static RAM. Motorola, Inc. 1998. http://motsps.com/products/memory/srams/synchronous/zbts/index.html

Rapid FPGA Prototyping of a DAB Test Data Generator Using Protocol Compiler

Klaus Feske[1)], Michael Scholz[2)], Günther Döring[1)] and Denis Nareike[3)]

1) FhG IIS Erlangen - Department EAS Dresden
Zeunerstr. 38, D-01069 Dresden, Germany,
e-mail: feske@eas.iis.fhg.de
2) Dresden University of Technology,
Dept. of Electrical Engineering
3) HTW Dresden, Dept. of Electrical Engineering

Abstract. To enhance efficiency in FPGA based rapid prototyping of digital telecommunication applications, we incorporated modeling and synthesis facilities of the Protocol Compiler into a proved design flow. This paper focuses on domain specific modeling styles and synthesis strategies aiming at design improvement, changeability and reuse. We discuss design and implementation of a DAB (Digital Audio Broadcasting) Test Data Generator which produces hierarchical structured data streams according to the DAB protocol. Summarizing the results, we improved design quality and efficiency.

1 Introduction

Conventionally, the controller design for structured data stream processing is not well supported by EDA tools. Furthermore the control part might occupy only 15% of the chip area, but designing the controller can take up to 75% of the development effort and debug time [SHM-96]. We are facing a bottleneck in the design process especially for telecommunication applications.

To fill the gap, we checked the qualification and usability of domain specific high level synthesis approaches [Sea-94, Syn-99, HoB-98] which are applicable in our FPGA based rapid prototyping design flow (Fig. 1). Moreover, we studied the modeling capabilities of this approaches to establish an adapted reuse methodology.

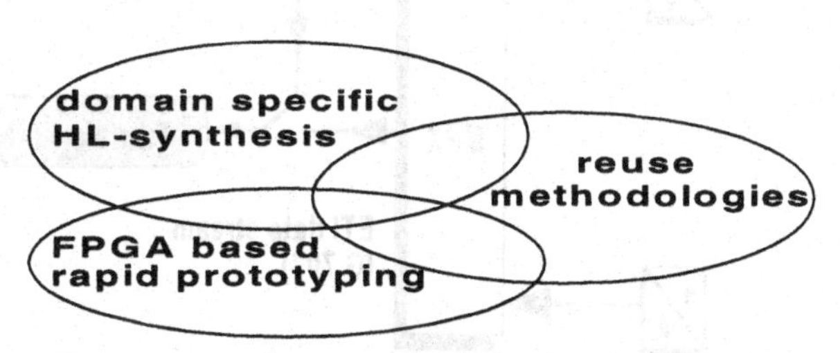

Fig 1. Ways to improve the design productivity

In the paper we explain the design of a Digital Audio Broadcasting (DAB) Test Data Generator (TDG) using an extended FPGA prototyping design flow. The task of the TDG is to produce hierarchical structured data streams according to the DAB protocol. Based on our previous work [FRK-98] and first experiences in utilizing the Protocol Compiler [SDF-99] the paper outlines high level modeling principles and related synthesis solutions in order to enhance efficiency in controller design and according to the requirements of the DAB test environment.

The paper is organized as follows. First we introduce the design requirements of the DAB test environment. Section 3 describes aims and details of domain specific modeling principles related to the rapid prototyping issues. Section 4 shows the synthesis results obtained for both, the particular protocol-components as well as the whole design. It also outlines first experiences made using reusable templates and protocol components. Section 5 describes our way to validate the design in a real DAB test environment. Finally, we summarize experiences concerning design quality and efficiency.

2 Requirements of Our Digital Audio Broadcasting (DAB) Test Environment

The digital radio system DAB is a broadband system, which transmits multiple audio and data services within a common program block: the ensemble transport interface ETI.

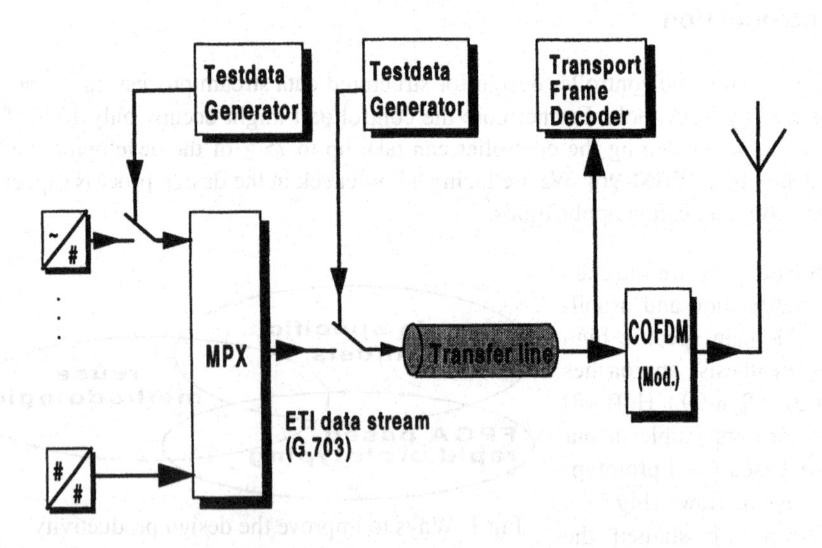

Fig. 2. Test Data Generator in a DAB collection network (simplified)

Digitized and preprocessed audio signals and data services are put together using a multiplexer (MPX, Fig. 2). Specific requirements of the DAB collection network are

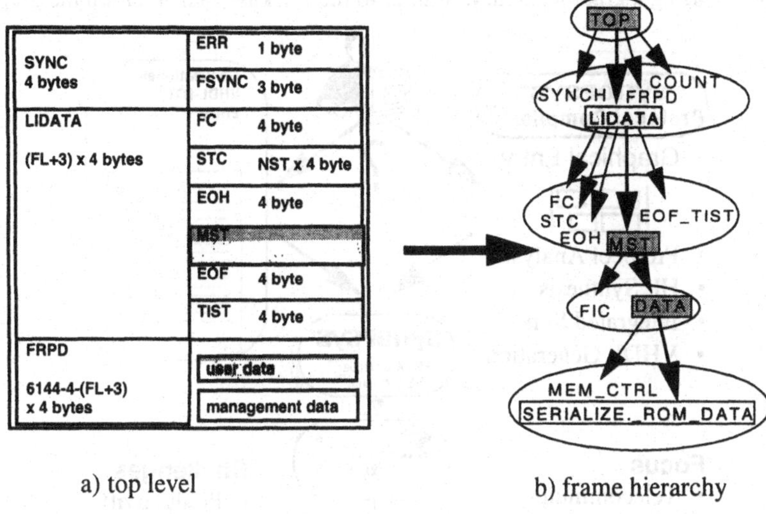

a) top level

b) frame hierarchy

Fig. 3. ETI frame structure

the reconfigurable multiplexing of different components which are provided by different sources, the coordination of the service providers and the data generation of the common signalling channel (FIC).

The ETI data stream [ETS-97] is a structured compilation consisting of three hierarchically arranged parts: the synchronization block (SYNC), the logical interface data (LIDATA) and the frame padding field (FRPD) (Fig. 3). In DAB project, our institute is responsible for the multiplexer design. Consequently, a suitable DAB test data generator represents an effective facility to support the test of complex DAB collection networks. It can be used either as a signal source at the multiplexer input or to emulate its output signal, useful for tests of DAB transfer lines.

For our work we have to examine the entire structure of the ETI data stream: Within the LIDATA area both the component characterizing and the main stream data are transmitted. It provides length and number of the subchannels. Additionally there are a header and a data checksum included. The timestamp area transmits timing information for the receiver. Finally the FRPD field adds the required number of padding bytes for a fix frame length of 24 ms.

3 Rapid FPGA Prototyping Exploiting Domain Specific Modeling Principles

As shown in Fig. 4 our design flow conventionally used starts at RT-level with a VHDL description of the design specification. As a new element in the chain of this flow, the Synopsys Protocol Compiler [Syn-99] is set on top of the process. In the domain of telecommunication applications it supports the design of controller circuits for processing data protocols. Regarding the protocol level, a high-level specification is composed using a graphical entry. Similar to the Backus-Naur-notation the graphical

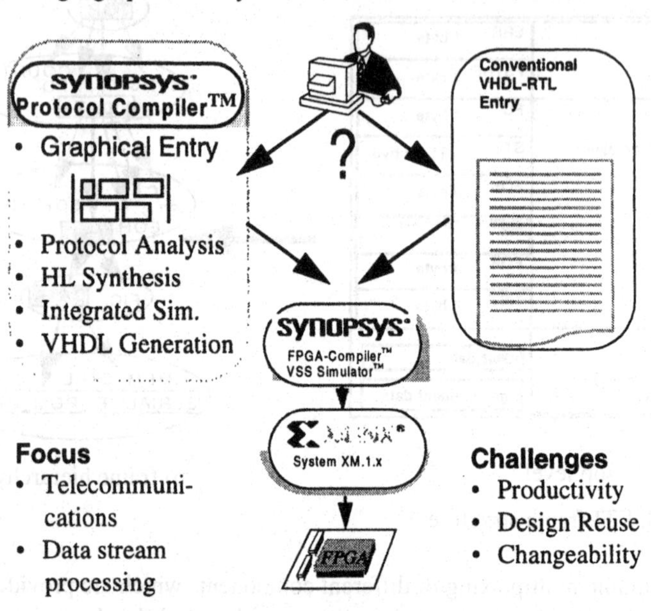

Fig. 4. Protocol Compiler in an experimental rapid prototyping design flow

symbolic format closely matches the high level specification (Fig. 5). Furthermore it allows a formal protocol design analysis, size and delay estimation, protocol simulation and protocol synthesis including high level optimization and HDL code generation.

- unspecified frame
- action (default, user)
- terminal frame
- reference frame
- epsilon operator
- qualifier operator
- if-frame operator
- repeat operator
- sequential operator
- alternative operator

Fig. 5. Types of frames and frame operators

The basic structure of the DAB TDG consist of several communicating and parallel working finite state machines. As an example Fig. 6 represents a model of the top design of the DAB TDG. It corresponds to the first level of the ETI frame hierarchy to be generated, using a sequential frame operator. Utilizing this modeling principle the successive action will be started not before its predecessors ends. The sequential frame operator '{}' includes a repeat operator '[]' and three reference frame operators which refer to the synchronizing function (SYNC), the component characterizing and mainstream data (LIDATA) and the subframe for padding bytes (FRPD) which completes the number of bytes transmitted to a total of 6144 bytes frame length. The alternative operator '| |' involves a concurrently working state machine symbolized by the reference frame F_Count_FRPD. According to the ETI frame structure as symbolized in Fig.4 b) the modeling details of the path LIDATA -> MST -> DATA -> SERIALIZE_ROM_DATA are given by the hierarchy of related reference frames in Fig. 7 and Fig. 8.

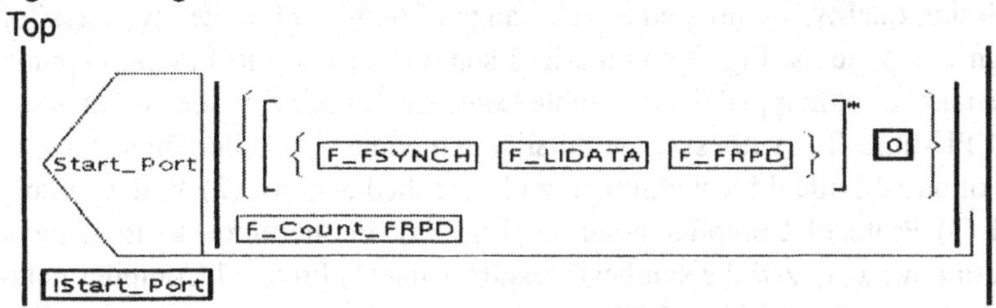

Fig. 6. Top level model using a sequential frame operator

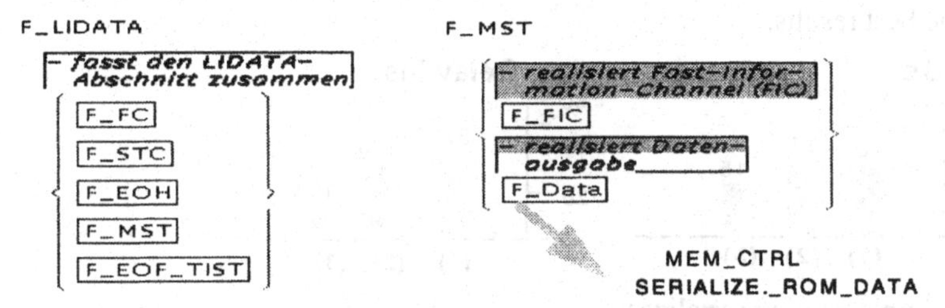

Fig. 7. Frame hierarchy of the reference frame F_LIDATA

Using these specification facilities, the embedding of DAB audio test data into the DAB frame structure and a way to expand the subchannels is shown in Fig 8. To add a new module, that inserts the additional data service into the data stream it is necessary to create a sequence. This has been accomplished by surrounding the repeat operator (including the reference frames F_Serialize_ROM_Data and F_Mem_ctrl) with a sequential frame operator. After that the appearing unspecified frame is replaced by a copy of the first module and only a few parameters have to be adjusted.

F_Data

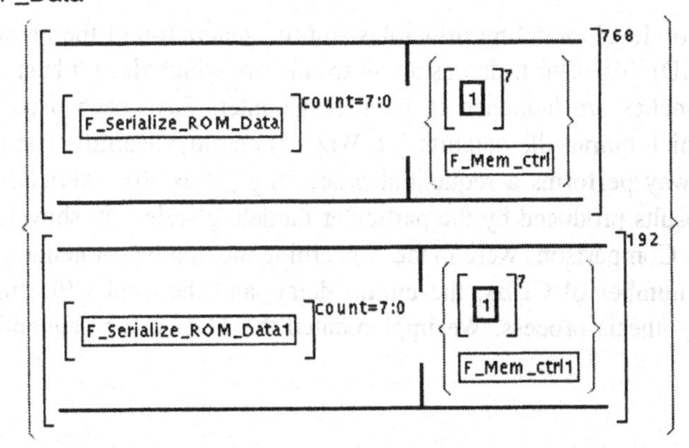

Fig. 8. Embedding test data into the ETI data stream

4 Results in Design Quality and Efficiency

4.1 Modeling Styles and Synthesis Results

Decisions in the design that has been made earlier significantly influence later steps in the design flow, e.g. logic optimization and technology mapping. Therefore, to improve the design quality, we utilized a wide range of high level modeling styles for relevant design components. Fig. 9 summarizes some results of the FPGA mapping comparing the number of mapped Configurable Logic Blocks (CLBs), the circuit delay and the total CPU-time. To synthesize the bit slicing module we studied three ways to serialize data of an individual bit-width: using (1) unrolled actions, (2) VHDL procedure call and (3) Protocol Compiler notation (Fig. 9a). Furthermore, to implement CRC computation we analyzed the synthesis results using (1) Protocol Compiler notation, (2) Protocol Compiler, option LFSR counters, and (3) sequential processing (Fig. 9b). Exploiting the high level optimization capabilities of the Protocol Compiler we obtained the best results.

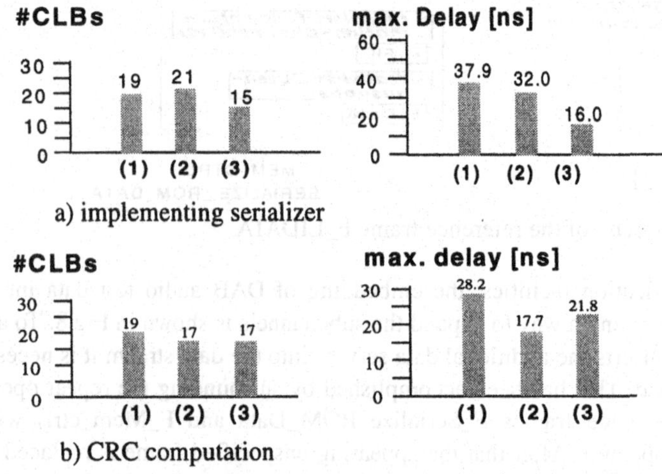

Fig. 9. Synthesis results for data stream processing components

It was found that three top level modeling principles suit the generation of the hierarchical frame structure [SDF-99]. Our task was about to find out which does it best. In the first variant, all branches are launched in parallel. Counters have been used to schedule the actions, which output the databits (1). Way (2) employs RunIdle frames [Syn-99] and the third way performs a sequential processing (3) as characterized in Fig. 6, section 3. The results produced by the particular modeling styles are shown in the bar charts of Fig. 10. Comparisons were made concerning the number of generated VHDL statements, the number of CLBs, the circuit delay, and the total CPU time needed for the whole synthesis process. We implemented the TDG using sequential processing (3). It fits into a Xilinx FPGA XC4013E using 421 out of 576 CLBs and provides two independent data channels for audio data and for digital data services.

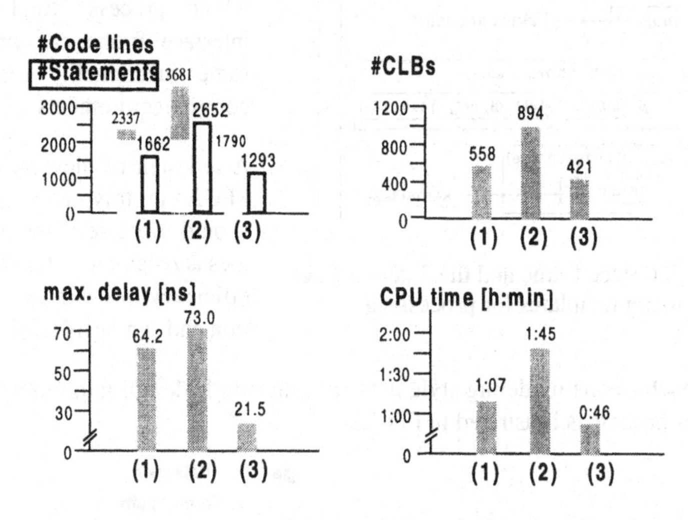

Fig. 10. Synthesis results: using (1) counters,
(2) RunIdle frames, (3) sequential processing

4.2 Enhanced Efficiency Using Protocol Templates

The processing of serial data streams is quite similar for different communication protocols. So we decided to built up a library containing reusable parts. This includes Protocol Templates, task-specific components and procedures for data processing.

Templates are organized like subroutines of programming languages. All parameters are formal parameters and will be replaced by actual values during execution [Syn-99]. They should be used if repetitions of tasks occur, but this is not only restricted to one design. A set of templates will form a class, covering a specific field in protocol processing. We have still been developing principles to retrieve templates from libraries according to system level descriptions and timing diagrams.

Components are more complex, interfaced subdesigns. They facilitate a hierarchically structured design process. One problem is that it is often necessary to provide additional logic to connect these interfaces. To obtain area optimized results, the interface should fit to the actual needs. The hierarchical notation supports dividing the components into an interface and a kernel module. The CRC (Cyclic Redundancy Check) generator implementation (Fig. 11) is taken as an example.

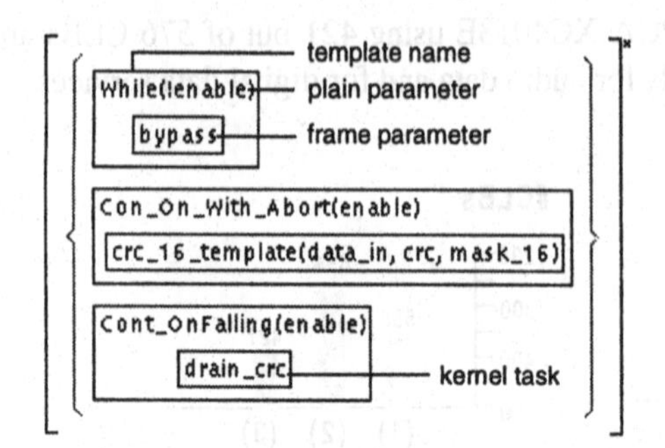

Fig. 11. interface frame and the 3 kernel tasks using templates for processing

This is a much-reduced interface, the signal ***enable*** controls the whole process. Replacing this interface frame can provide the same functionality for different control requirements.

A redesign of the TDG using the FPGA prototyping design flow should evaluate these methodologies and demonstrate the enhanced efficiency in description, verification and synthesis results.

The approach to this modeling style is to split the whole design into multiple components. This process is illustrated in Fig. 12.

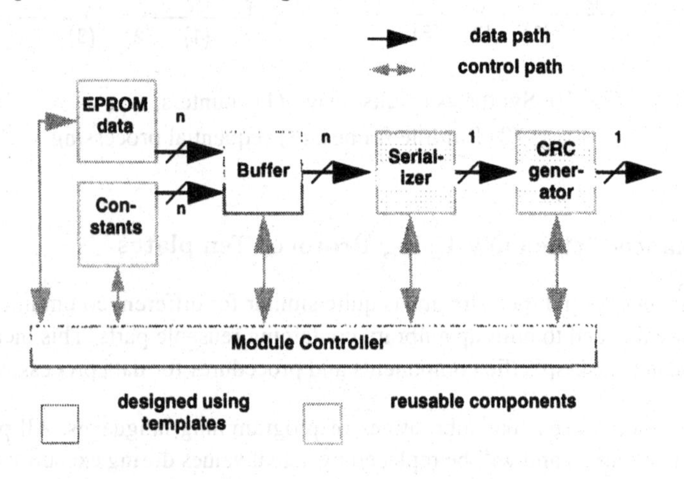

Fig. 12. structural approach to component based design

Obviously this is a completely new approach to the TDG design, it only requires to create the specific Module Controller and to fit the interface to the ETI data module. The several parts of this controller are completely independent and can be partitioned using the according control-style attribute (Fig 13).

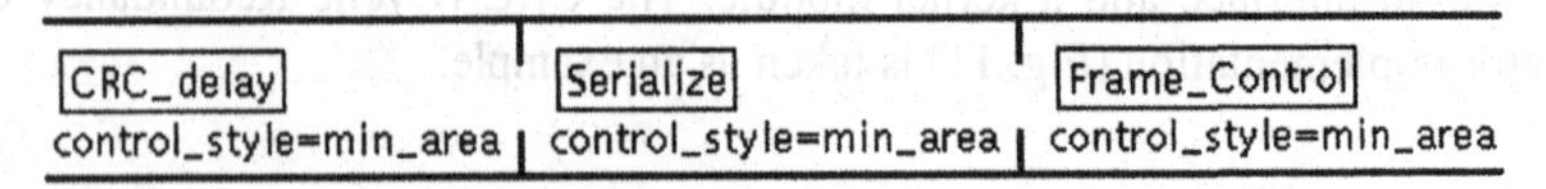

Fig. 13. Partitions of Module Controller

First implementation separates control and data path at the top level. This causes a large FSM. Distributing data and control path at different levels can optimize the resulting FSM with regard to area and speed of the FPGA implementation. Fig. 14 demonstrates how area optimization affects the number of used CLBs and the maximum frequency the FPGA can be operated.

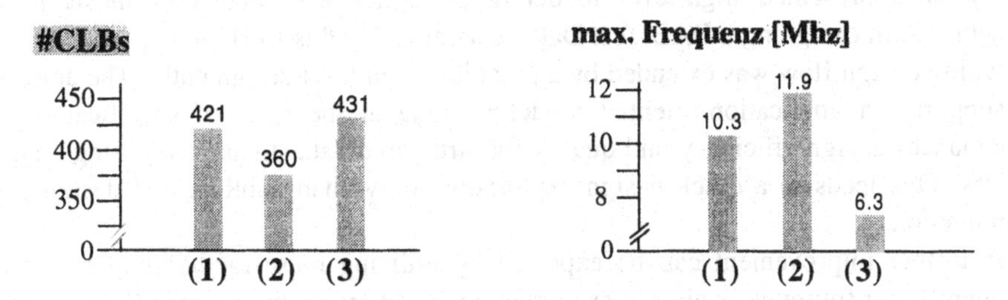

Fig. 14. Synthesis results for TDG: (1) using sequential processing
(2) redesigned, optimized FSM
(3) redesigned, separate data path & control part

5 Design Validation in a DAB Test Environment

Fortunately we are able to test the prototype in a real DAB environment, which allows a expressive testing. As symbolized in Fig. 15 a HBD3 converter translates the ETI data into a HDB3 compatible form. The first level tester checks the frame synchronization word and the correct frame length.

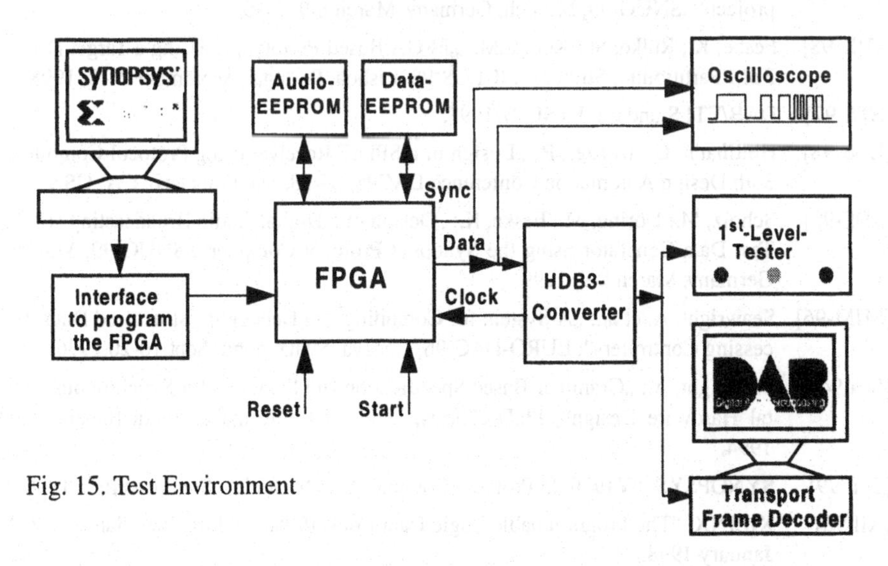

Fig. 15. Test Environment

The Transport Frame decoder (TFD) allows to examine all details of DAB/ETI data streams, such as frame-, CRC- and subchannel errors, and to detect the data and time stamp information. The TFD message viewer visualizes the recognized main stream content.

6 Conclusions and Future Work

We have presented high-level modeling principles and related synthesis results achieved in designing a DAB Test Data Generator. For this work a FPGA rapid prototyping design flow was extended by a graphical high level design entry. The approach supports an application-oriented modeling style at the level of specification and enhances design efficiency and quality for structured data stream processing controllers. This leads to a quick design exploration, easy changeability, and design cycle reduction.
A further improvement can be expected by utilizing reuse methodologies. Consequently, our future work aims at extending our rapid prototyping design flow by inserting a library of reusable protocol templates and components. First steps in this direction we discussed in chapter 4.2. We intent to adapt it to other projects in telecommunications and networking area [Bau-99].

Acknowledgment

This work has been partially supported by DFG Project SFB 358.

References

[Bau-99] Baumgart, A.: „Experiences made in using Protocol Compiler in a ATM ASIC project.", SNUG'99, Munich, Germany, March 8-9, 1999.

[FRK-98] Feske, K.; Rülke, St.; Koegst,M.: „FPGA Based Prototyping Using a Target Driven FSM Partitioning Strategy", ICECS'98, Lisboa, Portugal, September 7-10, 1998.

[ETS-97] DAB/ETI Standard; ETSI, 07/1997.

[HoB-98] Holtmann, U.; Blinzer, P.: „Design of a SPDIF Receiver using Protocol Compiler", 35th Design Automation Conference DAC98, 06/98, San Francisco, CA, USA.

[SDF-99] Scholz, M.; Döring, M.; Feske, K.: „Design of a Digital Audio Broadcasting (DAB) Test Data Generator using the Synopsys Protocol Compiler", SNUG'99, Munich, Germany, March 8-9, 1999.

[SHM-96] Seawright, A. et al.: „A System for Compiling and Debugging Structured Data Processing Controllers", EURO-DAC'96, Geneva, Switzerland, Sept.16-20,1996.

[Sea-94] Seawright, A.: „Grammar-Based Specification and Synthesis for Synchronous Digital Hardware Design", Ph.D. Thesis, Univ. of California at Santa Babara, June 1994.

[Syn-99] SYNOPSYS: "V1999.05 Protocol Compiler User's Guide", Synopsys Inc., 1999.

[Xil-98] XILINX: "The Programmable Logic Data Book 1998.", Xilinx, Inc., San Jose, CA, January 1998.

Quantitative Analysis of Run-Time Reconfigurable Database Search

N. Shirazi, W. Luk, D. Benyamin and P.Y.K. Cheung

Department of Computing
Imperial College of Science, Technology and Medicine
180 Queen's Gate, London SW7 2BZ, UK

Abstract. This paper reports two contributions to the theory and practice of reconfigurable search engines based on hashing techniques. The first contribution concerns technology-independent optimisations involving run-time reconfiguration of hash functions; a quantitative framework is developed for estimating design trade-offs, such as the amount of temporary storage versus reconfiguration time. The second contribution concerns methods for optimising implementations in Xilinx FPGA technology, which achieve different trade-offs in cell utilisation, reconfiguration time and critical path delay; quantitative analysis of these trade-offs is provided.

1 Introduction

As the volume of information stored in databases continues to expand, high performance database searching has become an important and necessary activity. For this reason, two database searching algorithms have been mapped to the Splash 2 FPGA-based custom computing machine [1]. These algorithms are a text searching algorithm that can process 20 million characters per second, and a fingerprint matching algorithm that can search through 250,000 fingerprint records per second. FPGA-based search engines have also been reported for biological databases [2], [4].

This paper reports two contributions to the theory and practice of reconfigurable search engines based on hashing techniques, which have been used in both text searching and fingerprint matching on Splash 2 described above to efficiently search large amounts of data. Our first contribution concerns technology-independent optimisations involving run-time reconfiguration of the hash functions for such search engines (Section 3). A quantitative framework is developed for estimating design trade-offs, such as the amount of temporary storage versus reconfiguration time (Section 4). Our second contribution concerns methods for optimising implementations in Xilinx FPGA technology, which achieve different trade-offs in cell utilisation, reconfiguration time and critical path delay (Section 5); quantitative analysis of these trade-offs are provided (Section 6). Before discussing these two aspects, we shall motivate in the next section the use of FPGAs in implementing hash functions.

2 Implementing Hash Function in FPGAs

An effective way of performing a database search is to use a hash function to map a word to a pseudo-random value which references a lookup table (LUT), indicating whether a given word is in the user dictionary. The lookup tables are generated by passing the user dictionary through the same hash functions that are used at run time. FPGA-based computing platforms are well suited for this type of application, because a hash function and LUT can be more efficiently implemented in an FPGA and an attached memory rather than a general-purpose processor. The efficient implementation of the hash function is due to: (1) the size of the hash function not always being 32 or 64 bits in size, and (2) the irregular bit-level operations that are performed.

The hash function involves computing the XNOR or XOR of each bit of the input word with the current hash value according to the hash function mask, and then performing an n-bit circular shift. Figure 1 shows what happens when the word "the" is passed through a 22-bit hash function. These hash functions are set up to use a single bit to represent the inclusion of a word in the search list. Cascading independent hash functions has the effect of reducing the probability of a false match. A 22-bit hash function will be used in our examples in Section 5, since it is sufficient for many uses [1].

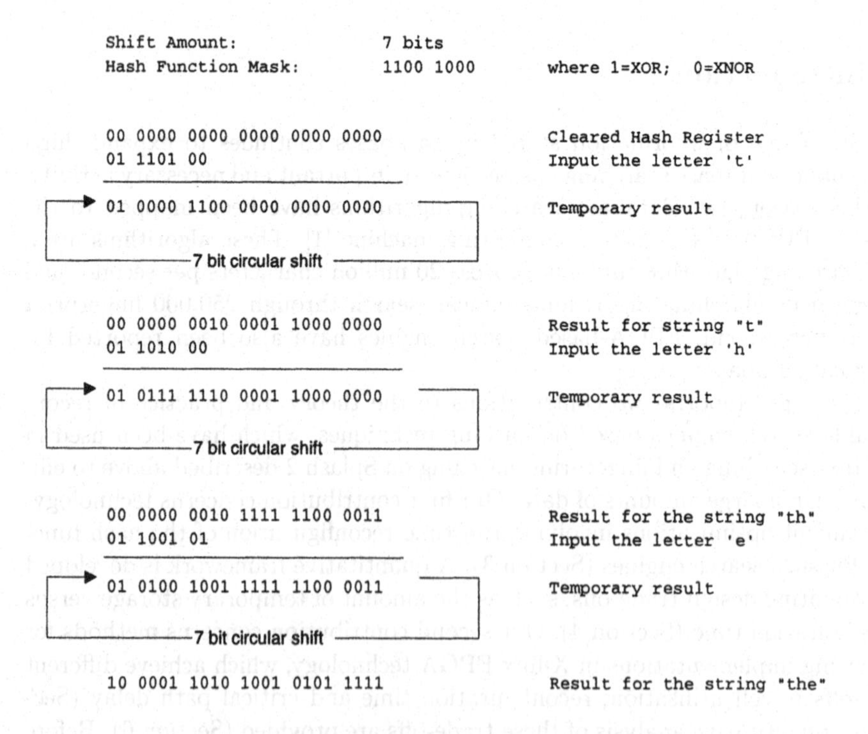

Fig. 1. 22-bit hashing example of the string "the".

3 Optimization Using Run-Time Reconfiguration

This section describes the use of run-time reconfiguration to optimise FPGA-based search engines. The system concerned should be partially and incrementally reconfigurable: the smaller the size of the reconfigured region, the lower the reconfiguration time.

Instead of cascading the hash functions in a pipeline manner, independent hash functions can be used sequentially. The sequential method has the advantage that it may result in a shorter execution, since additional hash functions are not required to test for a false match if a match does not occur.

Run-time reconfiguration can be used to change the hash function parameters and to switch between different hash functions. The two hash function parameters that can be changed at run time are the mask and shift values. An example reconfiguration state machine for the sequential method is shown in Figure 2. Since the hash functions are known at compile time, the combined reconfiguration method [5], which involves incrementally reconfiguring from one processing state to another, can be used.

A single bit of temporary storage is required for each input word to indicate whether the word is a match or not. The virtual pipeline method [3], which involves overlapping computation and reconfiguration, can be used to store the temporary data required for the next configuration. A framework will be introduced in the next section for quantitatively assessing various design trade-offs, such as the amount of temporary storage versus the length of reconfiguration time. Device-specific aspects will be covered in Sections 5 and 6.

4 Technology-Independent Design Analysis

Under what circumstances is it worth using run-time reconfiguration within an application? We attempt to answer this question by quantifying the trade-offs

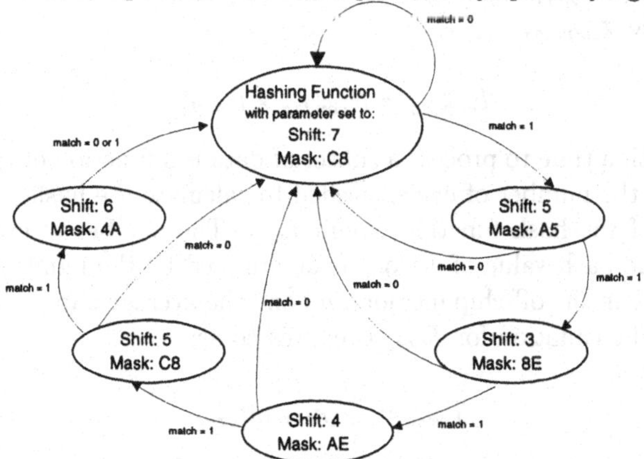

Fig. 2. Example of a reconfiguration state machine for the hash function used in the text searching application. Note that the initial state is at the top of the diagram.

between implementing the text searching application by using multiple FPGAs or a single partially reconfigurable FPGA. For this analysis, the design parameters that we will address are total execution time, FPGA area within a single FPGA or multiple FPGAs, and the amount of memory required for temporary storage to support data recirculation [3].

For this application, the input data set are divided into distinct parts to minimize the amount of temporary storage, and also to control the frequency of reconfiguration. We divide the total number of words, w, into l subsets of words, so each subset contains w/l words. This data subset is processed using a particular hash function, and one bit per word is used to indicate if a match has occurred. The indicator bit is stored along with the corresponding word in temporary memory. The temporary data are recirculated and processed by the next hash function, and the match indicator bit is updated on each iteration. This cycle is repeated until all the hash functions have been used. Once the cycle is completed, all temporary data are discarded and the next data subset is processed. The frequency of reconfiguration is controlled by the size of the data subset, because reconfiguration occurs once after each data subset has been processed.

In the equation describing the total execution time, we do not take into account the possibility that, if a match does not occur, the cycle is terminated and the remaining hash functions in the sequence are not used. Instead, we assume the worst case and execute all the hash functions in the reconfiguration state diagram.

The total execution time to process one subset of data, T_{subset}, is the sum of the reconfiguration time, T_{config}, and the execution time, T_{exec},

$$T_{subset} = T_{config} + T_{exec} \tag{1}$$

The configuration time is a product of the number of cycles required for reconfiguration, N_{config}, and the speed the FPGA can perform a single reconfiguration cycle, t_{config},

$$T_{config} = N_{config} \times t_{config} \tag{2}$$

The execution time to process a subset of data is a function of the size of the data set, w/l, the number of cycles needed to calculate the hash value, and the critical path of the hash function circuit t_{exec}. The number of cycles required to calculate the hash value of a word is determined by the number of cycles to access the LUT in the off-chip memory, m, and the average number of characters per word, c. The equation for T_{exec} is expressed as:

$$T_{exec} = \frac{w}{l}(m + c)t_{exec} \tag{3}$$

From Equation 1, the total execution time for one subset of data is given by:

$$T_{subset} = N_{config} \times t_{config} + \frac{w}{l}(m + c)t_{exec} \tag{4}$$

Since the execution order of the hash functions is not important, a number of them can be executed in a pipeline manner or by sequentially reconfiguring between them. The total execution time varies with the number of hash functions, h, and the number of these hash functions that are executed in parallel, p. The ratio h/p is the number of hash functions that are being computed at one time. This ratio, together with the number of data sets to be processed l and the execution time for a single subset of data (Equation 4), result in the equation for the total execution time, T_{total}:

$$T_{total} = \frac{hl}{p}\left(N_{config} \times t_{config} + \frac{w}{l}(m+c)t_{exec}\right) \quad (5)$$

where h denotes the number of hash functions
p denotes the number of pipeline stages
l denotes the number of subsets of the complete input data set
N_{config} denotes the number of cycles needed for reconfiguration
t_{config} denotes the duration for a single reconfiguration cycle
w denotes the number of words
m denotes the number of cycles needed to access a LUT in off-chip memory
c denotes the average number of characters per word
t_{exec} denotes the critical path of a hash function

Equation 5 is only valid for $w > 1$, $l \leq w$, $l > 0$ and $p \leq h$. The amount of temporary memory storage needed is based on the number of words in a subset of data, and the average number of characters per word of the subset. Also, an additional bit is added per word to indicate if a match has occurred. The total number of bits needed for temporary storage is given by:

$$Mem = \frac{w}{l}(b \times c) + \frac{w}{l} \quad (6)$$

where b is the number of bits per character, typically of value 7 or 8. The total circuit area is determined by the number of hash functions that are computed in parallel, p, and the size of each hash function, a,

$$Area = p \times a \quad (7)$$

Since the circular shifter is the primary component of our hash function, the above equations will be used in Section 6 to estimate the speed and size of shifter implementations for Xilinx 6200 FPGAs presented in the next section. Their efficiency will be calculated according to the functional density metric proposed by Wirthlin and Hutchings [6]: see Equation 8 below. Since we are also interested in characterising the efficient use of temporary memory, an additional metric, given by Equation 9, is obtained by replacing the area term by the amount of temporary memory used:

$$D_{Area} = \frac{1}{(Area)(T_{total})} \quad (8)$$

$$D_{Mem} = \frac{1}{(Mem)(T_{total})} \quad (9)$$

5 Device-Specific Mapping

This section introduces several shifter designs in Xilinx 6200 FPGA technology which supports partial run-time reconfiguration. The reconfiguration time of a circuit can be greatly reduced by taking into account the mechanism used to program an FPGA, as well as device-specific optimizations that may be available. For example, the Xilinx 6200 FPGA has a facility called 'wildcarding' that allows simultaneous configuration of multiple FPGA cells with the same data. Wildcarding can be performed on all 64 rows of the chip to allow an entire column to be configured in one configuration write cycle. However, wildcarding can only be performed on 4 columns at a time, therefore only 4 cells in a row can be configured at one time. Due to this limitation, it is advantageous to align components that are going to be reconfigured vertically along the columns of the FPGA in order to take maximum advantage of wildcarding.

A compact variable circular shifter has been implemented in Xilinx 6200 technology (Figure 3). The component is made up of an array of multiplexors that circularly shift the data n-bits; this design uses only the four nearest neighbour connections available in each FPGA cell. The clock period for a 22-bit variable circular shifter is 105 ns; the critical path delay is due to the control lines routed diagonally across the array to each of the multiplexors. Without taking routing into account, this component requires n^2 FPGA cells, or 484 cells in the case of a 22-bit shifter. If routing is taken into account, the area of the 22-bit variable circular shifter grows to 638 cells. Since the component is constructed as a variable circular shifter, it only takes a single cycle to change the shift amount. The relevant statistics are summarised in the top row of Table 1.

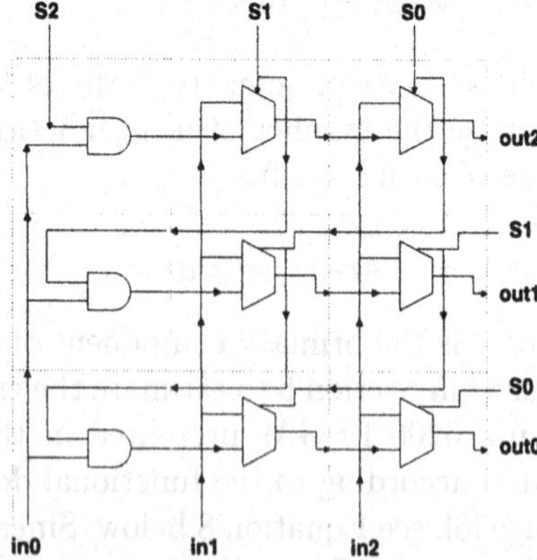

Fig. 3. Circuit diagram of a 3-bit Variable Circular Shifter. S0, S1 and S2 are signals controlling the shift amount; only one of them is high at any time.

Three different implementations of the circular shifter have been developed to reduce their critical path delay. The first method uses constant propagation

Table 1. Trade offs between different implementations of a 22-bit circular shifter.

Circuit	Clock Period (ns)	Area (cells + routing)	Reconfiguration Time (number of cycles)
Variable	105	638	1
Fixed	29.5	88	245
Square Hybrid	37	484	44
Rhombus Hybrid	55	484	4

techniques. The variable n-bit circular shifter is converted to an n-bit constant circular shifter, by treating the variable shift amount as a constant. The propagation of this constant converts the array of multiplexors into simple wire connections. For example, a 22-bit fixed circular shifter with a 7-bit shift amount has been designed by propagating the constant shift value into the circuit. The size of the component is reduced from 638 cells to 88 cells and its critical path from 105 ns to 29.5 ns. However, this shifter requires much random routing to perform the shift, and hence the time needed to reconfigure it is large.

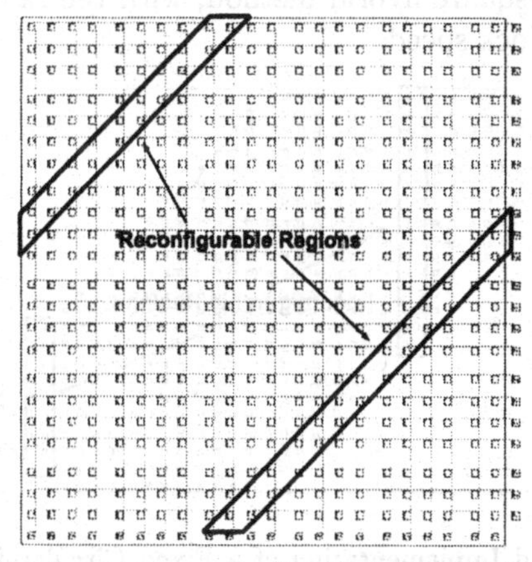

Fig. 4. Square Hybrid Implementation of a Fixed Circular Shift Component on a Xilinx 6216 device. The reconfigurable regions are enclosed by the two boxes.

The second method, known as the square hybrid method, is a hybrid between the variable implementation and the fixed version of the circular shifter. A two-dimensional array of buffers are used to control the routing of the component. The input data enter the component from the bottom and perform a corner turn and exit from the right. The shift amount is determined by the location of the corner turn in the array of FPGA cells; the corner turn is implemented by diagonally connecting the bottom inputs to the outputs on the right at different positions, resulting in two diagonally-placed reconfigurable regions (Figure 4). The reconfiguration time for this method is $2n$ cycles, where n is the size of the circular shifter, since it takes two clock cycles to make a new connection and remove the previous one.

The third method, known as the rhombus hybrid method, is similar to the square hybrid method except that the layout has been rearranged to enable fast partial reconfiguration. As previously discussed, the Xilinx 6200 FPGA supports wildcarding more effectively along a column of the chip rather than a row. The layout of the square hybrid implementation has been skewed into a rhombus shape, so that the previous diagonal reconfiguration regions are aligned into two columns (Figure 5). The number of cells used in the rhombus hybrid method is the same as that for the square hybrid method. However, due to its irregular shape, it is harder to take advantage of the unused triangular-shaped areas on either side of the component. Also, the input data have to propagate diagonally across the component; since the FPGA only has Manhattan-style routing, this increases the critical path. The advantage of using this method is that the reconfiguration time is reduced to a constant value regardless of the size of the component, thanks to wildcarding. Wildcarding can be performed since the reconfiguration involves changing the reconfigurable region to identically configured cells. This reconfigurable region consists of buffers that either perform a corner turn or pass data from the left to the right. As shown in Table 1, the reconfiguration time for the rhombus hybrid method is an order of magnitude less than that for the square hybrid method, with the same circuit area and a modest reduction in clock speed.

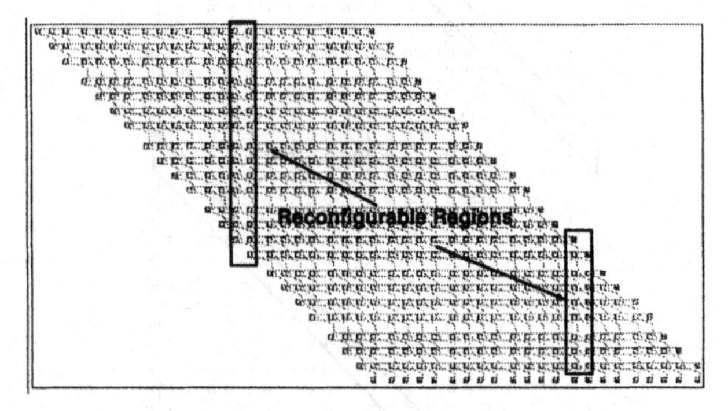

Fig. 5. Rhombus Hybrid Implementation of a Fixed Circular Shift Component on a Xilinx 6216 device. The reconfigurable regions are enclosed by the two boxes.

6 Design Analysis: Xilinx 6200 Specific Cases

Three test cases are used to illustrate the trade-offs between execution time and the amount of temporary storage required, using the equations developed in Section 4. The first test case assumes that there is a large amount of temporary storage available. The second test case examines the other extreme case and assumes that only a very small amount of temporary storage is available. In reality, there is usually a moderate amount of temporary storage available, either off chip or within the FPGA, and the third case examines this possibility.

Case #1. If a large amount of temporary storage is available, then we do not need to subdivide the input data set. Since the input data set is not subdivided, the amount of reconfiguration is minimized and reconfiguration only occurs after the complete data set has been processed. To specify this case, the value of l is set to 1 in Equation 5, which indicates that the input data set is one complete set. A typical number of words to be searched is 10×10^6, hence $w = 10 \times 10^6$ and the average number of characters per word is set to 5, so $c = 5$. The number of hash functions used to statistically ensure that false matches do not occur is 8, so $h = 8$, and we set $p = 1$ which indicates sequential execution of hash functions. Using these parameters, along with the area and speed data given in Table 1, we calculate the total execution time and functional densities with respect to the number of cells and the amount of temporary storage for each of the circuits described in Section 3. The results are shown in Table 2.

Table 2. Case #1: Total Execution Time and Functional Densities when $l = 1$.

Circuit	Execution Time (sec)	D_{Area} $(1/(cells \times sec))$	D_{Mem} $(1/(bits \times sec))$
Variable	50.4	0.31×10^{-4}	0.48×10^{-10}
Fixed	14.16	8.03×10^{-4}	1.72×10^{-10}
Square Hybrid	17.76	1.16×10^{-4}	1.37×10^{-10}
Rhombus Hybrid	26.4	0.78×10^{-4}	0.92×10^{-10}

Since reconfiguration time is very small compared to execution time, the circuit with the smallest critical path, the Fixed Circular Shift Component, is the fastest design. In this case, the critical path delay is the dominating factor in the overall computation time. The Fixed Circular Shift Component also has the maximum functional density with respect to both area and temporary storage size of the four different implementations. The disadvantage of using this method is that, in this case, 48 MBytes of temporary storage is required.

Case #2. If limited temporary storage is available, frequent reconfiguration minimizes the amount of temporary storage needed. In this case, we examine the extreme case by reconfiguring to the next hash function after every word. This is done by dividing the input data set into the smallest possible unit, a single word, by setting $l = w$. When $l = w$, Equation 6 is reduced to $Mem = (b \times c) + 1$, and therefore only 41 bits needs to be stored between reconfigurations. This can be achieved using a single register on the FPGA, and off-chip temporary storage is not required.

Using the same parameters in Case #1, we calculate the total execution time and functional densities with respect to the number of cells and the amount of temporary storage for each of the circuits (Table 3). In this case, since reconfiguration time is large compared to computation time, the reconfiguration time of the circuit is the dominating factor in the overall execution time. An implementation that does not incorporate run-time reconfiguration may be expected to have the fastest overall computation time. However, this is not the case in this example, since the Rhombus Hybrid method uses run-time reconfiguration but still has the overall fastest execution time. The small reconfiguration time is due to the optimizations discussed in Section 3.

Table 3. Case #2: Total Execution Time and Functional Densities when $l = w$.

Circuit	Execution Time (sec)	D_{Area} $(1/(cells \times sec))$	D_{Mem} $(1/(bits \times sec))$
Variable	58.8	2.7×10^{-5}	4.15×10^{-4}
Fixed	592	1.9×10^{-5}	0.412×10^{-4}
Square Hybrid	148	1.4×10^{-5}	1.65×10^{-4}
Rhombus Hybrid	44	4.7×10^{-5}	5.54×10^{-4}

Table 4. Case #3: Total Execution Time and Functional Densities when $w/l = 16$.

Circuit	Execution Time (sec)	D_{Area} $(1/(cells \times sec))$	D_{Mem} $(1/(bits \times sec))$
Variable	50.9	0.31×10^{-4}	2.99×10^{-10}
Fixed	50.3	2.26×10^{-4}	3.03×10^{-10}
Square Hybrid	25.9	0.79×10^{-4}	5.88×10^{-10}
Rhombus Hybrid	27.5	0.75×10^{-4}	5.54×10^{-10}

Case #3. A more realistic scenario occurs when there is limited amount of temporary storage and the optimal implementation has to be found. For example, if there is enough left-over chip area for 100 bytes of temporary storage on the FPGA, we need to know which one of the four circuits would have the fastest execution time. The same parameters in Case#1 yield a value of l to be approximately $1/16^{th}$ of w. Again, total execution time and functional densities for each of the circuits are calculated and are shown in Table 4. We find that the circuit with the fastest execution time, given these parameters, is the Square Hybrid Circular Shift circuit.

For $w/l = 16$, a plot of execution time versus the number of words is shown in Figure 6. If data partitioning is kept at $w/l = 16$, and fewer than 10^7 words are processed, we note from Figure 6 that the Rhombus Hybrid method is the

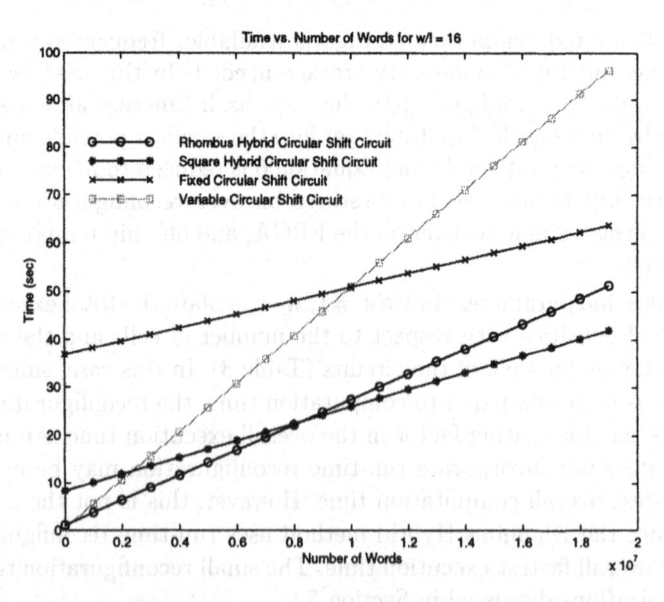

Fig. 6. Graph of Execution Time versus Number of Words for $w/l = 16$.

circuit with the fastest execution time. In this case we can reconfigure the FPGA to use this circuit until the cross over point is reached where the Square Hybrid circuit is the faster circuit.

The test cases do not cover the use of multiple FPGAs executing the text search in parallel as a pipeline. However, by changing the value of p, this case can be explored using the same method described in this section.

Equations 5, 6 and 7 enable us to find the circuit which is the most appropriate under given constraints, such as the availability of FPGA resources, by quantifying the trade-offs between the amount of temporary storage and execution time for this application. Other applications such as image processing can also be explored using these methods.

7 Summary

This paper describes how run-time reconfiguration can be used to optimise database search engines. Both technology-independent and device-specific aspects are covered by our framework, which supports quantitative analysis of design trade-offs in performance and resource usage. Current and future research includes extending our framework to exploit contextual information in specific applications such as multimedia processing, to target more recent FPGAs such as Xilinx Virtex devices, and to explore various circuit optimisations such as on-line arithmetic [4].

Acknowledgements. The authors are indebted to John Gray, Douglas Grant, Hamish Fallside, Tom Kean, Stuart Nisbet and Bill Wilkie for their constructive comments. The support of Xilinx Inc., the UK Overseas Research Student Award Scheme and the UK Engineering and Physical Sciences Research Council (Grant number GR/24366, GR/54356 and GR/59658) is gratefully acknowledged.

References

1. D. Buell, J. Arnold and W. Kleinfelder, *Splash 2, FPGAs in a Custom Computing Machine*, IEEE Computer Society Press, 1996.
2. E. Lemoine and D. Merceron, "Run time reconfiguration of FPGAs for scanning genomic databases", *Proc. FCCM95*, IEEE Computer Society Press, 1995, pp. 90–98.
3. W. Luk, N. Shirazi, S.R. Guo and P.Y.K. Cheung, "Pipeline morphing and virtual pipelines", *Field Programmable Logic and Applications*, LNCS 1304, Springer, 1997, pp. 111–120.
4. E. Mosanya and E. Sanchez, "A FPGA-based hardware implementation of generalized profile search using online arithmetic", *Proc. ACM Int. Symp. on FPGAs*, ACM Press, 1999, pp. 101–111.
5. N. Shirazi, W. Luk and P.Y.K. Cheung, "Run-time management of dynamically reconfigurable designs", *Field Programmable Logic and Applications*, LNCS 1482, Springer, 1998, pp. 59–68.
6. M.J. Wirthlin and B.L. Hutchings, "Improving functional density through run-time constant propagation", *Proc. ACM Int. Symp. on FPGAs*, ACM Press, 1997, pp. 86–92.

An On-Line Arithmetic Based FPGA for Low Power Custom Computing

Arnaud Tisserand, Pierre Marchal and Christian Piguet

Centre Suisse d'Electronique et de Microtechnique CSEM
Jaquet-Droz 1. CH-2007 Neuchâtel. SWITZERLAND
{arnaud.tisserand,pierre.marchal,christian.piguet}@csem.ch

Abstract. This paper describes the study of a new field programmable gate array architecture based on on-line arithmetic. This architecture, called Field Programmable On-line oPerators (FPOP), is dedicated to single chip implementation of numerical algorithms in low-power signal processing and digital control applications. FPOP is based on a reprogrammable array of on-line arithmetic operators. On-line arithmetic is a digit-serial arithmetic with most significant digits first using a redundant number system. The digit-level pipeline, the small number of communication wires between the operators and the small size of the arithmetic operators lead to high-performance parallel computations. In FPOP, the basic elements are arithmetic operators such as adders, subtracters, multipliers, dividers, square-rooters, sine or cosine operators.... An equation model is then sufficient to describe the mapping of the algorithm on the circuit. The digit-serial communication mode also significantly reduces the necessary programmable routing resources compared to standard FPGAs.

1 Introduction

High-performance and low-power implementations are essential in some applications such as digital control or signal processing for portable devices or embedded systems. The performed algorithms are quite complex and hence required fast and numerous computations. However, power consumption as well as size and weight have to be significantly reduced with respect to portable devices. Several implementation supports can be considered: software using general purpose processors or digital signal processors (DSP), or hardware implementations using application specific integrated circuits (ASIC) or field programmable gate arrays (FPGA). The choice relies on a trade-off between technical and financial parameters. Compared to ASIC solutions that often have too long time to market characteristics, and compared to DSP solutions that generally come up with several peripheral components, FPGA based solutions offer a very attractive alternative. Nervertheless, they still lack low-power considerations and require a non-necessary logical level description.

We present the study of a single chip architecture called *Field Programmable On-line oPerators* (FPOP) dedicated to numerical computations in low-power

* Patent pending

applications such as signal processing or digital control for portable or embedded systems. The FPOP architecture is close to a standard FPGA architecture. It is based on a reprogrammable array of on-line arithmetic operators. On-line arithmetic is digit-serial arithmetic with most significant digits first. A major difference between standard FPGAs and FPOP is the complexity of the reprogrammable cells. In standard FPGAs, the reprogrammable cells are logic operators designed to implement rather small logical functions, from 2-variable up to 5-variable functions, with a few registers. In FPOP, reprogrammable cells are arithmetic operators designed to implement adders, multipliers, dividers, square-rooters, sine or cosine operators. With respect to digital control and signal processing applications, FPOP integrates several analog to digital and digital to analog converters. The conversions and computations can overlap using on-line-arithmetic, and this can lead to shorter sampling period compared to DSP solutions.

Only high-level programming is necessary for the FPOP configuration. Contrary to the programming of standard FPGAs, numerical algorithms can be directly mapped on FPOP using a mathematical equation model because of its coarse grain architecture: the programmable cells are arithmetic operators. The FPOP architecture not only simplifies the synthesis of computational structures into the programmable operators, but it also reduces the placement and routing problems. Very fast prototyping of many complex solutions is allowed using FPOP because of the specific architecture and the simple programming flow.

The paper is organised as follows. First of all, section 2 presents on-line arithmetic and discuses the advantages and drawbacks of this arithmetic with respect to the target applications. Section 3 presents the FPOP architecture and its programming flow.

2 On-Line Arithmetic

Operands can flow through the arithmetic operators in a digit-parallel fashion or in a digit-serial fashion. Digit-parallel arithmetic leads to large operators and to numerous communication wires while digit-serial arithmetic allow the design of small operators with few communication wires. There are two possible directions in digit-serial arithmetic: least significant digits first (LSDF) or most significant digits first (MSDF). Although main operations such as addition and multiplication can be computed in both directions, some operations such as division, square root or comparison cannot be computed with LSDF. For simple operations such as additions and multiplications, serial arithmetic leads to slower circuits but it allows efficient pipelining at the algorithm level.

On-line arithmetic is a digit-serial arithmetic with most significant digits first. Using a redundant number system, it is possible to compute all standard scientific functions with MSDF: addition, multiplication, division, comparison, square-root, elementary functions $(\sin, \cos, \exp\ldots)$. The digit-level pipeline introduced by on-line arithmetic allows intrinsic sequential computations to be parallelised such as in a processor pipeline. Another advantage of on-line arithmetic is the real control of the precision. Indeed, in LSDF multiplication of 2

n-digit numbers, $2n$ digits are produced starting with the least significant part. Using an on-line multiplier only the n most significant digits can be produced, this leads to simpler control and more efficient computations because there are no breaks in the pipeline (using fixed-point arithmetic). On-line arithmetic has been introduced by Ercegovac and Trivedi in 1977 [1] and was further studied by several authors [2, 3]. On-line arithmetic has also been efficiently used by on of the author in a digital control application on standard FPGAs [4].

Table 1 presents the time and area complexity of the main arithmetic operators using parallel, on-line and LSDF arithmetic. The computation of the polynomials used for the approximation of the elementary functions can be performed using binomials $(ax + b)$ and the Horner scheme $P(x) = a_0 + a_1 x + a_2 x^2 + \ldots + a_d x^d = a_0 + x(a_1 + x(a_2 + x(\ldots(a_{d-1} + a_d x))\ldots)))$. For the evaluation of a degree-d polynomial, d binomials must be evaluated. The degree of the polynomial depends on the required accuracy and on the function, see [5, 6] for more details. All the complexity values reported in Tab. 1 are given for the best implementation of each arithmetic operator.

	parallel		on-line		LSDF	
	area	time	area	time	area	time
$\pm$	$O(n)$	$O(1)$	$O(1)$	$O(n)$	$O(1)$	$O(n)$
$\times$	$O(n^2)$	$O(\log n)$	$O(n)$	$O(n)$	$O(n)$	$O(n)$
$\div$	$O(n^2)$	$O(n)$	$O(n)$	$O(n)$	impossible	
$\sqrt{}$	$O(n^2)$	$O(n)$	$O(n)$	$O(n)$	impossible	
$ax + b$	$O(n^2)$	$O(\log n)$	$O(n)$	$O(n)$	$O(n)$	$O(n)$

Table 1. Time-area complexity of the main arithmetic operators.

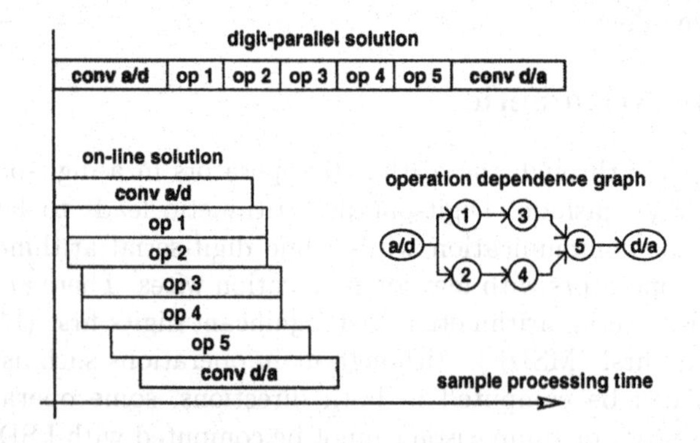

Fig. 1. Digit-parallel and on-line solution comparison.

In very accurate micro-systems, only digit-serial converters can be used [7]. These converters produce and take values in a digit-serial fashion with most significant digits first operators. The conversion from an on-line arithmetic value to an analog voltage can be achieved without intermediate conversion towards a standard number system by using two D/A converters $voltage(a) = voltage(a^+) - voltage(a^-)$, where $a^+ = \sum_{i=1}^{n} a_i^+ 2^{-i}$ and $a^- = \sum_{i=1}^{n} a_i^- 2^{-i}$. This solution can be extended to higher radices. Using digital serial converters and arithmetic operators, it is possible to overlap conversions and computations. This lead to a very efficient digit-level pipeline (see Fig. 1). This figure also shows how the combine use of on-line arithmetic and MSDF digit-serial converters can reduce the sample processing time.

3 FPOP

The reprogrammable on-line arithmetic architecture FPOP is based on architecture similar to standard FPGA architectures in which cells are arithmetic operators. FPOP takes advantage of both hardware performance and software versatility. It borrows hardware reprogrammability from FPGAs and software versatility from DSPs. Basic operators such as adders, multipliers and polynomiers are optimised blocks and may be assembled through programmable interconnections. Due to the digit-serial communication scheme between the operators, the programmable interconnection network has a reasonable size.

3.1 FPOP global architecture

FPOP is a heterogeneous cell array. As depicted in Fig. 2, it is composed of different kinds of blocks.

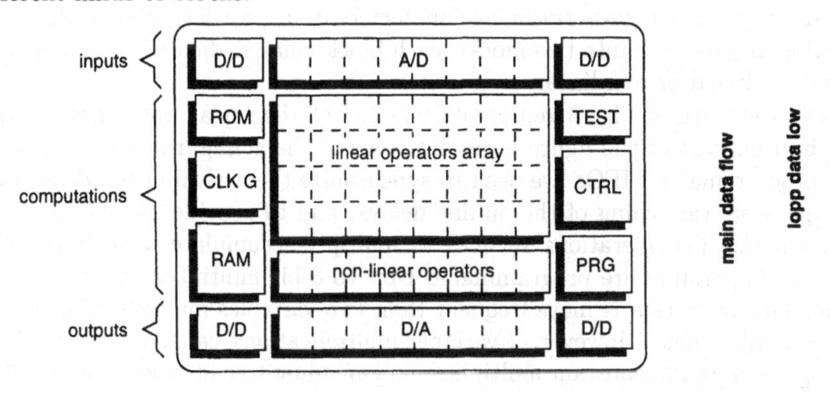

Fig. 2. Block scheme of the FPOP circuit.

3.2 Input and output blocks

The analog to digital conversion block (A/D) is devoted to interface analog measuring systems. In order to efficiently interface outside world to the internal on-line arithmetic operators, cyclic converters have been chosen, see section 2. As a matter of fact, this class of converters provides most significant digits first results and hence minimises the global on-line delay. Cyclic converters are also necessary to fulfil the accuracy requirements of high-precision applications. The resolution of cyclic converters can be programmed depending to the application requirements. It is also easy to incorporate in the final coding stage, the logic that generates the adequate input format for the on-line arithmetic operators as cyclic converters can directly produce redundant values.

The digital to analog conversion block (D/A) realises the interface with analog control system such as actuators, motors.... It performs a conversion from a digital value represented using a redundant number system to an analog value without any additional conversion delay compared to standard D/A converters, see section 2.

The digital to digital blocks (D/D) provide interfaces with standard binary digital inputs and outputs or with others FPOP circuits. These blocks perform the conversion between a standard binary notation and the internal redundant number system. This can be useful to connect microcontrollers or parallel converters. In order to connect several FPOP circuits together, these digital to digital blocks also provide a direct connection using the internal redundant number system. This avoids an unnecessary binary conversion.

3.3 Computation blocks

There are two kinds of computation blocks: the *linear operators matrix* for addition, multiplications and the *non-linear operators array* for divisions, square-roots and the elementary functions $(\sin, \cos, \exp, \log, \arctan)$. This decomposition has been imposed by the mathematical structure of our target applications. Most of the time, in signal processing and digital control applications, computations can be divided into 2 stages. The first stage computes sums of products, and the second stage computes specific functions on the result of the first stage. In order to achieve high-performance implementations, we decomposed the computation architecture into two blocks, each block being dedicated to one kind of operators, linear or non-linear.

The top part of Fig. 3 represents one cell of the linear operators matrix. The bottom part of the figure represent some of the configurations of the cell. The programmable FIFOs are used to synchronize the arrival of the digits into the operators (matching of the on-line delays) and to provide storage elements for memorization operations such as in multiply-accumulate operations. The multiplier operators are programmable 1-bit to 8-bit multipliers by a constant value. This operation is more frequent than two variables multiplication in the target applications. However, 2-variable multiplications can be performed by reprogramming two constant multipliers and an adder tree in order to implement a standard on-line multiplication operator. A constant multiplier can also be configured in order to compute squares. On-line arithmetic adder trees are used to sum the products produced in the left part of the cell. Several optimized nested adder trees are proposed for the addition of several numbers. Other small operators in the bottom part of the cell allow normalization or local comparison operations (saturation in loops). As the comparison of two on-line variables is a variable delay operator in on-line arithmetic, only limited comparison are integrated in the linear cells. This is sufficient for saturation functions such as in DSPs. Complete comparison operators are available in non-linear cells since there are less frequent in the considered applications.

The size of the operands can be defined during the cell configuration. In Fig. 3, several adjacent operators in a same row can be configured and connected in order to handle larger numbers. Each linear cell allows operations on 8-bit, 16-bit, 24-bit or 32-bit numbers. Some of the possible configurations are represented on the bottom part of Fig. 3 where x, y, z are variables and a is a constant value.

In all operations, r denotes the accumulator when placed in the right hand side of the operation.

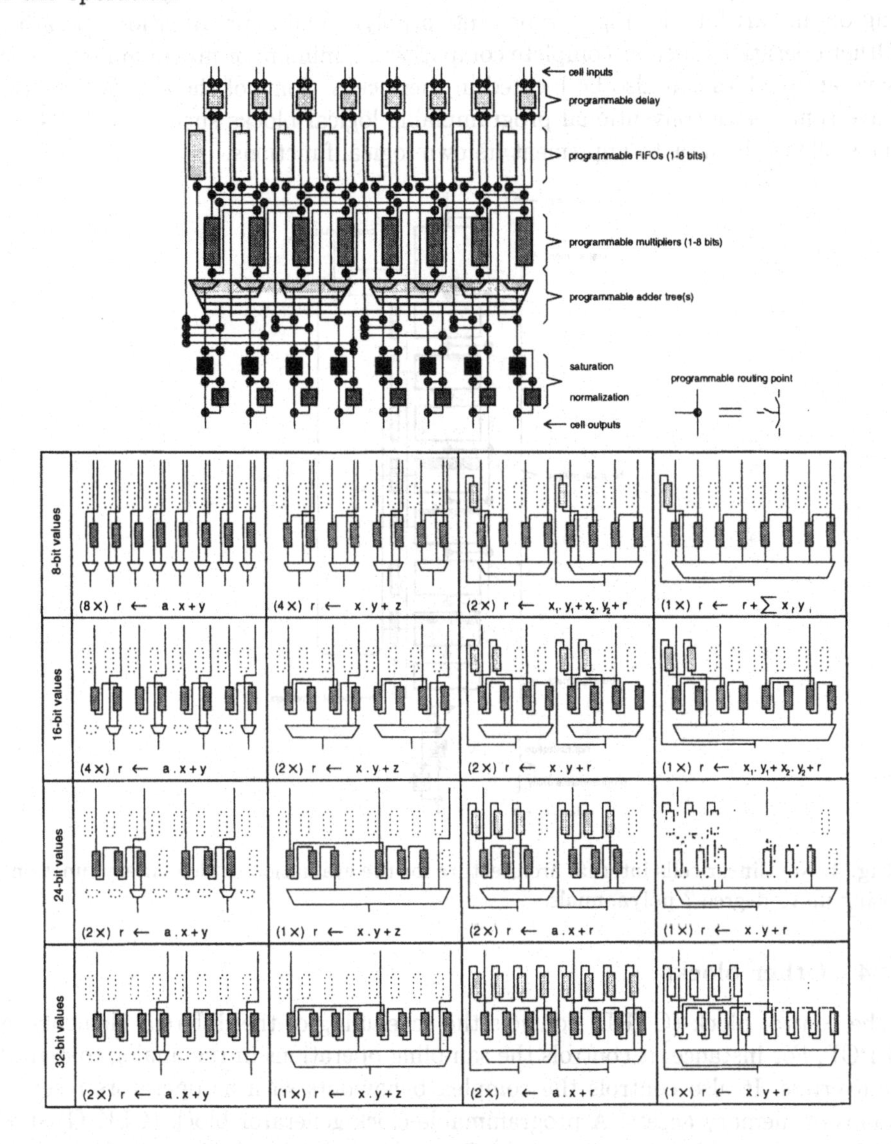

Fig. 3. A linear cell programmable architecture (top) and some of its possible configurations (bottom).

The non-linear operators block is composed of a single row of quite complex operators. Signal processing or digital control applications often leads to apply some non-linear operators after a cascade of weighted additions or sum of products (convolutions, dot products...). Non-linear functions such as division or square-root can be evaluated in this row of operators with specific optimized operators. Table-based or CORDIC-like [5, 8] methods can also be implemented in these cells in order to evaluate elementary functions (sin, cos, exp, ln, arctan, ...).

We also plan to integrate operators dedicated to the Ercegovac's E-method [6] which is a very efficient method for the evaluation of elementary functions using on-line arithmetic. Fig. 4 represents a polynomial approximation operator. Other operations such as complete comparison, minimum or maximum of two or several variables can also be located in these cells. Some of these cells can also have some more conventional programmable logical blocks such as in FPGAs. This allows the user to implement its own logical functions.

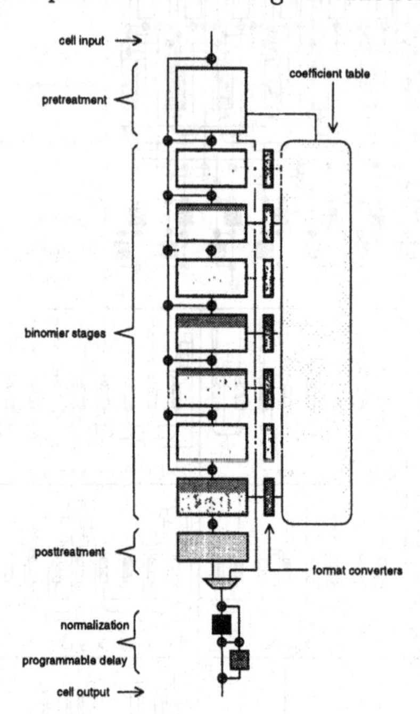

Fig. 4. Non-linear cells internal architecture for the evaluation of elementary functions using up to degree-7 polynomials.

3.4 Other blocks

The control block (CTRL) controls the scheduling of the different elements in FPOP. For instance, it controls the sampling operations in the analog to digital converters. It also controls the complex behaviours such as operators reset in loops or memory access. A programmable clock generator block (CLK G) simplify the control by providing the different clock signals used in all the circuit. This block is usefull as there are different kinds of cycles in the circuit (in the converters, in the internal loops, in the iterations of the algorithm).

The ROM stores the general coefficients used to approximate elementary functions (sin, cos, exp, log...). The RAM is provided for the user applications requiring a given precision on some precise ranges.

The configuration block (PRG) of the FPOP circuit is used during the configuration phase. It sends, in a sequential and/or in a parallel fashion, the configuration data to each programmable cell. It can also be used to test the circuit trough the test block (TEST).

3.5 FPOP programming

The main advantage at the programming level of FPOP is its ability to use high-level description for its configuration. Since all operators are arithmetic operators, an equation model of the algorithm is sufficient to perform the FPOP configuration. This allows using descriptions very close to C or Matlab programs as the input source code of the configuration. The programming tool, proposed in Figure 5, is still at the development phase.
The parser reads the source code and generates the following results:

- The list of all formal signals used in the algorithm with their names, their types and their size corresponding to the required accuracy.
- The list of all formal operations used in the algorithm with their function such as addition, multiplication, division, comparison, min..., and with their specific parameters such as size, on-line delay, internal constant values....
- The control structure of the algorithm: loops, tests, function calls....

From these data, the resource allocation can be performed. In this stage, formal operations are assigned to FPOP operators. The configurations of the used operators are build. Then internal size, internal constant value coding, internal routing schemes... are fixed. The result of the resource allocation stage is a netlist including the configuration of the operators and the formal connection between the operators extracted from the signal assignments. This netlist can also be input from a schematic tool, in which the user interconnects black boxes corresponding to the arithmetic operators.

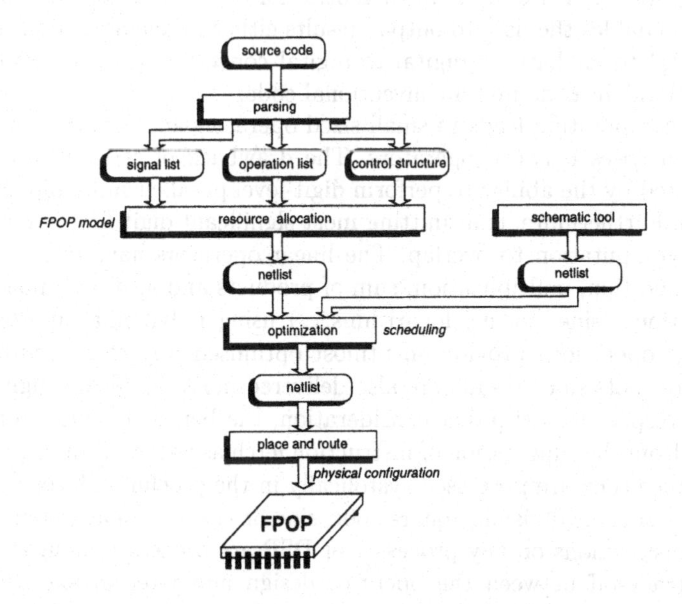

Fig. 5. FPOP configuration process.

An optimisation phase is performed in order to reduce the on-line delay of the generated design by using some transformation rules. For instance, the addition of several numbers can be optimised into a single specific adder with a very small on-line delay. The result of this stage is an optimised netlist.

The last stage in the configuration process is the physical mapping of the configuration netlist into FPOP architecture. This stage is close to the "place and route" phase in the configuration of standard FPGAs. In the FPOP case, the routing problem is simpler than with FPGAs because of the small number of connections between the operators. This is due to the digit-serial transmission mode and to the fact that there is only routing between arithmetic operators. Finally, the complete configuration file can be downloaded into the FPOP chip.

4 Conclusion

A novel architecture dedicated to complex numerical computations in low-power signal processing and digital control applications has been presented. In order to efficiently achieve computations, on-line arithmetic has been chosen. On-line arithmetic is a serial arithmetic with most significant digits first using a redundant number system. The architecture of the circuit is characterised by a hierarchical structure composed of three levels. A set of analog to digital converters and a set of digital to digital converters occupy the input level. Both types of converter properly encode inputs into a redundant code necessary for on-line arithmetic. The second level is devoted to on-line arithmetic computations. This part is itself composed of a matrix of linear operators (adders and multipliers) and a 1-D array of non-linear operators (dividers, square-rooters, comparators). It also includes RAM and ROM memories for the storage of parameters. The third level enables the user to output results either in analog or digital manner. Both digital to analog and digital to digital converters provide output results from a redundant code into a conventional code.

Serial computation leads to small sized operators and limited width of communication wires between operators. The drawback of serial computations is compensated by the ability to perform digit-level parallelism by pipelining computations. Furthermore, transmitting most significant digit first enables conversion and computation to overlap. The linear operators have been designed to compute addition, multiplication, sum of products and also to evaluate elementary functions (sine, cosine, logarithms...) using polynomial approximations. Non-linear operations provide an utmost optimised way to compute divisions and square roots since results are also delivered most significant digit first.

With respect to low-power consideration, the benefit of serial computation proceeds from the suppression of instruction fetch as well as data acquisition and storage since they are processed systolically in the pipeline. Moreover, complex algorithms such as division, square root, sine or cosine... that require hundreds or more instructions on any processor or DSP are processed alongside the data flow. A trade-off between the operator design and internal encoding scheme may be used to additionally improve the power consumption. The processing speed also has an impact on the efficiency of the solution. Generally speaking, regulation applications and digital signal processing applications generate fixed computations loops. The length of the pipeline directly provides the computation delay. It is hence possible to adapt the computation speed to the required loop length. This may be realised by adapting the clock frequency and hence the power supply. So, the FPOP architecture can lead to very efficient power×delay implementations (gain about 100).

The coarse grain structure of FPOP offers a high-level programming, since an equation model is sufficient to describe the mapping of the algorithm on the platform. The programming tools is also at the development phase.

A first prototype of the operators has been designed in Silicon on Insulator (SOI) 1-micron technology. This technology has been chosen for its advantages for portable electronics as well as for its radiation tolerance ability needed for aerospace applications. Basic cells have been integrated. As the technology is at the development phase, the circuits have not been finished yet.

References

1. M.D. Ercegovac, "On-line arithmetic: an overview.", in *SPIE, Real Time Signal Processing VII*, SPIE, Ed., 1984, pp. 86–93.
2. M.J. Irwin and R.M. Owens, "On-line algorithms for the design of pipeline architecture", in *Proceedings of the 4th IEEE Symposium on Computer Architecture.* 1979, IEEE Computer Society Press.
3. J.C. Bajard, J. Duprat, S. Kla, and J.M. Muller, "Some operators for on-line radix-2 computations", *Journal of Parallel and Distributed Computing*, vol. 22, pp. 336–345, 1994.
4. A. Tisserand and M. Dimmler, "FPGA implementation of real-time digital controllers using on-line arithmetic", in *Field Programmable Logic and Applications*, Springer, Ed., 1997, pp. 472–481.
5. J.M. Muller, *Elementary Functions, Algorithms and Implementation*, Birkhauser, Boston, 1997.
6. M.D. Ercegovac, J.M. Muller, and A. Tisserand, "FPGA implementation of polynomial evaluation algorithm", in *Field Programmable Gate Arrays for Fast Board Development and Reconfigurable Computing*, SPIE, Ed. SPIE, Oct. 1995, vol. 2607, pp. 177–188, Philadelphia, Pennsylvania.
7. B. Ginetti, *CMOS RSD Cyclic A-to-D Converters*, PhD thesis, Université Catholique de Louvain, Mar. 1992.
8. J. C. Bajard, S. Kla, and J. M. Muller, "BKM: A new hardware algorithm for complex elementary functions", in *Proceedings of the 11th IEEE Symposium on Computer Arithmetic*, E. E. Swartzlander, M. J. Irwin, and J. Jullien, Eds., Windsor, Canada, June 1993, IEEE Computer Society Press, Los Alamitos, CA.

A New Switch Block for Segmented FPGAs

M. Imran Masud and Steven J.E. Wilton

Department of Electrical and Computer Engineering
University of British Columbia,
Vancouver, B.C., Canada,
{imranm|stevew}@ece.ubc.ca
http://www.ece.ubc.ca/~ stevew

Abstract. We present a new switch block for FPGAs with segmented routing architectures. We show that the new switch block outperforms all previous switch blocks over a wide range of segmented architectures in terms of area, with virtually no impact on speed. For segments of length four, our switch block results in an FPGA with 13% fewer transistors in the routing fabric.

1 Introduction

An FPGA architecture consists of programmable logic elements and a programmable routing fabric. In commercial architectures, the routing consumes most of the chip area, and is responsible for most of the circuit delay. As FPGAs are migrated to more advanced technologies, the routing fabric becomes even more important [1]. Thus, there has been a great deal of recent interest in developing efficient FPGA routing architectures.

FPGA routing architectures consist of two components: fixed wires (tracks) and programmable interconnect between these tracks. A typical architecture is shown in Figure 2; the logic elements are surrounded by horizontal and vertical *channels*, each channel containing a number of parallel tracks.

At the intersection of each of these horizontal and vertical channels is a programmable interconnect block, often called a *switch block*. Each switch block programmably connects each incoming track to a number of outgoing tracks. Clearly, the flexibility of each switch block is key to the overall flexibility and routability of the device. Since the transistors in the switch block add capacitance loading to each track, the switch block has a significant effect on the speed of each routable connection, and hence the speed of the FPGA as a whole. In addition, since such a large portion of an FPGA is devoted to routing, the chip area required by each switch block will have a large effect on the achievable logic density of the device. Thus, the design of a good switch block is of the up-most importance.

Figure 1 shows three previous switch block architectures that have been proposed. In each block, each incoming track can be connected to three outgoing tracks. The topology of each block, however, is different.

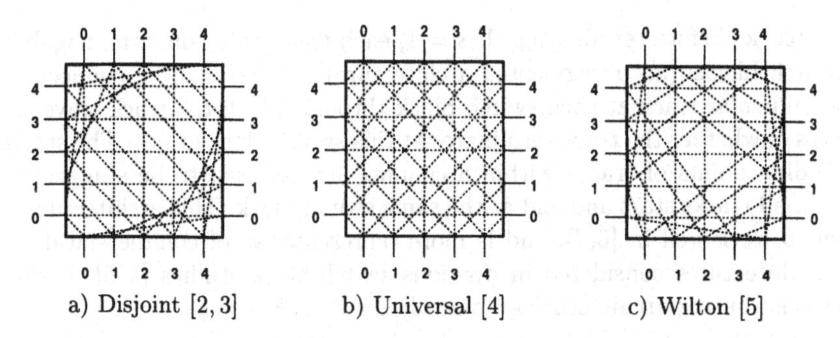

Fig. 1. Previous switch blocks.

Each of the blocks in Figure 1 was developed and evaluated assuming an architecture with only single-length wires (i.e. wires that only connect neighbouring switch blocks). Real FPGAs, however, typically have longer wires which connect distant switch blocks. Such a routing architecture is called a *segmented architecture*, and it is known that such architectures lead to a higher density and speed than an architecture with only single-length wires. Although each of the switch blocks in Figure 1 can be used in segmented architectures, this may not lead to the best density and speed. In particular, [6] showed that the Wilton switch block, while providing the best routability when used in a single-length architecture does not work as well as the Disjoint block in segmented architectures.

In this paper, we present a switch block designed for a segmented architecture. We show that it leads to significantly denser FPGAs than any other proposed switch block over a wide range of segmented architectures. This is important, since all commercial FPGAs rely on segmented routing of some sort, and we are unlikely to see any future FPGAs with only single-segment wires.

This paper is organized as follows. Section 2 describes the architecture we are targeting. Section 3 describes the new switch block, and Section 4 compares it to existing switch blocks.

2 Architectural Assumptions

We assume an island-style FPGA, in which each logic block is surrounded by vertical and horizontal routing channels, as shown in Figure 2.

Each logic block is assumed to be a cluster of four 4-input lookup tables and flip-flops. The logic cluster has 10 inputs and 4 outputs; each output can be fed-back to any of the lookup tables within the logic block. It is assumed that the four flip-flops are clocked by the same clock, and that this clock is routed on a dedicated FPGA routing track. Each of the other logic block inputs and outputs can be programmably connected to one-quarter of the tracks in a neighbouring channel.

Each routing channel consists of W parallel fixed tracks. We assume that all tracks are of the same length s. If $s > 1$, then each track passes through $s - 1$

switch blocks before terminating. If $s = 1$, each track only connects neighbouring switch blocks (an unsegmented architecture). A track can be connected to a perpendicular track at each switch block through which the track travels, regardless of whether the track terminates at this switch block or not. The starting and ending points of tracks within a channel are assumed to be staggered, so that not all tracks start and end at the same switch block. This architecture was shown to work well in [6, 7], and is more representative of commercial devices than architectures considered in previous switch-block studies [3–5]. Figure 2 shows this routing architecture graphically for $W = 8$, $s = 4$.

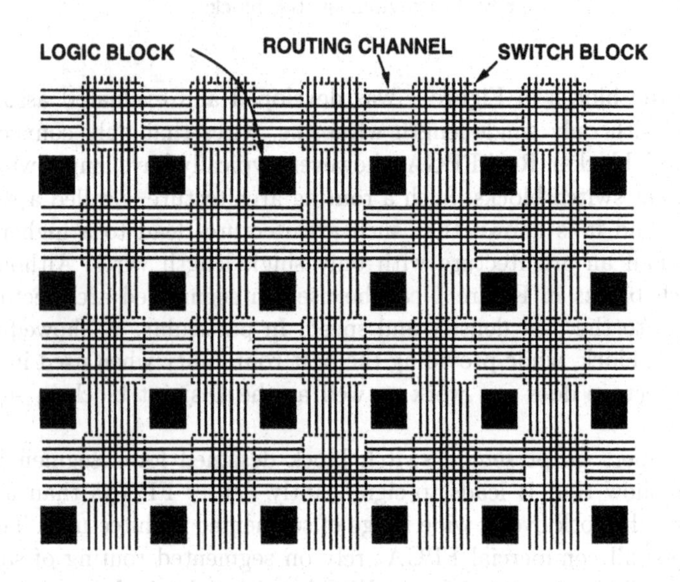

Fig. 2. Segmented Routing Arch. ($W = 8, s = 4$)

3 Switch Block Architectures

3.1 Previous Switch Blocks

Figure 1 shows three previously proposed switch blocks. In each case, each incoming track can be connected to three outgoing tracks. The difference between the blocks is exactly which three tracks each incoming track can be connected. In the Disjoint block, the connection pattern is "symmetric", in that if the tracks are numbered as shown in Figure 1, each track numbered i can be connected to any outgoing track also numbered i [2, 3]. This means the routing fabric is divided into "domains"; if a wire is implemented using track i, *all* segments that implement that wire are restricted to track i. It is known that this results in reduced routability compared to the other switch blocks. In the universal block, the focus is on maximizing the number of simultaneous connections that can

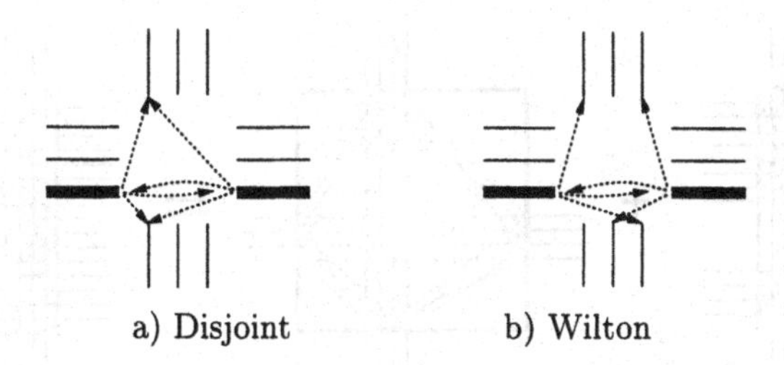

Fig. 3. Wire terminates at switch block

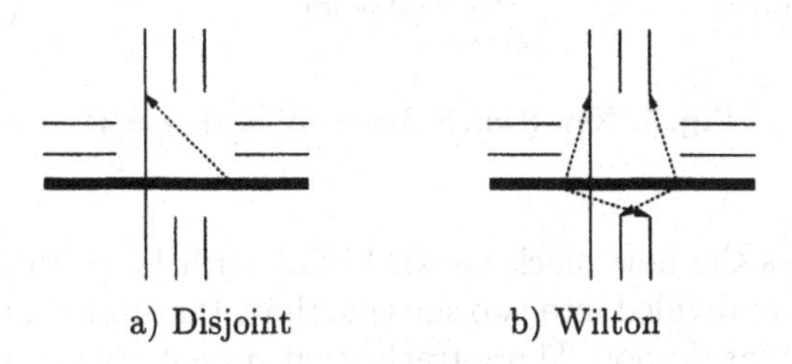

Fig. 4. Wire passes through switch block

be made using the block [4]. This does not take into account interactions between neighbouring switch blocks. The Wilton switch block is similar to the Disjoint switch block, except that each diagonal connection has been "rotated" one track [5]. This eliminates the "domains" problem, and results in many more routing choices for each connection.

The next section will show that the Wilton block is the most efficient for single-length routing architectures. It is not, however, the best choice in an FPGA with longer segments. This is because it requires more switches than the Disjoint block in such an architecture. Consider a track that terminates at a switch block. Figure 3(a) shows that for a Disjoint block, two horizontal wire segments require 6 switches to connect straight across and diagonally up and down (uni-directional switches are assumed). Figure 3(b) shows the same thing for the Wilton switch block; again, 6 switches are required. Now consider a track that passes through a switch block (and hence has a length greater than 1). In the Disjoint switch block, 5 of the 6 switches are now redundant, as shown in Figure 4(a). In the Wilton switch block, however, only two are redundant. Thus, when a wire does not terminate at a switch block, the Disjoint switch block requires fewer switches that the Wilton block, and hence is smaller and faster. In [6], it is shown that this has a significant effect on the overall speed and density achievable in the device.

3.2 New Switch Block

In this section, we propose a new switch block that combines the routability of the Wilton block and the implementation efficiency of the Disjoint block.

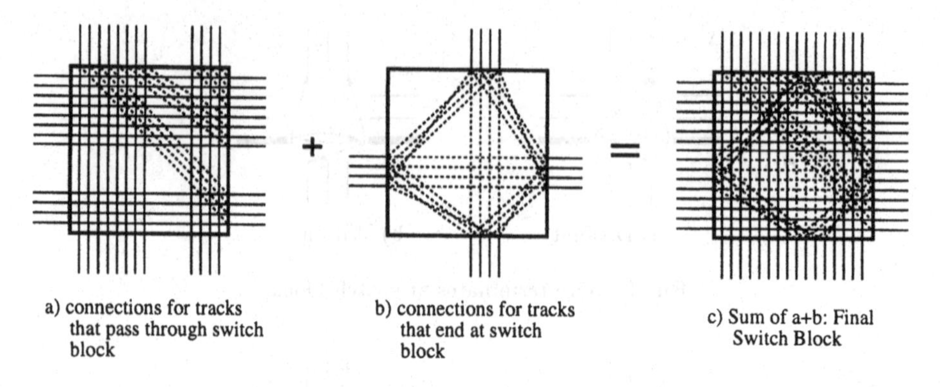

a) connections for tracks that pass through switch block
b) connections for tracks that end at switch block
c) Sum of a+b: Final Switch Block

Fig. 5. New Switch Block ($W = 16, s = 4$)

Figure 5 shows the new block for an FPGA with $W = 16$ and $s = 4$. The incoming tracks are divided into two subsets: those that terminate at this switch block and those that do not. Those tracks that do not terminate at the switch block are interconnected using a Disjoint switch pattern, as shown in Figure 5(a). Because of the symmetry of the Disjoint block, only one switch is required for each incoming track. The tracks that do terminate at the switch block are interconnected using a Wilton switch pattern, as shown in Figure 5(b). The two patterns can then be overlayed to produce the final switch block, as shown in Figure 5(c). Clearly, this pattern can be extended for any W and s.

Compared to the Wilton switch block, the new block requires fewer transistors. In the Wilton switch block, each track that does not terminate at the switch block requires 4 switches, as shown in Figure 4(b). The new switch block, however, only requires a single switch for each of these tracks. For the tracks that *do* terminate at this switch block, each block requires the same number of switches. Thus, we would expect the new switch block to be significantly smaller than the Wilton block in segmented routing architectures. As s increases, the number of tracks that terminate at each switch block decreases, meaning the new switch block is even more area efficient, compared to the Wilton block.

Compared to the Disjoint block, the new block has improved routability. As described above, the Disjoint block partitions the routing fabric into W subsets; all segments that make up a connection must be routed using the same subset. In the new switch block, the number of subset is reduced to s. Since there are fewer subsets, each subset is larger, and thus there are many more choices for each routing segment.

4 Experimental Results

In this section, we compare the proposed switch block to the existing switch blocks over a wide range of segmented architectures.

Nineteen large benchmark circuits were used. Each circuit was first mapped to 4-input lookup tables and flip-flops using Flowmap/Flowpack [8]. The lookup

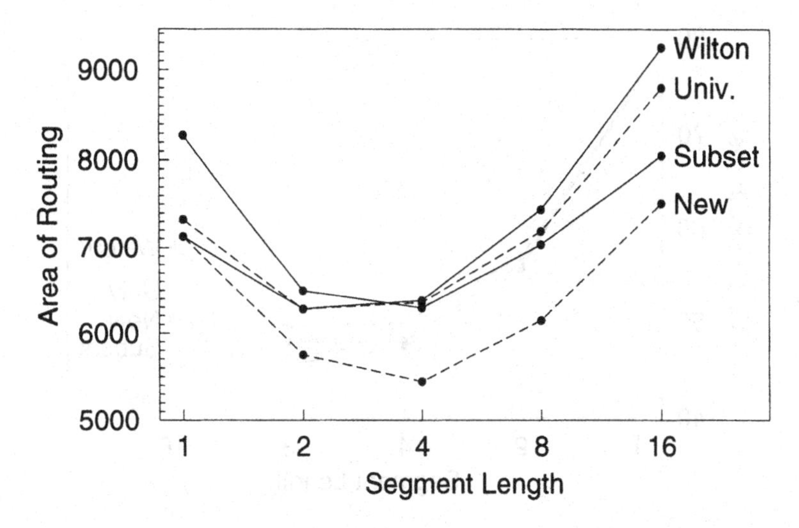

Fig. 6. Area results

tables and flip-flops were then packed into logic blocks using VPACK [6] (recall each logic block contains four lookup tables and four flip-flops). VPR was then used to place and route each circuit [6]. For each circuit and each architecture, the minimum number of tracks per channel needed for 100% routability was found; this number was then multiplied by 1.2, and the routing was repeated. This "low stress" routing is representative of the routing performed in real industrial designs. Detailed area and delay models were then used to estimate the efficiency of each implementation [6].

Figure 6 shows area comparisons for each of the four switch blocks as a function of s (segmentation length). The vertical axis is the number of minimum-width transistor equivalents per tile in the routing fabric of the FPGA, averaged over all benchmark circuits (geometric average). Previous switch block papers use the number of tracks required to route each circuit as an area metric; our metric is more accurate since it includes the effects of different switch block sizes. In addition to the transistors in the programmable routing, each tile contains one logic block with 1678 miminimum transistor equivalents [6], so the entire tile area can be obtained by adding 1678 to each point in Figure 6. As the graph shows, the new switch block performs better than any of the previous switch blocks over the entire range of the graph, except for $s = 1$, in which the new switch block is the same as the Wilton block. The best area results are obtained for $s = 4$; at this point, the FPGA employing the new switch block requires 13% fewer transistors in the routing fabric.

Figure 7 shows delay comparisons for each of the four switch blocks. The vertical axis is the critical path of each circuit, averaged over all benchmark circuits. A 0.35 μm process was assumed. Clearly, the choice of switch block has little impact on the speed of the circuit. If $s = 4$, the proposed switch block results in critical paths that are about 1.5% longer than in an FPGA employing

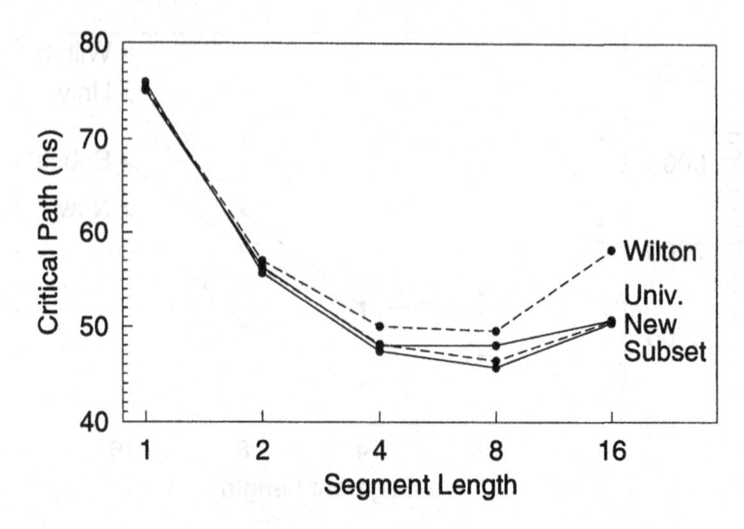

Fig. 7. Delay desults

the Subset switch block. However, this is the average over 19 circuits; in 9 of the 19 circuits, the proposed switch block actually resulted in faster circuits.

5 Conclusions

In this paper, we have presented a new switch block for FPGAs with segmented routing architectures. This new switch block combines the routability of the Wilton block with the area efficiency of the Disjoint block. Experimental results have shown that the new switch block outperforms all previous switch blocks over a wide range of segmented architectures. For segments of length 4, our switch block results in an FPGA with 13% fewer routing transistors. The speed performance of FPGAs employing the new switch block is roughly the same as that obtained using the previous best switch block.

Acknowledgments

This work was supported by British Columbia's Advanced System Institute, the Natural Sciences and Engineering Research Council of Canada, and UBC's Centre for Integrated Computer Systems Research. The authors wish to thank Dr. Vaughn Betz for his helpful discussions and for supplying us with the VPR place and route tool.

References

1. J. Rose and D. Hill, "Architectural and physical design challenges for one-million gate FPGAs and beyond," in *Proceedings of the ACM/SIGDA International Symposium on Field-Programmable Gate Arrays*, pp. 129–132, Feb. 1997.

2. Xilinx, Inc., *The Programmable Logic Data Book*, 1994.
3. G. G. Lemieux and S. D. Brown, "A detailed router for allocating wire segments in field-programmable gate arrays," in *Proceedings of the ACM Physical Design Workshop*, April 1993.
4. Y.-W. Chang, D. Wong, and C. Wong, "Universal switch modules for FPGA design," *ACM Transactions on Design Automation of Electronic Systems*, vol. 1, pp. 80–101, January 1996.
5. S. J. E. Wilton, *Architectures and Algorithms for Field-Programmable Gate Arrays with Embedded Memory.* PhD thesis, University of Toronto, 1997.
6. V. Betz, *Architecture and CAD for Speed and Area Optimizations of FPGAs.* PhD thesis, University of Toronto, 1998.
7. V. Betz and J. Rose, "FPGA routing architecture: Segmentation and buffering to optimize speed and density," in *Proceedings of the ACM/SIGDA International Symposium on Field-Programmable Gate Arrays*, Feb. 1999.
8. J. Cong and Y. Ding, "FlowMap: an optimal technology mapping algorithm for delay optimization in lookup-table based FPGA designs," *IEEE Transactions on Computer-Aided Design of Integrated Circuits and Systems*, vol. 13, pp. 1–12, January 1994.

PulseDSP - A Signal Processing Oriented Programmable Architecture

Gareth Jones

Systolix Ltd., 4th Floor India Buildings, Water Street, Liverpool L2 0QT, United Kingdom
gareth@systolix.co.uk

Abstract. Classical digital signal conditioning algorithms, such as FIR filtering, involve many simple independent calculations repeated in a fixed order. This makes them particularly appropriate for implementation using a parallel processing approach. The PulseDSP architecture is a programmable array developed to exploit this inherent parallelism. The architecture is a systolic array of simple processing elements. Data is passed between processing elements using a programmable network of serial data channels. Each processing element performs basic fixed operations on the data before passing it the next element. Algorithms are implemented by structurally describing the calculation as a signal flow then mapping it to the array on a one-to-one, operation-to-processor basis. This approach can provide a significant improvement in performance over standard DSP processor and FPGA implementations, particularly when large datawidths are required.

1 Introduction

The PulseDSP architecture has been designed to exploit the inherent parallelism of many standard DSP algorithms. Specifically it is optimised for simple repetitive signal conditioning algorithms such as Finite Impulse Response (FIR) filters rather than more decision based algorithms such as MPEG. In real applications, these simple signal conditioning algorithms tend to be located at the front of the signal chain where the data is initially digitised. This means they have higher performance requirements as the raw signal data is being processed at the full sample rate. The lack of a flexible platform with the ability process these high rate signals has been a limiting factor in the push to move back the boundaries of the analogue domain, particularly in the communications market.

The implementation of these simple algorithms is inefficient for a general purpose DSP processor as they require only a few simple instructions and make no use of the rest of the processors capabilities. The processor makes tradeoffs in order to perform across a general purpose instruction set. When implemented on an FPGA the arithmetic operators must be built using low level elements. As datawidth requirements become large this is a significant overhead, reducing the performance of the final solution. Dedicated programmable digital filters chips [1] overcome some of these problems but this is at the expense of flexibility. The user is presented with a fixed filter structure and can only modify limited elements such as coefficient values and filter length. These products cannot really be considered a generic solution even within the limited field of filtering.

The PulseDSP architecture has been developed to overcome these limitations by directly supporting the implementation of datapath type DSP algorithms. There are two main features of the architecture that differentiate it from a tradition programmable logic array and help support these designs. Firstly the basic elements of the array operate at the word level, rather than the bit level. The whole architecture is based around processing fixed word length blocks of data using processor elements optimised for this purpose. Secondly data is transmitted between elements as fixed packets comprising signal data and error information. This communication is all handled transparently to the user allowing them to work with signals and operate on them directly. Both these features make the architecture an application specific programmable array, but ensure the best performance. The resulting array is flexible device but it does not have the same degree of generality as a traditional FPGA.

2 Architecture

During development of the PulseDSP the focus was on producing an architecture optimised for DSP. This resulted in clear choices for the overall structure of the array, the basic processing elements and the communication between them. Figure 1 shows an overview of a 6 by 6 element array, this array is for illustration purposes. The smallest array actually developed is 12 by 12 elements.

2.1 Array Structure

The PulseDSP architecture is a systolic array of simple processing elements. A systolic array is an array of communicating elements where data is passed between elements synchronous to a global clock. One of the main reasons for choosing a systolic architecture is that, because of the synchronous nature of the transfers, it is very easy to predict the performance of such a device. All the elements work on a fixed word size so the matched performance also lends itself to this type of arrangement. The array uses bit serial techniques both for the transfer of data between elements and for the arithmetic processing. Each processing element performs a basic fixed operation on the data before passing it the next element. The data is passed between processing elements using a programmable network of serial data channels and is passed as a packet that contains both the signal data word and error control information.

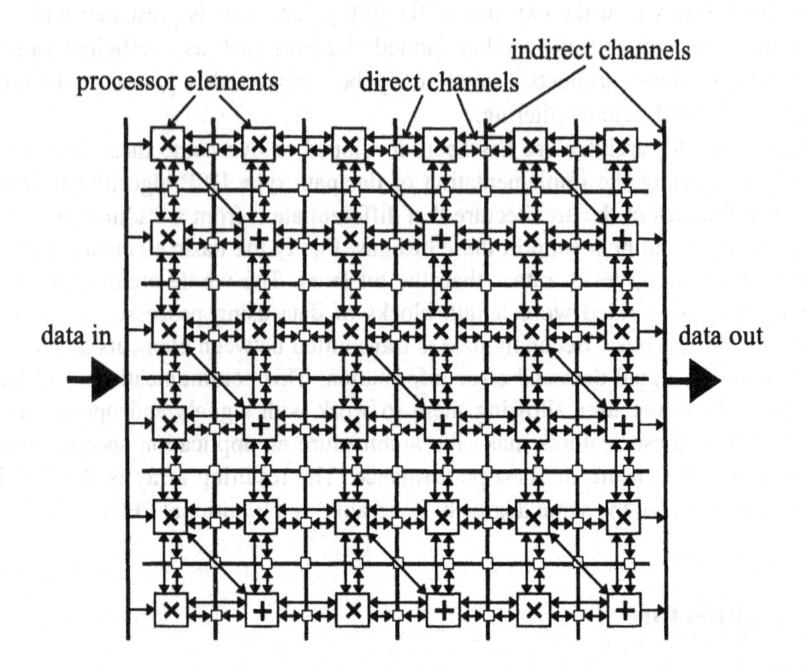

Fig. 1. The basic structure of a 6 x 6 PulseDSP array. Data enters from *left* the flows through the elements of array until it emerges on the *right*. Two types of element are shown, the multiplier element and the adder element, these occur in the ratio of 3 multipliers to 1 adder

A processor element, see figure 2, is either an adder or a multiplier. Coefficient values and data sources are programmable but not the basic function is not. Once again this is to ensure the optimum performance. The two basic functions are distributed over the array in the ratio of three multipliers to one adder, which was determined by analysing target algorithms. The multipliers and adders are arranged to combine conveniently in this ratio to form an efficient three multiply accumulate (MAC) block.

2.2 Processing Elements

The operations available are multiply two values or add up to four values. Each operation is a full calculation of N bits. For the multiplier the calculation is carried out on a stored coefficient and a signal or two signals. In the case of the adder only signal data can be processed.

In order to implement most DSP algorithms, processor elements must operate on delayed versions of the original signal data. As the array is a systolic array, all data processing is synchronised to a global clock. By having the processing element retransmit the original data, a single clock delayed version of the original is generated. By generating chains of delayed data values all operations can occur on local data thus removing the need for an external memory.

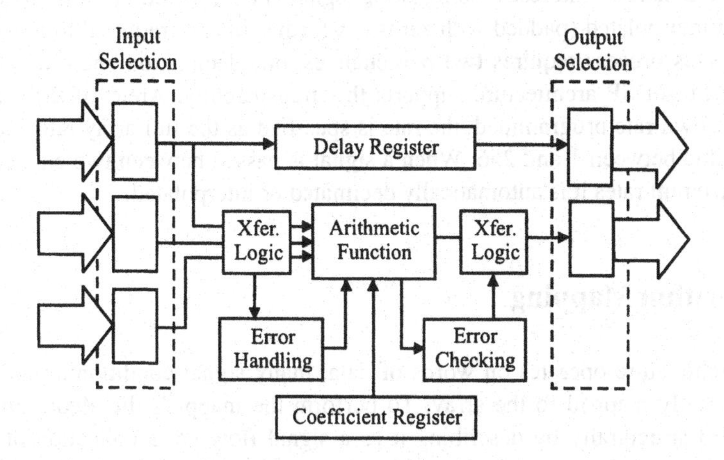

Fig. 2. This generic representation of a PulseDSP processor shows the main programmable elements, the coefficient register and the input/output selection. The transfer logic intercepts the data packet and strips out the error data for use by the arithmetic function. It also inserts the appropriate error flags based on the result of the arithmetic function. The transfer logic is also responsible for decimation or interpolation of the signal to achieve the required data rate

2.3 Communications

Data is transferred between elements using serial data channels, the control for these is embedded in each processing element. Data channels are either direct processor to processor connections or indirect connections through a programmable network of busses. The communication techniques employed on the PulseDSP array are essential to the overall performance of the architecture. Individual elements process the data serially, this means that by the time the final bit of the result has been calculated the rest of it has been transferred to the next element. This is not a problem unless the result overflows or underflows, in this situation the next stage must know to handle the data differently. The solution is to pass error flags along with the data to the next stage. These flags are used by the following stage to determine whether it is necessary to correct for the overflow or underflow situation. If this flag method were not used then the result would need to be stored locally to the element until the calculation was complete. At that point any errors could be dealt with and the result transmitted. This would both delay the signal data and increase the size of the array due to the need to temporarily store the result in the cell.

Very often a system will operate on a number of signals running at different rates. For instance in a multi-rate bandpass filter [2] the signal being processed is first mixed down with the signal from a local oscillator to lower the frequency of the band of interest. The signal is then decimated and filtered to remove any aliases. The signal is then bandpass filtered at the new lower rate, which will require fewer taps to obtain the specified band of interest. The resulting signal is then mixed up in frequency and the signal interpolated (padded with zeros). A further filter is required to remove any images. This process requires two rate changes, one decimation and one interpolation. The PulseDSP architecture supports this transparently. Areas of the array can have their own rate programmed, the rate is specified as the full array rate decimated by any value between 1 and 256. When a signal is passed between two elements running at different rates it is automatically decimated or interpolated.

3 Algorithm Mapping

As the architecture operates on words of data, many signal conditioning algorithms can be directly mapped to the array. To perform the mapping the algorithm is first represented structurally by describing it as a signal flow i.e. a collection of signals and operations on those signals. The Pulse Programming Language (PPL), a Systolix proprietary language, has been developed specifically for this task and allows user to textually describe a signal flow diagram. Examples of PPL descriptions are given in the examples in sections 3.2 and 3.3. The algorithm is then mapped from PPL to the array on a one-to-one basis with each operation mapped to a processor and each signal mapped to a data channel. . This approach makes the PulseDSP a highly intuitive platform for DSP designers to develop on. The mapping process is carried out automatically by the PulseDSP compiler, which takes PPL as its input.

3 Pulse Programming Language (PPL)

PPL provides a convenient language for defining algorithms to be implemented in the PulseDSP array. By using a proprietary language it is possible to restrict the user such that only structures that can be implemented on the architecture can be defined. This was considered appropriate for an application specific architecture and helps ensure that mapping to the array is guaranteed. This would have been a lot harder if a generic language such a VHDL was adopted. The language is also an intuitive and concise form for describing any algorithm that can be described as a signal flow graph.

PPL is a structural language consisting of modules, each of which contains a number of sampled signal definitions and expressions consisting of operators on those signals. When a signal is defined the user is able to operate on any point it the signals history, i.e. x[-n] is the nth previous sample. The result of each expression is always the most recent point in the signals history i.e. x[]. When implemented on the architecture all expressions are evaluated simultaneously, which means that all new signal values are also calculated simultaneously. It is possible to specify decimation and interpolation for individual expressions or modules this allows multi-rate systems to be defined. PPL also supports simple control structures that are available as an extension to the standard architecture however these are outside the scope of this paper.

3.2 Example - FIR Filter

The first example is a simple FIR filter [3]. The equation for which is given in equation 1. Following is the process by which the algorithm is implemented on the PulseDSP array. This process is fully automated by the PulseDSP compiler.

$$y(n) = \sum_{k=0}^{K-1} h(k)\, x(n-k) \tag{1}$$

As previously mentioned, a structural approach is used to implement algorithms. A structural representation of a signal flow can be simply mapped to the array. This process is illustrated in figure 3. Figure 3a shows a signal flow graph for the traditional direct form of an FIR filter. When building large accumulates on the PulseDSP architecture these are built up from smaller adders. This can be done as either a chain or a tree for lower latency. To take account of this the original direct form is modified to give the new construction shown in figure 3b, this can now be mapped directly to the array. The final mapping is shown in figure 3c.

The mapping provides the user with a netlist of processor elements and their programming. The final implementation stage consists of using a simple algorithm to place and route the netlist onto the array. The place and route algorithm can be reasonably simple as the complexity of the basic array elements means that the number of elements in an array will always be substantially less than a comparable programmable logic array. Additionally the algorithms that are likely to be implemented tend to have low fanout structure, this allows for the efficient use of direct connectivity. Another important factor is that data transfers are synchronous and guaranteed so there are no timing constraints to factor into the placement process which would make it a substantially harder problem.

PPL code for a simple 7tap FIR filter example

```
// Low pass FIR filter derived using FFT method with Rectangular
window (7 terms)
MODULE filter (x[], y[])
    Internal A1[], A2[]
    Y[] = 5.74608780697563E-02*x[-1] + 0.134951366142929*x[-2] +
0.19713963432779*x[-3] +A1[-1]
    A1[] = + 0.22089624291905*x[-4] + 0.19713963432779*x[-5] +
0.134951366142929*x[-6] +A2[-1]
    A2[] = 5.74608780697563E-02*x[-7]
End
```

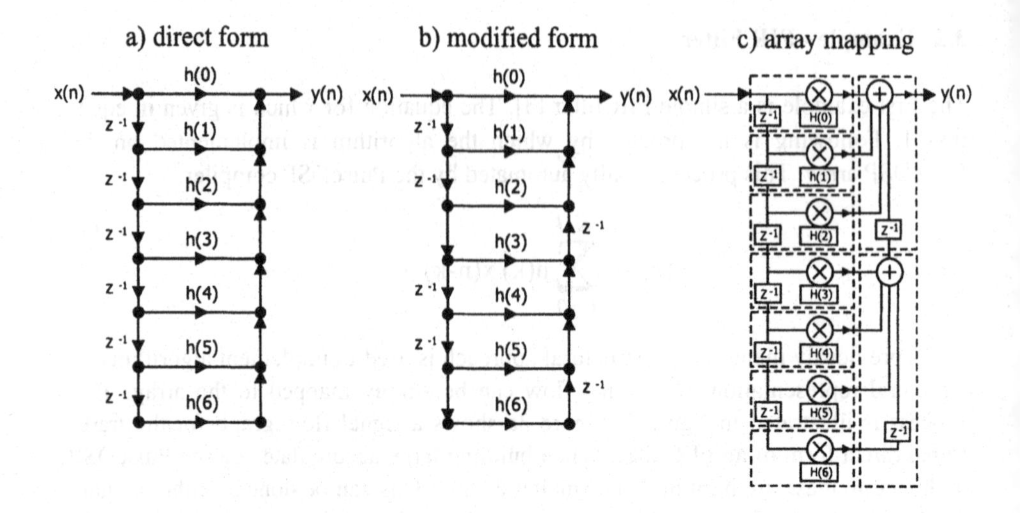

Fig. 3. *a* shows the standard direct form FIR filter with *b* giving the slightly modified form which is used for efficient mapping onto the PulseDSP architecture. C shows how the terms in b are directly mapped to processor elements on the array. A *dashed box* represents a single processor element

3.3 Example - Oscillator

The second example shows that the PulseDSP array is not just for filtering. Any system that can be defined as a difference equation can be implemented in the architecture. Consider a system as defined in equation 3, this is the z transform of the discrete sine function [4]. If an impulse is applied to this system it will oscillate with a frequency of ω. Obtaining the difference equation 5, this system can now be easily implemented in the PulseDSP array.

$$Y(z) = H(z)X(z) \tag{2}$$

$$\text{Where } h(n) = \sin(n\omega T) \text{ and } x(n)=\delta(n) \tag{3}$$

$$Y(z) = \frac{z\sin(\omega T)}{z^2 - 2z\cos(\omega T) + 1} \tag{4}$$

Giving the difference equation

$$Y(n) = \sin(\omega T)x(n-1) + 2\cos(\omega T)y(n-1) - y(n-2) \tag{5}$$

Figure 4 show the mapping process. Figure 4a shows equation 5 drawn as a signal flow graph and figure 4b shows this graph mapped to processor elements on the array. This figure also shows three elements being used to generate an impulse to start the oscillator. This oscillator could be used as a local oscillator in a modulation or demodulation system. By correlating the oscillator output with a another signal on the array, frequency analysis can be performed on that signal, this is the basis of the Goertzel algorithm [4][5].

PLL code for a simple oscillator

```
// Sine Oscillator
// 100KHz @ 1MHz sample rate
Module Oscillator (y[])
  Internal x[], d[]
  d[]= 1
  x[] = d[-1] - x[-1]
  y[] = 1.99999695327 *y[-1] - y[-2] + 0.001745328365898*x[-1]
End
```

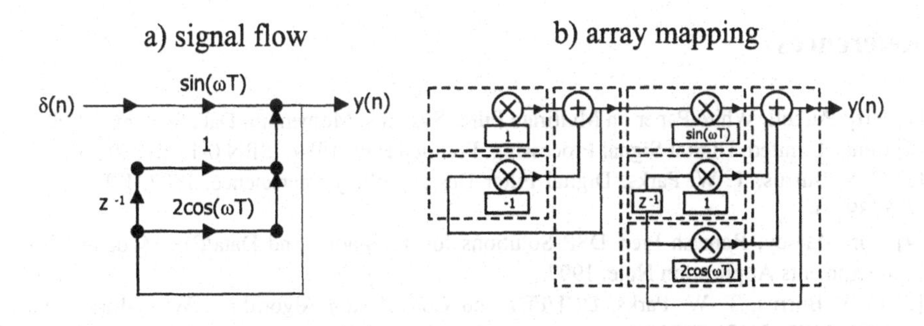

Fig. 4. *a* shows a signal graph representation of the oscillator with an impulse input to trigger the oscillation. *b* shows the mapping of *a* to the array and also illustrates the construction of an impulse generator from standard elements.

4 Performance

The simple nature of the basic processor element means that, on a standard 0.35μ CMOS process, arrays of over 3000 32 bit elements are possible. For the same process, estimates indicate that the sample clock, which determines the full 32 bit calculation time for each processor, can run at speeds of over 5MHz. A 3000 element array running at 5MHz would be capable of performing nearly 12 billions 32bit MAC calculation per second. These levels of performance allow the real time processing of signals for video, ultrasonic and wireless applications. By using parallel filters, higher rate signals can be handled. Multi-path filter structures are well supported in the architecture and the compiler. Using the quoted array as an example, it would be possible to perform a 112 tap 32 bit FIR filter on a signal with a sample rate of 100MHz.

5 Conclusions

The PulseDSP approach to the implementation of filtering and general signal conditioning algorithms provides a significant improvement in performance and ease of implementation over standard DSP and FPGA implementations, especially when large signal datawidths are required. The PulseDSP architecture can efficiently support abroad range of signal conditioning algorithms. Most filter structures can be implemented including many types of FIR and IIR filters. Multirate, multi-stage systems are supported and the programmable nature of the architecture means that adaptive systems can also be implemented. The array is not limited to signal conditioning applications, many other signal processing functions perform well on the architecture. Some of these include – Discrete Fourier Transforms, waveform synthesis, Taylor's series evaluation, complex mixing, correlation and convolution. The inclusion of control extensions to the architecture broadens this range even further.

References

[1] Jerry Purcell. White Paper on Multirate Filter Systems, Momentum Data Systems, 1999.

[2] Inmos Limited. Digital Signal Processing, Prentice Hall, 1989, ISBN 0-13-212804-7

[3] C. S. Burrus, T. W. Parks. Digital Filter Design, Wiley-Interscience, 1987, ISBN 0-471-82896-3.

[4] Tim Massey, Ramesh Iyer. DSP Solutions for Telephony and Data/Fax Modems, Texas Instruments Application Note, 1997.

[5] C. S. Burrus, T. W. Parks. DFT/FFT and Convolution Algorithms. Wiley-Interscience, 1985, ISBN 0-471-81932-8.

[6] Theodore S. Rappaport. Wireless Communications: Principals and Practice, Prentice Hall, 1996, ISBN 0-13-375536-3

[7] Chris Dick, Bob Turney, Ali M. Reza. Configurable Logic for Digital Signal Processing, Xilinx Technical Paper, 1999.

[8] Gregory Ray Goslin. A Guide to Using Field Programmable Gate Arrays (FPGAs) for Application-Specific DSP Performance, Xilinx Technical Paper, 1999.

FPGA Viruses

Ilija Hadžić, Sanjay Udani and Jonathan M. Smith
{ihadzic, udani, jms}@dsl.cis.upenn.edu

Distributed Systems Laboratory, University of Pennsylvania *

Abstract. Programmable logic is widely used, for applications ranging from field-upgradable subsystems to advanced uses such as reconfigurable computing platforms. Users can thus implement algorithms which are largely executed by a general-purpose CPU, but may be selectively accelerated with special purpose hardware. In this paper, we show that programmable logic devices unfortunately open another avenue for malicious users to implement the hardware analogue of a computer virus. We begin with an outline of the general properties of FPGAs that create risks. We then explain how to exploit these risks, and demonstrate through experiments that they are exploitable even in the absence of detailed layout information. We prove our point by demonstrating the first known FPGA virus and its effect on the current absorbed by the device, namely that the device is destroyed. We close by outlining possible methods of defense and point out the similarities and differences between FPGA and software viruses.

1 Introduction

SRAM-based programmable logic devices have been widely deployed wherever hardware performance and software flexibility are required concurrently. One of the most ambitious uses of the devices has been in the field of Run-time Reconfigurable Computing [16] where selected portions (or even the entirety) of algorithms are implemented in hardware, offering high performance while maintaining the flexibility of software systems. Various research and commercial platforms that utilize programmable logic as an accelerator or processing engine have been proposed[4, 5, 12, 15].

The majority of research on using the devices for computing has focused on the issues of mapping various well known algorithms to reconfigurable hardware[8], device technology[6], resource management[9], hardware-software codesign[11], and to some extent, programming models[10]. Very little attention - in fact none that we are aware of - has been paid to the security models for these devices. In fact, the connotation of *security* in the FPGA community has been framed in terms of protecting the intellectual property contained in a device's configuration, rather than the security and integrity of the system itself.

* This work was supported by DARPA under Contracts #DABT63-95-C-0073, #N66001-96-C-852 and #MDA972-95-1-0013, with additional support from the Hewlett-Packard and Intel Corporations.

Furthermore, PLD vendors assume that by keeping the architectural details and the format of the configuration data of their devices proprietary, the design contained in the configuration data can be secured[1].

In this paper we show that neither the system nor any associated intellectual property can be protected by practicing security through obscurity. We show by example that it is possible to deduce the architectural details necessary for constructing malicious configurations without knowledge of any proprietary information.

In the next section, we analyze the properties of FPGA devices and show how they can be exploited to create specific forms of attack. We also provide the definition of classes of attack that can be performed in run-time reconfigurable systems. In Section 3 we present an experiment in which a malicious FPGA configuration attacks a device at the transistor level and attempts to destroy it (in several experiments in our laboratory, the attempt succeeded). In Section 4, we outline the potential replicating mechanisms. In Section 5 we discuss possible methods for preventing and detecting attacks. We assess the impact of our results in Section 6, which concludes the paper.

2 Opportunities for Attack

Reconfigurable hardware has the interesting property that it can both change a system's behavior at the logic level as well its electrical properties. In general, no other architectural component has this property.

For example, by executing different programs, a processor changes its logic behavior, but the electrical properties of the system remain unchanged. Similarly, memory can be viewed as a lookup table for which the logic behavior is programmed by changing its content. On the other hand, reprogramming an FPGA device can change its electrical properties (*e.g*, power consumption, pin types, slew rate of output signals, etc.). A malicious user can exploit this property to cause damage (effecting a security attack) at the electrical signal level. This creates an entirely new class of destructive behavior than the attacks used by software computer viruses. Most interestingly, these attacks are centered on the physical destruction of the system (*e.g.* by overheating).

To classify the wide range of potential attacks to a system, we have created three categories based on the type of threat:

- *Level 0* (Electrical Signals): The attacker creates electrical conflicts either inside the device or at pins connecting the attacked device to other components of the system. The goal of this attack type is to physically destroy system components. We call this class of threat a Malicious Electrical Level Threat (*MELT*).
- *Level 1* (Logic Signals): The attacker generates signals which are electrically correct, but logically make no sense to other devices. For example, an FPGA device attached to a processor bus can generate a sequence of signals which do not represent any meaningful bus cycle causing unpredictable behavior

of the system. We call this class of threat a Signal Alteration Logic Threat (*SALT*).
- *Level 2* (Software Attacks): Finally, a virus may generate legitimate cycles which together compose an execution of a malicious task (*e.g.*, deleting data from the disk). This attack level is equivalent to the attacks performed by software viruses. We call this class of threat a Higher Abstraction Level Threat (*HALT*).

MELT represents an interesting, and most destructive, form of attack enabled by the addition of reconfigurable hardware. SALT attacks may cause unpredictable behavior in the system, but cannot directly cause physical damage (although they may indirectly cause damage, such as forcing a disk device or FLASH to operate until failure). It is less destructive than MELT, but its detection and prevention can be very difficult, mainly because any such prevention requires a rather complete model of the system in which the device is embedded. Finally, HALT attacks should be treated the same as malicious software code (*i.e.* software virus) and are thus not FPGA-specific. They are harder to detect since the device has no model for the valid behavior of systems scaffolded on top of it. In such instances, a defense should be based on establishing a trust relationship between the source of the configuration and the user executing it (*i.e.*, when configuring the FPGA device). A HALT attack is thus a classic security problem and therefore neither novel nor of particular interest to FPGA users.

The attacker's goal at the electrical level is to physically damage the system. To destroy a system component, the attacker must create high currents either inside the device or at its input/output pins. The latter case can be easily realized provided that the attacker is familiar with the board level architecture of the system. It is necessary to know which pins of the attacked device are supposed to be configured as inputs (*i.e.* connected to outputs of external devices) and configure them as outputs.

This will result in a potential conflict in logic levels creating a high current through the output transistors of both the device and whatever is connected to it, for example another device, as shown in Figure 1 (a).

Either device may be destroyed if a high current is applied for a sufficiently long period of time. In addition to knowing the board level architecture, the attacker must have some insight on the behavior of external signals, as high currents exist only if the attacked device outputs the logic complement of signals applied to it. Since the compiler is not aware of the board level system architecture, such a malicious configuration represents a legitimate design from the device's perspective and a defense using compiler techniques does not appear possible.

The second group of electrical level attacks attempts to create high currents inside the device by programming it with a configuration that creates a logic conflict in the internal connections. Most FPGA devices use pass gates to connect logic blocks to routing resources. This represents an opportunity for creating an internal conflict if two (or more) logic blocks are configured to drive the same routing resource as shown in Figure 1 (b).

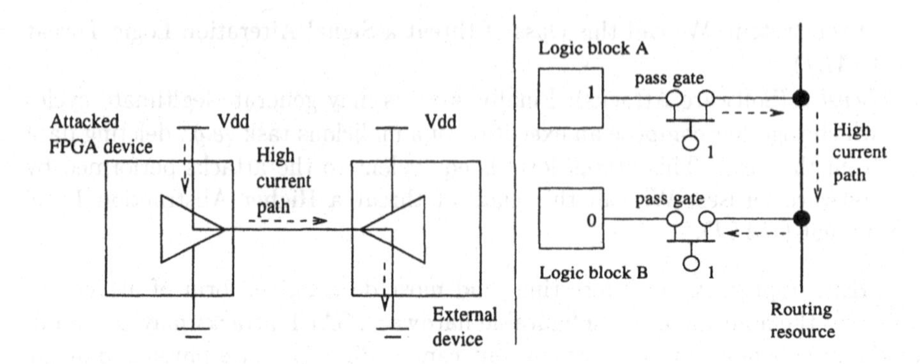

Fig. 1. Electrical conflict at a) I/O pins - left and b) logic elements - right

In contrast with the previous example, such a configuration cannot be generated by a compiler. However, it is in fact possible to modify the compiler output file (*i.e.*, the device configuration data) and create internal logic conflicts. In the next section we demonstrate the construction of such a configuration, using the Altera EPF8636ALC84-4 device[3] as an experimental platform (any vendor's devices will exhibit similar properties), using no Altera proprietary information and rendering the device inoperable.

3 Constructing a Destructive Configuration

In this section we describe an attack at the electrical level (MELT) that creates internal logic conflicts taking advantage of the fact that the interconnection between routing resources is achieved via pass-gates that connect multiple logic blocks to the same routing resource.

A connection of multiple logic blocks to the same routing resource via a pass gate represents a vulnerable point in FPGA device. In Altera Flex 8000 family, the logic elements (LE) from logic array blocks (LAB) in the same column share a column interconnect.

For example LE(1) in LAB(A1) in Figure 2 will share the column interconnect with LE(1) in LAB(B1). Therefore, if we could program the LE(1) in LAB(A1) and LAB(B1) to output complementary signals and connect both of them to the shared column interconnect, an internal conflict would be created. This conflict pattern can be replicated as many times as the device size allows increasing the device power consumption up to a level sufficiently high to overheat and destroy it. Column interconnects are the *only* vulnerable point in Altera's Flex8000 family, since each logic element has a dedicated row interconnect and internal conflicts among the logic elements in the same row are not possible.

To construct an internal conflict, we analyzed the configuration files of simple logic designs and compared the differences resulting from changing the logic element assignments and the logic functions. We identified locations in the configuration file which correspond to the logic element configuration and connections

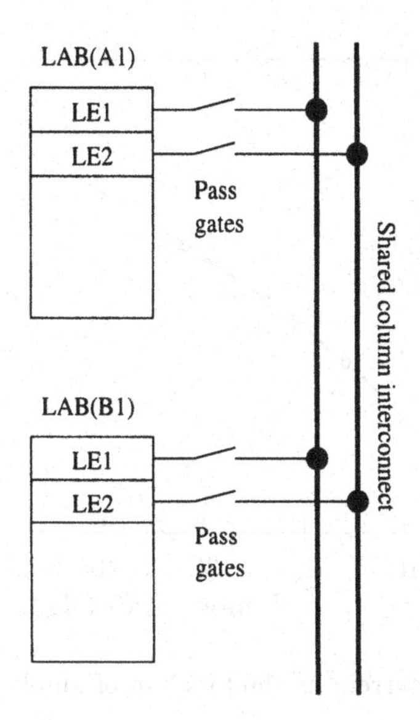

Fig. 2. Shared column interconnect in Altera Flex 8000/10K devices

to column interconnect. To create a conflict we used two designs, one that utilizes a 4-input NOR gate in row A of the device and one that utilizes a 4-input OR gate in row B[1]. We created a configuration with internal conflicts by mixing the logic elements and column interconnects from two previously described configurations. Since the configuration files for these devices do not use a global checksum or CRC, this cut-and-paste type of attack can be easily realized. Even if a global checksum or CRC existed, the device would not be secure since the attack would require only slightly more effort.

We have experimentally verified our claim by downloading configurations with internal conflicts into the device and measuring the supply current. No clock was applied so the measured results represent the quiescent supply current which is typically very low. The results of our measurements are shown in Figure 3. With only one conflict the quiescent current is greater than the maximum of $10mA$ specified by the datasheet[3] and it grows almost linearly with the number of conflicts. A small non-linearity appears due to the fact that the mobility of carriers in silicon drops with the rising temperature, causing some current flow to be reduced. Typically, the supply current will have an overshoot after the device is configured and will fall as the device heats up until it reaches steady state. Despite this negative temperature feedback, quiescent current can grow arbitrarily and the upper limit is determined only by device size – that is, the number of possible logic conflicts.

[1] The logic function used is arbitrary as long as the functions in row A always output the complement of functions in row B and are sufficiently "complex" so as to prevent the compiler from placing the logic in I/O blocks

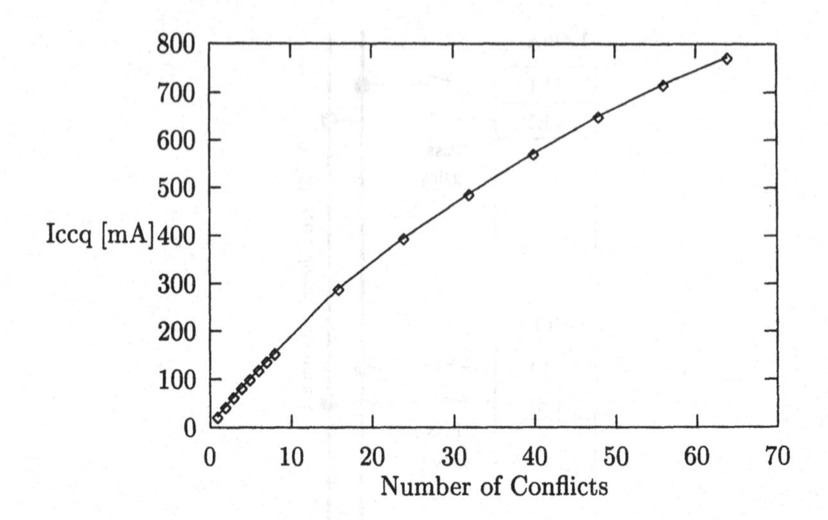

Fig. 3. Quiescent current as the function of number of logic conflicts

Although it is not possible to guarantee the physical destruction of a device, the attacker would normally try to make a device operate in the unsafe region and increase the probability of its destruction. As an illustration we will determine the critical current at which the junction temperature exceeds the maximum of $T_J = 135^oC$ specified by the datasheet. We will show that this temperature is easily achieved with a relatively modest number of logic conflicts.

The thermal resistance of the PLCC84 package is $\theta_{JA} = 35^oC/W$[2] and assuming the ambient temperature of $T_A = 25^oC$, we can easily calculate the maximum power consumption:

$$P = \frac{T_J - T_A}{\theta_{JA}} = \frac{135 - 25}{35} = 3.143[W] \tag{1}$$

For the power supply voltage of $V_{dd} = 5V$, we calculate that the maximum allowed supply current is $I_{cc}^{max} = 629mA$. From the graph in the Figure 3, it is clear that this limit can be easily exceeded with the quiescent current (*i.e.*, without applying the clock) if the number of logic conflicts is greater than 50. During our experiments, we have observed extensive device heating. In several trials, this resulted in physical destruction of the device, manifested through the device's inability for further reconfiguration after being exposed to the high supply current.

In addition to being a threat to the attacked device, the high current drawn by a infected device/card in a system will potentially reduce the available supply current for other devices in the system. For systems designed to be physically small and densely packed, temperature may be a critical issue and increasing the ambient temperature could make the entire system operate in an unsafe region, leading to larger scale failures. Even if the system power requirements are met, this heat can lead to long term instability.

For example, in the case of cards plugged into a PCI bus, the PCI Bus Specification (version 2.1)[13] says that the *total* power drawn by a PCI card cannot exceed $25W$. If a card with several FPGAs on it were infected with the virus (thus drawing high current), that specification would be violated. Depending on the design of the bus, the system may then crash or it may work intermittently or even work normally. This uncertainty is unacceptable for most systems. As programmable devices are used in more cards, this problem will grow.

4 Replication Mechanisms

Besides damaging components of the system (physically or logically), a property of a virus is that it spreads to other systems. Spreading mechanisms can be classified as either software assisted replication or pure hardware replication.

The first mechanism is equivalent to the replication mechanism of software viruses, with the only difference being in the system component the virus attacks. While most software viruses target the hard disk, the FPGA virus would target the programmable logic devices in the system. If run-time reconfigurable computing platforms become more widely adopted, future computer systems will utilize a combination of a general purpose CPU and some amount of reconfigurable hardware. An FPGA virus would then be a piece of code which carries a malicious FPGA configuration and whose replication mechanism is implemented in software. The attack could be performed either by directly programming the FPGA device once it has been found in the system or by replacing the FPGA configurations associated with other programs in the system which utilize the run-time reconfigurable logic. A hardware-library model, such as that proposed in [14], is especially vulnerable to such attacks.

The second replication mechanism provides limited opportunity for the virus to spread without software assistance. An FPGA device could theoretically store a configuration of an other smaller device in user memory and use it to create a reduced version of itself in other devices given a sufficiently large difference in device sizes. Although possible, this replication mechanism is difficult to realize in practice and is mentioned here only for completeness.

Logic conflicts inside the device can be generated either immediately upon downloading or some time after device configuration. Postponed logic conflicts can be easily generated by programming the short circuited logic blocks so that a control signal selects if they should output the same logic levels (no conflict) or opposite logic levels (logic conflict). The virus with hardware replicating mechanism would typically use postponed logic conflicts to avoid device destruction before it gets the chance to spread to other devices.

Software replication mechanisms have the same properties as classical software viruses and should be studied as such. On the other hand, hardware replication mechanisms are limited to the local system and can be easily disabled if the architecture on the board level is such that FPGA devices do not have access to configuration signals of other devices.

5 Detecting and Preventing the Attack

The design space for an attacker in systems with run-time reconfigurable hardware contains the design space for the attacker in CPU based systems.

Our goal is to study the defense methods only for attacks specific to reconfigurable hardware and to provide a safe environment on the electrical and logic signal levels (*i.e.*, protect from MELT and SALT attack types) so that the attacker's design space is reduced back to its software subset. This approach would then allow us to treat the FPGA viruses the same way we treat software viruses, which have been studied in work by others[7]. We now present the three main methods of defense against FPGA viruses, and discuss the advantages and disadvantages of each method.

5.1 Configuration File Verification

Before it is downloaded into a device, the configuration file can be analyzed for potential logic conflicts. For a logic conflict to exist either of the following must be true:

- two or more logic blocks are connected to the same routing resource
- an output pin is connected to the output of an external device

The goal of the configuration file analysis is to ensure that none of the above necessary conditions is satisfied. A correct place and route algorithm will ensure that the first condition is not satisfied. However, it is still possible to create a compiler that generates a malicious configuration or directly modify the configuration file as demonstrated in Section 3. The second condition cannot be addressed at compile time since the compiler is not aware of the board level architecture.

It is therefore necessary to analyze the configuration file against both conditions prior to downloading. Since the system is or can be made aware of the board level architecture, analysis against the second condition is also possible. For successful attack prevention, the system must know the format of the configuration file and search for the binary patterns that correspond to potential conflicts. Since the number of connections to a routing resource is finite, analyzing the configuration file is a viable solution.

The advantage of this method is that the attack can be prevented before the device is exposed to a malicious configuration. The major disadvantage is in the time necessary to perform the analysis, which adds to the download time. Users would also need to know the device configuration file format, something which device vendors are reluctant to make public.

5.2 High Current Detection

Electrical level attacks have the property that they must generate a high current either internally or at the I/O pins. In the case of internal logic conflicts the high current drawn causes a high current at the power supply lines. This current will be present in quiescent mode and it will typically be greater than the maximum allowed quiescent current[2].

The attack can be detected by measuring the supply current after the device has been configured but before applying the clock. Power can then be immediately disconnected from the device if the maximum is exceeded. There are many vendors (*e.g.* [17]) who carry current sensors which can be used to provide a digital signal when the current on a circuit exceeds a limit. The response time of these devices is typically on the order of a few microseconds. With the appropriate circuitry, these sensors could be used to guard not only power pins but also I/O pins from high current damage. Once the attack is detected, the device can be immediately cleared to prevent damage.

This detection method has neither of the disadvantages of the analysis method. The key disadvantage is in the additional protection circuitry and the existence of a short current pulse before the attack is detected. In addition, the current detection method could have problems detecting postponed attacks and attacks on partially reconfigurable devices, as these would typically happen after the clock has been applied and the supply current consists of both quiescent and switching current.

5.3 Avoiding Pass Gates

The final way of preventing electrical level attacks would be to remove pass gates from the architecture of FPGAs. Although it would completely eliminate the opportunity for an electrical level attack on the internal logic, this approach is not feasible for economic reasons. Replacing pass gates with logic (*i.e.* multiplexers and demultiplexers) would dramatically reduce the device density as well as achievable clock rates. Also, removing pass gates would not prevent external (I/O) level attacks.

6 Conclusion

We have presented a threat model for systems that utilize run-time reconfigurable hardware. This includes line cards and add-in boards with accessible FPGAs. We have developed the conceptual basis for, and experimentally demonstrated, that the threat is realistic.

We make a constructive contribution by outlining various methods for making reconfigurable hardware safe against some classes of attacks, particularly those we called MELTs (electrical level). For future work, we plan to extend the study to SALTs (logic signals level), provide demonstrations for other major FPGA families, and provide a more complete study of the threat model.

While research in the FPGA community is currently focused on stimulating adoption of FPGAs by demonstrating the potential of reconfigurable systems,

[2] We have experimentally verified this claim only for Altera Flex8000 devices, but we believe that other device families also have this property.

once reconfigurable computing goes mainstream the new threats to system security we have identified must be thoroughly understood. The goal of our demonstration was not to favor or disfavor and particular device vendor, but to point out the security threat shared by all FPGAs.

References

1. P. Allke. *Configuration Issues: Power-up, Volatility, Security, Battery Back-up.* Xilinx Inc., November 1997. Application Note 092, Version 1.1.
2. Altera, Corporation. *Altera Device Package Information - Data Sheet*, 7 edition, March 1998.
3. Altera, Corporation. *Flex 8000 Programmable Logic Family - Data Sheet*, 9.11 edition, September 1998.
4. Annapolis Micro Systems Inc., http://www.annapmicro.com. *Information on the Web.*
5. J. M. Arnold, D. A. Buell, and E. G. Davis. Splash 2. In *Proceedings of the 4th Annual ACM Symposium on Parallel Algorithms and Architectures*, pages 316–324, June 1992.
6. B. Borriello, *et. al.* The Triptych FPGA architecture. *IEEE Transactions on VLSI Systems*, 3(4):491–501, 1995.
7. F. Cohen. Computer Viruses, Theory and Experiments. *Computers and Security*, 6:22–35, 1987.
8. P. Graham and B. Nelson. A Hardware Genetic Algorithm for the Traveling Salesman Problem on SPLASH 2. In *Proceedings of FPL'95*, pages 352–361, September 1995.
9. J. Burns, *et. al.* A Dynamic Reconfiguration Run-Time System. In *Proceedings of FCCM'97*, April 1997.
10. E. Lechner and S. A. Guccione. The Java Environment of Reconfigurable Computing. In *Proceedings of FPL'97*, pages 284–293, September 1997.
11. G. McGregor, D. Robinson, and P. Lysaght. A Hardware/Software Co-design Environment for Reconfigurable Logic Systems. In *Proceedings of FPL'98*, pages 258–267, September 1998.
12. P. I. Mackinlay., *et. al.* Riley-2: A Flexible Platform for Codesign and Dynamic Reconfigurable Computing Research. In *Proceedings FPL'97*, pages 91–100, September 1997.
13. T. Shanley and D. Anderson. *PCI System Architecture.* Addison Wesley, 3rd edition edition, 1995.
14. D. Smith and D. Bhatia. Race: Reconfigurable and adaptive computing environment. In *Proceedings of FPL'96*, September 1996.
15. Virtual Computer Corporation, http://www.vcc.com. *Information on the Web.*
16. W. H. Mangione-Smith, *et al.* Seeking Solutions in Configurable Computing. *IEEE Computer Magazine*, pages 38–43, December 1997.
17. Zetex Semiconductors, http://www.zetex.com/sensors.htm. *Current Sensors.*

Genetic Programming Using Self-Reconfigurable FPGAs*

Reetinder P. S. Sidhu[1], Alessandro Mei[2], and Viktor K. Prasanna[1]

[1] Department of EE-Systems, University of Southern California,
Los Angeles CA 90089, USA
sidhu@halcyon.usc.edu, prasanna@ganges.usc.edu

[2] Department of Mathematics, University of Trento
38050 Trento (TN), Italy
mei@science.unitn.it

Abstract. This paper presents a novel approach that utilizes FPGA *self-reconfiguration* for efficient computation in the context of Genetic Programming (GP). GP involves evolving programs represented as trees and evaluating their fitness, the latter operation consuming most of the time.
We present a fast, compact representation of the tree structures in FPGA logic which can be *evolved as well as executed* without external intervention. Execution of all tree nodes occurs in parallel and is pipelined. Furthermore, the compact layout enables multiple trees to execute concurrently, dramatically speeding up the fitness evaluation phase. An elegant technique for implementing the evolution phase, made possible by self-reconfiguration, is also presented.
We use two GP problems as benchmarks to compare the performance of logic mapped onto a Xilinx XC6264 FPGA against a software implementation running on a 200 MHz Pentium Pro PC with 64 MB RAM. Our results show a speedup of 19 for an arithmetic intensive problem and a speedup of *three orders of magnitude* for a logic operation intensive problem.

1 Introduction to Self-Reconfiguration

1.1 Problem Instance Dependence and Hardware Compiling

Building logic depending on a single problem instance is the key advantage of reconfigurable computing versus ASICs. That essentially means that a good application for reconfigurable devices should read the input of the problem (the *instance*), compute *instance dependent* logic, i.e. logic optimized for that particular instance, and load it into a reconfigurable device to solve the problem. Applications which produce *instance independent* logic to be loaded onto a reconfigurable device are simply not exploiting the power of reconfiguration. In that case the logic mapped is static, depends only on the algorithm used, and is not conceptually different from ASIC approach.

A large class of applications developed for reconfigurable devices can thus be modeled in the following way (see Figure 1(a)). A process M reads the input problem instance. Depending

* This work was supported by the DARPA Adaptive Computing Systems Program under contract DABT63-96-C-0049 monitored by Fort Hauchuca. Alessandro Mei is with the Department of Mathematics of the University of Trento, Italy.

on the instance a logic E, ready to be loaded, is computed such that it is optimized to solve that single problem instance. This process is usually executed by the host computer. Let T_M denote the time to perform this.

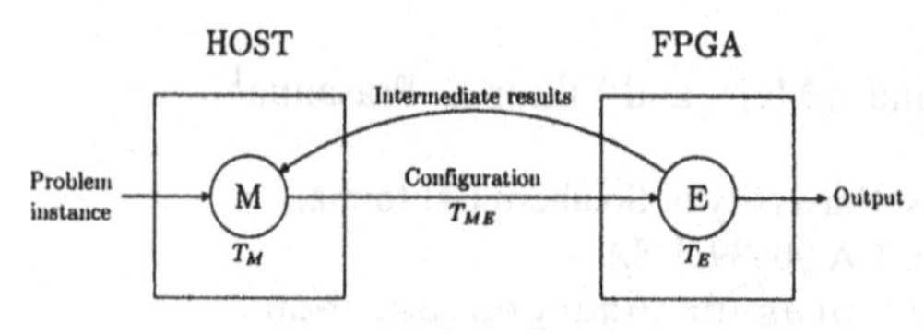

(a) Mapping and execution on a conventional reconfigurable device.

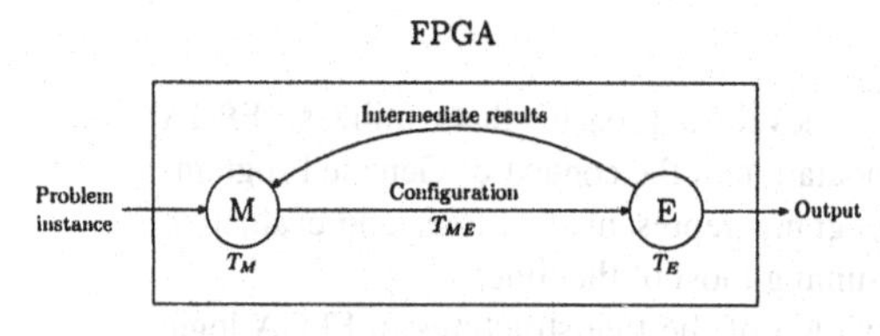

(b) Mapping and execution on a self-reconfigurable device.

Fig. 1. Problem instance dependent mapping.

After reconfiguring the device, E is executed. Let T_{ME} denote the time to reconfigure. The time T_E required for the execution includes the time needed for reading the inputs from the memory and producing the output and/or intermediate results sent back to the mapping module. Therefore, the time required by the execution of a single iteration of the computation described above is $T_I = T_M + T_{ME} + T_E$. This process can be iterated. The intermediate results returned by E can be used by M to compute and map new logic toward the final solution of the problem instance.

A large number of applications fit this model. In some of them, a small amount of parallelism can be obtained by running M and E in parallel. However, the best speedup that can be obtained this way is a factor of 2. Thus, we can suppose that only one of the two modules runs at a given time, without loss of generality. Of course, this factor cannot be ignored in performance analysis.

Self-Reconfiguration is a novel approach to reconfigurable computing presented in [12]. It has been shown to be able to dramatically reduce T_M and T_{ME} with respect to classical CAD tool approach. Since M has to be speeded up, the basic idea is to let fast reconfigurable devices to be able to execute it (see Figure 1(b)). In case a single FPGA is being used, the FPGA should be able to read from a memory the problem instance, configure itself, or a part of it, and execute the logic built by it to solve the problem instance. Evidently, in this case M is itself a logic circuit, and cannot be as complex and general as CAD tools.

Letting FPGA system execute both M and E on the same chip gives the clear advantage that CAD tools are used only *once*, in spite of classical solutions where they are needed for computing a logic for each problem instance. This is possible since the adaptations, needed to customize the circuit to the requirements of the actual input, are performed dynamically by the FPGA itself, taking advantage of hardware efficiency.

Another central point is that the bus connecting the FPGA system to the host computer is now only used to input the problem instance, since the reconfiguration data are generated locally. In this way, the bottle-neck problem is also handled.

These ideas have been shown to be realistic and effective by presenting a novel implementation of a string matching algorithm in [12]. In that paper, however, a simpler version of the above model was introduced which consists of a single iteration of the map-execute loop. Nevertheless, speedups in mapping time of about 10^6 over CAD tools were shown.

Since self-reconfiguration has been proved to be very effective in reducing mapping and host to FPGA communication time, we expect that mapping and communication intensive applications can get the maximum advantage from this techniques. A very important example of this kind of an application is Genetic Programming. Section 3 briefly introduces GP and shows how GP applications can fit our model, proving this way they can be dramatically speeded up by using reconfigurable computing enhanced with Self-Reconfiguration.

1.2 Previous Work Related to Self-Reconfiguration

The main feature needed by an FPGA device to fulfill the requirements needed by the technique shown in the previous section is self-reconfigurability. This concept has been mentioned few times in the literature on reconfigurable architectures in the last few years [5][4].

In [5], a small amount of static logic is added to a reconfigurable device based on an FPGA in order to build a self-reconfiguring processor. Being an architecture oriented work, no application of this concept is shown. The recent Xilinx XC6200 is also a self-reconfiguring device, and this ability has been used in [4] to define an abstract model of virtual circuitry, the Flexible URISC. This model still has a self-configuring capability, even though it is not used by the simple example presented in [4]. The concept of self-reconfiguration has also been used in the reconfigurable mesh [6]—a theoretical model of computation—to develop efficient algorithms. However, there has been no demonstration (except in [12]) of a practical application utilizing self-reconfiguration of FPGAs to improve performance. This paper shows how self-reconfiguration can be used to obtain significant speedups for Genetic Programming problems.

Devices like the XC6200 can self-reconfigure and are thus potentially capable of implementing the ideas presented in this paper. However, moving the process of building the reconfigurable logic into the device itself requires a larger amount of configuration memory in the device compared to traditional approaches. For this reason, multi-context FPGAs are better suited since they can store several contexts (see [10], for example, where a self-reconfiguring 256-context FPGA is presented).

2 Multicontext FPGAs

As described in the Introduction, the time required to reconfigure a traditional FPGA is very high. To reduce the reconfiguration time, several such *multicontext* FPGAs have been recently proposed [11][10][13] [7][3].

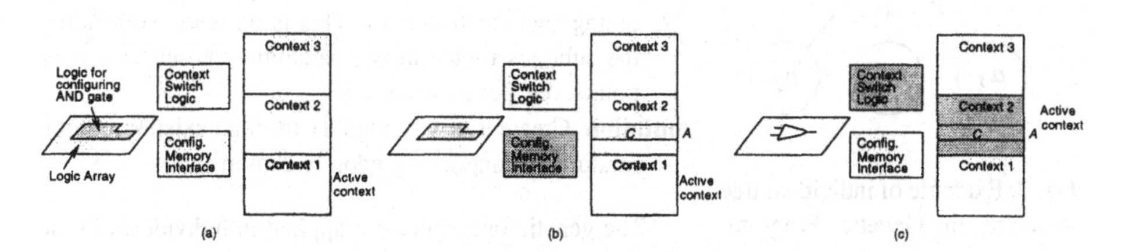

Fig. 2. Self Reconfiguration and context switching in a Multicontext FPGA.

These devices have on-chip RAM to store a number of configuration contexts, varying from 8 to 256. At any given time, one context governs the logic functionality and is referred to as the *active* context. Switching contexts takes 5–100 ns. This is several orders of magnitude faster than the time required to reconfigure a conventional FPGA (≈1 ms). For self-reconfiguration to be possible, the following two additional features are required of multicontext FPGAs:

- The active context should be able to initiate a context switch—no external intervention should be necessary.
- The active context should be able to read and write the configuration memory corresponding to other contexts.

The multicontext FPGAs described in [11][10][13] satisfy the above requirements and hence are capable of self-reconfiguration. Figure 2 illustrates how a multicontext FPGA with above features can modify its own logic. As shown in Figure 2(a), the active context initially is context 1 which has logic capable of configuring an AND gate. Figure 2(b) shows this logic using the configuration memory interface to write bits corresponding to an AND gate at appropriate locations in the configuration memory corresponding to Context 2. Finally the logic on the context 1 initiates a context switch to Context 2 which now has an AND gate configured at the desired location.

3 Introduction to Genetic Programming

Genetic Programming [8] is an adaptive and learning system evolving a *population* of individual computer programs. The evolution process generates a new population from the existing one using analogs of Darwinian principle and genetic operations such as mutation and sexual recombination. In Genetic Programming, each *individual* is obtained by recursively composing *functions* taken from a set $F = \{f_1, \ldots, f_{N_{\text{func}}}\}$, and *terminals* from $T = \{a_1, \ldots, a_{N_{\text{term}}}\}$. Each of the individuals has an associated *fitness* value, usually evaluated over a set of *fitness cases*.

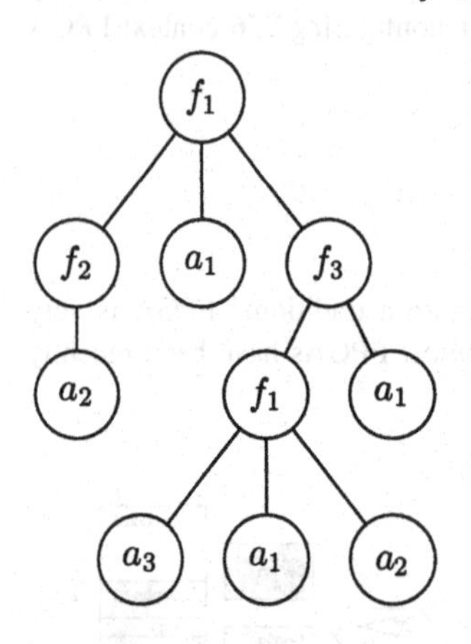

Fig. 3. Example of individual tree structure in Genetic Programming.

A natural way of representing an individual is thus as a *tree*, where a leaf contains a terminal and an internal node a function whose arity is exactly equal to the number of its children (see Figure 3). The evolution process, starting from a randomly generated population of individuals, iteratively transforms it into a new population by applying the following genetic operations:

reproduction Reproduce an existing individual by copying it into the new population.

crossover Create two new individuals by genetically recombining two existing ones. This is done by exchanging the subtrees rooted at two randomly chosen crossover points, one per parental tree.

mutation Create a new individual from an existing one by randomly changing a randomly chosen subtree.

The genetic operations are applied to individuals in the population selected with a probability based on their fitness value, simulating the driving force of Darwinian natural selection: survival and reproduction of the fittest. Computing the fitness value of each individual is a central computational task of GP applications, usually taking around 95-99% of the overall computation time.

It is thus not surprising that the main effort aimed to speedup a Genetic Programming application is focused on the fitness evaluation. For example, in [2] an FPGA is used to accelerate the computation of the fitness value of a population of sorting networks achieving much faster execution.

It is worth noting that the reconfigurable computing application presented in [2] nicely fits our model shown in Figure 1(a). Indeed, M is the process responsible for managing and storing the population, computing the logic E to fitness test each individual, mapping it onto the device, and reading the output value. This operation is repeated for each individual in the population and for each generation in the evolutionary process, resulting in a considerable mapping and host to FPGA communication overhead. Our performance evaluation (see Section 7) shows that reconfiguration time (T_{ME}) is greater than the fitness evaluation time (T_E) and thus self-reconfiguration is essential.

Moreover, in [2] only a rather specific application is shown to benefit from FPGA computing, and it is not clear how the same approach can be extended to an arbitrary GP application.

This paper presents important improvements toward in directions. First, it is shown how a generic GP application can be mapped onto an FPGA system, taking advantage of the massive parallelism of contemporary devices in several ways. Second, how Self Reconfiguration can dramatically speed it up, by handling long mapping and reconfiguration times, and by allowing the evolution phase, as well as the fitness evaluation phase, to be mapped onto the FPGA. The FPGA executes the complete GP algorithm and does not require any external control.

We begin by describing the mapping of the program trees onto FPGA logic in the following section. Section 5 presents the proposed operation of a GP algorithm on FPGAs. The two GP problems used as benchmarks are discussed in Section 6 and the results obtained are presented in Section 7. We summarize the contributions of this paper in Section 8.

4 Tree Template

Before execution begins, a number of *tree templates* are configured onto various contexts of the FPGA. Each template holds the tree representing an individual program throughout its lifetime. As evolution progresses, the nodes of the tree template are appropriately configured to represent the program—the interconnection remain fixed. By configuring the nodes to reflect the actual program, efficient execution results through pipelined execution of all nodes of the tree in parallel (see Section 5.2). By employing a template with static interconnect, fitness evaluation is speeded up and implementation of the mutation, reproduction and crossover operators is simplified (see Sections 5.2 and 5.3). The template tree is a complete binary tree of height k (having $n = 2^k - 1$ nodes). Number of levels of the tree is restricted to the number of levels of the template tree. (Restricting the number of levels is a common technique used in GP implementations to limit tree size.) Below we discuss its mapping onto FPGA logic cells and interconnect.

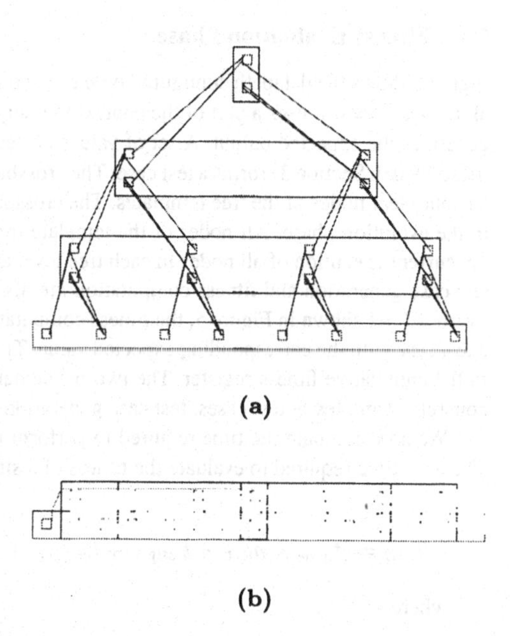

Fig. 4. Compact tree layout using hierarchical interconnect structure.

We map the nodes of the tree along a single row or column of logic cells. The sequence of nodes is the same as obtained through an in-order traversal of the binary tree. The width of each node is a power of 2 while its height is arbitrary—it depends upon the complexity of the functions in the function set. All nodes have the same size.

Figure 4(a) shows a complete 15 node binary tree. Also shown in Figure 4(b) is the mapping of the template tree edges onto wires of the interconnect of the Xilinx XC6200 FPGA architecture[1]. The compact mapping of the tree structure is possible because the interconnect of the XC6200 FPGAs, like that of most other FPGA architectures, is hierarchical. Moreover, most newer generation FPGA architectures (including the Xilinx XC4000 and Virtex, and the Atmel AT40K) have richer and more flexible interconnects than the XC6200. Thus the tree template can be easily mapped onto such FPGAs.

5 Operation

5.1 Initialization

A number of tree templates (equal to the required population size) are configured on one or more contexts of the FPGA. These templates are then initialized with trees generated using standard GP techniques [1]. The size of the nodes in the template is chosen to accommodate the largest area occupied by a function implementation. The size of the template itself is chosen to accommodate the desired maximum number of levels in the trees. Also configured is logic required for the fitness evaluation and evolution phases (explained in the following two sections).

5.2 Fitness Evaluation Phase

Figure 5 shows the datapath configured onto each context. The test case generator iterates through all the test cases. It uses a part of the context memory to store the test cases. For each case it also generates the expected output. A set of values corresponding to the members of the terminal set (described in Section 3) forms a test case. The crossbars are required to map these terminal set values onto leaf nodes of the tree templates. The crossbars are configured using self-reconfiguration in the evolution phase. All nodes of the template trees process the test case in parallel. There is concurrent execution of all nodes in each tree level and pipelined execution along each path. The test case generation and fitness computation are also pipelined and thus do not incur additional overhead. As shown in Figure 5, the fitness computation logic compares the output of the tree for a test case with the corresponding expected value. The resulting measure of fitness is accumulated in the cumulative fitness register. The two benchmark GP problems described in Section 7 give concrete examples of test cases, test case generation logic and fitness computation logic.

We now compute the time required to perform fitness evaluation using the above approach. The total time required to evaluate the fitness of a single generation is:

$$T_{FE} = (t_{\text{node}} \times n_{\text{tests}} + t_{\text{testgen}} + t_{\text{crossbar}} + kt_{\text{node}} + t_{\text{fitcomp}} + t_{\text{fitreg}}) \left\lceil \frac{n_{\text{trees}}}{n_{\text{tcontext}}} \right\rceil \quad (1)$$

where

$$k = \text{number of levels in the tree templates}$$
$$t_{\text{node}} = \text{latency of a node}$$

[1] It should be noted that the XC6200 is used purely for illustration and the proposed mapping in no way depends on any XC6200 specific features. Its choice here was motivated by the authors' familiarity with its CAD tools rather than any architectural considerations.

$$
\begin{aligned}
n_{\text{trees}} &= \text{total number of trees to be evaluated} \\
n_{\text{tcontext}} &= \text{number of trees per context} \\
n_{\text{tests}} &= \text{total number of fitness tests} \\
t_{\text{testgen}} &= \text{test generator latency} \\
t_{\text{crossbar}} &= \text{crossbar latency} \\
t_{\text{fitcomp}} &= \text{fitness computation latency} \\
t_{\text{fitreg}} &= \text{fitness register latency}
\end{aligned}
$$

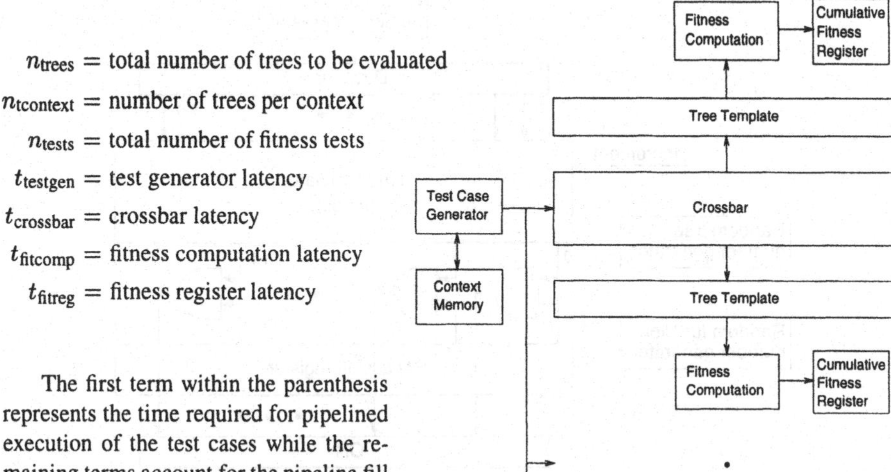

Fig. 5. Datapath for fitness evaluation. All nodes of multiple trees are evaluated in parallel.

The first term within the parenthesis represents the time required for pipelined execution of the test cases while the remaining terms account for the pipeline fill time. The sum within the parenthesis is the time in each of the $\left\lceil \frac{n_{\text{trees}}}{n_{\text{tcontext}}} \right\rceil$ contexts.

5.3 Evolution Phase

Evolution involves modifying some of the programs, letting a portion of the programs die (based on their fitness), and generating new programs to replace the dead ones. In this phase, self-reconfiguration is utilized to modify some of the trees and generate new ones. The modification is achieved through the genetic operations of mutation and crossover while reproduction is used to create new programs. Below we discuss how the ability of self-reconfigurable FPGAs to modify their own configuration is used to implement the genetic operators that manipulate the trees. It should be noted that the evolution phase consumes only 1–5% of the total execution time. Hence the discussion below is qualitative in nature and illustrates how self-reconfiguration elegantly implements the three genetic operations on the FPGA, without any external intervention.

Figure 6 shows the major logic blocks required to perform evolution. This logic operates in bit-serial fashion and is configured on a separate context. The random number generators shown can be efficiently implemented as discussed in [9]. The tree template shown has the same number of nodes as other templates but node contents differ. Each node of the template stores its own offset address. For e.g. the root node stores $\frac{n}{2}$. These can be efficiently stored in each node using $\log_2 n$ flip-flops. The offset—added to a tree base address—is used by the configuration memory interface to access a node of that tree. Each node also has an active bit which when set causes it shift out its address in bit-serial fashion. The logic shown solves in an elegant manner the problem of subtree traversal which is used for all the three operations as described below.

Reproduction The active bit of the root node of the tree template is set. Next the increment signal is applied for k (number of levels in template) clock cycles. In response to the increment signal all active nodes (while remaining active themselves) set the active bits of their child nodes. Thus after k clock cycles, all nodes are active. Next, one bit of the shift register is set (rest are 0). The offset address of the corresponding node is read out and is used to read a node from the source tree and write it into the destination tree template. Next the shift register advances one location and another node gets copied. In this manner, after n shifts, a tree is reproduced.

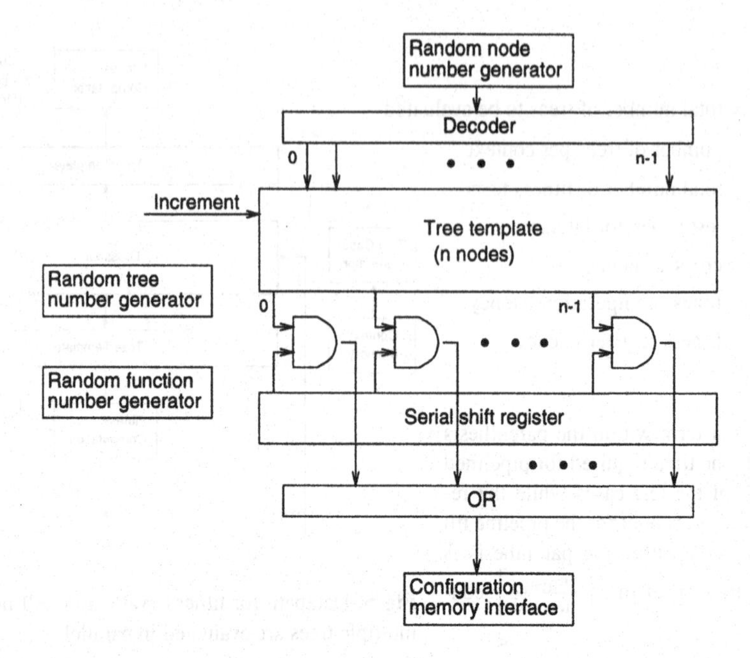

Fig. 6. Logic for the evolution phase. The tree template shown is used to map the node numbers onto the linear configuration memory address space. Each node stores its own offset address.

Crossover The problem is to swap (randomly chosen) subtrees of two (randomly chosen) parents. The key operation is the subtree traversal which is elegantly done. The output of the random node number generator is used to set the active bit of one node which forms the root of the subtree. Next (as in reproduction), the increment signal is applied for k clocks after which the active bits of all the nodes of the subtree are set. The shift register (with a single bit set) is then shifted n times. On each shift, the address of the corresponding node (if active) is read out. In this manner, after n shifts, the subtree is traversed. The crossover operation requires four such traversals, two for each of the subtrees. In the first two traversals, the subtrees are read into scratchpad configuration memory. In the next two traversals, they are written into each others' original locations.

Mutation Mutation involves replacing a randomly selected subtree with a randomly generated subtree. The subtree selection and traversal are performed as for crossover above. For each node visited during the traversal, its current configuration is replaced by the contents corresponding to the output of the random function number generator.

6 Implementation

We evaluated the performance of our approach in the following manner. We chose two GP problems as benchmarks. Both problems were selected from Koza's *Genetic Programming*[8]. For each we implemented the fitness evaluation logic (as discussed in Section 5.2) onto a conventional FPGA. This implementation was used to obtain the minimum clock cycle time and the maximum number of trees that fit on a single context. Using this information and Equation 1 the time required by our approach to perform fitness evaluation was computed. Next, a software implementation of the two benchmarks was run and the time spent on fitness evaluation measured.

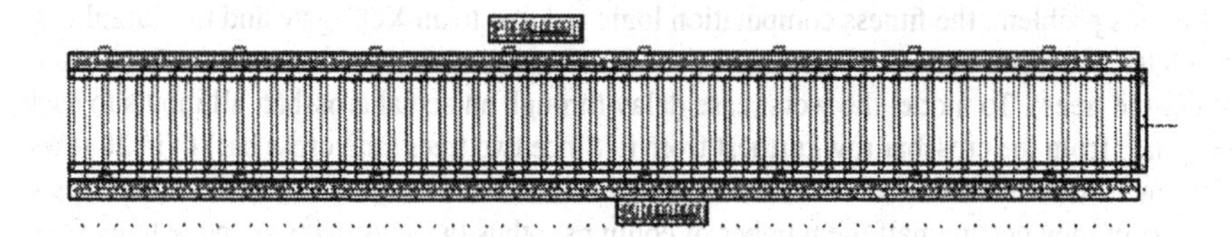

Fig. 7. A section of the layout for the multiplexer problem with 127 node tree templates.

The speedup obtained using our approach was then computed from the fitness evaluation times of both the approaches and Equation 1.

The choice of the two problems was motivated by the nature of their function sets (explained in Section 3). For one of the problems (multiplexer), all the members of its function set were bit-level logic functions. For the other (regression) all function set members were arithmetic functions operating on integers. Clearly, FPGAs would provide a greater speedup over a microprocessor for the former problem compared with the latter. Typically, the function set of a GP problem contains a mix of arithmetic and logic functions. Therefore, performance evaluation of the chosen problems would yield an estimate of the range of speedups that can be obtained over a microprocessor.

The following two GP problems were chosen:

Multiplexer The problem is to evolve a program that exhibits the same behavior as a multiplexer having 8 data inputs and 3 control inputs. The test cases are all 2^{11} possible inputs. The corresponding expected values are the boolean outputs a multiplexer would produce. The function set consists of logic functions `and`, `or` and `not`, and the `if` function which is essentially a multiplexer with 2 data inputs and a control input. The functions are much simpler than needed by many GP applications. But real problems such as evolution of BDDs also employ such simple functions. The terminal set has 11 members—the 8 data inputs and the 3 control inputs.

Regression the problem is to evolve a function that "fits" a number of known (x, y) points. The x coordinates are used as the fitness cases while the corresponding y is the expected value. We use 200 test cases (n_{tests}=200). The function set consists of `add`, `subtract` and `multiply`. The terminal set consists of the input value x and integer constants in the range $[-1, 1)$.

7 Performance Evaluation

The Xilinx XC 6264 was used as the target FPGA. Required logic was specified in structural VHDL and translated to EDIF format using velab. XACT 6000 was used for place, route and configuration file generation. Software implementation of the benchmarks was carried out using the lil-gp kernel [14]. The resulting executable was run on a PC with a 200 MHz Pentium Pro processor and 64 MB RAM. The population size for both approaches was fixed at 100 individuals.

7.1 Multiplexer

Area Requirements Figure 7 shows the layout on a Xilinx XC 6264 (128 × 128 logic cells) of two trees (each having 127 nodes) and the associated (simulated) crossbar for the multiplexor problem—it is similar to Figure 5 (except for the test case generator which appears on the right side). For this problem, the fitness computation logic reduces to an XOR gate and the cumulative fitness register is just a counter controlled by the XOR gate output (these appear on the top and bottom of the trees). To model the worst case delay through an actual crossbar, all inputs to each tree originate from the crossbar row furthest from it. Since two trees (and cross bar) fit in 20 rows, the 128 row FPGA can accommodate 12 127 node trees on it. The layout for the 63 node trees is similar except they occupy half the number of columns—thus twice as many 63 node trees fit on to the FPGA.

From the above mapped logic, the minimum clock cycle time (t_{clk}) for both tree sizes was determined which is shown in Table 1. It should be noted that the major component of $t_{clk}$is the crossbar—the critical paths through the 127 and 63 node trees were just 14.31 ns and 12.47 ns. Thus an efficient crossbar can provide even further improvements. Also shown are the number of clock cycles required which are computed using Equation 1 for n_{trees}=100 and $n_{tcontext}$=12 and 24. It should be clear from Figure 7 that all the times in Equation 1 (including t_{node}) are equal to t_{clk}. Finally multiplying by t_{clk}yields the time T_{FE} required to fitness evaluate a single generation of 100 trees using the proposed approach. It should be noted that T_{FE} is for fitness evaluation of all trees on all contexts.

Table 1. Area requirements for the multiplexor problem for tree templates having 127 and 63 nodes. Each context has 128 × 128 logic cells.

Structure	Area (in logic cells)	
	$n=127$ nodes	$n=63$ nodes
Tree template	127 × 3	63 × 3
Crossbar	127 × 11	63 × 11
Test case generator	1 × 11	1 × 11
Fitness logic	12 × 3	12 × 3
Number of trees per context ($n_{tcontext}$)	12	24

Table 2. Time required to fitness evaluate 100 trees using proposed approach.

	Area (in logic cells)	
	$n=127$ nodes	$n=63$ nodes
Clock cycle (t_{clk})	48.96 ns	37.08 ns
Clock cycles	18531	10290
Time taken (T_{FE})	907.3 μs	381.6 μs

Table 3. Fitness evaluation times for a generation of 100 individuals.

Approach	T_{FE}	
	$n=127$ nodes	$n=63$ nodes
Proposed	907.3 μs	381.6 μs
Software	930 ms	440 ms
Speedup	**1025**	**1153**

Time Requirements To obtain T_{FE} for the software implementation, it was executed for a population size of a 100 individuals. This experiment was conducted twice with the maximum nodes per tree restricted to 127 and 63 thus ensuring that the tree size limits are the same as in our approach. Each time, execution was carried out for a 100 generations and the total time spent on fitness evaluation was noted. From this, the average fitness evaluation time per generation was obtained which is shown in Table 3. As can be seen, the proposed approach is almost *three orders of magnitude* faster than a software implementation (for fitness evaluation).

7.2 Regression

Area Requirements Regression requires much greater area compared to the multiplexer problem due to the (bit-serial) multiply operation—each node requires 4 × 16 logic cells. Table 4 shows the area requirements. Note that since the terminal set consists of just one variable (x), in contrast to 11 for the multiplexer, the crossbar reduces to 124 × 1 logic cells. The other terminal set members (integer constants) are implemented by embedding them in the corresponding terminal nodes. Two 31 node trees and the associated circuitry fit into 35 rows of the XC 6264. Thus 6 trees can be accommodated. The fitness computation and accumulation logic consists of bit-serial comparator and adder.

Table 4. Area requirements for the regression problem for a tree template 31 nodes. Each context has 128 × 128 logic cells.

Structure	Area (in logic cells)
	$n = 31$ nodes
Tree template	124 × 16
Crossbar	124 × 1
Test case generator	1 × 16
Fitness logic	20 × 2
Number of trees per context (n_{tcontext})	6

Table 5. Time required to fitness evaluate 100 trees using proposed approach.

	Area (in logic cells)
	$n = 31$ nodes
Clock cycle (t_{clk})	28.86 ns
Clock cycles	115073
Time taken (T_{FE})	3321.0 μs

Table 6. Fitness evaluation times for a generation of 100 individuals.

Approach	T_{FE}
	$n = 31$ nodes
Proposed	3321.0 μs
Software	62.9 ms
Speedup	**19.0**

Time Requirements Operands are 16-bit values and all operations are performed in a bit-serial fashion. Latency t_{node}=33 clock cycles due to the multiply (only 16 MSB used). As can be seen from Table 5, the latency (in number of clock cycles) is higher but the clock cycle time is lower (since the "crossbar" is smaller and remains fixed) compared to the multiplexor. Clock cycles are computed for n_{tests}=200. Table 6 shows that the proposed approach achieves a speedup of 19 (in fitness evaluation) over a software implementation for the regression problem. This is a significant speedup for a single FPGA considering the arithmetic intensive nature of the problem.

8 Conclusion

We have demonstrated dramatic speedups—of upto three orders of magnitude—for fitness evaluation, the GP phase that consumes 95-99% of execution time. This speedup is achieved due to the fast, compact representation of the program trees on the FPGA. The representation enables parallel, pipelined execution of all nodes in parallel and also concurrent execution of multiple trees.

It should be noted that self-reconfiguration is essential for the above speedup. In the absence of self-reconfiguration, the evolution phase would be performed off-chip, and the resulting trees would have to be reconfigured onto the FPGA doing which would consume about 1 ms per context (much more if configuration done over a slow I/O bus). As can be seen from Section 7 our approach fitness evaluates a *several* contexts of trees in less than 1 ms. Since the reconfiguration time is greater than the execution time, the speedups obtained would be greatly reduced.

Self-reconfiguration eliminates external intervention and the associated penalty by allowing the chip to modify its own configuration and thus perform the evolution phase on-chip. We have also shown an elegant technique for performing the evolution phase using self-reconfiguration.

References

1. BANZHAF, W., NORDIN, P., KELLER, R. E., AND FRANCONE, F. D. *Genetic Programming – An Introduction; On the Automatic Evolution of Computer Programs and its Applications.* Morgan Kaufmann, dpunkt.verlag, Jan. 1998.
2. BENNETT III, F. H., KOZA, J. R., HUTCHINGS, J. L., BADE, S. L., KEANE, M. A., AND ANDRE, D. Evolving computer programs using rapidly reconfigurable FPGAs and genetic programming. In *FPGA'98 Sixth International Symposium on Field Programmable Gate Arrays* (Doubletree Hotel, Monterey, California, USA, 22-24 Feb. 1998), J. Cong, Ed.
3. DEHON, A. Multicontext Field-Programmable Gate Arrays. http:/-/HTTP.CS.Berkeley.EDU/ãmd/CS294S97/papers/dpga_cs294.ps.
4. DONLIN, A. Self modifying circuitry - a platform for tractable virtual circuitry. In *Eighth International Workshop on Field Programmable Logic and Applications* (1998).
5. FRENCH, P. C., AND TAYLOR, R. W. A self-reconfiguring processor. In *Proceedings of IEEE Workshop on FPGAs for Custom Computing Machines* (Napa, CA, Apr. 1993), D. A. Buell and K. L. Pocek, Eds., pp. 50–59.
6. JANG, J. W., AND PRASANNA, V. K. A bit model of reconfigurable mesh. In *Reconfigurable Architectures Workshop* (Apr. 1994).
7. JONES, D., AND LEWIS, D. A time-multiplexed FPGA architecture for logic emulation. In *Proceedings of the 1995 IEEE Custom Integrated Circuits Conference* (May 1995), pp. 495–498.
8. KOZA, J. R. *Genetic Programming: On the Programming of Computers by Means of Natural Selection.* MIT Press, Cambridge, MA, USA, 1992.
9. LAVENIER, D., AND SAOUTER, Y. Computing goldbach partitions using pseudo-random bit generator operators on a fpga systolic array. In *Field-Programmable Logic and Applications, Eighth International Workshop, FPL '98* (1998), pp. 316–325.
10. MOTOMURA, M., AIMOTO, Y., SHIBAYAMA, A., YABE, Y., AND YAMASHINA, M. An embedded DRAM-FPGA chip with instantaneous logic reconfiguration. In *1997 Symposium on VLSI Circuits Digest of Technical Papers* (June 1997), pp. 55–56.
11. SCALERA, S. M., AND VÁZQUEZ, J. R. The design and implementation of a context switching FPGA. In *IEEE Symposium on Field-Programmable Custom Computing Machines* (Apr. 1998), pp. 495–498.
12. SIDHU, R. P. S., MEI, A., AND PRASANNA, V. K. String matching on multicontext FPGAs using self-reconfiguration. In *FPGA '99. Proceedings of the 1999 ACM/SIGDA seventh international symposium on Field programmable gate arrays* (1999), pp. 217–226.
13. TRIMBERGER, S., CARBERRY, D., JOHNSON, A., AND WONG, J. A time-multiplexed FPGA. In *Proceedings of IEEE Workshop on FPGAs for Custom Computing Machines* (Napa, CA, Apr. 1997), J. Arnold and K. L. Pocek, Eds., pp. 22–28.
14. ZONGKER, D., PUNCH, B., AND RAND, B. lil-gp genetic programming system. http://GARAGe.cps.msu.edu/software/lil-gp/lilgp-index.html.

Specification, Implementation and Testing of HFSMs in Dynamically Reconfigurable FPGAs

Arnaldo Oliveira, Andreia Melo, Valery Sklyarov

Universidade de Aveiro, Dep. Electrónica e Telecomunicações
Campus Universitário, 3810 Aveiro, Portugal
arnaldo@ua.pt, andreia@ua.pt, skl@inesca.pt

Abstract. This paper discusses methods and software tools that we have developed for the specification, verification, implementation and debugging of control circuits. The specification method that we have adopted is based on the use of Hierarchical Graph-Schemes. The circuit implementation model is a Hierarchical Finite State Machine, which supports the top-down decomposition of the control algorithms. The application input/output interface provides links with other external tools that perform synthesis and implementation tasks. Some of the utilities we have developed, such as the random control algorithm generator, allow many useful supplementary tasks to be handled and provide powerful assistance for experiments. In particular, the tools have been used to implement Hierarchical Finite State Machines in the Xilinx XC6200 dynamically reconfigurable FPGAs.

1. Introduction

Most digital systems can be decomposed into a datapath and a control unit. Control units establish the sequence of operations performed in the datapath. They are often modelled as Finite State Machines (FSMs) and implemented using either hardwired circuits or microprogramming techniques. Since control units are unique for each project and usually very irregular, the problem of their automatic synthesis is important. Traditionally, the specification of FSMs has been based on state transition diagrams presented in the form of directed graphs, tables, etc. The specification technique considered in this paper uses Hierarchical Graph-Schemes (HGSs) [1], which provide a natural top-down description of the control unit behaviour with the aid of a notation that is similar to flowcharts. The hierarchical specification of control algorithms is very efficient because it simplifies the design of complex control units. In this methodology the algorithm specification is split into smaller sub-algorithms that can be individually designed and tested. A circuit model based on Hierarchical Finite State Machines (HFSMs) is convenient for implementing hierarchical control units. In addition, some HFSMs structures provide the physical separation of each sub-algorithm, thus simplifying the implementation of dynamically reconfigurable control

This work was partially sponsored by the grant FCT-PRAXIS XXI/BM/17678/98.

units, also known as virtual control circuits (VCCs) [2], [3]. The software we have developed provides an integrated environment that supports the design of control units using HGSs. Some extra steps of synthesis and hardware implementation are performed by external tools [2], [3] that can be invoked from the working environment. The current release is targeted to the Xilinx XC6200 FPGA family [4]. However, other circuit architectures can be accommodated through the development of appropriate modules (Dynamic Link Libraries – DLLs).

This paper is organised in seven sections. Section 1 is this introduction. Section 2 provides a brief review of HFSMs and VCCs. Section 3 presents the specification technique that we adopted. Section 4 shows the software tool we have developed that integrates all the steps required to implement HFSMs. Section 5 describes a template that makes possible the implementation of dynamically reconfigurable HFSMs in a Xilinx XC6216 FPGA. Section 6 presents the C++ class library and the device driver we used to access the Xilinx XC6200 FPGAs. The conclusion is in section 7.

2. Hierarchical Finite State Machines and Virtual Control Circuits

A hierarchically described control algorithm can be implemented in hardware using either hierarchical or non-hierarchical models of finite state machines (FSMs). In the case of a non-hierarchical implementation, the sub-algorithms which make up the specification are linked by the synthesis or implementation tool into a flat specification, and the resulting circuit presents the traditional structure of an ordinary FSM – fig. 1 a). This approach is based on macro substitution and consequently has some limitations. The most important constraint is that it is impossible to implement algorithms with recursive and/or cyclic invocations, which corresponds to one of the more interesting hierarchy applications. On the other hand, in the case of a hierarchical implementation, the circuit model can be based on HFSMs, which provides support for the run-time invocation of the sub-algorithms. The major difference between an ordinary FSM and a HFSM is the replacement of the state register with a stack memory. This component is used to store the current state of the HFSM and the states of the interrupted sub-algorithm(s) (caller(s) of the current sub-algorithm). However, this method is only recommended for complex control units where the amount of hardware used to manage the hierarchy is negligible when compared to the rest of the circuit.

There are two approaches to implementing HFSMs:

- The circuit synthesis for all sub-algorithms is performed at the same time. As a result, all these sub-algorithms will be incorporated into the same combinatorial scheme connected to the stack and other components – fig.1 b);
- The synthesis for each sub-algorithm is performed independently. As a result, the combinatorial scheme will be divided into autonomous segments in such a way that each segment implements only one sub-algorithm – fig.1 c).

The first approach is easier to realise, but we are losing the modularity of the specification during the implementation of the circuit. It is more appropriate to medium complexity algorithms where the synthesis tools can easily deal with not so large number of circuit states. In the second approach, the specification modularity and hierarchy are maintained in the implementation of the circuit. However, additional components to select and activate the correct combinatorial sub-circuit are required, so this approach is only recommended for control circuits of considerable complexity. The second approach has the following advantages:

- It allows incomplete specifications and future refinements to be dealt without significant changes in the circuit structure, and without redesigning pre-defined parts;
- It is easy to modify or extend the control unit behaviour after the design has been completed by changing or adding sub-algorithms;
- It is well suited to implementing virtual control circuits.

For very complex control circuits, a reduction in the hardware, and consequently in the power consumption can be achieved if the combinatorial circuits corresponding to each sub-algorithm can be updated on the fly. Circuits of this kind are called Virtual Control Circuits (VCCs) and may be implemented in field programmable devices with dynamic reconfiguration capabilities. The resulting circuit has virtual capabilities because the same hardware is used to implement different parts of a control algorithm. The reconfiguration can be performed statically or dynamically depending on the requirements, such as suspending the circuit operation or not.

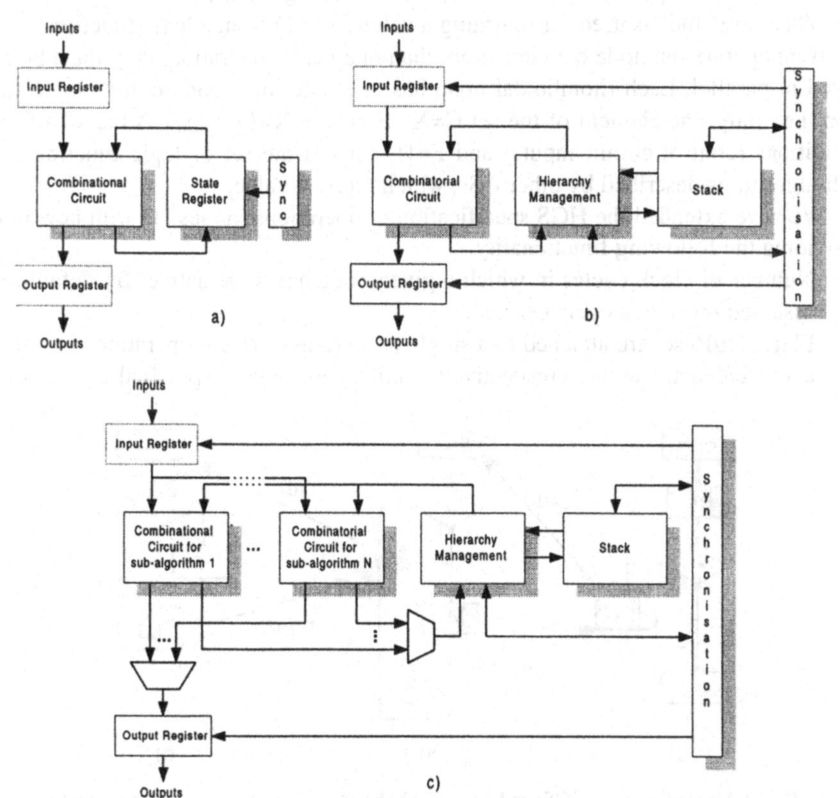

Fig. 1. High level structures of FSMs a) ordinary; b) hierarchical; c) hierarchical with separate implementations of the sub-algorithms.

3. Graphical Specification of HFSMs

HGSs are a formal graphical specification of hierarchical control algorithms. They are an extension to the Graph Schemes (GSs) considered in [5]. A HGS is composed of several relatively autonomous GSs, each describing a particular sub-algorithm. In a HGS there is one mandatory main graph that describes the main sub-algorithm. Optionally, it can invoke one or more logic functions and/or macro operations that are described by other GSs. A GS is a directly connected graph composed of different types of nodes, each with a specific function that can be recognised from its graphical shape. Rectangular (operational) nodes correspond to states and rhomboidal (conditional) nodes determine the algorithm flow, depending on the binary value of input signals. Fig. 2 shows an example of a HGS. There are four types of rectangular nodes:

- Operational nodes contain a subset of micro operations from the set $Y=\{y_1,...,y_N\}$ (which are control unit outputs) and/or a subset of macro operations from the set $Z=\{z_1,...,z_Q\}$ (specified by other GSs) [1];
- Dummy nodes are empty operational nodes, which introduce only a parameterisable time delay;
- Halt nodes suspend the execution of the control algorithm;
- An assign node is used for returning a value (0 or 1) from a logic function.

If an operational node contains more than one macro operation, they must be executed in parallel. Each rhomboidal node has only one logic condition or logic function, i.e., only one element of the set $C=X\cup F$, where $X=\{x_1,...,x_L\}$ is the set of logic conditions (control circuit inputs) and $F=\{f_1,...,f_I\}$ is the set of logic functions, i.e., sub-algorithms described by other GSs, which return a value.

We have extended the HGS specification of operational nodes [1] with new fields, providing the following functionality:

- Number of clock cycles in which a given node has to be active. By default, it is assumed one clock cycle per node;
- Flags Set/Reset are attached to a single or a group of micro operations in order to activate/deactivate them respectively, until a new value is specified.

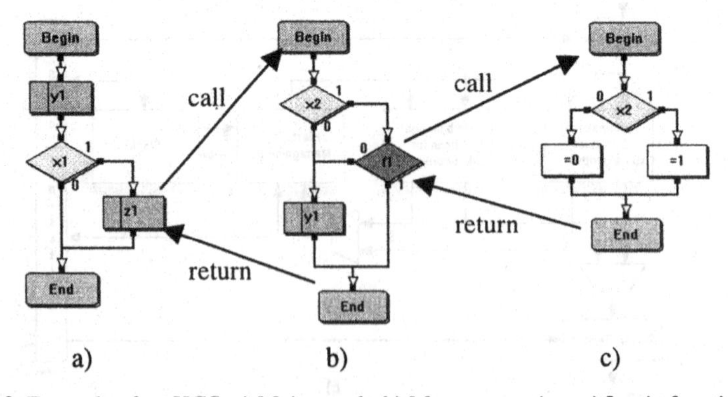

Fig. 2. Example of an HGS; a) Main graph; b) Macro operation; c) Logic function.

4. Integrated Design Environment for HFSM

To simplify the design of control units specified by HGSs, the software application, Graph Builder, was developed using Microsoft Visual C++ 6.0. The user interface of Graph Builder is shown in fig. 3. The main window is divided into menus, toolbars and local windows. In the left-hand frames of fig. 3 we can manage the hierarchy and in the right-hand window we can graphically build and modify the selected GS of the desired hierarchy level. Some embedded tools allow the verification, optimisation, random generation and test in hardware of a HGS. All fundamental abstractions, such as nodes, GSs and hierarchies have been implemented as C++ classes and stored in a DLL, so they can be directly used by other applications, such as synthesis tools. In addition, some synthesis tools can be invoked directly from within the design environment. This allows other applications to use this notation without the need for intermediate format or conversion utilities.

Verification should be performed in order to avoid specification errors that can lead to circuits with undesirable behaviour. Graph Builder automatically examines a list of specifications, and the items in each specification are checked sequentially or individually and marked with a symbol indicating either success or failure. Incorrect nodes are marked with identifying colours. The connections between nodes must be analysed in order to avoid the existence of unreachable nodes. The detection of infinite loops in a GS is very important because in this case the End node cannot be reached. All nodes that are along the path between the Begin and End nodes are marked if they cause an infinite loop.

Optimisation techniques are used to rebuild GSs in such a way that either their implementation in hardware will be simplified, or some pre-defined constraints, such as the number of inputs or outputs, will be satisfied. To achieve this we have to minimise the number of inputs and outputs. A software application was developed for implementing these optimisation techniques, and integrated in Graph Builder. Its main window is shown in fig. 4.

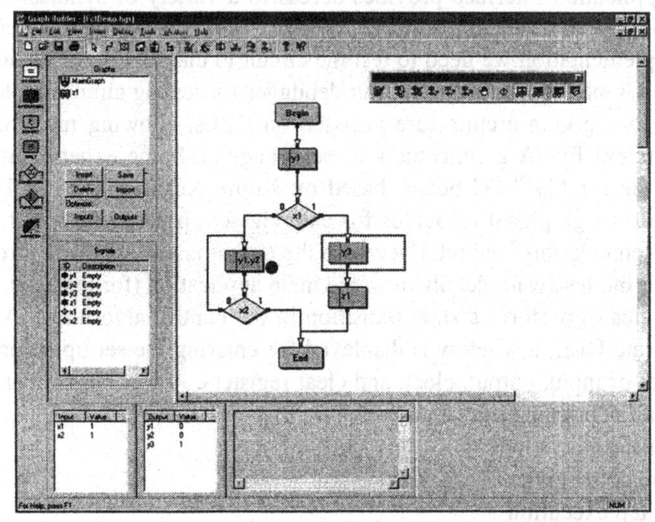

Fig. 3. The main window of the Graph Builder.

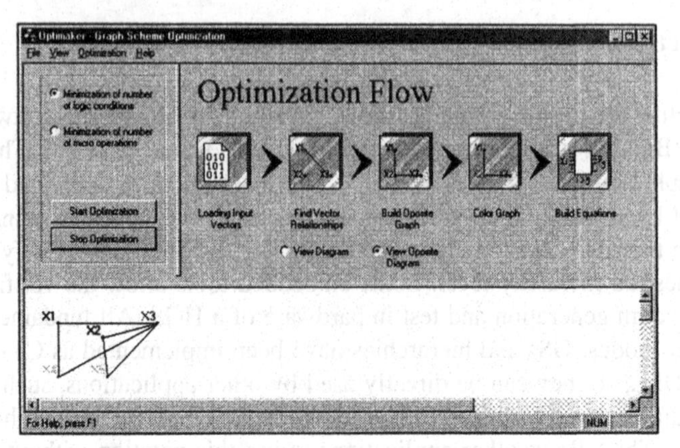

Fig. 4. The main window of the Optimaker.

The random HGS generator is an embedded tool for the rapid construction of arbitrary HGSs that can be used for testing purposes. Its interface is based on a wizard that performs a series of steps in which the user can choose:

- The type of GS to build, such as an entire hierarchy, a logic function, a macro operation or the main graph of a HGS;
- The minimum and maximum number of logic conditions and micro operations of the control algorithm;
- The maximum number of micro operations per node;
- The generation of parallel sub-algorithms.

The HGS generator calculates all the necessary parameters randomly based on the specified constraints. Then the nodes are grouped and connected in order to produce a consistent graph.

The interaction with external tools can be achieved either through intermediate textual descriptions of the HGS, or through a DLL that encapsulates the HGS notation. The application interface provides access to a variety of synthesis tools corresponding to different implementations and/or different device architectures.

After implementation we need to test the circuit to check if the behaviour is as expected. The Graph Builder has a built-in debugger for testing circuit implementations in FPGAs. Its working architecture is based on DLLs, allowing many implementations in different FPGA architectures to be debugged. Some experiments were performed on the FireFly™ PC board, based on Xilinx XC6216/XC6264 FPGA. Each DLL has its own graphical resources for entering set-up parameters that are specific to each implementation. The interface with the circuit is always made through a DLL that hides all the hardware details from the main application (for instance, the number of clock cycles to perform a state transition in the control algorithm). After loading the appropriate DLL, a window is displayed for entering the set-up parameters, such as the names of input, output, clock and clear registers. A specific toolbar is activated to perform all debugging operations, such as:

- Display the Setup window
- Start/Stop debugging
- Step by step execution

- Run/Break, which allows execution in a free running mode (performing sequential steps) until the user stops the circuit execution or a breakpoint is encountered
- Insert/Remove breakpoints
- Setting input binary values
- Watch input/output signals in different windows placed at the bottom of the screen (see fig. 3)
- Open a log window for displaying log and error messages and clear its contents

Debugging is performed graphically by marking the currently active node in the GS with two enclosing rectangles: one is used to activate the node that is predicted in software and the other indicates the node that is really active in hardware. If the circuit is executing as expected, both rectangles must be placed over the same node, otherwise it corresponds to an error. If after testing the circuit some errors were encountered or some improvements can be made, we can modify the initial circuit specification in the same application environment so there is feedback that allows us to close the design cycle.

5. Implementation of HFSMs in Dynamically Reconfigurable FPGAs

The device we have used to implement HFSMs is a Xilinx XC6216 partially dynamically reconfigurable FPGA installed in the FireFly Development System [6]. This is a PCI board that provides access to all the features available for the XC6216 FPGA. It also contains a Xilinx XC4013 FPGA (that establishes the PCI bus interface) and 512 Kbytes of local SRAM. The implementation is based on a parameterisable template we have designed that segments the FPGA into several individually reconfigurable windows (fig. 5) and provides a set of common capabilities that simplify the design of HFSMs. Some of these windows are used in fixed functions such as synchronisation and hierarchy management, while others are used to implement the parts that are reconfigurable and specific for each sub-algorithm. The template parameterisation is important for implementing algorithms of different complexities. This is achieved through the variation of one or more of the following parameters:

- Number of reconfigurable windows, which establishes the maximum number of sub-algorithms simultaneously loaded in hardware and ready for execution;
- Number of inputs and outputs of the control unit;
- Number of invocations (hierarchical transitions) per sub-algorithm;
- Number of hierarchy levels (stack depth).

The first template version allows up to 64 sub-algorithms to be implemented, with typically 100 nodes and 4 hierarchical transitions (calls) each. The maximum number of hierarchical levels is 256 and recursion is supported.

The HFSM model presented in section 2 was adapted to fit the XC6200 FPGA architecture. This does not provide gate primitives with 3 or more inputs or on-chip memory blocks (lookup tables) that allow a large set of two level logic functions to be implemented in a speed and area efficient way. This prevents the utilisation of traditional binary state encoding techniques. Moreover, every cell contains a flip-flop, which helps in the use of the one-hot state encoding. In fact, comparatively to other techniques, it has at least two advantages - a reduced power consumption and straightforward HGS synthesis. For performance and optimisation reasons, the reconfigurable area that implements the sub-algorithm contains both combinatorial and sequential parts. Only states with hierarchical transitions are saved on the stack, contributing to an improvement in the speed of the circuit. All other states are considered to be an internal attribute of the respective sub-algorithm.

Circuit reconfiguration is controlled by a software application (FireFly Manager) and triggered each time a required sub-algorithm is not already loaded in hardware. The trigger mechanism can be based on polling or on an interrupt. During the reconfiguration process, the FireFly Manager loads the sub-algorithm configuration bitstream in a window, eventually discarding an idle one. To simplify the reconfiguration process and decrease its duration, the inputs and outputs of the reconfigurable windows were fixed in predefined locations. The size of each window is important to guarantee that the functional and routing resources don't cross the window boundaries.

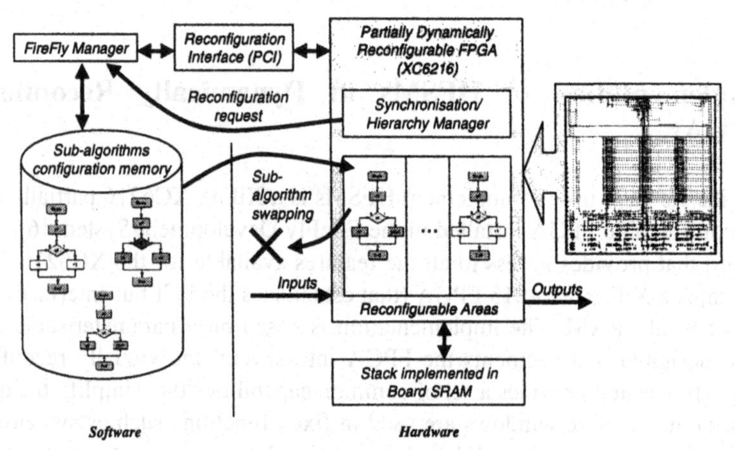

Fig. 5. Implementation structure.

The design flow used to implement a HFSM in the template is depicted in fig. 6, in which the shadowed blocks were developed by the authors. The starting point is the specification of the control algorithm using the Graph Builder application. The resulting specification can be placed in a single "HGS" file or spread in several "GS" files. Both allow a separate synthesis of each sub-algorithm. The subsequent steps should be performed individually for each sub-algorithm. The specification is then converted into structured VHDL code, which is used by VELAB [7] to generate the EDIF netlist of the circuit. The mapping, placement and routing procedures are performed by XACT6000 [8], which outputs three different file types. At the moment, only the configuration file (CAL) is relevant for our purposes. To allow the dynamic reconfiguration of an FPGA window, this file is processed, filtered and stored in binary format to decrease the configuration time. The sub-algorithms and template configuration files are used by the Graph Builder debugger and by the FireFly Manager application for debugging and running the circuit respectively. The template design flow is similar to the flow considered above except for the start point, which is a set of structural VHDL files.

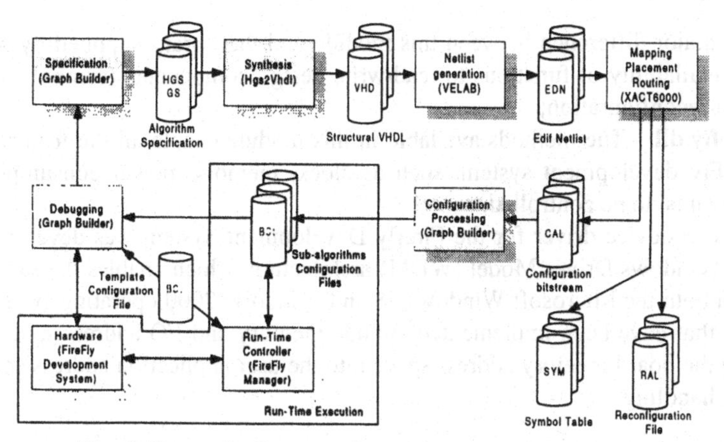

Fig. 6. Dynamically reconfigurable HFSM Design flow

6. Software Library and Device Driver for Hardware Access

A C++ class library was developed to simplify the access to the hardware mentioned in the previous section. It was written using Microsoft Visual C++ 6.0, runs on Windows 95/98/NT, and corresponds to an improved version of the RalLib and FireFly interface classes supplied by Xilinx and Annapolis Micro Systems respectively [8]. The structure of those libraries was changed in order to increase reusability. As an example, the functions used to read/write specific registers of the XC6200 FPGA family are now located in a class that encapsulates each device of this family, rather than in a class designed specifically for the FireFly development system. On the other hand, the library contains functions to simplify common tasks such as read/write components (symbols) in the FPGA and reconfigure them. The library is composed of several DLL files, which simplifies its inclusion and sharing among projects and makes future modifications or improvements easier. It uses several features of the Microsoft Foundation Classes (MFC) library such as, common data structures, serialisation, run-time type information, etc. A drawback of this approach is that it is difficult to port the code to other platforms. However, in our case it was impossible to avoid it, because the library includes Graphical User Interface (GUI) elements and MFC provides a convenient way to encapsulate them. Examples of GUI elements are dialog boxes to configure FPGA or board control registers. The library is divided between the following DLLs:

- **BoardInterface.dll** – Provides general classes to access boards with different types of interfaces, such as PCI, Parallel, etc;
- **Xc6200.dll** – Contains classes that abstract most of the features of the Xilinx XC6200 family, such as read/write of cells or control registers. It also supplies reusable graphical controls (dialog boxes) which can be utilised to modify FPGA registers;
- **CalFile.dll** – This module is used primarily to read CAL files [8], to process them if required, and to download them into a XC6200 device;
- **RalLib.dll** – Contains functions to read and process SYM and RAL files [8], which are used for high-level hardware access and reconfiguration respectively.

The major difference between this module and the library supplied by Xilinx is the availability of functions to read/write design components (symbols) and perform reconfiguration;

- **Firefly.dll** – The methods available in this module enable all the features of the FireFly development system, such as clock, memory, power consumption and interrupts, to be controlled.

Finally, a device driver for the Firefly Development System was developed. This uses the Windows Driver Model (WDM) architecture, which enables the same driver to run on both the Microsoft Windows 98 and Windows 2000 operating systems. The facilities that have been implemented include memory and I/O address space access, mapping the board memory address space into the user application address space, and interrupt handling.

7. Conclusion

The template we have discussed is a framework that simplifies the development of HFSMs in dynamically reconfigurable FPGAs, and offers a set of common capabilities for a variety of different applications. The proposed approach is based on the hierarchical specification of the control algorithms and the separate synthesis of each sub-algorithm. The software we have developed provides an integrated environment for building, debugging and modifying control algorithms graphically, thus it making feedback to the initial specification possible. The debugging architecture based on DLLs that we have adopted adds the flexibility for debugging a variety of circuit implementations in different FPGA architectures.

References

1. Sklyarov, V., "Hierarchical Graph-Schemes", Latvian Academy of Science, Automatics and Computers, Riga, 1984, N 2, pp. 82-87.
2. Sklyarov, V., Monteiro, R. S., Lau, N., Melo, A., Oliveira, A., Kondratjuk, K., "Integrated Development Environment for Logic Synthesis Based on Dynamically Reconfigurable FPGAs". In: Field-programmable Logic and Applications-8th Int. Workshop, FPL'98, Hartenstein, R. W., Keevallik, A. (Eds.), August 1998, pp. 19-28.
3. Sklyarov, V., Lau, N., Oliveira, A., Melo, A., Kondratjuk, K., Ferrari, A. B., Monteiro, R. S., Skliarova, I., "Synthesis tools and design environment for dynamically reconfigurable FPGAs", SBCCI98 XI Brazilian Symposium on Integrated Circuit Design, Búzios, Rio de Janeiro, Brazil, September 1998, pp. 46-49.
4. Xilinx, "XC6200 Field Programmable Gate Arrays, Product Description", (http://www.xilinx.com/partinfo/6200.pdf), April 1997.
5. S. Baranov, "Logic Synthesis for Control Automata". Kluwer Academic Publishers, 1994.
6. Nisbet, S., Guccione, S., "The XC6200DS Development System", Proceedings of FPL'97, London, September 1997, pp. 61-68.
7. Xilinx, "Velab: VHDL Elaborator for XC6200 (V0.52)", http://www.xilinx.com/apps/velabrel.htm.
8. Xilinx, Series 6000 User Guide, 1997.

Synthia : Synthesis of Interacting Automata Targeting LUT-based FPGAs

George A. Constantinides[1], Peter Y. K. Cheung[1], and Wayne Luk[2]

[1] Electrical and Electronic Engineering Dept., Imperial College, London, U.K.
george.constantinides@ieee.org, p.cheung@ic.ac.uk
[2] Department of Computing, Imperial College, London, U.K.
wl@doc.ic.ac.uk

Abstract. This paper details the development, implementation, and results of *Synthia*, a system for the synthesis of Finite State Machines (FSMs) to field-programmable logic. Our approach uses a novel FSM decomposition technique, which partitions both the states of a machine and its inputs between several sub-machines. The technique developed exploits incomplete output specifications in order to minimize the interconnect complexity of the resulting network, and uses a custom Genetic Algorithm to explore the space of possible partitions. User-controlled trade-off between logic depth and logic area is allowed, and the algorithm itself during execution determines the number of sub-FSMs in the resulting decomposition. The results from MCNC benchmarks applied to Xilinx XC4000 and Altera FLEX8000 devices are presented.

1. Introduction

Finite State Machine (FSM) decomposition is the implementation of a large FSM as a network of smaller interacting FSMs, a problem studied since the late 1960s [1]. In VLSI architectures, FSM decomposition is useful for a number of reasons. By controlling the topology and manner in which the machine is decomposed, it is possible to aim for an implementation with characteristics such as: high speed, as decomposition can usually lead to a reduction in logic depth [2]; low area, as state-encoding heuristics will often cope more efficiently with each of the smaller sub-FSMs than with the large lump-FSM [3]; reduction in interconnect complexity [4], and I/O minimization [5].

This paper describes one approach to FSM decomposition, which specifically targets LUT-based FPGA implementations. The technique partitions the states of a large FSM between several sub-machines, while also partitioning its inputs. Output don't-care conditions are exploited in order to minimize the interconnect complexity of the resulting network, and a custom Genetic Algorithm is used to explore the space of possible input partitions. A user-controlled area/speed trade-off is allowed, and the algorithm itself determines the number of sub-FSMs in the decomposition.

2. Background

Feske et al. [3], have developed an approach to FSM decomposition which operates at the State Transition Graph level. Decomposition strategies that partition the STG allow a wider solution space to be searched by the following phases of synthesis [2]. They proceed by breaking the STG into a number of sub-STGs, each containing an extra 'wait' state representing all states outside the sub-STG's domain. The sub-FSMs pass messages to each other when a transition occurs which would cross a sub-STG boundary, indicating which state to enter. Unlike most techniques for FSM decomposition, [3] not only allows an arbitrary number of sub-FSMs to be formed, but decides on that number itself. The simplicity of the messages passed means that complex don't care conditions, arising from the association of particular inputs with particular groups of states, are not exploited. The authors report improvements compared to a one-hot implementation: reduction in circuit depth (29%) and number of logic blocks (38%) when circuits were mapped to a Xilinx XC4000 FPGA.

Yang et al. [4], have developed an FSM decomposition to minimize the complexity of VLSI interconnection. The first step is to partition the set of inputs between n sub-FSMs, each containing all the states of the lump FSM. The communication between sub-FSMs necessary for each sub-FSM to calculate its correct (next-state, output) combination is then found. Outputs are partitioned between sub-FSMs, leading to possible redundancy in states. This is followed by a state-minimization on each sub-FSM. Because the messages passed between sub-FSMs are only functions of machine inputs, the logic to generate them is combinational and simple compared to the sequential logic for the sub-FSM. This technique exploits more complex don't care conditions, leading to a significant reduction in interconnect complexity. This is important for FPGA design, as routing resources tend to be limited. However, as developed in [4], little advantage is taken of incompletely specified FSMs. In addition, the user is required to specify the number of sub-FSMs in advance of algorithm execution. In a later paper [8], the authors apply their technique to FPGA implementations, claiming a significant speedup at the cost of a 44% increase in the number of logic blocks.

3. *Synthia*

3.1 Decomposition Topology

The *Synthia* system performs a partitioning of both inputs and states between sub-FSMs. The state partitioning is performed by adapting the linear partitioning approach described in [3]. The constraint that each lump-FSM state is assigned to exactly one sub-FSM is retained. However, the constraint that every input must be available to every sub-FSM [3], is relaxed. Instead, our approach introduces two types of messages which may be passed between sub-FSMs. Message from one sub-FSM can either instruct another sub-FSM to enter a particular state, or inform another sub-FSM of the status of its inputs. It is worth noting that the two different types of message are never needed simultaneously, since if a given sub-FSM is in its wait-state, then it is only waiting for next-state type messages, whereas if it is not in the wait-state, it will never receive next-state type messages. Shown in Fig. 1(b) is a single sub-FSM.

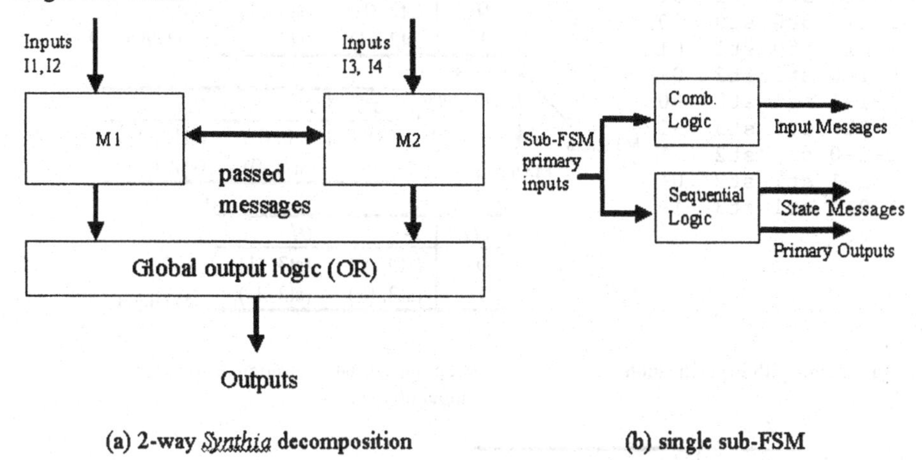

(a) 2-way *Synthia* decomposition (b) single sub-FSM

Fig. 1. Synthia decomposition topology

3.2 Input Messages

Yang [4] uses *n*-dimensional state-transition matrices in order to find what messages to be pass between *n* sub-FSMs. Here we extend the idea of state-transition matrices to better cope with two phenomena not present in [4]: in our decomposition, states belong to only one sub-FSM and incompletely specified outputs may be present. Assuming that the input and state partitioning has already been performed using the method described in section 3.3, an (*n*+1)-dimensional matrix may then be built for each sub-FSM *M*. One dimension corresponds to the state of *M*. The other *n*-dimensions correspond to possible input cubes seen by each sub-FSM, and the data value at each coordinate represents a next-state/output combination. Consider the state-machine specification shown in Fig. 2(a). Let us assume that st0 and st1 are contained in sub-FSM M1, the other states belonging to other sub-FSMs. Further, assume that the first two inputs are sent to sub-FSM M2, and the final one to sub-FSM M3. A 4-D state-transition matrix is constructed for M1 shown in Fig. 2(b).

It may now be possible to merge axis headings on each of the (*n*-1) dimensions not representing M1, while still retaining all necessary information, thus reducing interconnect complexity. We aim to minimize interconnect complexity for two reasons: routing resources are scarce in FPGAs, and small sub-FSMs are likely to make heavy use of fast local FPGA interconnect, whereas interconnect between sub-FSMs is likely to make more heavy use of slower, non-local routing resources.

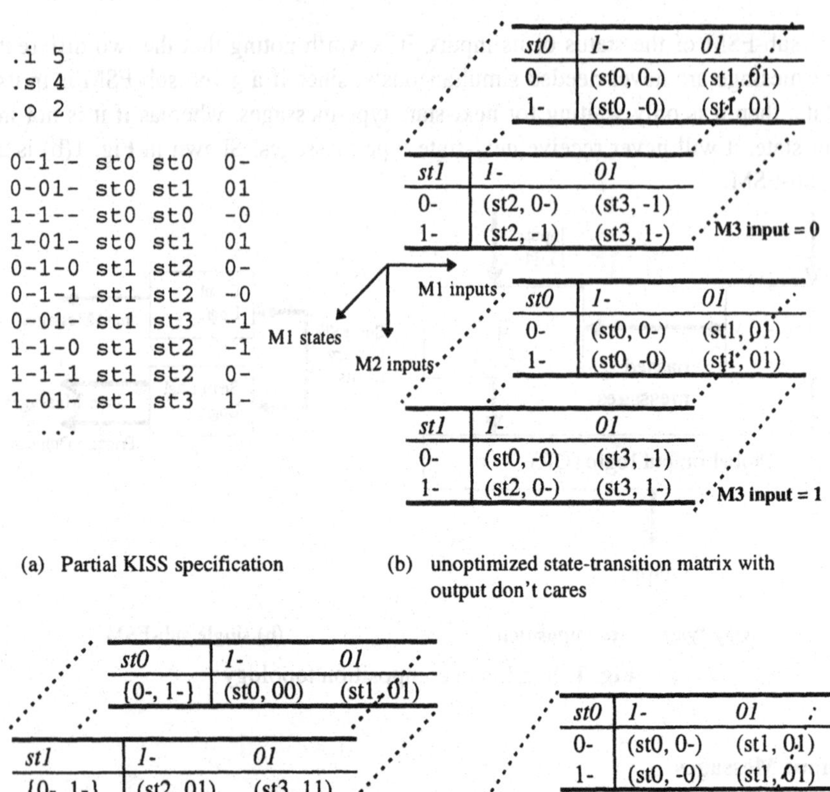

(a) Partial KISS specification

(b) unoptimized state-transition matrix with output don't cares

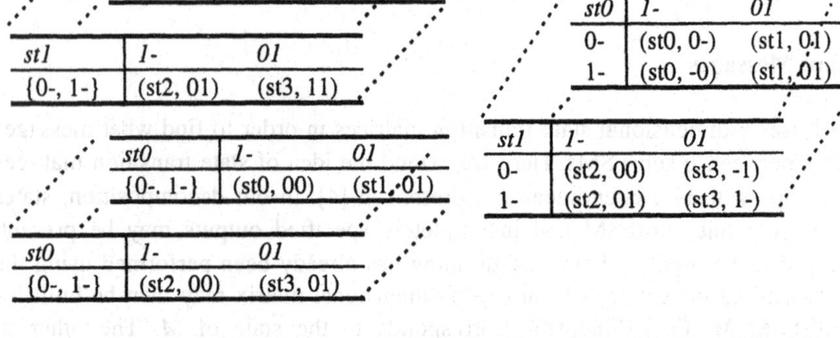

(c) one optimization of Fig. 2(b)

(d) another optimization of Fig. 2(b)

Fig. 2. State-transition matrices and their optimization

When there are don't care conditions on the output specifications, it is possible that there is no single unique solution to the problem. For example, the reduced matrix could take the form of Fig. 2(c) or 2(d) equally. This situation arises because don't care conditions may be expanded into concrete '0' or '1' specifications as a result of one merge, which would conflict with the concrete value arising from another possible merge. A heuristic has been implemented, as detailed in section 3.4, to select which merges to implement.

3.3 Primary Input and State Partitioning

A Genetic Algorithm (GA) is used to search the space of possible input partitions. We may represent the mapping between primary inputs and sub-FSMs by a vector **I**, indexed by primary input number, and with entries in the range 0 to (n-1), indicating to which sub-FSM each input is assigned. When viewing the problem in this way, a clear demarcation already exists between alleles in a chromosome - the index of the vector **I**. With larger allele size there is not full genetic control over the search-space, whereas with a smaller size (say bits, more typically used in a GA) no extra information is present. The unusual properties of the GA applied are: valid allele values are 0 to (n-1), n may vary during algorithm execution, and mutation consists of randomly choosing an allele value between 0 and (n-1).

Rather than extending the GA approach to the partitioning of states, certain *a-priori* information may be used effectively: states that are successors or predecessors of other states in the STG are more likely to benefit from incorporation within a single sub-FSM than any two states chosen at random. Such sub-FSM networks are likely to have fewer inter-FSM transitions, and in addition, the results of [3] indicate that particular inputs tend to be associated with particular sub-graphs of the STG. Indeed, the algorithm described in [3] is a heuristic incorporating elegantly this prior knowledge. The approach taken in that paper has been modified to incorporate the two different message types, and the presence of an extra phase - that of input assignment.

3.4 Algorithms

At the core of the algorithm employed is the integration of the two phases: an STG-level heuristic for state-partitioning, and a GA for input partitioning. The solution taken is to interleave the two algorithm phases in the following way. The top-level algorithm is shown in pseudo-code below.

```
OptimizeFSM() {
        ga_optimize( start_its );
        do {
          foreach sub-FSM B {
            do {
              gain = Optpart( B, best_chromosome );
            } while( gain > 0.0 );
          }
          cleanup_popul( popul, decomp );
          ga_optimize( popul, next_its );
        } while( newCost < oldCost );
}
```

The algorithm starts with an initial state partition of one state per sub-FSM (a one-hot encoding on the lump-FSM). After performing a GA optimization with respect to that state-partition, the main loop is entered. `Optpart` works on one block of

M0	-00	011	010	001
-00				
011	T			
010	T	T		
001	F	F	F	

M1	--0	001
--0		
001	T	

M2	000	100	110	111
000				
100	F			
110	F	T		
111	T	F	F	

Fig. 3. Three compatibility tables

Table 1. *Synthia* run-times

Ckt	Run-time
bbara	4.01s
beecount	1.44s
cse	15m 17s
shiftreg	2m 57s
dk14	6m 47s
dk27	1m 54s
dk512	5m 31s
ex6	11m 59s
opus	5m 42s
s208	13m 45s

the state-partition at a time, trying to suck-in any state which has a predecessor or successor state within another block. If this results in a positive gain, the move is retained. If this the results in an empty block, all references to that block in the chromosome **I** are set to point to the new block containing that state. (A similar function is performed by `cleanup_popul`). Two, possibly different, numbers of iterations are used in the GA - one inside and one outside the main loop body. This is because it is quite likely that the population of input partitions before entering the state-partitioning heuristic is a good *first-guess* for the GA after exiting the heuristic. This insight was confirmed by the results collected.

The delay and area of the resulting network is estimated from the optimized state-transition matrices. After constructing an unoptimized state-transition matrix (as discussed in section 3.2), the algorithm optimizes the matrix by merging axis headings as much as possible. The heuristic iteratively merges axis headings, two at a time. It is a non-backtracking approach: the merge chosen at any given step is the one judged 'most-likely' to result in the largest interconnect complexity reduction, and once that choice is made it will not be changed at a later stage. For each axis in the given state-transition matrix, a *compatibility table* is constructed. This table indicates which axis headings are mergible with which others (to be determined by searching the matrix entries). The example compatibility tables are shown in Fig. 3. Note that these are lower triangular matrices with binary entries (the compatibility relation is bi-directional and Boolean). The heuristic proceeds by counting the number of possible merges on each axis (3 for M0, 1 for M1 and 2 for M2). The axis with the greatest number is chosen for merging, in the hope that some merges will still be possible after the current one. Then the heuristic picks the heading H_1 with the largest number of possible merges (in this case either -00, 011 or 010). Finally, the pair to merge is completed by choosing the heading with the next largest number of possible merges H_2, subject to the constraint that H_1 and H_2 are mergible. After merging the two axis headings, and adjusting their entries (replacing don't cares by '0' or '1') as necessary, the heuristic is ready for its next iteration.

After optimization, the matrices form a complete specification for the network of sub-FSMs, which is then passed to a function in order to calculate a cost estimate.

This proceeds in four stages: input message encoding, state message encoding, STG construction for each sub-FSM and finally delay and area estimation.

The message assignments are arbitrary minimum length encodings. The delay and area of each sub-FSM are estimated separately from its neighbours for speed of execution. This is done using SIS [9] procedures for LUT-based FPGAs, as developed by Murgai [10], using a JEDI [11] minimum length state encoding on each individual sub-FSM. The worst-case logic depth (and therefore the estimated worst-case delay) of the resulting network occurs when more than one sub-FSM is involved in a transition. The maximum number of sub-FSMs that may be involved in a transition is two. Hence if all the logic depths of all sub-FSMs are written as d_0, d_1, ..., d_n, with $d_0 \geq d_1 \geq ... \geq d_n$, then the worst-case logic depth is bounded above by d_0+d_1, and below by d_0. The returned delay estimate is $d_0 + 0.5d_1$. The area estimate returned for the network is simply the sum of all area estimates for the sub-FSMs. The two costs (area and logic depth) are combined in a linear manner with a user-controlled factor in order to return a single cost-function value.

4. Experimental Results

The algorithm described was implemented and integrated within the SIS logic synthesis package [9]. The implementation takes an STG as input and produces a file of hierarchical VHDL source-code describing the network of interacting FSMs. A diagram illustrating a typical design-flow when using *Synthia* is shown in Fig. 4.

The format for describing FSMs in SIS is the KISS format. This can be acquired either directly from design specification, or through extraction from an HDL description. Synopsys software [12] has such an extraction feature, 'extract'. The

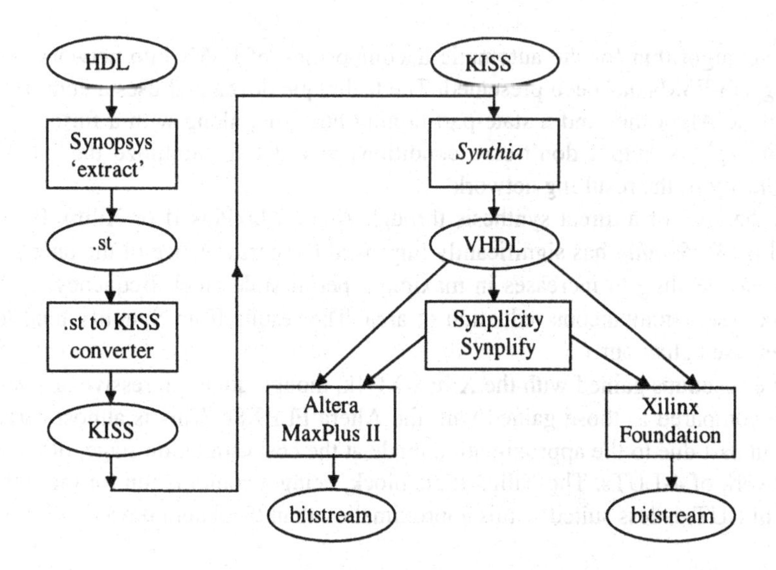

Fig. 4. *Synthia* design flow

state-machine format used by Synopsys is another abstract description, '.st' for which we have written a converter to KISS.

FSMs from a subset of the MCNC benchmark set were decomposed using the algorithm. The resulting VHDL code was compiled by Xilinx Foundation tools [13] targeting Xilinx XC4003EPC84-1, and Altera MaxPlus II [14] targeting EPF8282A-2. In addition, the VHDL code was synthesised by Synplify [15], and the resulting network passed through MaxPlus II and Xilinx M1.
The overall number of Altera LCs and Xilinx CLBs were collected, alongside the reported maximum clock frequencies. For the purposes of comparison, each benchmark was also encoded in a one-hot style by JEDI, automatically written as VHDL code, and compiled as above by the FPGA vendor tools. These results are shown graphically in Fig. 5. In addition, *Synthia* run-times on a Pentium II 233 running Linux are reported in Table 1.

The results shown an average speedup of 29% for XC4k using Xilinx Foundation, and 49.7% for FLEX8k using MaxPlus II. These figures change to speedups of 37% and 55% respectively, when counting negative speedups as 0 (i.e. when either one-hot or *Synthia* is used, whichever is faster). Also shown is an average area reduction of 10% for XC4k using Xilinx Foundation and an average area increase of 16% for FLEX8k using MaxPlus II (becoming a 15% and 12% reduction when modified as above). Also illustrated are the results for XC4k using Synplify, showing an average area increase of 39.13% and clock rate reduction of 27.74% (becoming 0% and 0% when modified as above). Finally the results for FLEX8k using Synplify, show an average area increase of 59% and clock rate reduction of 14.67% (becoming a 0% area reduction and a 1.5% speedup when modified as above).

5. Conclusion

A novel algorithm for the automatic decomposition of FSMs into networks of interacting sub-FSMs has been presented. The technique developed uses a combination of a Genetic Algorithm and a state-partitioning heuristic, along with a further heuristic which exploits output don't-care conditions in order to minimize the interconnect complexity of the resulting network.

In the case of a direct synthesis through Altera MaxPlus II or Xilinx Foundation (see Fig. 4), *Synthia* has significantly improved the performance of the later stages of synthesis, leading to increases in maximum permissible clock frequency, and in the Xilinx case, simultaneous reduction in area. The results from Synplify tend to show the reverse behaviour.

The speedups gained with the Xilinx XC4k, though quite impressive, are relatively small compared to those gained with the Altera FLEX8k. This is almost certainly at least in part due to the approximation made at the cost estimation stage, of a circuit as a network of 4-LUTs. The Xilinx logic block, being a combination for more than one type of LUT, is less suited to this approximation than the Altera device.

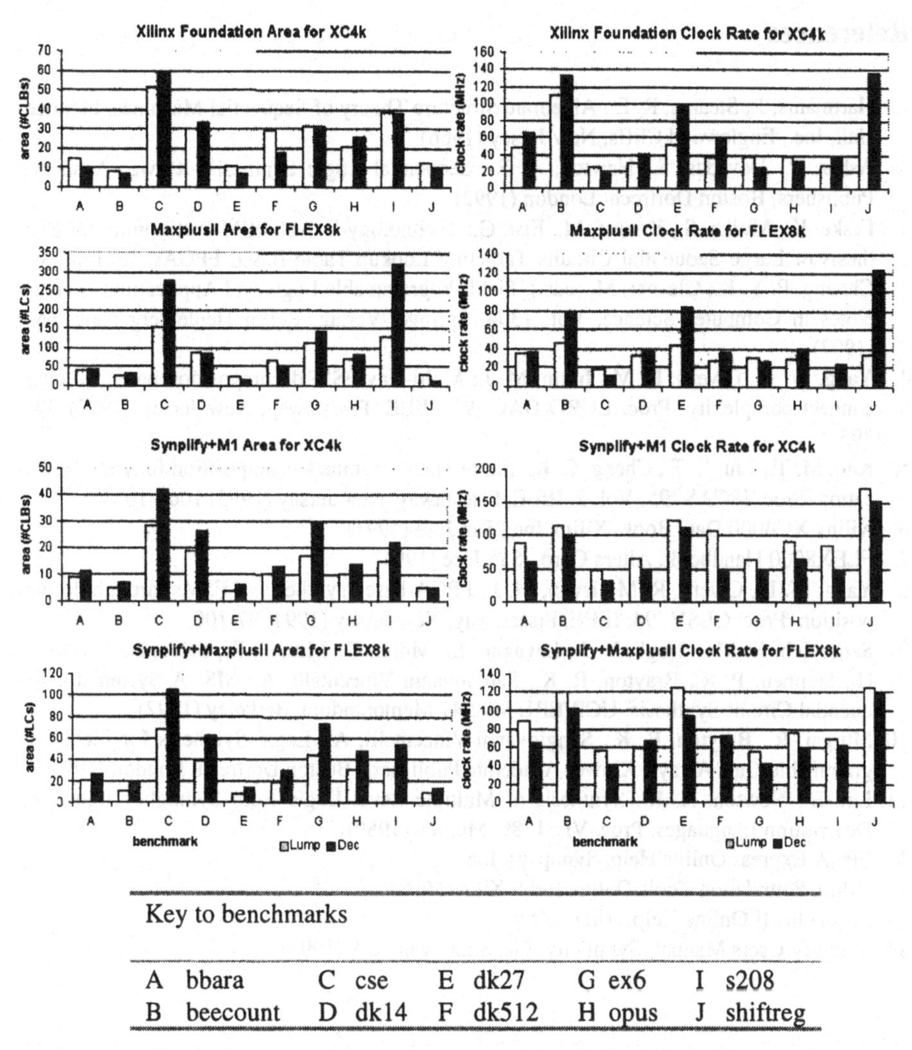

Key to benchmarks									
A	bbara	C	cse	E	dk27	G	ex6	I	s208
B	beecount	D	dk14	F	dk512	H	opus	J	shiftreg

Fig. 5. Experimental results

The question of choosing an efficient message encoding for inter-FSM communication stands out as a missing part of this work. This question is related to the ongoing research topic of hierarchical synthesis of sequential circuits.

Acknowledgments

The Authors would like to acknowledge the support of Alan Marshall, Nick Wainwright and John Lumley of Hewlett Packard Laboratories, Bristol, UK. In addition, donations of software tools were gratefully received from Xilinx and Altera.

References

1. Hartmanis, J., Stearns, R. E.: Algebraic Structure Theory of Sequential Machines. Prentice-Hall, Inc., Englewood Cliffs, New Jersey (1966)
2. Ashar, P., Devadas, S., Newton, A. R.: Sequential Logic Synthesis. Kluwer Academic Publishers, Boston Dortrecht London (1992)
3. Feske, K., Mulka, S., Koegst, M., Elst, G.: Technology-Driven FSM Partitioning for Synthesis of Large Sequential Circuits Targeting Lookup-Table Based FPGAs. In: Luk, W., Cheung, P. Y. K., Glesner, M. (eds.): Field-Programmable Logic and Applications. Lecture Notes in Computer Science, Vol. 1304. Springer-Verlag, Berlin Heidelberg New York (1997)
4. Yang, W. L., Owens, R. M., Irwin, M. J.: Multi-way FSM decomposition based on interconnect complexity. Proc. EURO-DAC '93. IEEE, Piscataway, New Jersey (1993) 390-395
5. Kuo, M. T., Liu, L. T., Cheng, C. K.: Finite State Machine Decomposition for I/O Minimization. Proc. ISCAS '95, Vol. 2. IEEE, Piscataway, New Jersey (1995) 1061-1064
6. Xilinx XC4000 Data Book, Xilinx Inc., San Jose (1991)
7. FLEX8000 Handbook, Altera Corp., San Jose (1994)
8. Yang, W. L., Owens, R. M., Irwin, M. J.: FPGA-based synthesis of FSMs through decomposition. Proc. GLSV '94. IEEE, Piscataway, New Jersey (1994) 97-100
9. Sentovich, E. M., Singh, K. J., Lavagno, L., Moon, C., Murgai, R., Saldanha, A., Savoj, H., Stephen, P. R., Brayton, R. K., Sangiovanni-Vincentelli, A.: SIS: A System for Sequential Circuit Synthesis. UCB/ERL M92/41 Memorandum, Berkeley (1992)
10. Murgai, R., Brayton, R. K., Sangiovanni-Vincentelli, A.: Logic Synthesis for Field Programmable Gate Arrays. Kluwer Academic Publishers, Boston Dortrecht London (1995)
11. Lin, B., Newton, A. R.: Synthesis of Multiple Level Logic from Symbolic High-Level Description Languages. Proc. VLSI '89, Munich (1989)
12. FPGA Express Online Help. Synopsys, Inc.
13. Xilinx Foundation Tools Online Help. Xilinx, Inc.
14. Max+Plus II Online Help. Altera Corp.
15. Synplify Users Manual. Synplicity, Inc. Sunnyvale, CA (1998)

An FPGA-based Prototyping System for Real-Time Verification of Video Processing Schemes

Holger Kropp, Carsten Reuter, Matthias Wiege, Tien-Toan Do, and Peter Pirsch

Laboratorium für Informationstechnologie, Universität Hannover,
Schneiderberg 32, 30167 Hannover, Germany
kropp@mst.uni-hannover.de
http://www.mst.uni-hannover.de/Forschung/Projekte/RPS/RPS.shtml

Abstract. A real-time prototyping environment for complete video processing schemes is presented. To realize a real-time processing, a commercial FPGA-based prototyping system is extended by a special video interface, efficient pipelined FPGA macros, and a modified design flow. Reductions of 48% in terms of FPGA resources, and 80% of compilation time are achievable. The feasibility of the prototyping environment is shown for a complete H.263 video codec.

1 Introduction

Today's video processing schemes, for example CCITT visual telephony (H.263) [1] or MPEG [2], have a steadily increasing computational complexity. They consist of various algorithms, e.g., transformations, filters, motion estimation, or variable length coding. If the processing has to be performed under real-time constraints, these schemes as well as their algorithms have to deal with high throughput rates. Their implementation often demands application-specific integrated circuits (ASICs). Since development costs for those ASICs are high, video processing schemes should be verified and optimized before implementation. For this purpose, a flexible real-time prototyping of a whole video processing scheme is mandatory.

Traditionally, software simulation is used for verifying video processing schemes as well as ASIC designs. However, due to the enormous throughput rate and huge amount of operations, a software simulation is only possible for video sequences of a few seconds. The simulation time for longer sequences takes usually hours or days. Advantages in VLSI technology and, in particular, the increasing complexity and capacity of programmable logic devices have been making hardware emulation possible. The underlying key of such a prototyping system is the use of reconfigurable field programmable gate arrays (FPGAs), like SRAM-based XC4000 Xilinx's FPGAs [3] or Altera's Flex10K [4]. Hardware emulation can considerably reduce the time taken in analysis and verification of image processing algorithms and circuits, compared to software simulation. Furthermore, a prototyping system can operate in the target environment under real-time constraints, so that influences of the modification of algorithm parameters on the quality of processed video sequences, can be subjectively analyzed and evaluated. Thus, algorithms and circuits can be optimized before implementation.

In recent years, a few prototyping environments for signal processing applications were published. Examples for video processing applications can be found in Luk et al.

[5] or SPLASH 2 [6]. With these rapid prototyping systems, single video processing algorithms can be emulated, e.g., Median filter, fast Fourier transform, or edge detection. The *'Graphical RApid Prototyping Environment' (GRAPE-II)* [7] is a heterogeneous prototyping systems and consists of commercial DSP processors (TMS320C40) and FPGAs (Xilinx). Implementations can take place on DSPs, FPGAs or a combination of both. GRAPE II is used in the field of audio processing.

The applications for these prototyping environments are single algorithms for video and audio processing. More complex are complete video processing schemes, because they are composed of different algorithms. So called low-level algorithms, e.g., filters and transformations, are characterized by their deterministic number of input and output data. The processing is independent of the values of the data. To increase the throughput, an intensive parallel processing of low-level algorithms could be applied. In contrast to that, the data of medium- or high-level algorithms, e.g., quantization or variable length coding, are symbols, that do not directly represent the picture content. Processing of the symbols depends on their values. Compared to low-level algorithms, only a restricted parallel processing is possible.

Algorithm parameters have mutual influence and make it difficult to predict the impact of parameter variations on image quality. To support a designer in his decision among architectural alternatives, algorithm, and circuit parameters, the *'Video and Image Processing Emulation System'* (*VIPES*) has been developed. It provides a real-time emulation of complete video processing schemes. The aim is the implementation and real-time emulation of a H.263 codec on the FPGA based platform *VIPES*. Therefore, high throughput and a compact realization are necessary. In order to occupy as few FPGAs as possible, the implementation from single arithmetic operations up to complete algorithms has to be adapted to the underlying FPGA architecture.

The paper is organized as follows. The prototyping environment *VIPES* as well as some sophisticated FPGA modules will be presented in section 2. Results for an H.263 codec are given in section 3, while section 4 provides concluding remarks.

2 VIPES

The *'Video and Image Processing and Emulation System' (VIPES)*, provides an FPGA based platform for real-time emulation of complete video processing schemes. Thereby, a subjective assessment of image quality is possible and the impact of different algorithm or circuit parameters can be determined and optimized. *VIPES* consists of a video input and output interface, the FPGA based rapid-prototyping system (RPS) *'System Realizer M250'* from Quickturn Design Systems [8], a software environment, and a library of pre-implemented circuits in form of FPGA macros and Verilog interface modules. A simplified representation is shown in Fig. 1.

The RPS contains 80 XC4013 FPGAs and permits the realization of designs with a size up to 250,000 gate equivalents. The main components of an XC4013 FPGA are the configurable logic blocks (CLBs) and routing resources. The CLBs contain SRAM-based look-up tables (LUTs) for the implementation of boolean equations, as well as flip flops, which can be used for intensive pipelining.

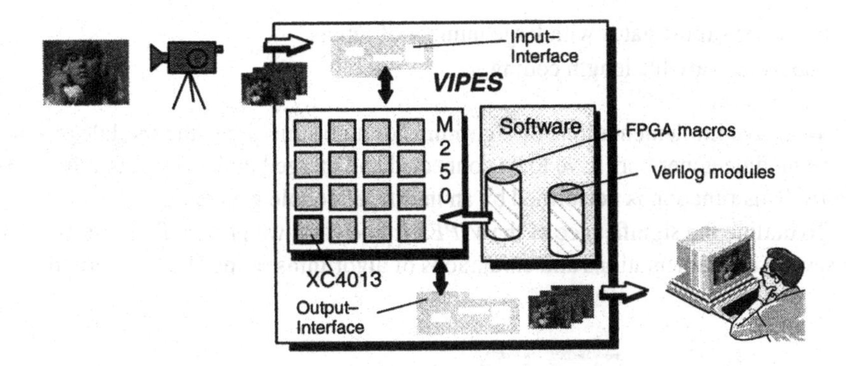

Fig. 1. Simplified representation of VIPES

The mapping of a circuit onto the *'System Realizer M250'* is performed in two steps on different levels by Quickturn's emulation software (*QUEST* [8]). On the higher system level a gate level netlist, e.g. Verilog netlist, is imported. The netlist is checked, flattened, and partitioned into subnetlists of gate primitives, that fit into single FPGAs. Thus, the actual number of realizable gate equivalents depends essentially on partitioning. Afterwards, the subnetlists are converted into FPGA netlists (XNF netlists). On the lower FPGA level these XNF netlists are partitioned and placed by Xilinx tools. Thereby, gate primitives and flip flops are instantiated by elements from *QUEST's* FPGA-library. This approach leads to poor resource utilization, low throughput, and long compilation time, because the FPGA library consists only of elements of at most a few CLBs.

To derive efficient implementation of circuits and sub circuits, one has to consider a compact mapping of a circuit's architecture onto CLBs, as well as CLBs' interconnection with a minimum delay. For this purpose, we have been developing relationally placed macros (RPMs) for important basic arithmetic operations and algorithms. RPMs are structures, where the placement of CLBs is pre-determined. These modules are integrated in the *VIPES* library. *QUEST* designates so far no integration of user defined RPMs. To overcome this disadvantage, methods were developed, which enable us to merge optimized RPMs in the designflow [9]. We substitute in the Verilog netlist appropriate modules (here for instance a multiplier) by RPMs adapted to required word widths, and number of pipeline stages. A simplified description shows Fig. 2. Thus, circuit mapping is accelerated, since no synthesis and no FPGA partitioning and placement of these pre-implemented macros are necessary.

For a multitude of video processing algorithms like filtering or DCT, basic arithmetic operations are essential. In order to derive fast and compact modules for different applications, we have developed several module generators that create RPMs, for example combinatorial or pipelined arithmetic operations. The following module generators are available:

– pipelined adders, incrementers, subtractors, and decrementers
– pipelined array multipliers [10]
– rounding units

- units for simple gates with large number of inputs
- tables for variable length coding .

To derive implementations of algorithms or algorithm parts, the modules from the above module generators have to be connected and merged into a netlist that fits on one FPGA. This function is performed by an interface module generator.

To outline the significance of the *VIPES* framework, we present first results derived by several implementations and emulations of algorithms of the H.263 standard.

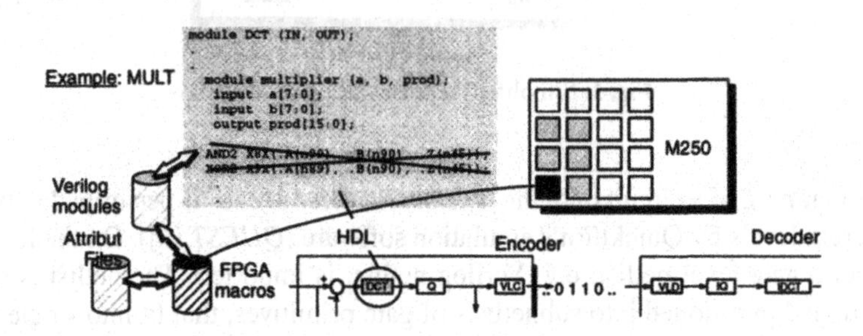

Fig. 2. Substitution of single modules in VIPES

3 Design Example: H.263

As application example for our FPGA-based prototyping environment we have chosen the video coding scheme compliant to H.263-standard [1]. This coding scheme is applied for video telephony applications. The H.263 codec is divided in one encoder and decoder part, respectively. Both consist of low-level as well as of medium- and high-level algorithms. Low-level algorithms are the discrete cosine transformation (DCT), their inverse (IDCT), and the motion estimation (ME). For these algorithms, different realization forms are possible. Nevertheless, for real-time processing of such algorithms, a large number of arithmetic operations have to be performed. Unfortunately, these operations, especially multiplications, are costly parts in terms of FPGA resources. Therefore, compact arithmetic modules like multipliers, adders, subtracters, etc. are required.

In the H.263 codec, also some medium-level algorithms like the quantization (Q), and their inverses (IQ) are applied. High-level algorithms are the Huffman-based variable length coding (VLC) and decoding (VLD) as well as the arithmetic coder (AC) or decoder (AD). For the medium- and high-level algorithms, beside arithmetic operators, further FPGA modules for whole algorithms are required. At the current status of *VIPES*, not all algorithms required for H.263 have been implemented. The results of those which have been implemented are shown in Tab. 1 in terms of the number of required FPGAs and the achievable emulation frequency. If a well known architecture was used, this is indicated.

Table 1. Implementation results for different algorithms

Algorithm	*Architecture*	*#FPGAs*	$f_{EMUL.}$
DCT	fast DCT [11]	12	5.5 MHz
IDCT	fast inv. DCT [11]	12	5.5 MHz
Quantization	-	1	6.5 MHz
inv. Quantization	-		6.5Mhz
Run & Variable Length Coding	-	1	6.5 MHz
Motion Estimation	TSS, NTSS, FSS, NFSS [12]	9	4.0 MHz
Arithmetic Coding	SAC [1]	5	3.6 MHz
Arithmetic Decoding	SAC [1]	7	3.0 MHz

With respect to H.263 video coding standard, CIF images are processed with a frame rate of 25 Hz. This leads to a real-time constraint of approximately 27 MHz sample frequency.

The most significant algorithms for H.263 are the DCT and the IDCT, respectively, because these two use identical operations and require a large amount of FPGA resources. We selected an appropriate architecture from Zhang and Bergmann [11]. A two-dimensional implementation needs 26 additions, 32 subtractions, and 22 multiplications. We have implemented this 2D-FDCT using 25 pipeline stages. In a first approach the FDCT was described in Verilog, synthesized by Synopsys into a gate-level netlist and mapped onto 23 FPGAs of the RPS. The achievable emulation frequency was about 5 MHz. Only components of the QUEST FPGA library were used. Based on this architecture we re-implemented the FDCT with pipelined FPGA macros taken from our FPGA library. To derive best results concerning pin and FPGA usage we performed manual partitioning and optimized word widths for each arithmetic unit. The number of necessary FPGAs could be lowered by 48% to 12 FPGAs. The achievable emulation frequency was slightly improved to 5.5 MHz. Measurement of CPU time for compilation indicated a reduction of approximately 80%, due to the use of pre-implemented XNF netlists. Because of the architectures' throughput of eight pixels per clock cycle, this corresponds to a sample frequency of 44 MHz. Therefore, it is possible to emulate the DCT and IDCT with H.263 compliant video sequences in real time.

The quantization (Q) and its inverse (IQ) require 1 multiplier, 2 incrementers, and 2 decrementers as well as some flip flops and glue logic, each. These modules were generated by the module generators listed in section 2. They fit into one XC4013 FPGA. The designs' time critical path is in range of 57 ns, this corresponds to a frequency of 17 MHz. But due to the limitation of an internal logic analyzer, the highest emulation frequency is 6.5 MHz. The modules for run- and variable length coding (RLC/VLC), fit into one FPGA and the maximal emulation frequency could be achieved. The RLC requires only 12 CLBs, whereas the VLC needs 410 CLBs, because large look-up tables have to be realized. The architectures for motion estimation, arithmetic coding, and arithmetic decoding were described in hardware description language Verilog without any of our FPGA macros. Due to that, they need a relatively large amount of FPGAs. The achievable emulation frequencies of the designs are about 3-4 MHz. In order to reduce the number of FPGAs and to improve emulation frequency we currently work on new FPGA macros to derive more efficient modules.

For the implementation of the whole H.263 codec with *VIPES* up to 80 XC4013 FPGAs are available. Based on the values of Tab. 1 and the assumption that the algorithms VLC, VLD and the controlling require approximately 6 FPGAs, the implementation of the H.263 codec will fit into the RPS.

4 Conclusion

A real-time emulation system (*VIPES*) for the emulation of complete video processing schemes has been presented. *VIPES* is based on a commercial FPGA emulator and related software. To meet real-time constraints for complete schemes, the emulation environment has been extended by dedicated video interfaces, efficient flexible FPGA macros, and a modified design flow. Using the FPGA macros and the modified design flow, it is possible to reduce the number of FPGA resources by 48%, the compilation time by approximately 80%, as well as to increase the throughput rate.

For future work, it is planed to implement additional tools with consideration of our experiences. These tools should mainly support the implementation of low-level algorithms concerning word width analysis, partitioning, and automatic FPGA macro generation and integration.

Acknowledgements

The work presented is supported by the Deutsche Forschungsgemeinschaft, DFG, under contract number Pi–169/7.

References

1. ITU-T Draft Recommendation H.263, "Video coding for low bitrate communication," 1995.
2. ISO-IEC IS 13818, "Generic Coding of Moving Pictures and Associated Audio," 1994.
3. Xilinx Inc., *The Programmable Logic Data Book*, 1996.
4. Altera Co., *Data Sheet FLEX10K*, 1995.
5. W. Luk, P. Andreou, A. Derbyshire, F. Dupont-de-Dinechin, J. Rice, N. Shirazi, D. Siganos, "A Reconfigurable Engine for Real-Time Video Processing," *8th Int'l Works. on Field-Programmable Logic and Applications*, Estonia, 1998.
6. P.M. Athanas, A.L. Abbott, "Image Processing on a Custom Computing Platform," *4th Int'l Works. on Field-Programmable Logic and Applications*, Prague, Sept. 1994.
7. R. Lauwereins, M. Engels, M. Ade, J.A. Peperstraete, "Grape-II: A System Level Prototyping Environment for DSP Applications," *IEEE Computer*, vol. 28, no. 2, pp. 35-43, 1995.
8. Quickturn Design Systems Inc., *System Realizer User's Guide*, Version 5.1, 1997.
9. H. Kropp, C. Reuter, P. Pirsch, "The Video and Image Emulation System VIPES," *Proc. 9th Int'l Works. Rapid System Prototyping*, June 1998, pp. 177-175.
10. H. Kropp, C. Reuter, T.T. Do, P. Pirsch, "A Generator for Pipelined Multipliers on FPGAs," *Proc. 9th Int'l Conf. on Signal Processing Appl. and Tech., Vol. 1*, Sept. 1998, pp. 669-673.
11. J. Zhang, N.W. Bergmann,"A New Fast DCT-Algorithm for Image Compression," *IEEE Int'l Works. on Visual Signal Processing and Communications*, Melbourne 1993, pp. 57-60.
12. S. Kappagantula, K.R. Rao,"Motion compensated interframe image prediction," *IEEE Trans. on Communications*, vol. COM33, no. 9, pp. 1011-1015, Sept. 1985.

An FPGA Implementation of Goertzel Algorithm

Tomas Dulik

Technical University of Brno,
Faculty of Electroengineering and Computer Science,
Department of Computer Science, Bozetechova 2, 612 66 Brno,
Czech Republic
dulik@dcse.fee.vutbr.cz
http://www.fee.vutbr.cz/~dulik

Abstract. Field Programmable Gate Arrays (FPGAs) has already proven to be the best solution in cases, where hardware flexibility and/or reprogrammability was required. Recently, with the growth of their capabilities and speed, a new kind of completely different applications has arosen. This paper describes a FPGA implementation of Goertzel algorithm and its application - a multichannel DTMF (Dual Tone Multi Frequency) decoder, that can outperform a middle-class DSP processor based system by an order of magnitude.

1 Introduction

Goertzel algorithm is used for Discrete Fourier Transformation (DFT) coefficients computation. Its core consist of bank of IIR filters, whose resonance frequencies are equal to the discrete frequencies of the transformation. For example, a two-point DFT that computes weights of frequencies f_1, f_2 can be computed by Goertzel algorithm with two IIR filters with resonance frequencies f_1 and f_2, respectively. The filters have both poles on the unit circle, so they amplify harmonics with frequency equal to the resonant frequency of the filter and attenuate all the other harmonics. The signal flow of one IIR Goertzel filter is on Fig. 1.

This core must be used repeatedly for all the analyzed frequencies. As this seems to be a disadvantage when implementing in software, this approach gives a great opportunity for parallel fashion of computation in dedicated FPGA hardware.

The two taps of the IIR filter are formed by two registers, an adder and a constant multiplier. Since all of these functions can be implemented effectively in Xilinx FPGAs that have the RAM (ROM) feature (their logic resources can be configured as distributer RAM or ROM cells), it is obvious that if only few discrete frequencies are needed, the computation can be made fully parallel by placing one Goertzel IIR core per frequency point on the chip.

The second part of the algorithm, that is triggered after every N samples are passed to the IIR filter, is the evaluation (N determines the accuracy of the

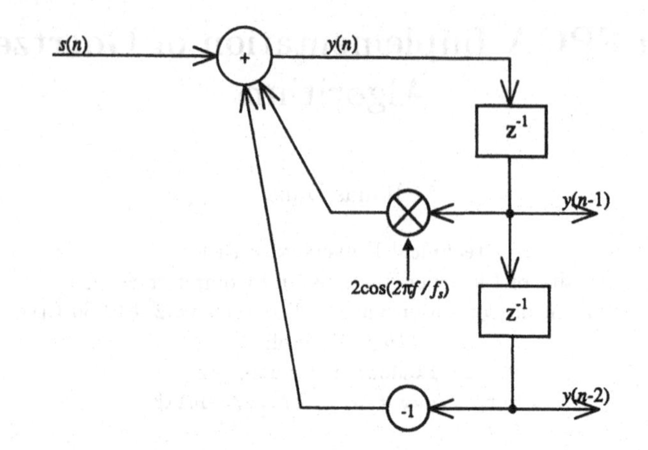

Fig. 1. The Goertzel algorithm IIR core. The $s(n)$ stands for sequence of input signal samples, while $y(n)$ forms the output. The constant $2\cos(\frac{2\pi f}{f_s})$ is dependent only on the sampling frequency f_s and analyzed frequency f.

analyze and its value depends on application demands – e.g., a DTMF decoder would have $100 < N < 300$). It must be done only if there is a need for exact DFT magnitude of the frequency and can be skipped, if the goal is just a frequency detection. Also, there is no need for implementing the magnitude computation in the FPGA - the required data-throughput of this operation is always at least 100 times lower than that one of the IIR filter. Therefore, it is better to concentrate on accelerating the IIR filter computation in the FPGA, while the magnitude computation can be done by a cheap microcontroller or DSP processor. The magnitude M can be computed as

$$M = \left| y(n) - y(n-1).e^{-j2\pi\frac{f}{f_s}} \right| . \tag{1}$$

where $y(n)$ are the output samples, f is the frequency analyzed and f_s is sampling frequency.

The function of one Goertzel IIR filter is demonstrated on Fig. 2. In the upper half of it, there are two filter responses to different harmonics - on the left, the frequency of the harmonic is equal to the resonant frequency of the filter (941 Hz) and thus the filter starts to oscillate on that frequency with the amplitude of the oscillation growing, on the right the frequency is different (1209 Hz). In the lower half of the picture, there is the frequency response of the filter, that was computed point by point using the magnitude evaluation for different input frequencies. On the left, the length of each input sequence was 200 samples, on the right it was 400 samples. The sampling frequency was 8000 Hz – the same as in the most of digital telephone systems – and the filter resonant frequency is taken from the DTMF system, which defines signals for dialing numbers in modern telephones.

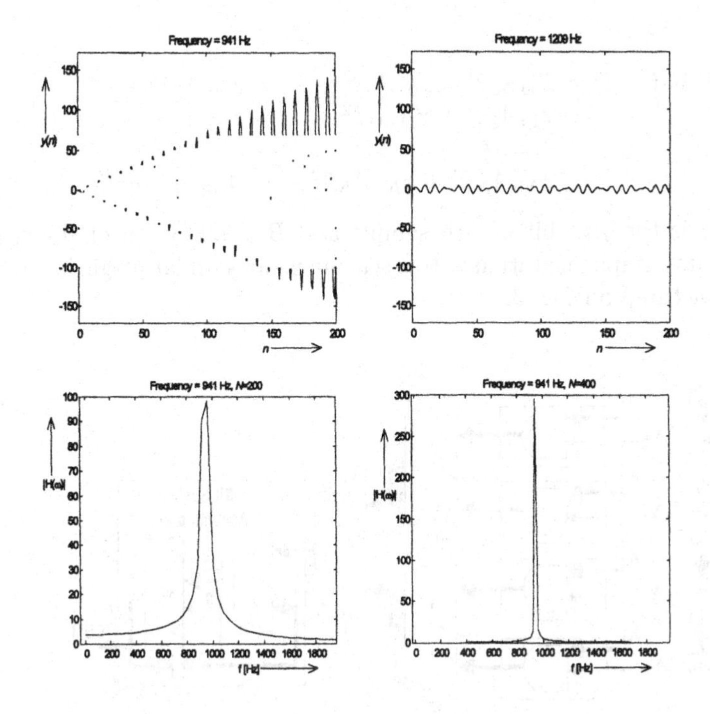

Fig. 2. Output responses of one Goertzel IIR filter

2 FPGA and Constant Coefficient Multiplier

As multipliers always create bottleneck in any FPGA design, there have always been research efforts to reduce the overhead – computation time and occupied chip area. E.g., Prof. Mintzer in [2] proposed to use distributed arithmetics for constant coefficient multipliers. His approach became standard in all modern FPGA designs because it claims the least FPGA resources of all the other methods introduced before that one, while keeping the computation time as low as 1 CLB delay.

2.1 Distributed Arithmetics

Distributed arithmetics can be used anywhere where a quick and space efficient *Multiply and Accumulate (MAC)* operation is needed. The *MAC* operation on sequence of N samples x_n and N coefficients A_n is defined as:

$$MAC(x, A) = \sum_{i=0}^{N} x_i A_i = x_0 A_0 + x_1 A_1 + \ldots + x_N A_N . \quad (2)$$

This can be decomposed to the bit level:

$$\begin{aligned} MAC(x, A) = {} & x_{00}A_0 2^1 + x_{01}A_0 2^2 + \ldots + x_{0B}A_0 2^B + \\ & x_{10}A_1 2^1 + x_{11}A_1 2^2 + \ldots + x_{1B}A_1 2^B + \\ & + \ldots + \\ & x_{N0}A_N 2^1 + x_{N1}A_N 2^2 + \ldots + x_{NB}A_N 2^B \,. \end{aligned} \tag{3}$$

where x_{ij} is the j-th bit of i-th sample and B is bit width of the sample (number of bits). Implementation of this schema using standard logic is straightforward, as pictured on Fig. 3.

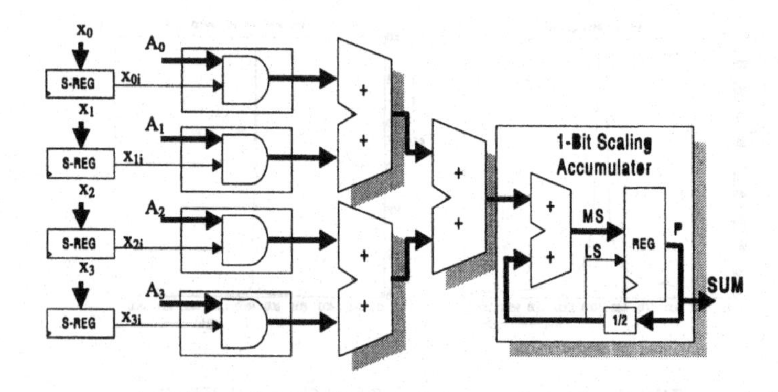

Fig. 3. The *MAC* operation using standard logic (picture taken from [4])

The computation of the *MAC* is done in mixed serial-parallel way, with parallel adders and serial multipliers (the input bits are shifted into an array of AND gates, whose other inputs are the bits of A_n. However, this can be further optimized when noticing fact, that the whole block of 4 AND arrays and following adders is in fact a 4-input logic function, that can be easily implemented in a ROM look-up table 16xB bits. When considering the possibility of using ROM feature of a FPGA, it is obvious that this operation can be done very effectively (see Fig. 4).

The 4 input look-up table easily maps to the Xilinx's XC4000 series FPGAs, taking one CLB per 2 bits of the table bit width. The operation of this circuit can be further parallelized by computing 2 or more bits a time, but there is always the trade-off with chip area occupied. The Scaling Accumulator block accumulates the partial results, scaling the them down by factor 2 each step. It also includes sign extension logic (for deeper description see [1]).

3 The IIR Core of Goertzel Algorithm and FPGA

Since the IIR core determines the overall throughput of the whole algorithm – it must be computed every sample, while the final evaluation is triggered only once per N samples, where $N > 100$ – we focus to the core algorithm only. The

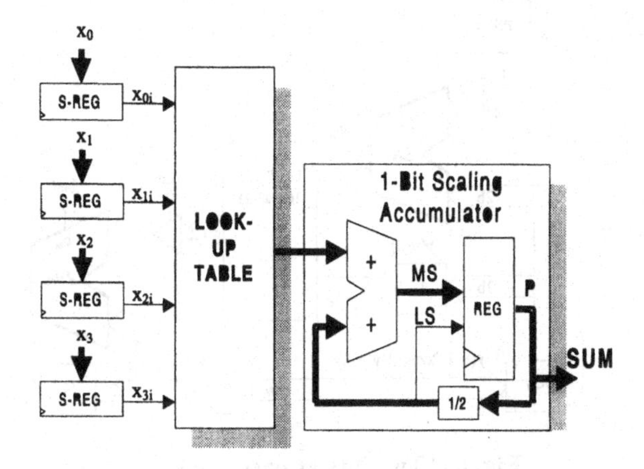

Fig. 4. The *MAC* operation using FPGA's ROM lookup table (taken from [4])

IIR core of Goertzel algorithm must complete following operation for getting one output sample $y(n)$:

$$y(n) = y(n-1)2\cos(2\pi f/f_s) - y(n-2) + s(n) . \tag{4}$$

This equation is in fact a *MAC* operation with coefficients $2\cos(\ldots)$, -1 and $+1$. However, it would be inefficient to implement it the way presented on Fig. 4 – one lookup table input would stay unused. It is more efficient to choose another solution – either computing 2 bits at once, or even 4 bits. The final choice depends on application circumstances, e.g. number of frequencies analyzed.

3.1 "Two bits at once" version

The possible solution is shown on Fig. 5:

This is typical example of distributed arithmetics – LUT (look-up table) block with the RFA (Registered Full Adder) computes the sum of products $y(n-1)2\cos(\ldots) + k.y(n-2)$, where k is constant< 0, e.g. -256. This constant makes the term $y(n-2)$ being aligned properly, because the format of numbers used is fractional integer with the point in the MSB position. Input samples enter the PSCs (Parallel to Serial Converter), which converts them to 2-bits-wide serial stream (from the LSB). If needed, the PSC block together with SA (serial adder) can carry out conversion to the two's-complement code too. The serial adder then adds 1 to the inverted value of every negative input samples. Otherwise, the SA only adds input samples to the intermediate result $y(n-1)2\cos(\ldots) + k.y(n-2)$. The input marked "last" is part of the sign extension logic (for closer description see [1]).After N input samples are proceeded, the result can be read from the register of RFA and all the registers can be reset to allow the next evaluation period to start.

The whole unit runs in pipelined fashion, which enables using the fastest clock rates possible. The width of output samples $y(n)$ and intermediate results $y(n-1)$

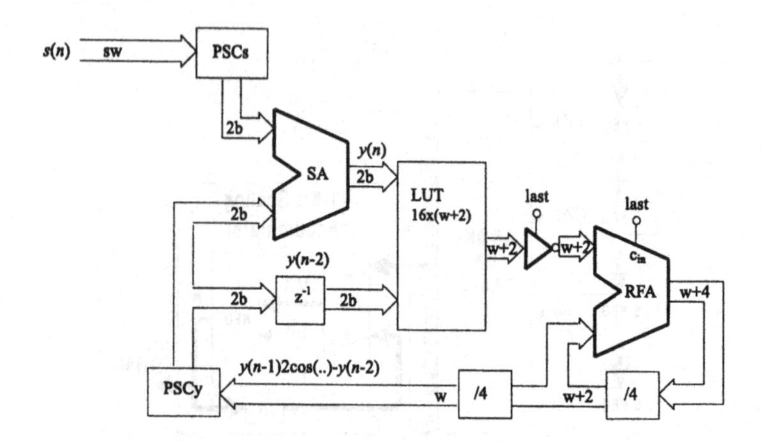

Fig. 5. "Two bits at once" version

and $y(n-2)$ is dependent on the width of the input samples sw - it must be wide enough to prevent the overflow. By simulating the unit in MATLAB I found, that for 8 bit input samples and $N = 200$ is $w = 13$ enough – when applying the resonant frequency to the input of the filter, the output magnitude never exceeded $2^{12} - 1$ during the computation of the 200 samples. The throughput of this solution can be found by counting the number of cycles needed to complete computation of one output sample. This is 7 cycles in this case. Using internal clock of 32 MHz, it would be possible to use sampling frequencies of up to $32.10^6/7 = 4.6$ MHz. Using more expensive, quicker chips, it would be possible to achieve at least 2 times better results. The chip area occupied is summarized in following table:

Table 1. Nr. of CLBs–"Two bits at once"

Block	Nr. of CLBs
PSC	4
SA	2
LUT 16x15	8
INV15	8
FA	10
z^{-1}	2
SUM	34

3.2 "Four bits at once" version

The second solution reduces the distributed arithmetics to standard serial-parallel multiplier using LUT, as can be seen on Fig. 6.

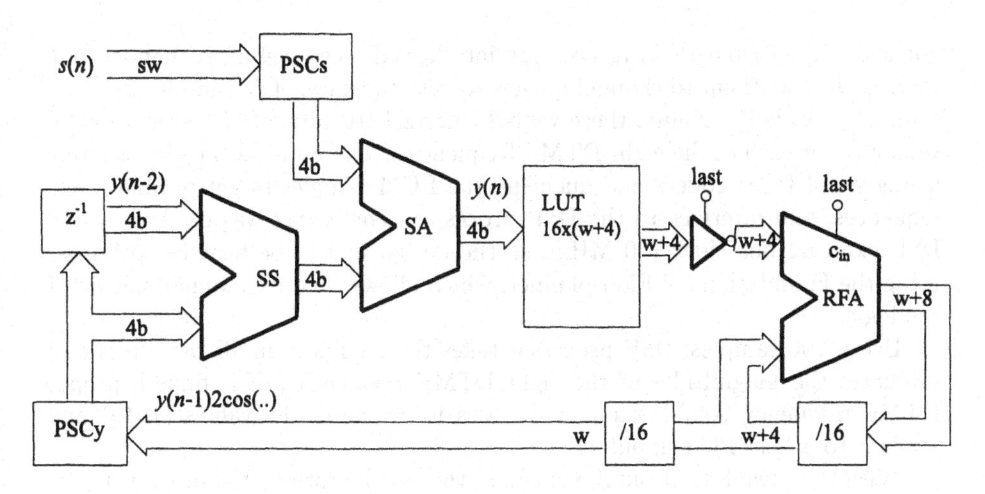

Fig. 6. "Four bits at once" version

The function of this unit is similar to the one on 5, except the SS (Serial Subtracter) block, that performs the subtraction of the terms $y(n-1)2\cos(\ldots)$ and $y(n-2)$. The number of cycles needed for processing one 8 bit input sample is 4. This is almost 2 times less than in the previous version. As for the number of occupied CLBs, it is summarized in the table:

Table 2. Nr. of CLBs–"Four bits at once"

Block	Nr. of CLBs
PSC	4
SA	4
LUT 16x15	4
INV15	9
FA	12
z^{-1}	4
SUM	46

This means, it would be possible to place 12 of the Goertzel IIR filters to the 4013 chip, and that is still quite enough when considering the throughput is almost 2 times better than with the previous version.

4 The DTMF Decoder Application

The application of multichannel DTMF decoder, that used the Goertzel algorithm, was built as single board with DSP320C32-60 and Xilinx XCS30-4 (576 CLBs) with external SRAM memory 128kx8, that served as input sample queue (samples from PCM are coming interleaved – channel after channel, but we need to sort them to channel queues to get sequences of N samples for each channel). Inside the Xilinx, there were 8 Goertzel IIR filters ("Two bits at once version") for each of the eight DTMF frequencies. There was also logic for interfacing serial PCM codecs and queueing the PCM samples to get normal input sequences, and interface to the DSP processor. The Xilinx was clocked by the DSP external bus clock (30 MHz), so the design had to be heavily optimized using the Foundation 1.5 Floorplanner, which allows effective manual placing of the logic.

Every 200 samples, DSP processor takes the results from all the filters and evaluates the magnitudes of the eight DTMF frequencies. If it finds a proper DTMF frequency combination, it decodes it to one of the values (1..16) and sends it to a host PC computer.

With the "two bits at once" version in this configuration, it is possible to decode as much as $30.10^6/(7.8000) = 535$ channels (Xilinx internal clock=30 MHz, nr. of clocks per sample=7 and sampling frequency is 8 kHz), so the Xilinx would have to include 16 PCM32 (32 channel PCM) interfaces. Because this would take about 60 CLBs, it was impossible to fit it in the XCS30. Therefore the number of channels was limited to 256. To use the real power of the FPGA Goertzel core, we would have to use bigger chip or codecs that support the PCM with 64 channels per frame.

5 Conclusion

The FPGA technology can bring an unexpected increase of throughput, if used in a DSP system. E.g., a multichannel Dual Tone Multi Frequency (DTMF) decoder, using FPGA configured as a Goertzel detector (with the "Four bits at once" version) could handle 1000 telephone channels. In comparison with a classical DSP processor TMS320C32-60, that can handle only 32 channels, this is 32-times increase of throughput when the price has grown only 1.5 times (for deeper analysis see [1]). Therefore, it is obvious that the FPGA chips can bring a revolution to certain areas of DSP applications.

References

1. Dulik, T.: DSP unit using FPGAs. Diploma project, TU Brno (1998)
2. Mintzer, L.: Mechanization of DSPs. Handbook of DSP (1987), 941–973
3. Texas Instruments: Modified goertzel algorithm for DTMF using teh TMS320C80. Applicatio Report SPRA066 (1996)
4. Goslin, G. R.: A guide to using FPGAs for application-specific DSP performance. Xilinx application notes (1995)

Pipelined Multipliers and FPGA Architectures

Mathew Wojko
Dept. of Electrical and Computer Engineering
University of Newcastle
Callaghan, NSW 2308, Australia
mwojko@ee.newcastle.edu.au

Abstract. This paper studies pipelined multiplication techniques for implementation on FPGAs with emphasis on the utilisation of FPGA hardware resource. Performance of multiplier implementations are measured for commercially available FPGA architectures where two inherent issues are introduced and investigated. These being the imbalance of critical interconnect delay between general routing and static carry interconnects, and the amount of FPGA logic area used and its poor utilisation. For each of these issues suggestions are proposed and investigated.

1 Introduction

Multiplication on FPGAs is considered an expensive operation. For high throughput multiplier implementations where the result is calculated in a pipelined parallel fashion, large Logic Cell (LC) counts are required. The advent of fast-carry logic has allowed multiplier implementations to achieve sustainable speeds for many DSP operations. However for increasing bit size multiplication operations, high LC counts still provide poor FPGA device utilisation limiting FPGAs from a wider range of DSP or arithmetic intensive applications.

Several techniques have been proposed to provide better utilisation of the FPGA resources when implementing multiplication. Mintzer [6] noted that for fixed coefficient multiplication, by using distributed arithmetic and storing coefficient look-up information in the Look-Up Table (LUT) element of LCs within the FPGA, reduced logic implementations were achieved. Run-Time Reconfiguration (RTR) has been employed to update the content of LUTs to provide a new multiplicand value. Previous work has shown reconfigurable multiplier implementations on the XC6200 series FPGA [1, 7]. Reconfiguration of LUT content is performed through the SRAM configuration interface using pre-computed configuration data. A novel approach named the Self-Configurable multiplication technique was presented by the author and ElGindy [10] for the Xilinx XC4000 series by performing the LUT reconfiguration on chip. On new input, the configuration data is computed on chip and stored into the LUTs in parallel. Other work to improve hardware efficiency for multiplier implementations suggests embedding arithmetic specific Flexible Array Blocks (FABs) capable of implementing a 4×4 bit multiplier, within a conventional FPGA structure [2].

However, each of these techniques results in a compromise. For the fixed coefficient technique, coefficient values cannot be updated. Reconfiguration of stored values requires additional time where the circuit is typically off-line while configuration takes place. Furthermore, for an FPGA with multiple logic cell types, a segmentation in hardware utilisation between the cell type occurs. This paper presents two alternatives for performing multiplication on FPGA architectures.

2 Implementation and Analysis of Existing Techniques

In this section we review the pipeline performance of three multiplication techniques implemented on the Xilinx XC4036EX-2 [5] and ALTERA EPF10K70RC-2 [8] devices. The parallel array, parallel add (vector based) multiplier and the Wallace carry-save multiplier are reviewed (refer to [3,4,9] for a description of the techniques). For the implementations, all techniques were pipelined at every stage of computation to achieve the highest throughput possible. Designs were targeted at the physical level, such that where possible specific architectural features (such as static carry) were utilised. Design mappings were placed and routed with high emphasis on reducing interconnect delays between successive pipeline stages. Figure 1 below compares the LC count requirements for each technique on each device while Figure 2 compares the maximum clocking speeds achieved.

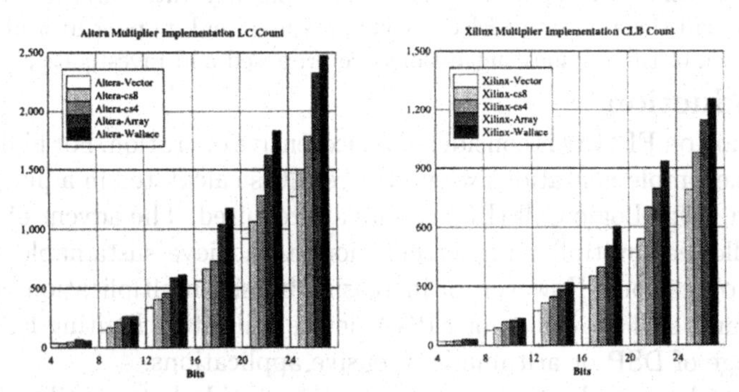

Fig. 1. Multiplier implementation LC Count for Xilinx and Altera Devices.

Comparing the LC counts for each FPGA, we see the same general trend across each architecture for each technique, i.e. the logic requirements scale quadratically with the input bit size of the multiplier. We can observe the relative differences of how each technique maps to an FPGA architecture containing 4-input LUTs and static carry. The Wallace technique saves carry values from propagating within pipeline stages at the expense of additional logic, the vector and array techniques propagate carry the bit length of the input operands making use of the static carry interconnect. The parallel array technique can be viewed to require additional cells for pipeline buffering.

However, the trends observed for the maximum cycle speed for each technique in Figure 2 are different between FPGAs. This result is due to the difference in interconnect strategies employed in both architectures. Xilinx utilises a hierarchical scheme in which interconnect paths vary in CLB span lengths, while the Altera architecture uses row and column interconnects that span the whole device. For Xilinx, Wallace implementation results are seen to converge with vector and array for increasing input bit sizes. For Altera there is a distinct difference between them. This is because the vector and array based techniques make high use of the static carry interconnect, thus reducing the routing demand placed on the general purpose interconnect. Since the Wallace technique does not propagate any carry values, it makes no use of the carry interconnect and thus places a higher demand on the general routing interconnect. Results show that the Altera architecture is able to sustain more consistent routing delays for increasing bit input sizes through the use of spanning interconnects than the Xilinx hierarchical length interconnect architecture. This has been verified by analysis of routing floor-plans produced by the routing tools used. As routing channels become congested, determining alternate paths to establish routes requires more effort for the Xilinx interconnect than Altera.

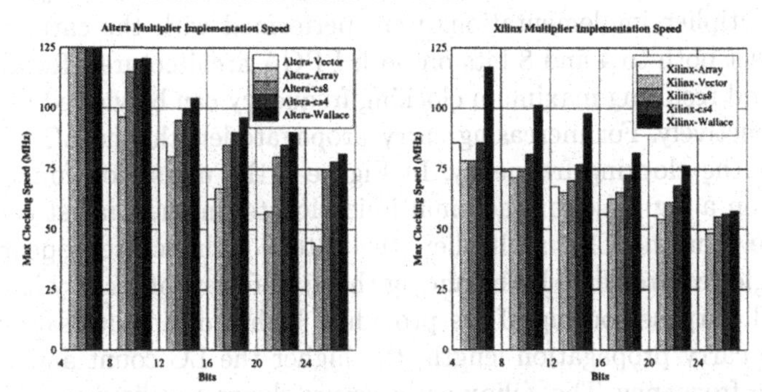

Fig. 2. Multiplier implementation delay for Xilinx and Altera Devices.

3 Carry Save/Propagate Hybrid Multiplier

The Carry Save/Propagate (CSP) multiplication technique was developed as a hybrid between pipelined vector and carry save multiplication techniques to provide a configurable balance between LC count, maximum clocking frequency and interconnect type utilisation (i.e. static carry and general purpose). By making the carry propagation length within the multiplier a design parameter, the CSP technique establishes a continuum of multiplier instances and implementation results between the extreme cases of parallel add vector based and Wallace multiplication techniques.

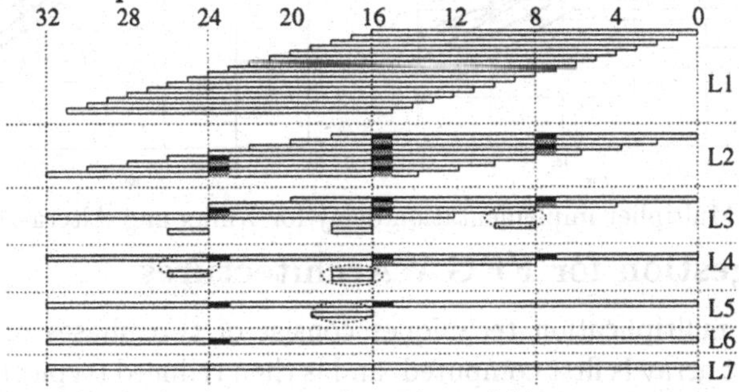

Fig. 3. 16-bit CSP multiplier example with an 8-bit carry propagation length.

In Figure 3 an example of the bit operations within a 16-bit CSP multiplier of propagation length 8 is shown. The initial array of partial products is shown at the first level (L1). These are then added in parallel to produce $\frac{n}{2} = 8$ results producing 16 carry values at the 8-bit wide defined boundaries. The CSP technique maintains carry values in two ways. At each level they are registered as either a carry-in for the next stage of addition, or to be accumulated with other carry values for later addition to the partial product values (denoted respectively as black or grey). Two sets of values progress throughout the operation of the CSP multiplier, these being the accumulated values of the partial products, and the accumulated carry-save values. Add operations are performed in each set to reduce the number of values. At certain stages, additions using accumulated carry values occur (denoted by the dotted oval). The addition operations continue until the height of the array of values is reduced to one and no carry values remain as in L7.

CSP multiplier implementations were performed with the carry propagate parameter set both to 4 and 8 bits on both FPGA architectures. Results for the LC count and resulting maximum clocking frequency can be viewed in Figures 1 and 2 respectively. For increasing carry propagate length, the LC count is reduced as is the clocking frequency. In Figure 4 the utilisation function $\frac{freq}{\#LCs}$ is graphed on a log scale for each multiplication technique against each FPGA architecture. For the CSP multiplier, the Altera architecture appears able to leverage logic requirements by balancing the use of the static carry interconnect and general purpose routing. Thus providing higher device utilisation, i.e. the shorter the carry propagation length, the higher the LC count and the higher the clocking frequency. The Xilinx architecture shows a utilisation increase, but not one significant enough to show an effective balance between the use of interconnect types for the required LC count.

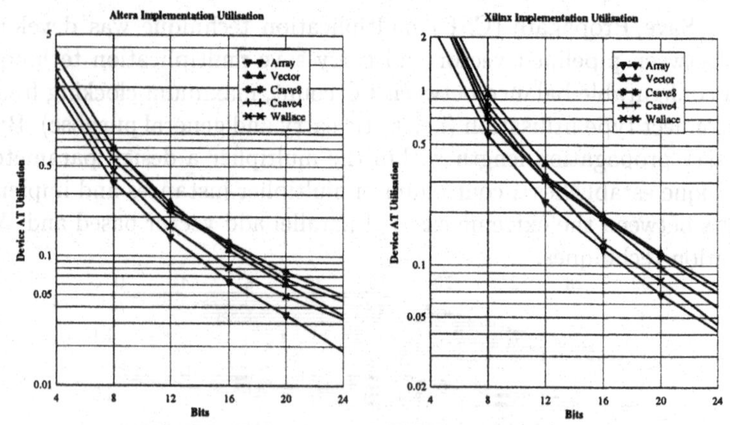

Fig. 4. Multiplier implementation delay for Xilinx and Altera Devices.

4 A Suggestion for FPGA Architectures

Most parallel multiplication techniques consist of two phases of operation; a partial product array is first computed, and is then reduced (typically) in parallel to provide the multiplication result. Considering how a multiplication technique maps to either the Xilinx or Altera FPGA architectures, there are two examples of poor utilisation. In the generic 4-bit input LC schematic in Figure 5(a), the LC contains a 4-bit input LUT, fast static interconnected arithmetic logic and a register. This gives a typical description of the LC functionality present in both the Altera FLEX10k and Xilinx XC4000 architectures. For the first phase of multiplication, (the bit-cyclic convolution) for an n-bit multiplier n^2 2-bit AND operations are performed. This requires n^2 LCs to calculate the partial product array. Implementing a 2-bit AND operation on a 4-bit input LC provides extremely poor utilisation of the cell. For the second phase, the partial product results are reduced by parallel addition. Pairs of values are added together to reduce the number of partial products by half between successive stages. Again, we see that each LC implements a Full Adder (FA) which again only utilises two of the four inputs of the LC plus the static carry interconnect. It is evident that the potential bandwidth of the LC within an FPGA is not exploited and is only at half of its capacity. Due to this fact, high LC counts are required for multiplier implementations, providing poor resource utilisation.

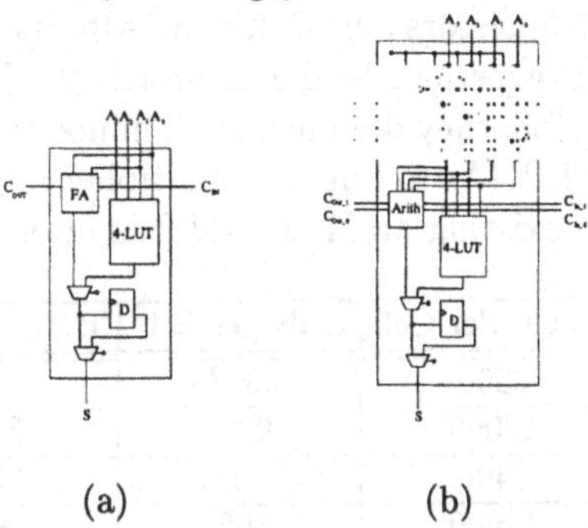

Fig. 5. Common and suggested Logic Cell architectures.

Figure 5(b) shows the suggested LC architecture that ensures the potential bandwidth of each LC is fully utilised such that reduced LC count multiplier implementations can be achieved. Added to the LC are four static interconnects with pass transistors and AND gates, and an enhanced arithmetic unit accepting four bit inputs, two carry-in values and producing a single bit result and two carry-out values. All static fixed interconnects are intended exist between rows (or columns) of LCs and not to be considered as general inputs to the LCs. For multiplier implementations, the first phase of partial product values are computed by utilising the top four static busses. Inputs to the LC are either be routed onto the static bus, or propagated through the ANDs gate as input to the LUT (by configuring pass transistors in the cell). This mechanism provides the ANDing phase for the multiply operation where four bit values can be broadcast to a row of LCs such that a set of 4 n-bit wide partial products can be generated internally within one row of LCs. These values can be added to provide a $4 \times n$ multiplication result. The addition is performed by the ARITH unit shown in the figure. The unit accepts four bit values which it adds to a 2-bit carry input and produces a 3-bit output. Bit zero is output as the bit result for the cell while bits two and three are propagated to the next cell as carry-in bits. A carry-chain requiring two bits is established to add the four n-bit values within a single row of LCs. As a result, to implement a $4 \times n$ bit multiplier using the modified LC architecture, $n + 5$ LCs are required. On average an $n \times n$ bit multiplier can be implemented with LC counts four times less than conventional FPGAs by using this modified LC. [1]

[1] While it is acknowledged the Xilinx Vertex architecture enables partial product values to be generated within cells by the use of an additional AND gate, halving

To evaluate the effectiveness of the modified LC, the suggested changes are integrated into both the Altera and Xilinx devices and are compared with a previous study [2] on multiplier implementations. In this study, multiplier implementations using FABs are compared against those on both the Xilinx 4000 and Altera FLEX architectures. Results are presented comparing transistor counts and relative counts per bit^2 for multiplier implementations on each. These results are reproduced below in Table 1 where new results are presented for the modified LC changes integrated into the FLEX and XC4000 architectures implementing vector based multipliers. Viewing the results, we can see respective improvements of factors 3.79 and 3.91 transistors per bit for the Altera and Xilinx architectures using the modified cell suggestion. While these results do not achieve the density provided by using FABs, they do illustrate a minor architectural change that can be made to existing FPGAs to improve multiplier implementation efficiency while not disrupting the existing architectural framework of the FPGA.

Implementation	Trans. Per Cell	Cells per Bit^2	Trans. per Bit^2	Relative Area
FABs	2300	0.0625	144	1
Altera 10k	12000	0.26	3169	22
Xilinx 4000	3400	1.14	3880	27
Altera 10k Modified	12600	0.066	832	5.8
Xilinx 4000 Modified	3550	0.28	944	6.9

Table 1. Relative bit area requirement comparison.

References

1. J. Burns A. Donlin J. Hoggs S. Singh M. de Wit. A dynamic reconfiguration run-time system. In *FPGAs for Custom Computing Machines*, April 1997.
2. S. D. Haynes and P. Y. K. Cheung. A reconfigurable multiplier array for video image processing tasks, suitable for embedding in an FPGA structure. In *FPGAs for Custom Computing Machines*, April 1998.
3. K. Hwang. *Computer Arithmetic: Principles, Architecture, and Design*. John Wiley and Sons, 1979.
4. K. Hwang. *Advanced Computer Architecture: Parallelism, Scalability, Programmability*. McGraw-Hill, 1993.
5. Xilinx Incorporated. Xilinx XC4000 data sheet, 1996.
6. Les Mintzer. FIR filters with field-programmable gate arrays. *Journal of VLSI Signal Processing*, 6(2):119–127, 1993.
7. B. Slous T. Kean, B. New. A multiplier for the XC6200. In *Sixth International Workshop on Field Programmable Logic and Appications*, 1996.
8. ALTERA Corporation. FLEX 10000 data sheet, June 1996.
9. C. S. Wallace. A suggestion for fast multipliers. *IEEE Transactions on Electronic Computers*, EC-13:14–17, Feb. 1964.
10. M. Wojko and H. ElGindy. Self configurable binary multipliers for LUT addressable FPGAs. In Tam Shardi, editor, *Proceedings of PART'98*, Newcastle, New South Wales, Australia, September 1998. Springer.

LC counts for multiplier implementations. The use of single bit carry lines does not allow as high utilisation as the modified LC cell does.

FPGA Design Trade-Offs for Solving the Key Equation in Reed-Solomon Decoding [1]

Emanuel M. Popovici[1], Patrick Fitzpatrick[1], and Colin C. Murphy[2]

[1] National Microelectronics Research Centre, Cork, Ireland
popovici@nmrc.ucc.ie, fitzpat@ucc.ie
[2] Department of Electrical Engineering and Microelectronics, National University of Ireland, Cork, Ireland
cmurphy@rennes.ucc.ie

Abstract. Reed-Solomon codes are widely used in communications as well as in data storage for the correction of errors due to channel noise. In this paper we present a comparison between implementations of the Berlekamp-Massey algorithm and the Fitzpatrick algorithm. Both algorithms were synthesised and implemented on an FPGA and compared in terms of area, speed and routability. The modules can be used as part of a core-based design for Reed-Solomon decoders.

1 Introduction

Reed-Solomon (RS) codes are constructed and decoded using a Galois Field $GF(q)$ [1][2]. The central computation in the decoding of RS codes is the determination of the error locator polynomial $\sigma(x)$ and the error evaluator polynomial $\omega(x)$ from Berlekamp's key equation

$$\omega \equiv \sigma h \bmod x^{2t},$$

where h is the syndrome polynomial and t is the number of errors that can be corrected. Once the error locator polynomial is found, its roots are determined in order to find the locations of the errors. The usual technique is to use a Chien search which consists simply in the computation of $\sigma(\alpha^i)$ for all $i = 0, 1, ..., q-2$, where α is a primitive element of GF(q), and checking for a result equal to zero. Use of this method is feasible since the number of elements in the field is finite. The next step is to find the error values. The method currently used is called Forney's algorithm [2]. In this paper we focus on implementations of the key equation solver using F and BM.

1.1 Solving the key equation

Algorithm F described in [3–5] for solving the key equation has the same computational complexity as the best known algorithm currently in use, namely the

[1] The research presented in this paper has been supported by Silicon Systems Limited and Forbairt under Applied Research Scheme Grant HE/97/302

Berlekamp-Massey algorithm [1,2,6,7]. The advantages of F over BM derive from its construction via a new approach to the solution of the key equation using Gröbner bases of polynomial modules [5].
We present a comparison of F and BM and outline the development of appropriate hardware structures. The computations involved in the two algorithms are presented in Figure 1 (*cf.* [1,2,5,6] for more details)

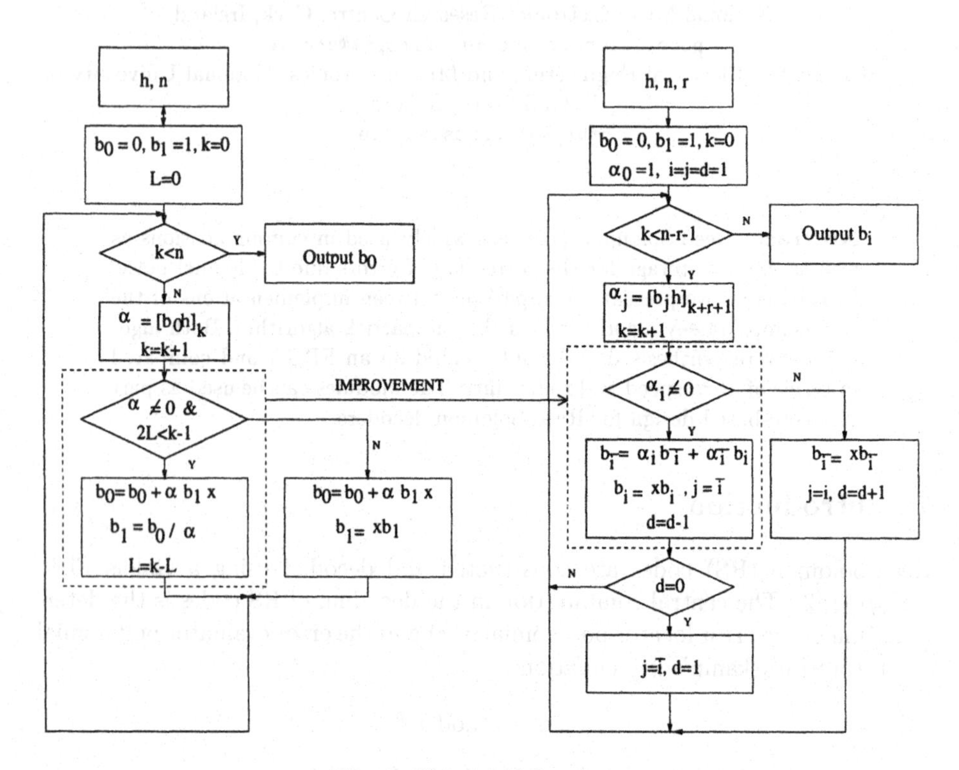

Fig. 1. BM and F flowcharts

Both algorithms construct a sequence of pairs of polynomials b_0, b_1 with coefficients in $GF(2^m)$. F is controlled by two tests for zero, while BM requires two tests for zero and an integer comparison test. Also, F presents advantages from the VLSI perspective since division is an expensive task.

2 Implementation architectures

The architectures presented here can be used in various applications depending on specific requirements. Our designs are fully parameterized. The adoption of design re-use has resulted in the availability of a variety of implementation options. The modules presented here can be used as part of core-based design of an RS decoder.

The arithmetic operations are performed in the finite field $GF(2^m)$ where in this particular application we choose $m = 8$. To add two elements we simply add their vector representations. Multiplication can be implemented in a combinatorial logic circuit or by using a sequential structure (see [1]). For inversion we use a look-up table while a division can be represented as inversion followed by a multiplication.

2.1 Serial architectures

A typical shift register structure for F is presented in Figure 2. The syndrome register S, used to store the polynomial h, has $2t + 1$ stages while registers B_0 and B_1 have $t + 1$ stages, where t is the maximum number of errors that can be corrected. The register S has an extra symbol to facilitate synchronization of multiplication with b_j for the calculation of α_j. The control unit supplies the control signals for the multipliers and the shift registers [8].

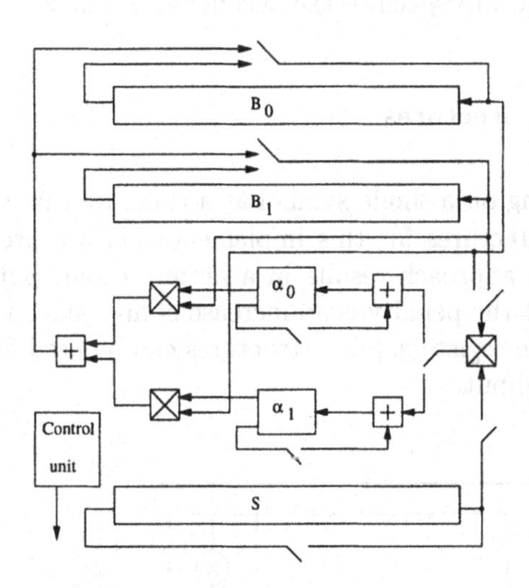

Fig. 2. F algorithm implementation

The architecture of BM [2] is presented in Figure 3. The syndrome register S has $2t+1$ stages, register B_0 has $t+1$ stages and register B_2 has $t+2$ stages. Again, the register S has an extra symbol for the reasons mentioned above.

We work on a single symbol at a time instead of entire polynomial. The required area rises linearly with t, the number of errors. The main advantage of the structure presented in Figure 2 compared with that of Figure 3 is that F is division-free [8]. This leads to better results in terms of area, speed and routability since division is both time and area consuming. Since solving the key

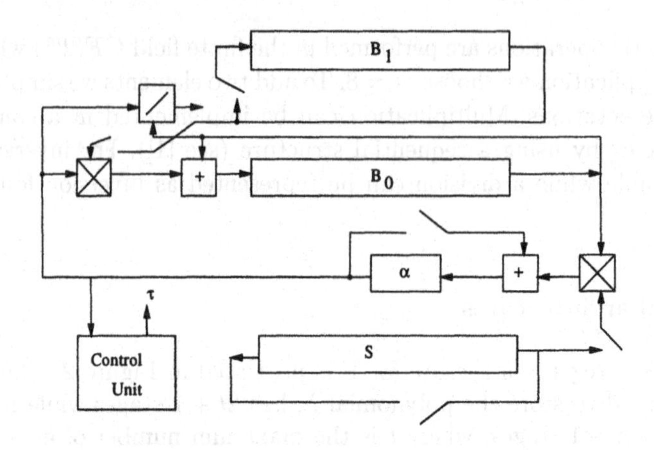

Fig. 3. BM algorithm implementation

equation is the most computationally intensive task, this approach has a positive impact on both cost and speed of the decoder as a whole.

2.2 Parallel architectures

Rather than working on a single symbol at a time, we can work on entire polynomials. The architectures for this implementation are presented in Figure 4 and Figure 5. This approach results in a better throughput than in the serial implementation but the penalty is an increase of area since we need a multiplier for each stage of the registers. Such structures can be used for applications that require high throughput.

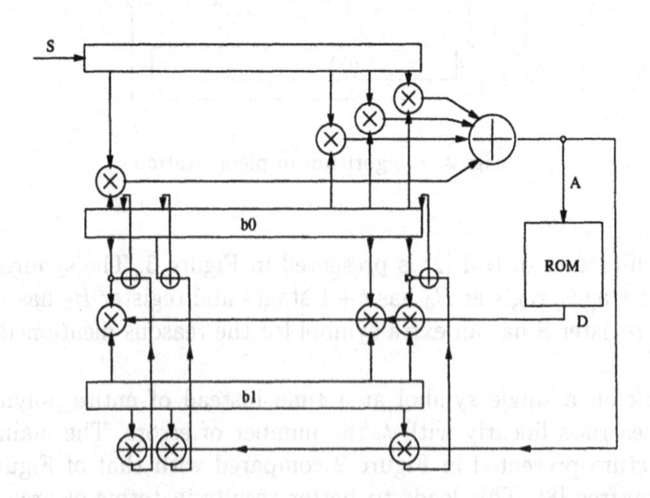

Fig. 4. BM parallel architecture

For the BM architecture, the inversion is achieved using a ROM table. For $m = 8$ the size of the memory is 256 words. Multiplication is carried out in a combinatorial fashion in order to achieve a high throughput, but we note that this can cause a bottleneck from the point of view of routability.

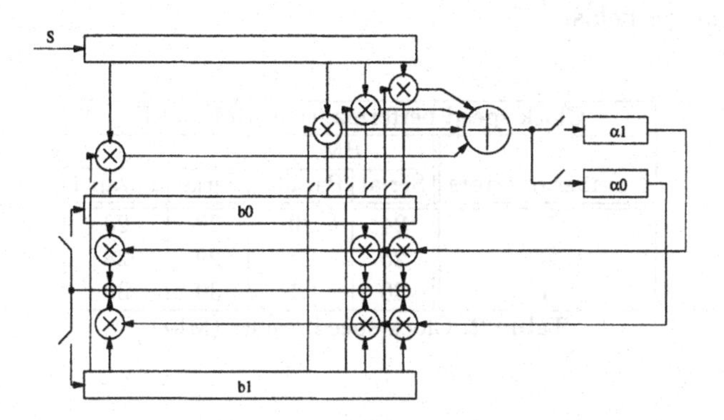

Fig. 5. F parallel architecture

Since F does not use division, it is more suitable for VLSI implementation. However, inversion is needed for the implementation of the Forney algorithm used in computing the error values. In F the structure used for multiplication is sequential giving smaller area and higher clock speed.

As can be seen, both algorithms use a third register in order to calculate the α_i. The syndrome value enters serially for each iteration.

3 Conclusions

Both algorithms have been implemented using the FPGA compiler provided by Synopsys and Altera Max Plus II software. For hardware testing we used an EPF10k100GC503 device provided by Altera.

Area performance of BM and F				
	BM		F	
Number of errors	Serial	Parallel	Serial	Parallel
7	945	1951	838	1744
5	935	1480	758	1320
3	924	793	641	928

Table 1. Area results given in logic cells

From Table 1, we can see that F requires less area than BM. Note that for a parallel implementation the values given for the area required by F include the syndrome register.

As we can see from Table 2, F works at higher clock rate. In our implementations BM has a better throughput compared with F (for the 'parallel' case) because it was implemented using the combinatorial option for the $GF(2^m)$ arithmetic, whereas F uses a sequential oriented implementation for same arithmetic. The combinatorial approach leads to routing and area problems as we migrate to larger fields.

Clock speed performance of BM and F				
	BM		F	
Number of errors	Serial	Parallel	Serial	Parallel
7	24	10	29	28
5	26	11	29	28
3	26	11	30	30

Table 2. Clock speed results (Mhz)

The conclusion is that division-free F can be used for high speed applications. If there is need for a high throughput then a specific architecture can be chosen. The highest throughput is achieved for a combinatorial approach, but this is too expensive for higher order fields. The sequential approach leads to better results in terms of routability and area but with loss of latency. The serial implementations are attractive from an area point of view but they became too slow as latency increases for a higher number of errors.

References

1. Lin S., Costello D. J.: *Error control coding: fundamentals and applications*, Prentice-Hall (1983).
2. Wicker S. B., Bhargava V. K.: *Reed-Solomon codes and their applications*, IEEE Press (1994).
3. Fitzpatrick P., Flynn J.: A Gröbner basis technique for Padé approximation, *Journal of Symbolic Computation*, 13, (1992) 133–138.
4. Fitzpatrick P.: New time domain errors and erasures decoding algorithm for BCH codes, *Electronics Letters*, 30, (1994) 110–111.
5. Fitzpatrick P.: On the key equation, *IEEE Trans. on Information Theory*, 41, (1995), 1290–1302
6. Fitzpatrick P., Jennings S. M.: Comparison of two algorithms for decoding alternant codes, to appear in Applicable Algebra in Engineering, Communication and Computing, 9, (1998), 211–220
7. Blahut R. E.: *Theory and practice of error control codes*, Addison-Wesley, (1983)
8. Popovici E. M., Fitzpatrick P.: Reed-Solomon decoders for the read-write channels, *IEE Systems on a Chip Colloquium Digest*, September, (1998), 9.1–9.5.

Reconfigurable Multiplier for Virtex FPGA Family

Juri Põldre[1] and Kalle Tammemäe[2]

[1] Tallinn Technical University, Tallinn, Estonia
jp@pld.ttu.ee
[2] Tallinn Technical University, Tallinn, Estonia
nalle@cc.ttu.ee

Abstract. This paper describes integer multiplier design optimizations for FPGA technology. The changes in partial product generator component enable to infer CLB fast carry logic for building Wallace trees. This change increases speed and gives better resource allocation.

1 Overview

As usable gate count grows an important factor in designing system on chip is to be able to reuse the intellectual property (IP) [4]. Contemporary ASIC technology allows more than 20 Mgates per die. To utilize these gates efficiently a variety of tools have emerged that use the knowledge about architecture of design to create regular structures for arithmetic operators. Such regular structures are used in applications like signal processing, data compression and cryptography. The regular structure generators are available commercially - like the Module Compiler from Synopsys [3]. A free arithmetic module generator is represented here by a tool from Norwegian University of Science and Technology [5]. Multiplier can be partitioned into three separate units: partial product generator, Wallace tree and final adder [1]. The Wallace tree consists of full-adder and half-adder cells. The cells are however connected in such a way that the longest carry chain is equal to $\log_2(\text{argument\_length})$. The structure also contains carry chains of intermediate lengths. The shortest common factor in these trees is two. This decomposition can be made in such manner that the carry propagation signal between adder cells is not required elsewhere.

Most FPGA providers include carry propagation circuits inside CLB [6]. These internal carry chains are optimized for building the fast ripple carry adders. The carry out signal of one stage is directly connected to carry in of the next within one CLB and cannot be accessed outside. If the Wallace tree is decomposed into length two carry chains we can map these chains directly to CLB.

Commercial structure generator tools allow to specify target technology. To use the FPGA library the tool creates vector versions of elements in target library. The tool tries to find the elements for datapath synthesis in target library. In case it finds none the respective cell is generated from more primitive cells. In case of FPGA we cannot directly infer full adder cell, but rather two full adder cells – a length 2 ripple-carry chain.

The other solution is to use some standard cell library and run technology translation on target netlist. This leaves us with non-optimal result because internal carry logic is not inferred. The penalty is in both area and speed.

With both the commercial and free module generators we do not have access to the routine that generates the tree. The output is gate-level VHDL or Verilog file. From that netlist it is rather hard to replace cells because we need topology info to decide which full adders to group into single CLB. The simplest solution here is to group all these cells into single CLB that have the carry chain used only internally.

The best solution can be achieved when generating the netlist directly for the target technology. We can create structural designs in VHDL using generate statements. Such design can also support carry in and carry out from Wallace trees what are otherwise connected to ground and removed by logic optimizer.

Using these carry signals we can create a serial-parallel multiplier with more degrees of freedom exploiting the bit widths of both arguments. Consider 1024×1024 bit multiplier for example. In first solution we can use carry-save accumulator and do it in 64 steps using 16×1024 multiplier as a building block. This can be done by both generators.

If we can access carries within Wallace tree we can further partition it into eight times more steps using 16×128 multiplier as a base block. The above-described block can now easily be included into behavioral design with arrays mapped to FPGAs internal RAM cells. Such multiplier design leaves us with large set of architectural solutions and fast synthesis time. In the following sections we will overview the structure of Wallace tree and generators. Then the methods described above are used to create the optimized netlist. First the structural netlist from technology translation is used directly. Then the necessary changes are introduced to group the cells together based on the carry chain connection constraint. Finally the structural VHDL generator is used. All designs are mapped to VIRTEX FPGA device and results are compared with respect to area and delay.

2 Wallace Tree and CLB

The Wallace tree compresses partial products into two components - Sum vector and Carry vector. The final result can be calculated by adding these vectors. If magnitude comparison is not needed then the intermediate results can be kept in carry-save format. The usual practice in modular exponent calculations is to use squaring method. The method breaks down the exponent into $2 * N$ multiplications, where N is the size of the argument. The $N \times N$ bit multiplication is further reduced by scanning second argument with some determined bit width, usually 8 or 16 bits. This core of these calculations in $k \times N$ multiplication what will be called $2kN$ times. During these calculations the result is kept in carry-save format and only added together at the end of exponent calculation. The similar practice is used in DSP multiply-accumulate cycles. In all these calculations Wallace Tree becomes the main resource and constraint.

The Wallace Tree consists of full-adder and half-adder cells. There are several ways to connect these cells. For this article the generalized 4to2 adder cells are used. Each cell has 4 primary inputs and 2 primary outputs, carry in and carry out. This cell adds two carry-save digits. It has carry in and carry out signals to expand the calculations. This cell is like full adder, but all IO lines are duplicated. Using these blocks the adder tree of any size can be built much as with ordinary full adders. The main difference with full adders is that there is no timing path from carry inputs to the primary inputs. All carries are introduced at lower level only. The Wallace tree for 16 arguments is depicted in fig. 1.

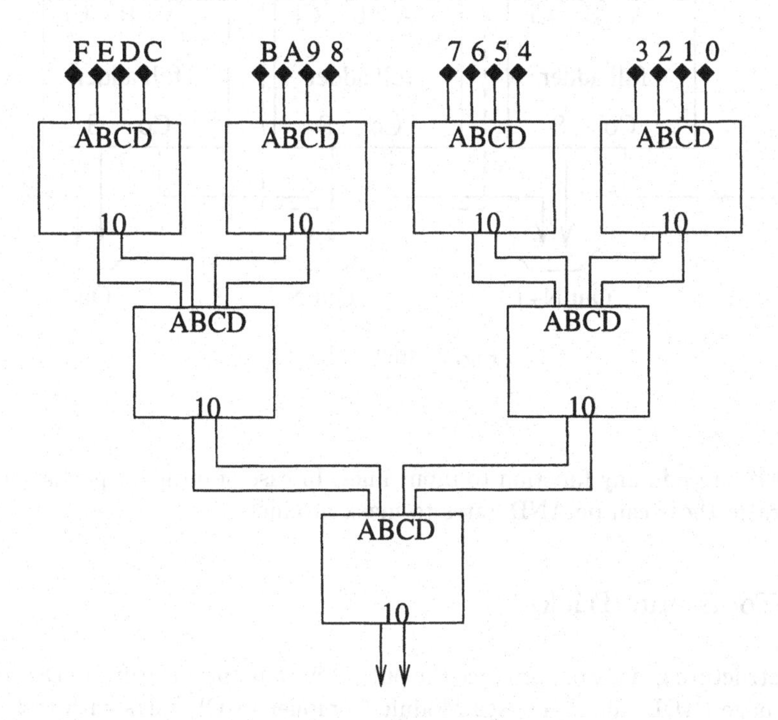

Fig. 1. Wallace tree

The expanded 4to2 adder cell is shown in fig. 2. The carry logic in FPGA CLB cell is depicted in fig. 3. Comparing these figures it is evident that the 4to2 adder can be grouped differently to use CLB fast carry logic. The only signal that is not available in CLB is the carry propagation between full adder cells - marked with the wide line in fig. 2.

The grouping is done in fig. 4. Notice the addition of M1 and M2 signals. As now we use the dedicated circuits to build carry chains then CLB resources are free. These can be used to include partial product generator in the cell. The M1

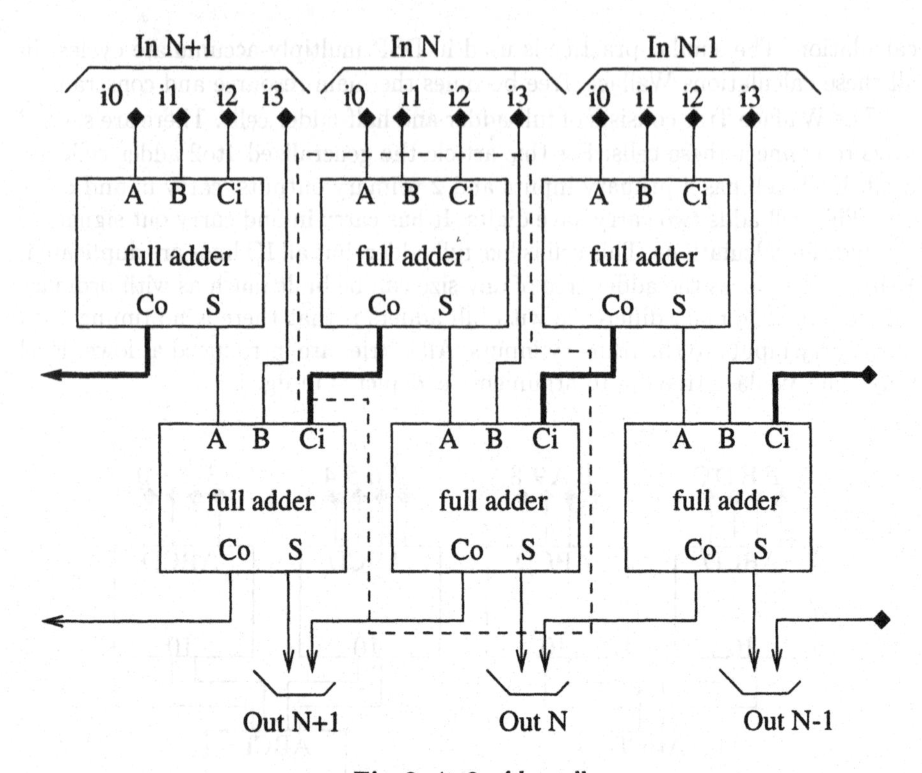

Fig. 2. 4to2 adder cell

and M2 can add any function to input lines. In case of ordinary partial product generator these can be AND-gates to mask products.

3 Tools and Tricks

The structure generators produce the output in structural Hardware Description Language (HDL) file. Synopsys Module Compiler (MC) starts with technology database. Let us first look at it.

3.1 Technology Library

The Synopsys has tool for adding components to technology library. With this tool it is possible to add two versions of full adder cell described in section 2 to technology library. The first is bare 4to2 cell and second is the cell with mask functions included. While adding these cells the designer has to be sure to set the area and delays less than respective full adder cells. After MC library creation the cell is included in datapath library. When building Wallace trees it is selected instead of plain full adder cell because area and delays are better.

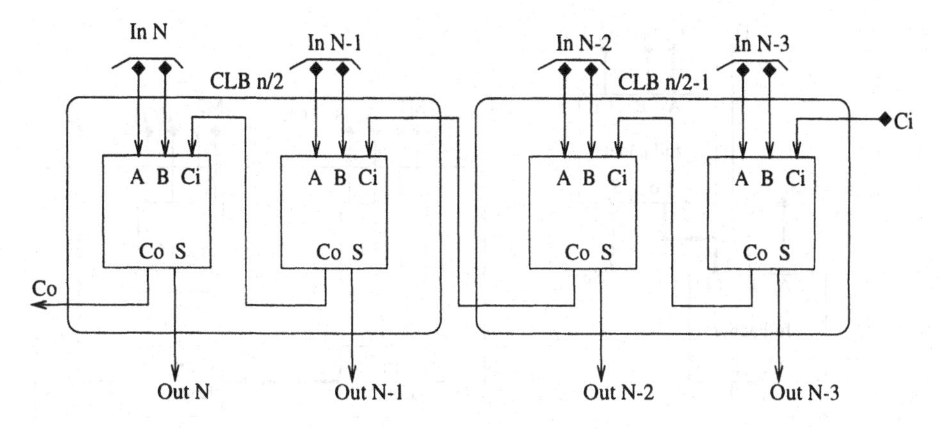

Fig. 3. FPGA CLB carry logic

3.2 Target Netlist Translation

In most cases the starting point is the produced HDL netlist. This netlist usually has instantiations of half and full adder cells. It is simple to change these cells to 4to2 sell as described in sec. 2. The algorithm is the following:

REPLACED ← NIL
for N in all nets
    if net is connected to carry out in full adder cell (A1) **and**
        A1 $\notin$ REPLACED **and**
        net is connected to carry in of another full adder cell (A2) **and**
        A2 $\notin$ REPLACED **and**
        net is **not** connected to any other nets **then**
            replace A1 and A2 with cell as in sec. 2
            REPLACED ← REPLACED ∪ A1
            REPLACED ← REPLACED ∪ A2
    end if
end for

The more complex algorithm would also study the other inputs of A1 and A2 for possibility to use additional free logic in LUT.

3.3 Structural Generator

VHDL language has the generate construct. It is a nice tool for creating structures. Using this it is possible to write generator that creates exactly the structure described in sec. 2. It is possible to keep the technology dependent part at bare minimum. In following section we will use this as a golden device to compare different solutions to it.

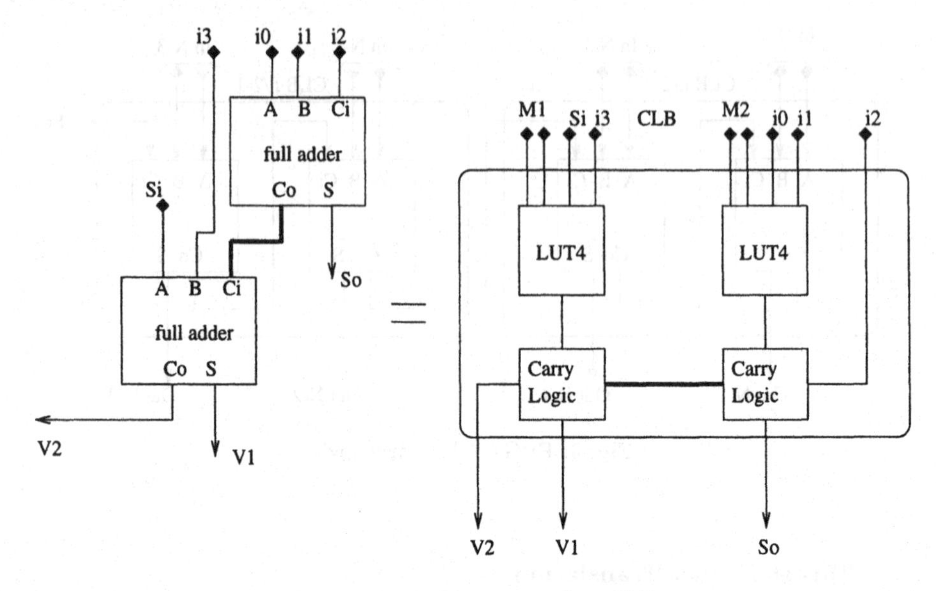

Fig. 4. 4to2 adder mapped to CLB

4 Results

The results of the experiments were not fully available for printing in proceedings but are provided in Web:
`http://www.pld.ttu.ee/~jp/FPL99/`

5 Conclusions

In this paper Wallace Tree component of integer multiplier for VIRTEX FPGA is described. The method of inferring fast carry logic is developed. Compared with current approaches the method gives better device resources utilization and yields faster multiplication.

References

1. David A. Patterson, John L. Hennessy Computer Architecture. A qualitative Approach Morgan Kaufmann Publishers, Inc., (1996)
2. Põldre, J., Tammemäe, K., Mandre, M.: Modular Exponent Realization on FPGAs. Proceedings of FPL98
3. Synopsys Module Compiler User Guide. Synopsys, Inc, ver. 1998.08
4. Michael Keating, Pierre Bricaud Reuse Methodology Manual Kluwer Academic Publishers, (1998)
5. `http://redback.fysel.ntmu.no/~pihl/iwlas98/index.html`
6. Product Specification. Virtex 2.5V. Xilinx, Inc., Nov. (1998)

Pipelined Floating Point Arithmetic Optimized for FPGA Architectures

Iakovos Stamoulis, Martin White, Paul F. Lister

Centre for VLSI and Computer Graphics,
University of Sussex, Falmer, Brighton,
BN1 9QT, England, UK
Email: M.White@sussex.ac.uk

Abstract. There are many methods to perform floating point arithmetic functions. Until recently, such circuits could not be efficiently implemented in FPGAs due to their small size and low speed. The limited resources of FPGAs, both in terms of logic functions and routing seriously limits the complexity of a design that uses floating point arithmetic. We will show alternative algorithms that take advantage of the features of FPGA architecture can yield impressive performance and very small area requirements. We also describe efficient architectures for designing pipelined floating point units for addition/subtraction, multiplication and division that were used for 3D computer graphics applications in FPGAs. By exploiting the intrinsic architecture of such devices, and overcoming the limitations of VHDL and synthesis tools, these pipelined arithmetic units allow the use of multiple floating point units on the same device

1. Introduction

FPGA are versatile devices, which until recently were very small. Advances in FPGA technology in terms of both size and architectural efficiency have enabled FPGAs to be used in a variety of new applications that were not possible before. Floating point arithmetic has been virtually impossible to implement on programmable devices due to the complexity of these operations and the large gate counts required [4]. In addition, floating point requires very wide data paths, which are not usually available in FPGAs where in many cases routing complexity is the limiting factor of designs.

VHDL has allowed designers to rapidly prototype and simulate designs. However, synthesis tools cannot always take full advantage of all the architectural features of FPGAs and results are often sub-optimal. Fortunately, this situation is changing and the recent introduction of many FPGA specific synthesis tools has pushed further the capabilities of these tools. [8] [9].

Knowledge of the target architecture enables the designer to choose the suitable algorithm that will match the architecture of the target technology and make optimisations on the design that are not possible by purely device independent VHDL code. Another challenge for the designer who intends to use FPGA devices is to make the best use of the limited available resources. This means that the designer must balance the routing, logic and registers of the design to match those of the target architecture.

In this paper, we present an investigation of methods for designing pipelined floating point units and assess their suitability for implementation in FPGA devices. We also show that further optimisations can be applied depending on the exact family of FPGA used according to the architecture features.

2. Floating Point Multiplication

Floating point multiplication is regarded as the easiest of the floating point operations mainly because it is not very much different to integer multiplication. The following are the typical steps required for multiplication [5] [2]: check if any operand is zero, add the exponents, derive sign, multiply the fractions, normalise the result and finally adjust and bias the exponent

The most critical component that directly affects the area and performance of the design is the integer multiplication of the fractions.

The sign is simply found by XORing the sign of the multiplicand and the multiplicator. This can be implemented in one LUT. The exponent is the sum of the two exponents. However, the final exponent must be biased and adjusted according to the result of the normalisation. An advantage of normalised numbers in multiplication is that, in order to normalise the number, only one bit shift is required. The same adder can perform the biasing of the exponent and the adjustment as a result of the normalisation. FPGAs have typically very efficient structures for implementing adders [6]. The availability of dedicated carry chains and the adder configuration of logic cells ensures that integer adder occupy very little area and are very fast as well.

Special attention must be paid to the design of the integer multiplier for the fractional parts. The multiplication of the fractions is an integer unsigned multiplication of two numbers.

Three methods are typically available:

- Using a LUT-based multiplier
- Use of a binary tree of partial products
- Use of Vendor provided parametrisable cores

We can obtain a LUT-based multiplier using the Synopsys DesignWare Design Analyzer libraries. This can be implemented as a Booth-code Wallace-Tree or an array of carry save adders depending on the constrains.

Synopsys generates a multiplier by implementing the logic by generic component, optimising it and then packing it into Look Up Tables, (LUTs). This is not an optimal solution as it does not take advantage of the architectural features of the target technology, and the large amount of logic cells required, make it impractical for an FPGA implementation.

For the experiments we used the Lucent ORCA 2CxxA family of FPGAs. This architecture has the advantage of configuring the logic cells that are called Programmable Functional Units (PFUs) into a special multiplication mode, which can significantly reduce the required resources for a multiplier [7].

The ORCA SCUBA is a Lucent specific program that produces designs optimised for the target architecture. The design produced by this program uses solely the logic cells in a 4x1-multiplying mode (FMULT41) and cascades them. Although, this method requires very little logic resources, the critical path is very long due to the extensive use of cascading with ripple carry chains.

A hand-made multiplier was also made which used the FMULT41 arrangement of the partial products but a binary tree of adders for the final result. This methods uses the special multiplication mode to compute the partial products, but an binary-tree arrangement of adders to compute the final product [6].

The designer can choose between a SCUBA or a hand-made depending whether his design is area or timing critical. In this application we used our own implementation to meet the tight timing constraints.

The design can be pipelined to a great extent in order to achieve the desired performance. The register intensive nature of FPGAs means that most of the pipeline registers will come for free as they will be hosted in the same logic cells as the preceding logic functions.

3. Floating Point Addition and Subtraction

Floating point addition and subtraction is one of the most challenging operations in floating point arithmetic. Shifters are the most expensive component in floating point addition and subtraction. The aim in this circuit is to reduce the area to a minimum while maintaining high speed operation. A key element in all circuits designed for FPGAs with area constraints is to adjust the design's routing and logic complexity balance to match those for the target FPGA architecture.

The steps are typically required in a standard floating point adder are: Check of zero operands, Subtract the exponents, Align the fractions by shifting the fraction with the smaller exponent, Add or subtract the fractions, Set the exponent of the result equal to the larger exponent and finally normalise and round the result

The first step in floating point addition is to subtract the exponents. This will determine a number of things. First, it will determine which of the two numbers is the biggest and if swapping of the significands is needed. Next, it will detect if underflow will occur. The amount of shifting will also be determined here. The circuit in Figure 1 is used to realise this function. It takes advantage of the very efficient adder/subtractor structures of FPGAs and the available three-state buffers to do the multiplexing. Although this arrangement would be considered irrational for ASIC implementation, it gives optimal results in FPGAs where it is both smaller and faster than traditional methods.

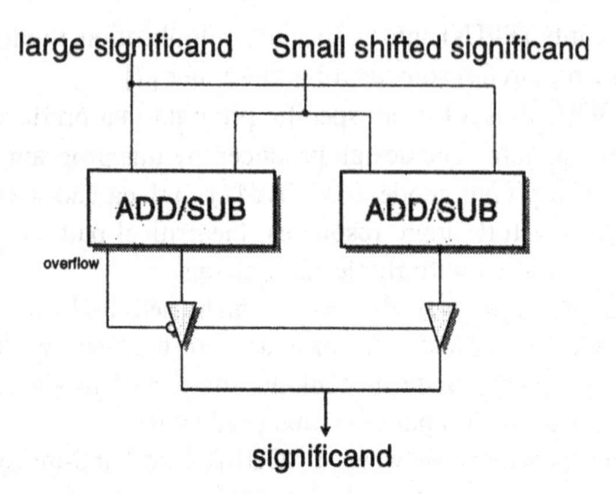

Figure 1 : Subtraction of significands

The next stage is to swap the significants and shift the smaller of the two. Swapping is performed using three states buffers placed right after the registers or memory cells where they are stored. Shifting is then done using LUT-based multiplexer components and not with tri-state buffers as this would limit the design's routability and cascading three states buffers is not efficient in FPGAs as shown in Figure 2.

The next stage is to add or subtract the fractions. The desired operation is derived from the signs of the two numbers. There is a special case, when subtraction of equal exponent numbers is desired, where the result will be in two's complement. The problem is tackled with a method similar to that employed for the exponents.

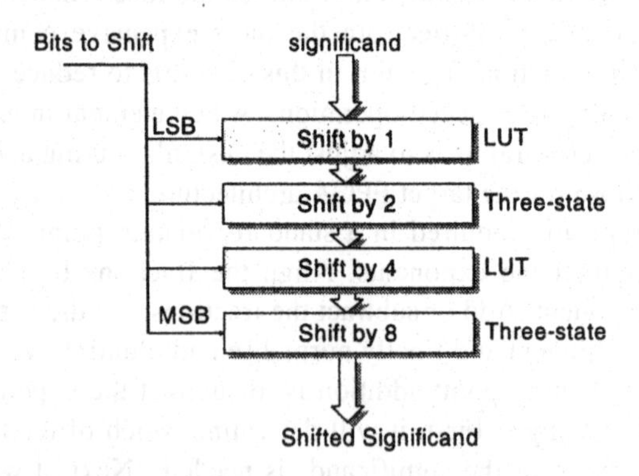

Figure 2: Shifting of Smaller Significand

The result is then normalised and the exponent is biased. The rounding stage was omitted from this implementation as the design is mainly area critical and in our application, rounding errors are not accumulative and therefore do not affect significantly the result.

When synthesised to our reference FPGA device, this design uses 67 logic cells, which is just 7.5% of an ORCA2C-40 device. Although, the maximum speed depends on the routing delays, a reasonable placement it is capable of running at 10 MHz. Pipelining a design with three-state buffers in a FPGA needs careful placement of the registers in order to achieve best results and avoid the use of excessive extra logic.

4. Floating Point Division

Floating point division is a very complicated operation, which is usually implemented using iterative approximations using a multiplication circuit and the Newton-Raphson method. The implementation of multipliers and adders is well understood and a lot of literature and implementations exists, the design of dividers still remains a serious design challenge which is often viewed as "black-art" among system designers [3].

A Common method of implementing multipliers uses the Newton-Raphson multiplication-based algorithms such as functional iteration. However, these implementations are not very efficient in deeply pipelined applications.

Long methods based on successive subtraction or addition of the fractions, such the ones based on the work of Sweeney, Robertson and Tocher that are now commonly known as SRT-type of divisions [1] will give very good result as integer adder/subtractors can be very efficiently implemented in FPGA.

The following diagram shows the internal architecture of a design based on long division that used successive subtractor cells.

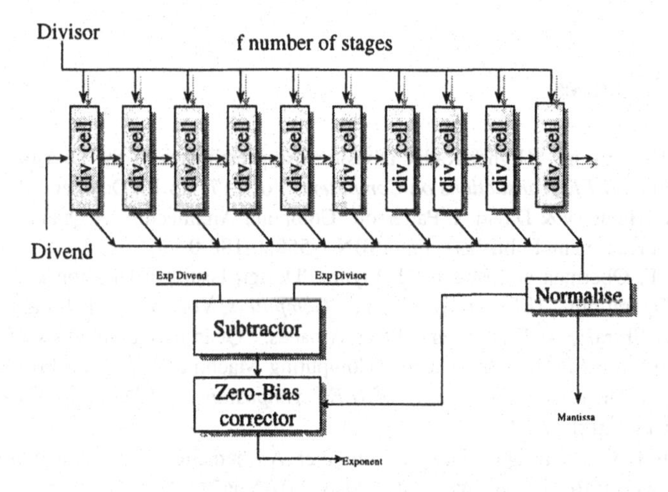

Figure 3: Internal structure of divider

We call the main subcomponent used in the division process DIV_CELL. The DIV_CELL cells subtracts the divisor from the dividend if possible. If this is not possible, then the overflow bit is set and the dividend is passed to the next stage unchanged using a multiplexer. Depending on the exact FPGA family used, the multiplexing can be performed with or without three-state buffers. In families where three-states are connected to long routing resources, such as ORCA2C, this would significantly limit the routability of the design. In newer FPGAs where three-states are part of the logic cells normal outputs, such as the Lucent ORCA3C, multiplexing using three-states would save significant amount of logic cells.

This method relies on the fact that subtraction circuits are very small and very fast, which is the case for most popular FPGA Architectures., e.g. Lucent ORCA2A, Altera FLEX etc, Xilinx XC4000 etc.

5. Conclusions

FPGA devices are increasingly used in new applications and for simulating ASIC. As the FPGA market is rapidly expanding, vendors are increasingly offering highly sophisticated and flexible architectures.

Floating point arithmetic has been avoided in FPGA as the limited speed and capacity of those devices prohibited such implementations. We show methods that allow an FPGAs to be configured with multiple floating-point arithmetic units within the same device, thus they can be used to compute complex mathematical formulas

Examples based on Lucent ORCA FPGA were given which were used in the TETRA 3D Graphics rendering pipeline, which was developed under the GraphMem Project. The GraphMem project is supported under the ESPRIT OMI programme of the European Commission (EP20488).

6. References

[1].Daniel E. Atkins, "Higher-Radix Division using estimates of the Divisor and Partial Remainders", *IEEE Transactions on Computers*, Vol C17, No.10, October 1968.

[2].John L. Hennesy & David A. Patterson "Computer Architecture, A Quantitative Approach" Morgan Kaufmann Publishers, Inc. ISBN 1-55860-188-0

[3].Stuart F. Oberman and Michael J. Flynn, "Design Issues in Division and other Floating-Point Operations", *IEEE Transactions on Computers*, Vol.46 No.2 February 1997.

[4].Nabeel Shirazi, Al Walters and Peter Athanas, "Quantitative Analysis of Floating Point arithmetic on FPGA based Custom Computing Machines." Virginia Polytechnic Institute and State University, *Proceeding of IEEE Symposium on FPGAs for Custom Computing Machines*, April 1995.

[5].Joseph J. F. Cavanagh, "Digital Computer Arithemetic - Design and implementation", McGraw-Hill Book Company, 1985, ISBN 0-07-Y66213-4.

[6].Iakovos Stamoulis, Nicky Ford, Martin White, Graham J. Dunnett, Paul F. Lister, "VHDL Methodologies for effective implementation in FPGA devices and subsequent transition to ASIC technology", Designer Track DATE '98, Paris 23-26 February 1998.

[7]. Lucent Technologies, "ORCA OR2C Field Programmable Gate Arrays", Users' Manual, January 1998.

[8].Synplicity, Inc. "Synplify, Simply Better Synthesis", User's Guide Version 3.0 with HDL Analyst", July 1997.

[9].Synopsys, Inc. "FPGA Express" Version 2.0.1, Users' Manual, P/N 01733-000 DB February 1998.

SL – A Structural Hardware Design Language

Samuel Holmström

Turku Centre for Computer Science, Åbo Akademi University,
Department of Computer Science, FIN-20520 Turku, Finland
Samuel.Holmstrom@abo.fi

Abstract. SL is a simple language designed to improve the productivity of hardware design. It is easy to use and it adopts reusable word-level and bit-level descriptions. This results in concise and easily grasped descriptions which give a good overview of the design. Used together with our SLAVE compiler, SL offers a fast way to produce complicated structural VHDL-code. This structural VHDL-code can then be used by Velab VHDL elaborator in order to produce a netlist for Xilinx XC6200 FPGA.

1 Introduction

Hardware description languages have made hardware design significantly faster compared to traditional schematic approaches. In many cases, a high level description is enough because the low level details of a circuit can be synthesized from a behavioral or RTL-level description. This is not always true, however. Sometimes the only possibility is to use a detailed structural description which in practice is just a massive textual description of a traditional schematic. These descriptions tend to be hard to maintain because it is not that easy to see the structure from, say, several hundreds of lines of code written in some hardware description language. This is also the case for structural VHDL, which is needed by Velab in order to produce netlists for the XC6200 family.

Our contribution is a new language, SL, that abstracts away from the many details. Of course, it is possible to use libraries of components containing all the details. However, it can be a very hard job to create these libraries. We believe that the key to success is a concise abstract notation – applied to design of circuits as well as for creating libraries of components.

There exists several other hardware description languages, for instance VHDL, Handel-C [1] and Lola [7]. Among the languages compared in Table 1, the language Pebble [6] is closest to SL. The language Handel-C has a simple syntax and it offers a convenient way to describe hardware on a high level [3], but we cannot use it for the Xilinx XC6200 family. The Lola Programming System for the H.O.T. Works Development System [5] offers an integrated environment for fast and efficient design. Unfortunately, quite often something goes wrong with the netlists. Therefore, we had to stick to structural VHDL, which we have been using together with the Velab VHDL elaborator. The complexity of VHDL, on the other hand, has led us to believe that a simpler approach will provide a better foundation on which to build abstractions and tools.

	SL	VHDL	Handel-C	Lola	Pebble
Family	simplified VHDL	Ada	C	Pascal	simplified VHDL
Syntax	simple	complicated	simple	simple	simple
Level	low	all	high	low/medium	low
Placement	yes	yes/no	no	yes	yes
Tools	Velab only	many	limited	integrated	VHDL tools
FPGA	XC6200	all	XC3195	XC6200	XC6200

Table 1. SL compared with other hardware description languages.

This paper is organised into four further sections. Section 2 provides an overview of the SL notation. Section 3 describes how to create basic components with SL. Section 4 uses a summation example in order to demonstrate the power of our notation. Section 5 provides some concluding remarks.

2 Language Overview

SL is an alias for *Small Language.* The primary objective for SL is to facilitate the development of efficient and reusable designs by offering a clear and concise notation which gives a good overview of the design. Because of its simple syntax, SL is both easy to learn and to use. In principle, an SL description is just a list of components like any other structural HDL-description. For the moment, XC6216 is the only FPGA supported by SL, but this limitation will be removed in the future.

SL can be regarded as a much simplified variant of structural VHDL. It reminds in some sense of Pebble, but SL does not have any support for run time reconfiguration because we wanted to keep its syntax as simple as possible. SL provides a means of representing component descriptions hierarchically. The basic features of SL are outlined below.

- A SL component is a block, defined by its entity and its architecture.
- The entity consists of an entity name and a number of directed port signals.
- The architecture consists of declarations, instantiations and assignments.
- Variables as well as port signals can be one or several bits wide.
- A primitive component has no architecture; the top level component has an architecture but does not have any port signals.

SL syntax rules for components are described in Fig. 1. In here, $\ll exp \gg$ means zero or one occurrences of exp, $\ll exp \gg^*$ means any number of exp, while $\prec alt_1, alt_2 \succ$ means that either alt_1 or alt_2 is used. This should not be confused with the placement information $< x, y >$ or the braces which are used when generating components. The expression $< x, y >$ simply describes the coordinates for where to place the component on the RPU. The coordinates can be anything from (0,0) to (63,63) for XC6216. In SL, there are only four reserved words: IN, OUT, VAR, and OPEN. The keyword OPEN is used for an unconnected port.

EntityName ≪ (≺IN, OUT≻ Signal ≪; ≺IN, OUT≻ Signal≫*) ≫ :
≪ **VAR** Signal; ≫
≪ ≺Generate, Component≻ ≫*
≪ SignalName ≪Vector≫ = SignalName ≪Vector≫; ≫

Signal ≙ SignalName ≪ Vector ≫ ≪ ; SignalName ≪ Vector ≫≫*
Vector ≙ ≺ [StartIndex..EndIndex], [Index] ≻
Component ≙ Label = ComponentName(Signal) Placement;
Generate ≙ Label = { Id @ [StartIndex..EndIndex] | Component}
Placement ≙ <ColumnIndex, RowIndex>;

Fig. 1. SL syntax rules

3 Basic Building Blocks

SL has a simple, block-structured syntax. All descriptions are built up of small blocks of components, where the primitive components are declared in a VHDL library provided by Xilinx. For convenience, we use the same names for primitive components as in the given library. By combining the primitives `XOR2` and multiplexer `M2_1`, we can create a one-bit full-adder element as follows:

```
ADD1(IN cin, a, b; OUT cout, sum):
  VAR temp;
  X1 = XOR2(temp, a, b)<0,0>;
  X2 = XOR2(sum, cin, temp)<2,0>;
  M1 = M2_1(cout, cin, b, temp)<1,0>;
```

The coordinates for the internal components in `ADD1` only describe the internal placement *relative* to the placement of `ADD1`. For instance, if `ADD1` is instantiated at top-level and placed at the position (8,4), then the *exact* placement of the internal components labeled `X1`, `X2` and `M1` will become (8,4),(10,4) and (9,4) respectively.

We can now build a 16-bit adder `ADD16` by using sixteen one-bit full-adders. We leave the carry-out port of the last one-bit adder unconnected. We could easily connect it by adding a new signal, say, `cout` to the `ADD16` list of port signals and replacing `OPEN` with this new signal name. However, we will use the adder `ADD16` as such in section 4. The SL description is as follows:

```
ADD16(IN cin, a[0..15], b[0..15]; OUT s[0..15]):
  VAR temp[0..15];
  A0 = ADD1(cin, a[0], b[0], temp[1], s[0])<0,0>;
  G1 = { i @ [1..14] |
         A = ADD1(temp[i], a[i], b[i], temp[i+1], s[i])<0,i>; }
  A15 = ADD1(temp[15], a[15], b[15], OPEN, s[15])<0,15>;
```

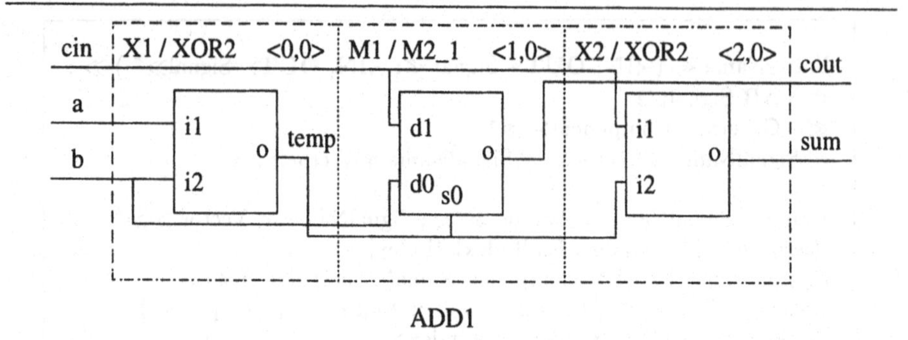

Fig. 2. Full-adder built up of basic components placed on RPU.

4 Example: Summation of 16-bit Words

This section demonstrates the power of SL by describing a small design that calculates the sum of eight 16-bit words. The circuit can be built as follows. First, the rightmost sixteen bits of a 8 x 16-bit register `K` and the content of the result's register `RG1` is sent to a 16-bit adder `AD1`. The result is then stored back to `RG1`. On the next clock cycle, the contents of `K` is rotated one bit to the right, and the addition starts over again. A counter `CTR`, built up of T flip-flops (`TFF`) and AND-gates, keeps track on the number of additions. The counter continues to count as long as the controller `CTL` allows it. When `CTL` is zero, no calculations are done. The calculations start when the register `EN` is activated or when `CTR` is greater than 0.

Some of the components needed for the summation of the eight 16-bit words are described in SL below. We reuse the 16-bit adder which was described earlier as well as some other components, e.g. the D-latch `FDC`, without repeating their descriptions here.

```
REG16(IN clk, clr, en, d[0..15]):
  G1 = { i @ [0..15] | R=REG1(clk, clr, d[i], en, OPEN)<0,i>; }

TFF(IN clr, clk, d; OUT q):
  VAR Qt, X_out;
  T = XOR2(X_out, Qt, d)<0,0>;
  FF = FDC(Qt, X_out, clk, clr)<0,0>;
  q = Qt;

CTR1(IN clr, clk, i1, i0; OUT q, a):
  VAR A_out;
  A = AND2(A_out, i1, i0)<0,0>;
  T = TFF(clr, clk, A_out, q)<1,0>;
  A = A_out;
```

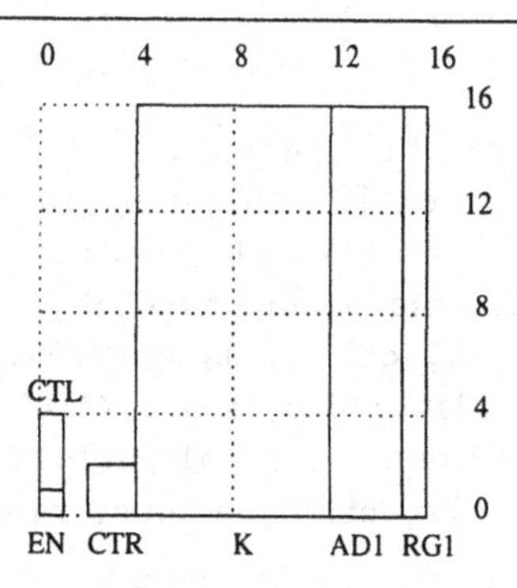

Fig. 3. Placement of the summation's components.

```
CTR3(IN clr, clk, d; OUT q[0..2]):
  VAR Qt[0..2], At;
  CO = TFF(clr, clk, d, Qt[0])<1,0>;
  C1 = CTR1(clr, clk, d, Qt[0], Qt[1], At)<0,1>;
  C2 = CTR1(clr, clk, At, Qt[1], Qt[2], OPEN)<0,2>;
  q = Qt;

ROT1(IN clk, clr, en; OUT q):
  VAR t[0..7];
  REG_0 = REG1(clk, clr, t[7], en, t[0])<0,0>;
  REG_1 = REG1(clk, clr, t[0], en, t[1])<1,0>;
  G1 = { i @ [1..6] |
         REG = REG1(clk, clr, t[i], en, t[i+1])<i+1,0>; }
  q = t[7];

R_ROT16(IN clk, clr, en; OUT q[0..15]):
  G1 = { i @ [0..15] | R = ROT1(clk, clr, en, q[i])<0,i>; }

CTRL(IN en, i[0..2]; OUT c):
  VAR t1, t2;
  o0 = OR2(t1, i[0], i[1])<0,0>;
  o1 = OR2(t2, i[2], t1)<0,1>;
  o2 = OR2(c, t2, en)<0,2>;
```

Now we can create the summation circuit easily by using the components above. The placement of the components is illustrated in Fig. 3. The SL description is as follows:

```
SUM:
  VAR clk, clr, pwr, ctl_c, enable,
      r1_q[0..15], a1_out[0..15], k_out[0..15], ctr_q[0..2];
  EN = FDC(enable, clr, clk, clr)<0,0>;
  CTL = CTRL(enable, ctr_q, ctl_c)<0,1>;
  CTR = CTR3(clr, clk, ctl_c, ctr_q)<2,0>;
  K = R_ROT16(clk, clr, ctl_c, k_out)<4,0>;
  AD1 = ADD16(clr, k_out, r1_q, a1_out)<12,0>;
  RG1 = REG16(clk, clr, ctl_c, a1_out, r1_q)<15,0>;
```

Compared with structural VHDL, descriptions are clear and much shorter in SL which makes it easier to overview the design and thus makes the structure easier to capture. We can describe the circuit above in full detail by less than 70 lines using SL, and then let our SLAVE (*Simple LAnguage to VElab*) compiler translate this into 260 lines of structural VHDL-code that the Velab VHDL elaborator can handle. More details, including complete SL-code and automatically generated VHDL-code for the example above, can be found elsewhere [4].

5 Conclusions and Future Work

We have presented the language SL and its use in hardware design. We were in some sense inspired by the Action System notation [2], which we want to use in order to enable a higher level formal description of hardware design and preserve the correctness from initial to concrete level of design. SL is our first step in that direction. We plan to extend SL to support reconfiguration and timing. We also plan to add a graphical user interface supported by SL as well as a simulator to make it possible to simulate the design before the place-and-route phase.

References

1. M. Aubury, I. Page, G. Randall, J. Saul, R. Watts: Handel-C Language Reference Guide. Oxford University Computing Laboratory. 1996.
2. R.J.R. Back, K. Sere: From Action Systems to Modular Systems. Software-Concepts and Tools (1996) 26-39. Springer-Verlag.
3. S. Holmström and K. Sere: Reconfigurable Hardware — A Case Study in Codesign. TUCS Technical Report No 175, May 1998.
4. S. Holmström: SL — A Structural Hardware Design Language for the XC6216. TUCS Technical Report No 287, June 1999.
5. H.O.T. Works Development System Details. (Home page visited at 25.2.1999) `http://www.vcc.com/Hotworks.html`
6. W. Luk and S. McKeever: Pebble - A Language for Parametrised and Reconfigurable Hardware Design. In: Proc. of the 8th International Workshop on Field-Programmable Logic and Applications, LNCS 1482, Springer-Verlag 1998.
7. N. Wirth: Digital Circuit Design for Computer Science Students. Springer-Verlag 1995.

High-Level Hierarchical HDL Synthesis of Pipelined FPGA-Based Circuits Using Synchronous Modules

R.B. Maunder, Z.A. Salcic and G.G. Coghill

University of Auckland, New Zealand
z.salcic@auckland.ac.nz

Abstract. In this paper we present an approach to high-level synthesis of digital circuits from synchronous modules. The synthesiser implemented takes as its input a functional description of the circuit in the form of a netlist using predefined functional modules with desired parameters, and produces an AHDL description as an intermediate circuit representation. The functional modules can be designs entered and produced using different design entry tools and design compilers. The synthesis allows hardware resource sharing, variable data path widths, variable bit resolutions, and various number representations (e.g. parallel, serial, stochastic) for different parts of a circuit. As a result of synthesis, pipelined circuit analysis ensures coherent dataflow through the circuit is produced. At the end, the overall control unit that controls data flow through the circuit is automatically generated. The synthesiser presents the first part of the implementation of a tool for the optimisation of circuit design for FPGAs as a target technology.

1. Introduction

The aim of the synthesiser is to take a high-level functional description of a circuit and produce an implementation of the circuit in AHDL. This intermediate language output may be further processed to physically instantiate the circuit on an FPGA device. Synthesis parameters may be chosen to guide the resulting circuit towards the planned hardware resource usage and circuit processing time constraints. The synthesiser has been designed to be used in conjunction with an optimisation algorithm that adjusts the synthesis parameters so that different design alternatives may be analysed and optimised for these desired criteria.

The types of circuits targeted by the synthesiser are those that are largely data-path oriented, with clearly defined functional blocks. Ideal applications would be the implementation of functions to be controlled by a separate microcontroller, or circuits such as digital filters that require a high degree of parallelism.

2. Design Entities

The circuit to be synthesised is described by its *functional description*, which is in the form of a netlist created with our graphical design entry tool. The netlist describes the circuit as a number of interconnected synchronous modules. The modules are synchronous as they are all clocked together and have their data flow controlled by a common *control unit*, which is also automatically synthesised by the synthesiser. Each *synchronous module*, or just *module*, is an abstraction of a circuit that performs a certain function, and can be implemented in various ways depending on its *module parameters* which are chosen by the synthesiser (or optimiser). In addition to each module's function being described in the netlist, the *accuracy*, or number of bits of resolution required is also indicated. The synthesised part of the circuit that is being controlled is the *circuit's data path*.

There are two different types of modules used by synthesiser: functional and auxiliary. *Functional modules* are those that perform a certain, predefined function such as multiply or filter and are used in the circuit's functional description. *Auxiliary modules* are used and generated by the synthesiser in its task of ensuring the circuit performs correctly. Examples of auxiliary modules are accuracy converters, delay modules and modules for converting between *data types* (such as parallel, serial or stochastic).

Each *module implementation*, which can be described by various and\or multiple methods such as VHDL, Verilog, AHDL, XNF, EDIF to name a few, has an interface to the synthesiser in AHDL known as the *module interface shell*. The module interface shell represents an unambiguous way of describing the module implementation. Each module's parameters are calculated for a specific target FPGA family, and when there are both logic elements and memory cells, these are treated as separate resources. Different module implementations reflect the different ways that the module may be implemented, based upon the same algorithm. A *module instance* is an unambiguously defined module; one that has been supplied with all its module parameters. These *module parameters* include the data type of the module, and the data path widths. Lastly there is the *physical module instance*, which is the actual on-device implementation. This differs from the module instance because, for example, a number of identical module instances can utilise *resource sharing*, and actually use the same hardware, or physical module instance, to perform their function.

3. Synthesiser: Functional Description and Implementation

3.1 Functional Circuit Description

The input netlist (generated by the graphical entry tool) describes the functionality of the circuit, suggests data types of modules to use if available, and recommends which modules should share their resources. The netlist also indicates the overall data type and accuracy of the circuit, and whether, or not it should be treated as operating on signed numbers. Other module-dependent parameters are provided in the netlist (such as the constant of a constant-multiplier). The compute time of each module is not known until a specific module implementation is chosen during synthesis and so a designer cannot use delay elements to store data in the circuit between subsequent initiations of the circuit. Feedback paths, however, provide the required delays because the synthesiser ensures fed back data is available at its destination module at the correct clock cycle in the subsequent circuit initiation. Feedback paths must be identified in the netlist, as it is not possible for the synthesiser to resolve the required functionality otherwise.

3.2 Synthesis Parameters

There are a number of synthesis variables that are associated with each module of the circuit. These affect the overall hardware resource usage and the time taken for the circuit to process its inputs (*compute time*). Also affected is the *average latency* of the circuit, which describes how many cycles (on average) must pass before new inputs can be processed by the circuit. Often this is given as the number of completions (or initiations) per cycle, known as the *initiation rate,* which is the reciprocal of the average latency. The combination of hardware resource usage and timing gives a solution point in the possible design space for the functional implementation.

Altering the module variables allows different solutions to be analysed. Generally, there will be a trade-off between hardware resource usage and circuit compute time. A more subtle trade-off can be obtained between circuit compute time and average latency. An optimisation algorithm can then search through the design space for a solution that conforms to various criteria.

The variables that can be modified for each module are its implementation version, the data type it operates on, and whether the synthesised module instance should share hardware resources with other modules by sharing a single physical module instance. The implementation version describes the method or algorithm employed by a module implementation to obtain its result - for example arithmetic series or look-up table. The minimum accuracy of a module is a design decision and cannot be reduced. Increasing it does not usually have an advantageous effect, and so it is not considered to be one of the variables. Table 1 shows the resulting module parameters for an 11-bit unsigned multiplier for different synthesis variables.

Data Type	*Module Version*	*Size (Logic Cells)*	*Size (Memory Cells)*	*Compute Time (Cycles)*	*Average Latency (Cycles)*
Parallel	A	181	0	3	3.00
Parallel	B	64	7168	1	1.00
Serial	A	65	0	22	33.00

Table 1. The effect of different synthesis variables on an 11-bit unsigned multiplier module.

3.3 Synthesiser Functional Description

The synthesiser begins its operation by loading a netlist, which is a functional description of the circuit to be synthesised. The library of predefined modules is loaded into the database, and all the module generators are initialised. The netlist is verified to ensure that all modules can be implemented. A global synthesis view is illustrated in figure 1.

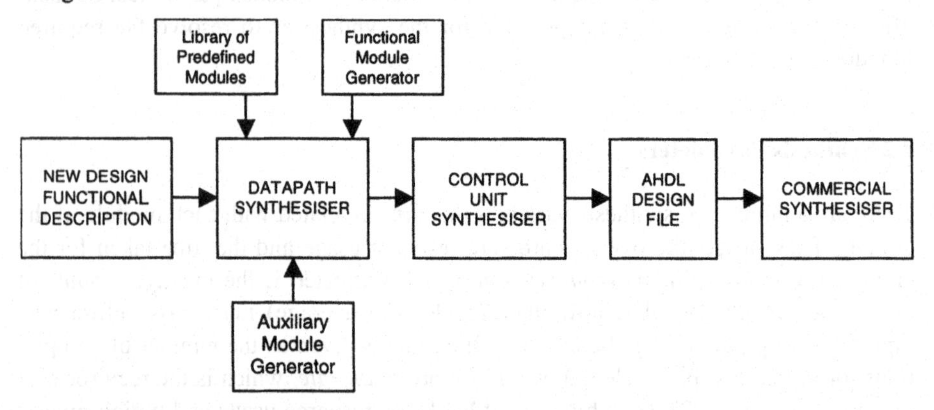

Fig. 1. Global View of the synthesiser flow.

To repair the circuit description, functionally equivalent module implementations are chosen that are closest to those defined in the netlist. These chosen implementations may be of a different data type or greater accuracy than the netlist recommends. Accuracy converters and data type converters are added by the synthesiser to ensure data type compatibility throughout the circuit. Multiplexers are added to perform resource sharing of physical module instances. Module instances that are resource shared and will not have a physical instantiation of their own are removed.

Next, the synthesiser must ensure, by the insertion of delay elements, that data flows through the circuit coherently such that all inputs to a module arrive at the same time. In addition, all resource shared modules are assigned the most efficient dataflow order.

Accuracy converters and data type converters are added in feedback paths where necessary. Delays are inserted in feedback paths to ensure coherent dataflow in them. The feedback delay elements are of a constant duration, because serial delays of different duration, for example, require different numbers of flip-flops, and the inserted module must be of an *exactly* defined function. This limits subsequent initiations to be spaced in time by a constant latency.

Circuits that do not have feedback paths are not constrained to having a constant initiation latency. In this case, a full pipelined analysis of the circuit is performed to determine a *control strategy* that indicates when circuit initiations should be made. The collision vectors for each module, which describe when initiations of each module can occur, are used to build up a set of initiations that cannot occur for the whole circuit; if they do, collisions of the data will occur, that is, the same hardware is expected to operate on two or more sets of inputs at the same time. This set of initiations that will cause collisions is known as a *forbidden latency set*. From the forbidden latency set, a collision vector for the whole circuit may be derived [7].

The circuit collision vector is used to derive a *modified state diagram*, which describes all the possible initiation latencies between allowable initiations [1]. The method of determining the control strategy from the modified state diagram is that of finding all the greedy cycles it contains. This is done by moving from state to state in the modified state diagram by choosing the lowest latency from each state [2]. The strategy of using greedy cycles does not always provide the maximal data throughput, but it is readily calculable whether the minimum achievable latency cycle has been found [3]. The result of the greedy cycle is a sequence of initiations that produces no collisions. Patel and Davidson have presented another method for ensuring the maximum data throughput of a pipeline is achieved by the addition of delays, but their approach may increase the compute time of the circuit [6].

Circuit hardware resource usage is estimated by summing the usage for each module that will be physically instantiated. Logic cell usage is kept separate from memory cell usage. The circuit compute time, average latency, and collision vector are obtained directly from the pipelined dataflow analysis stage. All other parameters can be derived from the circuit by the synthesiser to completely describe the circuit as a module for use in further designs.

As the final stage of the synthesis, the output files are generated. Predefined module implementations and associated files are copied to the output directory. (Associated files are those required by a module such as a lower level module or an initialisation file for on-device memory.) Further required functional and auxiliary modules are synthesised with their associated files. The data path of the circuit and the control unit are synthesised in AHDL. A report is generated that describes the circuit parameters, the controller, the required external control signals, and other aspects of the synthesised circuit. The circuit parameters are added to the synthesised circuit file to produce a reusable module in the form of a module interface shell.

3.4 Software Implementation

The synthesiser was implemented for 32-bit Microsoft Windows platforms in C++ using the Borland Database Engine. A subset of SQL (Structured Query Language), known as SQL92 (or ANSI-92 SQL) was also used so that the underlying database engine could be changed (to Oracle for example) without code changes. Advantage was taken of the Borland's Visual Component Library (VCL) to provide high-level structures to work with. Coding time and debugging time was reduced at the expense of slower code operation and reduced portability between platforms. The module generators were developed as application extensions, or DLLs (Dynamic Linked Libraries), which are loaded into memory by the synthesiser during its initialisation.

4. Example Design

A pipelined circuit was developed using the synthesiser as an example of its operation. The circuit performs the averaging of a group of pixel values within a 3×3 mask, as part of an image enhancement data path. The pixel values are fed into the circuit in a pipelined fashion, three adjacent pixels at a time. All pixel values are available as 8-bit parallel numbers. The circuit topology fed into the synthesiser, to implement the approximated mask that ignores the centre pixel value, is shown in figure 2. Results obtained from the synthesiser for various combinations of data types and resource sharing are given in table 2. Note that with resource sharing, if modules with different accuracies are shared, the highest accuracy is chosen for the implemented module, which in addition to the added multiplexer and delays can make resource sharing an unattractive option. It can also be seen from this table that the synthesiser's circuit-size estimation is fairly accurate. Figure 3 presents a timing diagram for one of the implemented circuit variations after compilation and full timing simulation. It shows the pipelined operation of the circuit and propagation delays introduced by the circuit, resulting in a finite settling time of the circuit's output.

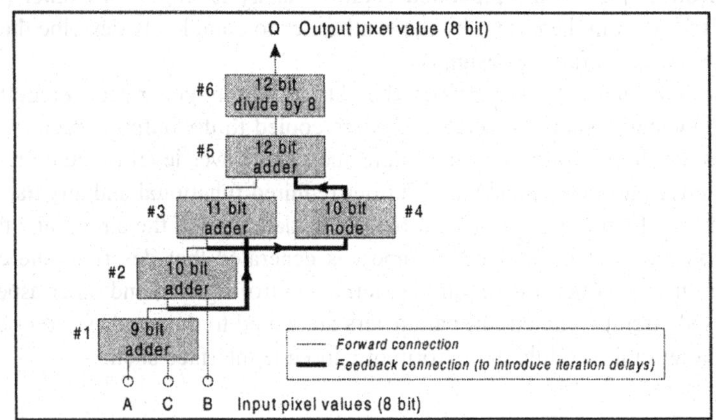

Fig. 2. Pixel averaging filter circuit using a 3 × 3 spatial mask approximation. The pixels A, B and C are fed into the circuit in a pipelined fashion.

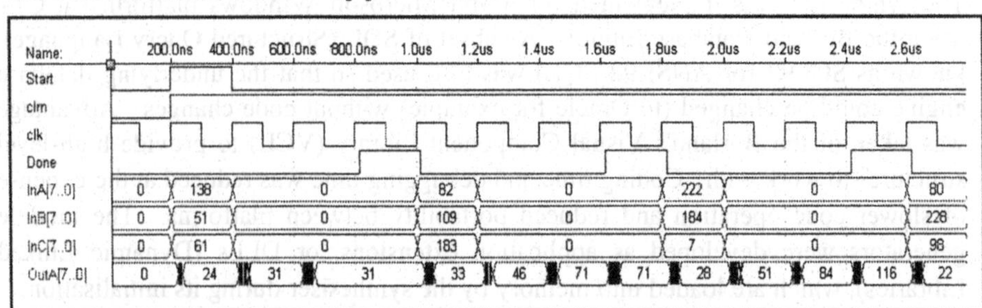

Fig. 3. Pixel averaging filter timing diagram for implementation with compute time of 3 cycles and constant latency of 4.00 cycles. The inputs are presented starting with the *Start* signal and then every fourth clock cycle. The outputs hold their correct values when the *Done* signal is high.

Parallel Module Numbers	Share *Modules*	*Actual Logic Cells*	*Estimated Logic Cells*	*Compute Time (Cycles)*	*Average Latency (Cycles)*
all	-	97	109	0	1.00
6	-	101	88	12	12.00
1,2,6	-	120	119	12	12.00
2,4,5,6	-	525	596	20	11.00
1,2,3,4,6	-	122	124	12	12.00
all	(1,2,3,5)	288	352	3	4.00
6	(1,2,3,5)	323	290	48	48.00
all	(3,5)	153	196	1	2.00
6	(1,2,5)	238	218	36	36.00
6	(1,2)(3,5)	171	158	24	24.00

Table 2. Results for various pixel-averaging circuit implementations using logic cell based modules only. The sizes shown are actual implementation hardware usage after HDL compilation (including controller circuit), and estimates from the synthesiser. The *Parallel Modules* column lists the module numbers of the modules implemented as parallel modules; the remaining modules are implemented as serial.

The circuit was modified to average all nine pixels to compare the resulting circuit parameters with a full integer divide rather than a shift operation. Table 3 shows the different resulting parameters for circuits consisting of only parallel modules for both the 8 pixel and 9 pixel averaging filter circuits. For logic-cell based dividers, the increased number of pixels averaged by the mask results in greater logic cell usage, compute time and average latency. For the memory-cell based dividers, both logic-cell usage and memory-cell usage increase but at no time expense.

Pixels Averaged	*Size (Logic Cells)*	*Size (Memory Cells)*	*Compute Time (Cycles)*	*Average Latency (Cycles)*
Logic-cell-based dividers:				
8	109	0	0	1.00
9	201	0	14	14.00
Memory cell based dividers:				
8	76	49152	1	1.00
9	115	106496	1	1.00

Table 3. Comparison between using divide-by-8 and divide-by-9 division modules in the pixel-averaging circuit with all parallel modules and no resource sharing.

5. Conclusion

This paper has presented an HDL synthesiser that was developed as a method of analysing and synthesising hierarchically designed circuits from high-level descriptions. Dataflow and pipeline design techniques were used by the synthesiser to obtain high data throughput rates for the circuit. The synthesiser was able to utilise modules designed with different tools, and combine them to produce the required circuit functionality. Variable width data paths were supported, as were different data types, and resource sharing. It was shown that by varying the module parameters, designs with different size and time requirements could be produced - giving different possible solutions in a circuit's design space. The synthesiser was developed to work in conjunction with an optimiser that is currently under development.

Proposed extensions to the synthesiser are:

- Integration with hardware-software co-design through the inclusion of software modules and associated algorithms for their processing.
- More accurate estimation of hardware resource usage through a 'compression factor' that accounts for the fitting processes of the AHDL compiler, depending upon the circuit topology.
- Estimation of the propagation delay of each synthesised module could be made- including the circuit's clock frequency in the optimisation goal. The circuit repair would have to sum sequential propagation delays and ensure module outputs stabilise before the latching edge of the clock signal, otherwise delay elements would be needed.
- The extension to the library of synchronous modules that utilise a wide range of data types. This would broaden any search performed of the design space.

References

[1] E. S. Davidson. The Design and Control of Pipelined Function Generators. In *Proceedings: International Conference on Systems, Networks and Computers,* pages 19-21, Oaxrepec, Mexico, January 1971.

[2] P. M. Kogge. *The Architecture of Pipelined Computers.* Hemisphere Publishing, 1981.

[3] D. Lewin. *The Theory and Design of Digital Computer Systems.* Thomas Nelson and Sons, 1980

[4] R. B. Maunder, G. G. Coghill and Z. A. Salcic: "Genetic Algorithm Optimisation of FPLD Circuit Synthesis from High-Level Modular Descriptions", in *Progress in Connectionist-Based Information Systems* (N. Kasabov et al. editors), Springer, 1997, pp. 678-681.

[5] R. B. Maunder, Z. A. Salcic and G. G. Coghill : "FPLD HDL Synthesis Employing High-Level Evolutionary Algorithm Optimisation", in *Field-Programmable Logic '97, Lecture Notes in Computer Science 1304* (W. Luk, P. Cheung and M. Glesner editors), Springer Verlag, 1997, pp.265-273

[6] J. H. Patel and E. S. Davidson. Improving the Throughput of a Pipeline by Insertion of Delays. In *Proceedings: Third Annual Computer Architecture Symposium,* number 76CH0143-5C, pages 159-163. IEEE, 1976

[7] H. P. Stone. *High-Performance Computer Architecture.* Addision-Wesley, 1990.

Mapping Applications onto Reconfigurable KressArrays

R. Hartenstein, M. Herz, T. Hoffmann, U. Nageldinger

Computer Structures Group, Informatik
University of Kaiserslautern
D-67653 Kaiserslautern, Germany
Fax: +49 631 205 2640
abakus@informatik.uni-kl.de
http://xputers.informatik.uni-kl.de

Abstract. This paper introduces a design space explorer for coarse-grained reconfigurable KressArray architectures - to enable the designer to find out the optimal KressArray architecture for a given application. This tool employs a mapper based on simulated annealing, and is highly configurable for a variety of different KressArray architectures. Using performance estimation and other statistic data, the user can interactively change the architecture, until it suits the requirements of the application.

1. Introduction

The research area of reconfigurable computing has experienced a rapid expansion in the last few years. In many application areas reconfigurable hardware systems have proven to achieve substantial speed-up over classical (micro)processor-based solutions [1]. The range of these application areas has been extending continuously, including image processing as well as DSP, encryption, pattern recognition and many more.

The implementation of such applications has been done mostly on systems based on fine-grained FPGAs, as they are widely available. However, it turns out that these devices do not suit well for computational applications due to several reasons [2] [3]. As an alternative to fine-grained FPGAs, several coarse-grained reconfigurable architectures have been developed [4][5][6][7][8], which try to overcome the above problems by providing multiple-bit wide datapaths and more complex operators in the processing elements.

An example for such architectures is the KressArray [7], which features arithmetic and logic operators on the level of the C programming language, making the mapping of applications much more simple than for FPGAs. For the first KressArray prototype, which is also known as rDPA (reconfigurable Datapath Architecture), a Datapath Synthesis System (DPSS) [9] based on simulated annealing has been implemented, which accepts input at a high level language. However, after experiments with different KressArray architectures, it has shown, that the available communication resources have a major impact on the efficiency of the mapping. Thus, a KressArray design space

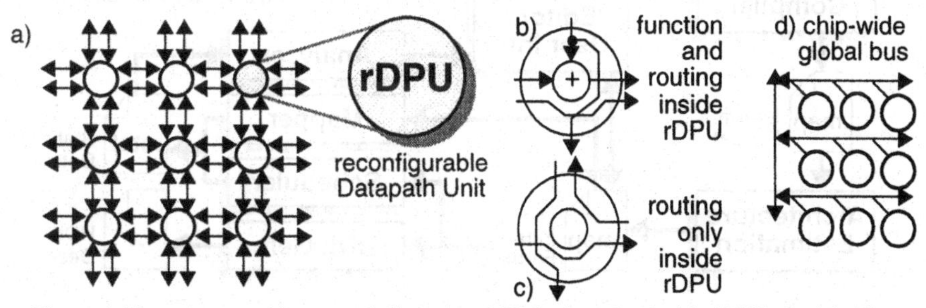

Figure 1: KressArray architecture: a) KressArray-III; b,c) routing inside rDPU; d) global bus

explorer has been implemented, which assists a designer to find the best KressArray architecture for a given problem. This paper concentrates on the simulated-annealing based mapper, which does a placement and routing of a given datapath onto a KressArray architecture. The mapper is highly parametrized in terms of available communication and function resources, which allows experimenting within a short time with a wide variety of different architectures for a given application, or, an entire application domain.

Section 2 briefly summarizes the KressArray having been published elsewhere [7] [9][10]. In section 3 the KressArray design space explorer will be introduced. In section 4 the configurable mapper tool is described briefly. The operation of this tool is demonstrated by 2 versions of an example in section 5. Finally, conclusions are drawn.

2. The KressArray

As an example of the KressArray architecture family [7][9][10], the current new prototype KressArray-III is illustrated in Figure 1a. It consists of a mesh of Processing Elements (PE), also called rDPUs (reconfigurable Datapath Units), which are connected to their four nearest neighbours by two bidirectional links with a datapath width of up to 32 bit, where "bidirectional" means that a direction is selected at configuration time, i. e. is fixed at run time. Nearest neighbour connections (NN links) are used to transfer operands to/from a rDPU, and, to route other data through a rDPU (Figure 1b), as well as for routing through only (Figure 1 c). Besides the NN links, a background communication architecture (called *back buses*) with one global bus (Figure 1 d) or multiple global buses and/or bus segments, i. e. semi-global buses (Figure 3 a and b) provides additional communication resources.

Currently the functionality of each PE consists of the integer operators provided by the programming language C. The mapper, however, also supports rDPUs with other operator repertories, such as e. g. specialized for accelerator usage in a particular application area. Execution inside PEs is transport-triggered, i. e. it starts as soon as all operands needed are available. The data to be processed and the results to be stored back can be transferred to and from the array in two different ways: over the global bus and over rDPUs ports at the edges of the array.

3. Design Space Exploration

An overview of the KressArray design space explorer is given in figure 2. The user provides a high-level description of the application using the high level ALE-X language [9]. The ALE-X compiler generates an expression graph, written in an intermedi-

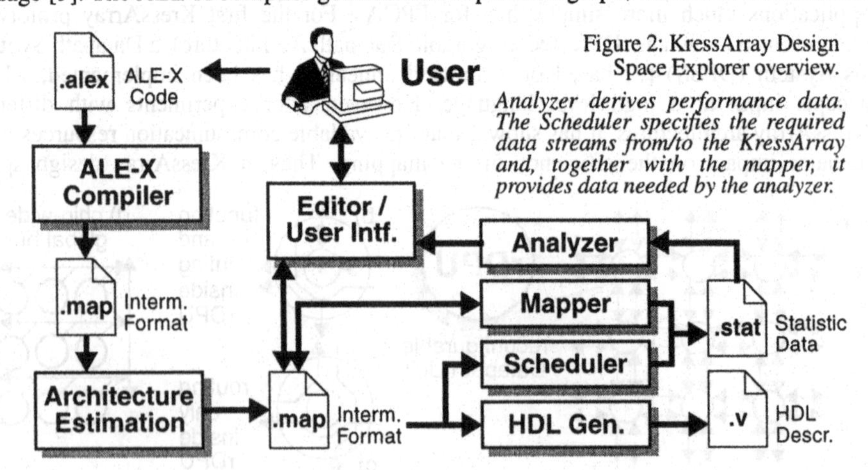

Figure 2: KressArray Design Space Explorer overview. *Analyzer derives performance data. The Scheduler specifies the required data streams from/to the KressArray and, together with the mapper, it provides data needed by the analyzer.*

ate format. This format also describes the properties of the rDPU operator repertory and the KressArray architecture, as well as the mapping added later on (explained later). In a next step, the expression tree is analysed and minimal requirements to the architecture are estimated. The intermediate file is then enhanced by these architecture definitions. In the following design process, the application is mapped onto the KressArray by the mapper tool, which generates a file in the same intermediate format as its input, allowing several consecutive mapping steps. A data scheduler determines an optimized array I/O sequence as well as a performance estimation. Both, mapper and scheduler, generate statistical data like usage of the communication resources and critical path information. From this data, an analyser derives possible enhancements to the architecture, which are presented to the user by an interactive editor. This editor is also used to control the design process itself. When a suitable architecture has been found, a HDL description (currently Verilog) can be generated from the mapping for simulation.

4. The Mapper

The mapper tool maps the application datapath onto the KressArray by placement and routing. The mapper can be seen as an extended version of the KressArray-I mapper [9], which now can handle different KressArray architectures. The mapping algorithm is based on simulated annealing, including a router for the nearest neighbour connections. For the mapper, different architectural properties may be specified:

- The size of the array.
- Areas with different rDPU functionality. For example, in figure 3c, the rDPUs in every second column of the array have a different function set available. The mapper has to consider this when placing the operators.
- The available repertory of nearest neighbour connections (see figure 1). Per side of the cell can be one or multiple unidirectional and/or bidirectional connections.
- One or multiple row and column buses (see figure 3a,b). These buses connect several rDPUs in a row or a column. The buses may be segmented (figure 3b).
- The length of routing paths for nearest neighbour connections.
- The number of routing channels through a rDPU (see figure 1b,c).
- Peripheral port location constraints to support particular communication architectures connecting the array to surrounding memory and other circuitry
- particular placements or groups (e. g. library items) can be frozen (modular mapping)

In addition to the structure of the array, parameters to control the annealing process may be specified, such as e. g. the starting temperature, the number of iterations, and the end temperature. Also, for each of these communication resources, a cost factor for a connection using this resource can be specified. As the mapper produces as output the same intermediate file format as the input, several annealing steps may be performed sequentially, e.g. a low-temperature simulated annealing after a normal mapping.

For each communication resource, as well as for the global bus, the costs for a connection can be specified, which together make up the cost function for the annealing.

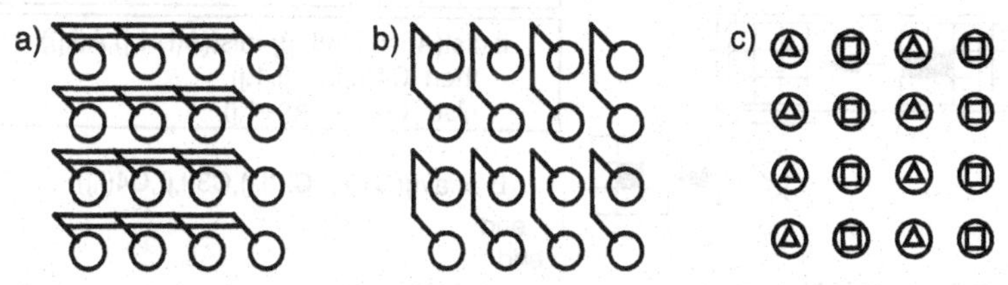

Figure 3: Architectural properties of KressArrays: a) example row buses; b) example column buses (segmented); c) different function sets in alternating columns

This way, specific resources can also be discouraged. Normally, one would assign the global bus a high cost factor, since those connections are slow per se, and, in contrast to routing through cells, by not supporting pipelining. If row or column buses are available, they would normally get a medium cost level, while nearest neighbour connections have the lowest cost, since being preferred. With this strategy, the mapper will eventually find a mapping, which does not use expensive interconnect. Such connect resources can then be removed for the next iteration step in the design space exploration process, leading to a more effective implementation of the KressArray.

5. Mapping Applications onto the KressArray

For reconfigurable computing, the use of fine grained FPGAs implies a large connectivity overhead because even each simple operation on dataword widths of 16 or 32 bit massively needs interconnect resources to be implemented on single bit PEs. In spite of approaches like carry chains in contemporary FPGAs[11], a coarse grained architecture involves a much smaller routing problem. That's why we have used a heuristic algorithms, which would require unacceptable computation times for fine grained FPGA mappings. For the KressArray, a simulated annealing approach has been chosen, which gives quite encouraging mapping results. Next subsection shows the efficiency of this approach by an example: mapping the datapath of an image processing filter.

5.1 A complex example: SNN Filter

The Symmetric Nearest Neighbour (SNN)-Filter is used as sharpening filter in object detection. In the following, a short introduction to this algorithm is given.The basic operation of this filter is shown in fig. 4. The algorithm calculates a pixel of the result image by considering a 3-by-3 neighbourhood of the original pixel. First, four neighbouring pixels of the central pixel in the original image are selected according to the illustration in fig. 4. The selections are based on the colour distance dist(a,b), which is calculated like shown at the bottom of the source code in the figure, involving colour separation of the pixels. After the selections, the resulting pixel is the average colour value of the four selected pixels.

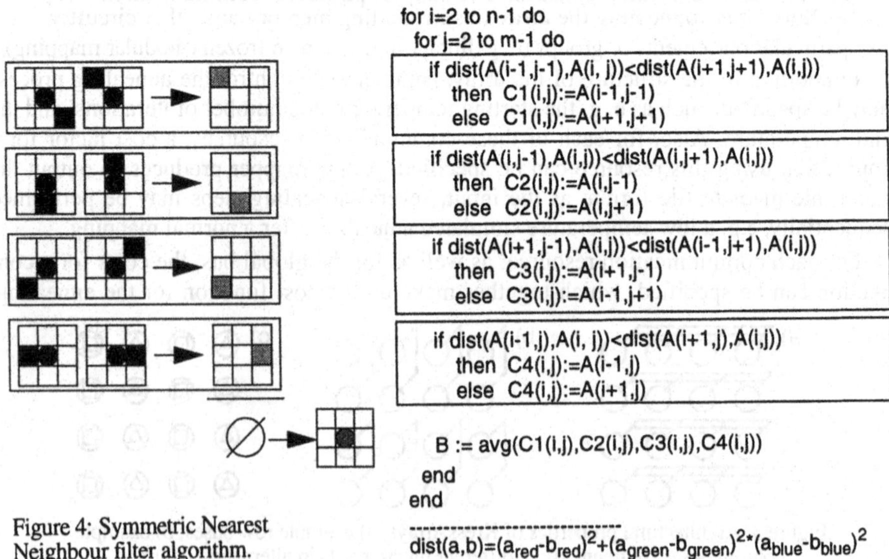

for i=2 to n-1 do
for j=2 to m-1 do
if dist(A(i-1,j-1),A(i, j))<dist(A(i+1,j+1),A(i,j))
then C1(i,j):=A(i-1,j-1)
else C1(i,j):=A(i+1,j+1)
if dist(A(i,j-1),A(i,j))<dist(A(i,j+1),A(i,j))
then C2(i,j):=A(i,j-1)
else C2(i,j):=A(i,j+1)
if dist(A(i+1,j-1),A(i,j))<dist(A(i-1,j+1),A(i,j))
then C3(i,j):=A(i+1,j-1)
else C3(i,j):=A(i-1,j+1)
if dist(A(i-1,j),A(i, j))<dist(A(i+1,j),A(i,j))
then C4(i,j):=A(i-1,j)
else C4(i,j):=A(i+1,j)
B := avg(C1(i,j),C2(i,j),C3(i,j),C4(i,j))
end
end
dist(a,b)=(a_red-b_red)^2+(a_green-b_green)^2*(a_blue-b_blue)^2

Figure 4: Symmetric Nearest Neighbour filter algorithm.

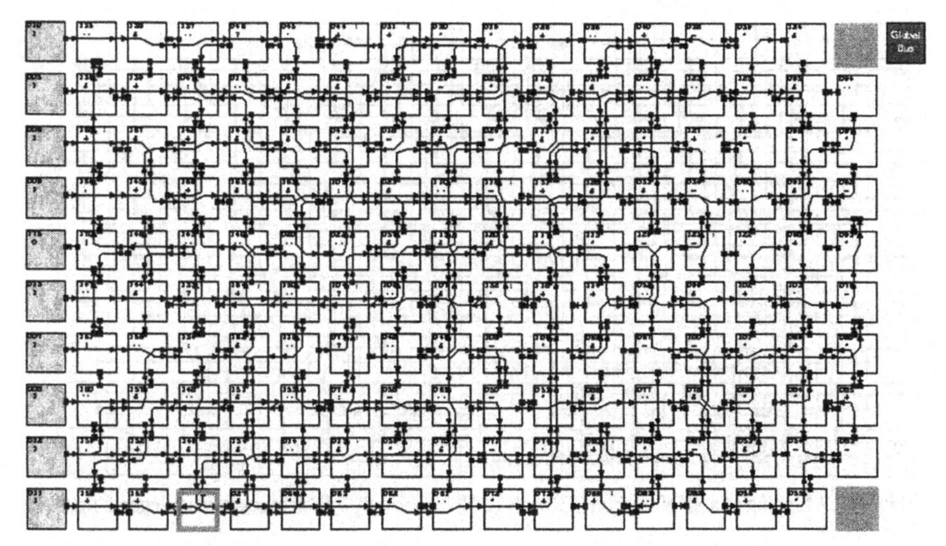

Figure 5: Mapping of a SNN filter on PEs with 8 NN links: 16 back buses (not shown, legend see figure 7).

The result of the first mapping is shown in fig. 5. The algorithm uses 157 PEs and has been mapped onto a 10-by-16 KressArray. In the chosen architecture each PE provides two bidirectional nearest neighbour connections to each of its four neighbour-PEs.

There are nine input ports and one output port, which were all put at the western side of the array, leaving the exact placement up to the annealer again (the output port is the fifth one from top). As the nearest-neighbour resources are used up in some places, additional global bus connections are necessary. These can be identified by the small bars in the upper right corner of the PEs. The mapping in fig. 5 uses 16 bus connections, which is a moderate value for a datapath of such a complexity. However, for the I/O, no global bus connections were needed. Also, the neighbour connections have a quite good utilization, which is also caused by the fact, that in this algorithm many output values of PEs are used multiple times. E. g. eight of the nine input values are used at four different PE inputs. The mapping took about 24 minutes on a Pentium-II 366-MHz PC.

The design space explorer enables the designer to map his algorithm on different architectures and to compare the resulting structures. Depending on the chosen KressArray architecture the mappings will differ in many points like the number of used PEs, the used communication resources, etc. With this information the designer can decide his compromise between complexity (and costs) of the base-architecture and the execution time.

In the example above an alternative architecture could be the same KressArray (10-by-16) with three nearest-neighbour connections *(NN links)* in each direction instead of two. Then the number of required PEs is still 157 because the number of operations remains the same and the operators provided by the PEs are also unchanged. But, due to the enhanced routing capabilities of the PEs, some connections which were realized by the global background bus *(back bus)* in the first mapping can now be routed using NN links. This explains why in the second mapping only seven bus connections are needed.

The SNN example raises an important question regarding the granularity of the configurable architecture. For the example mappings, the datapath description was fed directly in the DPSS system. However, the direct mapping is not always optimal. E.g. for the colour separation, two PEs were used, one for a shift operation and the next for

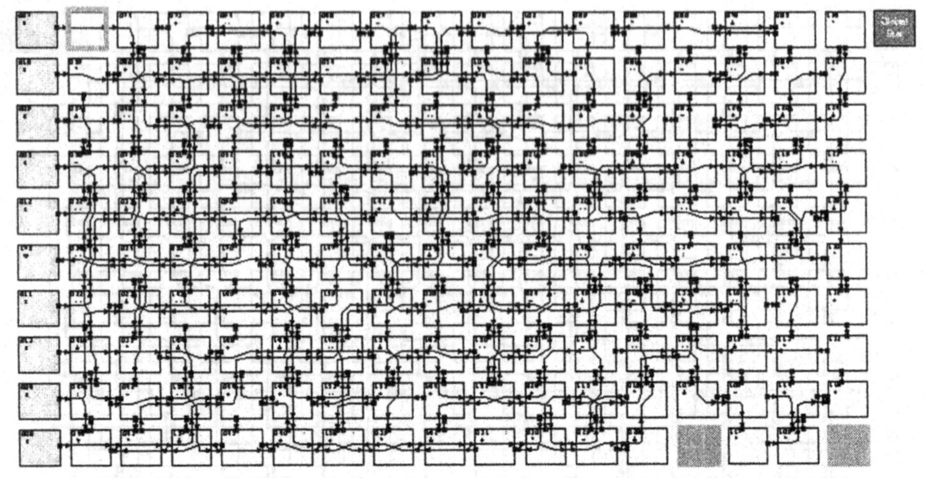

Figure 6: Mapping of a SNN filter on PEs with 12 NN links: only 7 back buses needed (legend: figure 7).

masking out the according value. It would not increase the complexity of the PEs to provide a function, which combines these two operations, resulting in less PE usage.

6. Conclusions

- rDPU not used
- used for routing only
- operator and routing
- port location marker

Figure 7: Legend.

We have introduced a design space exploration environment for the KressArray architecture family. The KressArray is a coarse-grained reconfigurable architecture, which allows much simpler application mapping than fine-grained FPGAs. A highly flexible configurable mapper has been introduced, which is capable of handling a wide variety of different architectures. Based on the mapper, the design space explorer allows the user to find the best suitable KressArray for a given application domain.

Literature

1. W. Mangione-Smith, et. al.: Seeking Solutions in Configurable Computing; IEEE Computer, Dec. 1997.
2. R. Hartenstein (invited paper): The Microprocessor is no longer General Purpose: why Future Reconfigurable Platforms will win; ISIS'97, Austin, Texas, U.S.A., Oct. 1997
3. R. Hartenstein (opening keynote): Next Generation Configware merging Prototype and Product; RSP'99, Int'l Workshop on Rapid Prototyping, Leuven, Belgium, June 3 - 5, 1998
4. E. Mirsky, A. DeHon: „MATRIX: A Reconfigurable Computing Architecture with Configurable Instruction Distribution and Deployable Resources", Proc. FPGAs for Custom Computing Machines, pp. 157-166, IEEE CS Press, Los Alamitos, CA, U.S.A., 1996.
5. E. Waingold et al.: „Baring it all to Software: Raw Machines", IEEE Computer 30, pp. 86-93.
6. C. Ebeling, D. Cronquist, P. Franklin: „RaPiD: Reconfigurable Pipelined Datapath", Workshop on Field Programmable Logic and Applications, FPL'96, Darmstadt, Germany, 1996.
7. R. Kress: „A Fast Reconfigurable ALUs for Xputers", Ph.D. thesis, Univ. Kaiserslautern, 1996.
8. A. Marshall et al.: A Reconfigurable Arithmetic Array for Multimedia Applications; FPGA'99, Int'l Symposium on Field Programmable Gate Arrays, Monterey, CA, U.S.A., Febr. 21 - 23, 1999
9. R. Kress et al.: A Datapath Synthesis System for the Reconfigurable Datapath Architecture; ASP-DAC'95, Makuhari, Chiba, Japan, Aug. 29 - Sept. 1, 1995.
10. J. Becker et al.: Parallelization in Co-Compilation for Configurable Accelerators; Asian South Pacific Design Automation Conference 1998 (ASP-DAC'98), Yokohama, Japan
11. N. N.: The Programmable Logic Data Book, Xilinx Inc., San Jose, California, 1998

Global Routing Models

Martin Daněk and Zdeněk Muzikář

Dept. of Computer Science and Engineering
Czech Technical University
Karlovo náměstí 13, CZ - 121 35 Praha 2, Czech Republic
e-mail: {danek,muzikar}@fel.cvut.cz

Abstract. A simplified FPGA global model supporting timing-driven placement and global routing is presented. A real Xilinx XC4000 family FPGAs environment was modelled this way and the model was used by timing-driven algorithms using an integrated approach [4]. The performance of this type of model compared to a complex global model [1], [2] was evaluated. The tests were based on real designs and MCNC benchmarks using an experimental system [3] interacting with Xilinx XACT design system.[1]

1 Introduction

To deal with FPGA placement and routing there are two possible approaches: algorithms can deal with a *detailed* or with a *global* FPGA model. The routing model is usually a graph representing realizable routes in the FPGA. A detailed model should be a one to one representation of the FPGA structure, whereas a global model reduces the number of edges and vertices used so as to lower placement/routing algorithm execution time. It is understandable that although global routing models are not as accurate as detailed models, they have to carry the information required by placement and global routing algorithms.

We have presented a complex global routing model [1], [2] inspired by Xilinx XC4000 [6] family FPGAs. This model supports integrated timing-driven placement and global routing algorithms [3], [4]. It describes both *placement* and *interconnection* possibilities of an FPGA environment and it supports *signal delay* estimation for any admissible net topology with a reasonable precision.

This model is implemented as an undirected graph and can be divided into four basic layers. The first layer contains configurable logic blocks, input/output blocks, magic boxes and single length line segments. The second layer consists of double length line segments, the third layer is formed by longlines, and the fourth layer shows global distribution network and global buffers. The model is rather complex (it reduces the number of objects with respect to a detailed model 30 times), but it preserves routing properties of a real device, and it has been proved to support both linear and Elmore-based signal delay estimation methods that can be used by timing-driven algorithms.

[1] This research is supported by the Czech Technical University grant no. IG 309907903 and by the Grant Agency of the Czech Republic grant no. GAČR 102/99/1017.

The rest of the paper is organized as follows: Section 2 describes the principles of an integrated timing-driven approach, Section 3 outlines the main idea of the simplified model, Section 4 describes the experiments and their results and in Section 5 we draw conclusions.

2 Integrated Timing-Driven Approach

Satisfying timing requirements of all signals involves optimizing the topology of block interconnections with respect to its delays. As it requires changes of block placement, the estimation of future interconnection paths during the placement process is necessary. A global routing simulation is a possible solution of this demand [5]. It aims to integrate the *placement* (or rather *replacement*) and *global routing* phases into a compound subproblem where both tasks are performed simultaneously.

The integrated timing-driven approach to a layout synthesis [3], [4] is based on the following assumptions:

- The layout technique is an iterative process where the (re)placement and global routing steps are overlapping subtasks.
- This iterative process of layout generation is controlled by a nonlinear complex cost function which takes into account both routability and timing requirements.

These principles place a crucial demand on the model used: it must describe logical *topology* of the FPGA device and it must provide *delay estimation* for all routed nets.

The complex model described in Section 1 satisfies these demands. As placement and global routing algorithms usually have quadratic or cubic complexity, it is a natural question whether it is possible to reduce further the amount of information contained in this model. This idea leads to the *simplified global model.*

3 Simplified Global Model

A direct way how to simplify a complex model is to replace the four levels or types of interconnect with just one type equivalent to single length line segments. The Elmore-based delay estimation cannot be held in this case. This model contains only one basic parameter - a modified linear delay per segment, which is based not only on a simple delay analysis, but it also reflects a statistically based distribution that describes the probability that a given type of line segment will be used. This model reduces the number of objects with respect to a complex model 3 times (e.g. ratio 90:1 with respect to a real device). It supports only a linear delay estimation method, which is in accordance with its simplicity.

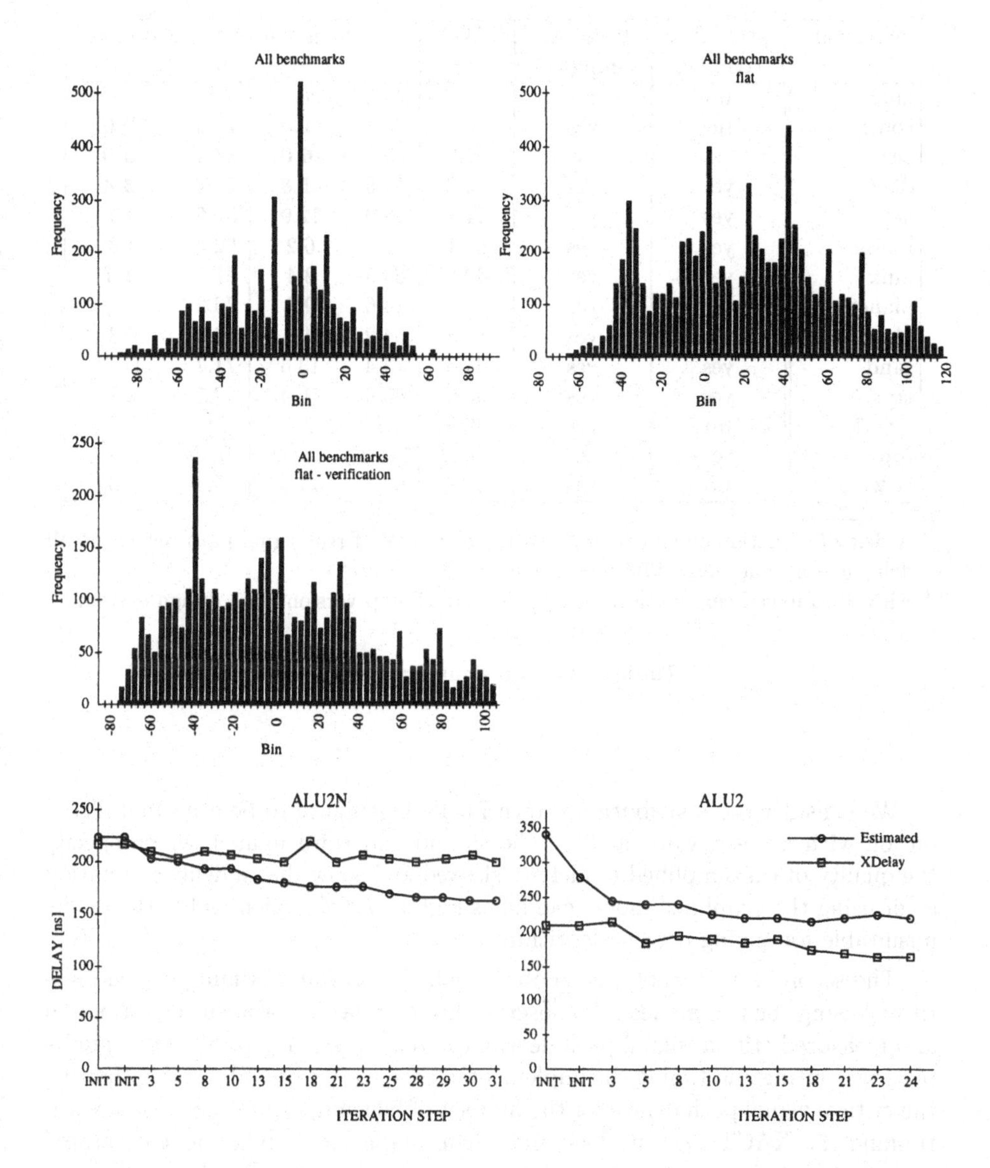

4 Experiments

We have performed two sets of experiments. The first set tells us about the quality of the routing topology of the two models. We took LCA files placed and routed solely by the Xilinx XACT design system, projected the delays generated by the XDelay program to the corresponding global routing model and compared them with the delays calculated by our delay estimator based on the linear delay model. The first were obtained for standard MCNC benchmarks and show the best relative delay estimation error histograms for the two models, the third histogram was generated using seven moderately-sized real world designs and is a sort of verification of delay parameters used in the simplified model.

benchmark	pessimistic estimation[a]		XACT[b]	complex model		simplified model	
	complex	simplified	ns	ns	%	ns	%
alu2	**no**	yes	166.1	200.0	20.4	178.1	7.2
comp	**no**	yes	76.6	75.5	**-1.4**	74.6	**-2.6**
cu	**no**	**no**	37.9	35.6	**-6.0**	34.5	**-8.0**
dk16	yes	yes	55.6	54.6	**-1.8**	53.7	**-3.4**
keyb	yes	yes	51.8	58.0	12.0	56.6	9.3
minm	yes	yes	88.1	93.6	6.2	92.3	4.8
mux	yes	yes	62.2	64.3	3.4	61.5	**-1.1**
planet	**no**	yes	68.3	74.6	9.2	73.2	7.2
sla	yes	yes	51.8	54.2	4.6	52.0	0.3
sand	yes	yes	64.6	73.4	13.6	71.7	11.0
styr	yes	yes	65.9	65.0	**-1.4**	67.7	2.7
term1	**no**	yes	99.9	109.3	9.4	108.4	8.5
top	yes	yes	109.2	115.3	5.6	107.5	**-1.6**
ttt2	**no**	yes	62.7	62.7	14.7	65.4	4.3

[a] A delay estimation should provide an upper bound of real signal path delays, so all delay estimations should be pessimistic.
[b] All values have been measured using the XACTStep version 5.2.1 design system.

Table 1. Critical signal path delays

We consider the distribution shown in the first figure to be a normal distribution with a mean value at 0%. The second and third figures, which specify the quality of the simplified model are skewed and show that a delay estimation error using the simplified model can be as high as 200%, which makes the model unsuitable for timing-driven algorithms.

The second set of tests was to observe the behaviour of timing-driven algorithms using the two models. We observed the correlation between the estimated and measured critical signal path delays, the reference being provided by guide-placing, routing and timing the implemented designs. We provide a table showing the critical signal path delays for the first set of benchmarks obtained both when running the XACT system alone and when guiding it by placement generated by our timing-driven algorithm. The two graphs describe the development of critical signal path delays during iterations. You can observe that when using the complex model the estimated delay is less than the measured delay, which we attribute to the impossibility to guide the Xilinx detailed router (ppr). It seems to us that the general strategy adopted by the Xilinx detailed router is to start routing using single length lines and only in following improvement stages to use double length lines and longlines. Our global router is strictly timing-driven and routing resources are chosen according to a complex non-linear function described in [4]. The different approaches together with the impossibility to guide the detailed router might explain why the correlation between the estimated and measured signal delays is not better when using the complex model.

5 Conclusion

We have constructed a relatively complex global routing model that reduces the number of objects in a ratio 30:1 and that still preserves routing properties of a real device. The model supports a reliable signal delay estimation. By simplifying this model we have got a model that reduces real world objects 90 times. We have checked that this model significantly decreases the quality of signal delay estimation and therefore it is not suitable for timing-driven placement and routing algorithms.

We have used both models in our timing-driven placement and routing algorithms and observed the correlation between the estimated and measured critical signal path delays. Better results were obtained using the simplified model, which we attribute to the loose coupling between our global router and the vendor detailed router in connection with less degree of freedom provided by the simplified model.

References

1. Daněk,M.: *FPGA Analysis*, Master thesis - CTU Prague, Faculty of Electrotechnical Engineering, Dept. of Computer Science and Engineering, May 1997.
2. Daněk,M., Servít,M.: *Xilinx XC4000 Global Routing Model and Signal Delay Estimation*, DMMS'97 Workshop, Budapest, 1997, pp 213-222.
3. Muzikář,Z.: *Timing-Driven Layout Synthesis for FPGAs*, PhD thesis - CTU Prague, Faculty of Electrotechnical Engineering, Dept. of Computer Science and Engineering, October 1997.
4. Servít,M.Z., Muzikář,Z.: *Integrated Layout Synthesis for FPGA's*, in Hartenstein,R.W., Servít,M.Z. (Eds.), Field-Programmable Logic, Springer-Verlag, 1994, pp. 23-33.
5. Tomkevičius,A., Muzikář,Z., Servít,M.: *Integrated Approach to Placement and Global Routing in Gate Arrays*, Research report DC-94-04, CTU Prague, Faculty of Electrotechnical Engineering, Dept. of Computer Science and Engineering, 1994.
6. *XILINX - The Programmable Logic Data Book*, San Jose, 1994.

Power Modelling in Field Programmable Gate Arrays (FPGA)

*Andrés Garcia, **Wayne Burleson, *Jean-Luc Danger

garcia@enst.fr, burleson@ecs.umass.edu, danger@enst.fr
*Ecole Nationale Supérieure des Télécommunications. Paris, France.
** University of Massachusetts. Amherst. MA. USA.

Abstract. This paper presents a power consumption model for FPGAs based on measurements. This model will permit us to optimize power consumption on FPGAs using existing architectures, as well as helping direct the design of new power-sensitive FPGA architectures.

1 Introduction

Power consumption is an important constraint in VLSI design. This factor is becoming increasingly critical because of the rapid growth of portable wireless and battery-powered applications. During the last years, integrated circuit designers have focused their efforts on increasing the clock frequency and gate density of systems at the expense of an increase in power consumption. The use of FPGAs is increasing because they now have adequate resources to satisfy high-speed systems and hardware cost constraints. FPGA implementation allows the construction of rapid prototypes reducing development times and board area. However, it is not at all clear whether FPGAs are an appropriate technology for truly low-power applications. This paper presents a model of power consumption of FPGAs based on measurements. The model shows the influence of the Netlist and it will be used to develop some power optimization techniques using commercial devices.

2 CMOS Power Consumption

The power dissipated in any logic circuit can be decomposed into two components: static and dynamic. The static power is the power dissipated while the clock is not changing (i.e. idle), and the dynamic power represents the power consumed during transitions. Static power represents a very small percentage of global power consumption of almost all CMOS devices so for this study it will be ignored. The majority of power consumption comes from the activity of the gate. This dynamic power is dominated by the charge and discharge of the load and parasitic capacitances during switching. If we just assume a purely capacitive model, the power consumed by a gate switching F times per second is:

$$P_{DYN} = C_L * V_{DD}^2 * F \tag{2.1}$$

For a CMOS device that contains several gates, the dynamic power consumption can be represented using the following equation:

$$P_{DYN} = V_{DD}^2 * F * \sum \alpha_i * C_i \tag{2.2}$$

Where Ci is the load capacitance of node i and αi is the activity rate in node i. Considering the equation (2.2), we can assume that the power consumption of CMOS devices depends on the following factors: Frequency, Load Capacitance, Activity rate, and Power Supply.

3 Brief Description of FLEX10K and XC4000E FPGAs

The Flex10K family is a coarse-grained FPGA based on CMOS SRAM programmable technology. The Flex architecture consists of a matrix of logic and memory blocks (embedded array block) within a network of continuous horizontal and vertical routing channels (called Fast Tracks) that traverse the entire device. **Figure 3.1** shows a simple view of the internal architecture of a Flex10K device. The Logic Array Block (LAB) contains eight Logic Elements (LE). Inside the LEs, a 4-input look-up table (LUT) allows the computation of any 4-input function (logic or arithmetic). Each LE contains a DFF (D-type Flip-Flop) that could be used or disabled as needed. The I/O cells contain a bi-directional I/O buffer and a DFF that can be used either as an input or output register.

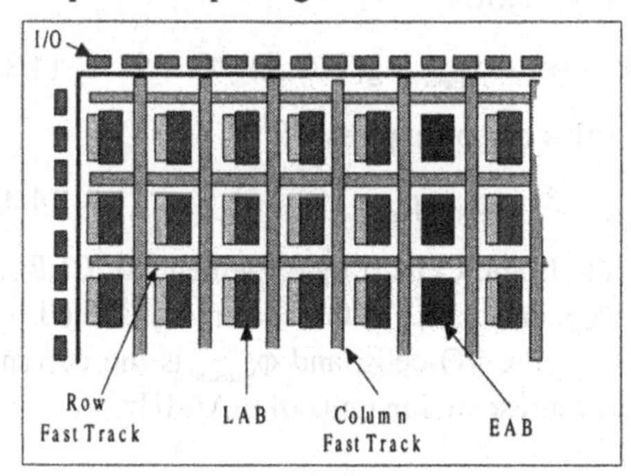

Fig. 3.1. The Flex10K architecture.

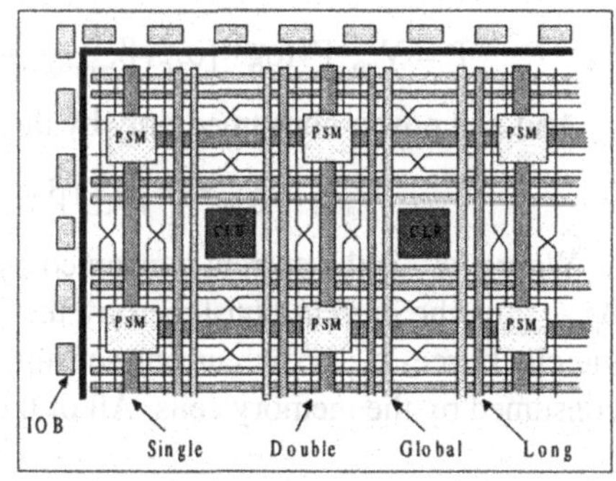

Fig. 3.2. The XC4000E architecture

Figure 3.2 shows the internal architecture of a Xilinx XC4000E. This device is a fine-grained FPGA based on CMOS SRAM programmable technology. It consists of a matrix of CLB (Configurable Logic Block) interconnected by a hierarchy of routing resources, and surrounded by a perimeter of programmable I/O blocks. Two 4-input LUT (F and G), one 3-input LUT (H), and two DFF compose the CLBs. The Function Generators F, G and H permit the CLB to be configured as Memory. The I/O blocks

(IOB) can be configured for input, output or bi-directional signals. Each IOB has two DFF, one to memorize the input signal and the other to memorize the outgoing signal.

In general, we can identify the following internal sub-elements of a FPGA: The Logic Cells, that are formed by LUTs and Registers; the Interconnect Resources; the I/O cells; the Clock buffer; and the Embedded or distributed Memory.

4 Study of Power Consumption in FPGAs

In this section we present a model of power consumption in FPGAs based on the previous analysis and on the measurements results. The following equation summarizes the parameters to be considered in the power analysis:

$$P = V_{DD}*F*tog*g(netlist, T°C, K) \quad \textbf{(4.1)}$$

Where F is the frequency, tog is the activity rate, and g is a function of the room temperature, the technology (K), and the internal resources used (netlist). This function is represented in mA/MHz. Load capacitance (C_L) considered in equation (2.1), and (2.2), is contained into the netlist factor. Our work only analyzes the power consumption from the netlist. The other factors will be considered constant. We propose to use simplest designs with a fixed toggling rate, and change only the number of internal resources in the design.

The idea is to increase the number of one internal element and keeping constant the others. This methodology (called incremental) allows to obtain a more accurate value of the power consumed by each internal element. Using equation (4.1), we can define the power consumption of the netlist as follows:

$$P = V_{DD}*F*tog*\,[\alpha_{LC}+\beta_{Interconnect}+\delta_{CLK}+\varepsilon_{I/O}+\varphi_{Memory}]*g(T°C,K) \quad (4.2)$$

And, the power consumed only by the netlist can be expressed:

$$P_{netlist} = V_{DD}*F*\,[\alpha_{LC}+\beta_{Interconnect}+\delta_{CLK}+\varepsilon_{I/O}+\varphi_{Memory}] \quad \textbf{(4.3)}$$

Where: α_{LC} is the current consumed by the Logic Cells (Logic Elements or CLBs), $\beta_{interconnect}$ is the current consumed by the interconnect, δ_{Clk} is the current consumed by the clock tree, $\varepsilon_{I/O}$ is the current consumed by the I/O cells, and φ_{Memory} is the current consumed by the memory cells. All of them represented in units of mA/MHz.

5 Measurement Methodology

In this section, we describe the incremental methodology used to determine the contribution of each internal element to the total Power Consumption. The devices used in this work are an Altera's Flex10K100-BGA-504, and a Xilinx's XC4003E-PC-84.

5.1Interconnect Resources and Look-Up Tables

Figure 5.1 shows a design used to estimate the power consumed by the interconnect. First at all, we measured the current consumed by two adjacent LUT. Then using the Floorplan Editor from MAX+PLUSII and the EPIC Editor from Foundation, we changed the distance between both LUTs to increase the number of interconnect resources.

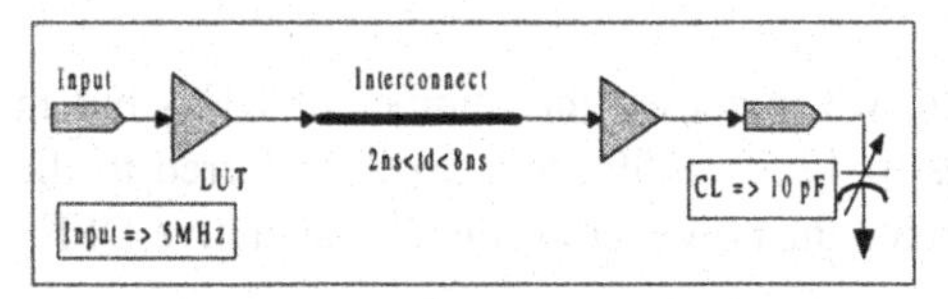

Fig. 5.1. Test of the interconnect

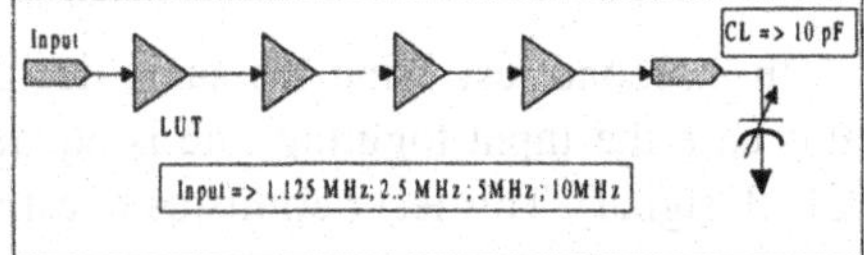

Fig. 5.2. LUT chain

Figure 5.2 shows the design used to estimate the power contribution of LUTs. In this case, we increased only the number of LUTs to obtain the increment of current.

5.2 Inputs and Outputs

Using a chain with a fixed number of LUTs (**fig. 5.3**), we tested the I/O cells configured as Output. In this case, we increased the number of outputs.

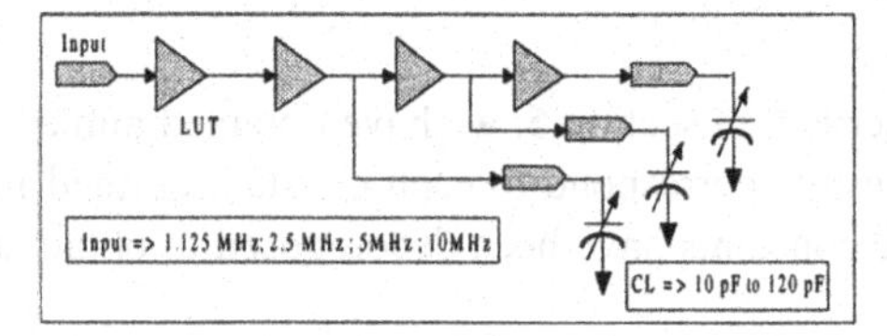

Fig. 5.3. Outputs

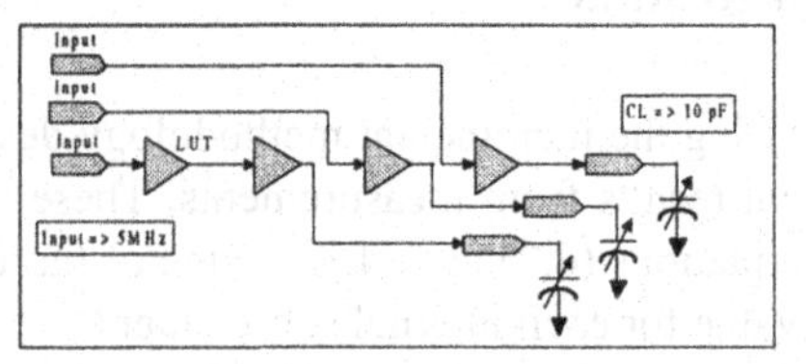

Fig. 5.4. Inputs

The load capacitance has been changed for the same exercise to verify its influence and to verify our results. Then, we increased the number of Inputs of the same design to estimate the influence of an I/O cell configured as Input (**fig 5.4**).

5.3Clock Tree and DFF

Figure 5.5 was used to obtain the power consumed by the Clock tree and the DFFs. The first test consists on an increment of DFFs with an input toggling rate equal to zero. In this case only the Clock tree will consume.

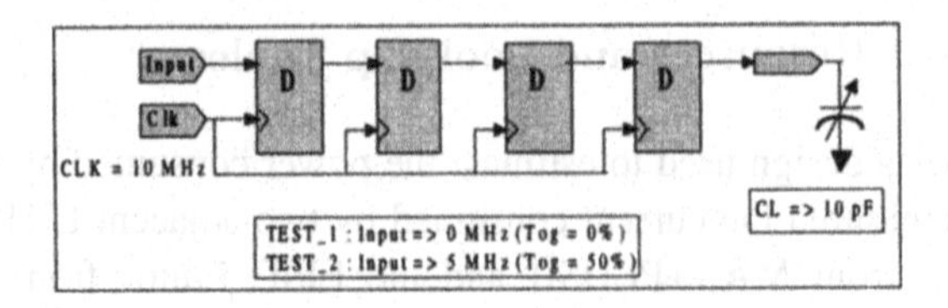

Fig. 5.5. Test of the Clock tree

In a second test using the same design, we increased the number of DFFs but in this case the input toggling rate is equals to 12.5%, 25%, and 50 % (referred to the Clock signal). This test permits us to estimate the power consumed by an active DFF.

5.4 Memory

We use simplest memory structures in order to measure its power contribution, in this case, we increased the size of the memory (increasing the size of the address bus) and keeping constant the size of the word. We used RAM and ROM structures. For the Flex10K device, we increase also the number of EAB. In XC4000 device, we increase the number of CLB configured as memory (64 CLBs correspond to 1 EAB).

6 Results

Using the incremental methodology described in section 5, we have obtained numerical results from measurements. These results correspond to each constant defined in equations (4.2) and (4.3). Some of these constants have been decomposed to obtain a value for each internal sub-element.

6.1Interconnect Resources and Loop-Up Tables

Equation (6.1) represents the power consumed by the interconnect:

$$P_{interconnect} = V_{DD}*F*\beta_{interconnect} \quad \textbf{(6.1)}$$

Where $\beta_{interconnect}$ can be decomposed for both Flex10K and XC4000E devices as follows:

$$\beta_{Flex} = \beta_a N_{half\_Fast\_Track} + \beta_b N_{full\_Fast\_Track} + \beta_c N_{Columns} \quad \textbf{(6.2)}$$

$$\beta_{XC4000E} = \beta_a N_{Direct\_Paths} + \beta_b N_{Single\_lines} + \beta_c N_{Double\_lines} + \beta_d N_{Long\_lines} + \beta_e N_{Global\_lines} + \beta_f N_{PMS} \quad \textbf{(6.3)}$$

Where N represents the number of resources contained in the netlist and β_i are constants in mA/MHz that correspond to each sub-element. The results obtained are presented in tables 6.1 and 6.2:

Table 6.1. Flex10K100 interconnect

Element	P (mW/MHz)	I (mA/MHz)
Half F.T.	0,115	0,023
Full F.T.	0,19	0,038
Column	0,18	0,036

Table 6.2. XC4003E interconnect

Element	P (mW/MHz)	I (mA/MHz)
Direct Paths	0,09	0,018
Single	0,07	0,014
Double	0,08	0,016
Long Lines	0,4	0,08
Global	0,35	0,07
Switch Box	0,01	0,002

Inspired in equation (6.1), we can represent the power consumed by look-up tables as follows:

$$P_{LUT} = V_{DD}\alpha_{LCELL} \quad \textbf{(6.4)}$$

Where α_{LCELL} can be decomposed in two components:

$$\alpha_{LCELL} = \alpha_a N_{LUT} + \alpha_b N_{DFF} \quad \textbf{(6.5)}$$

$$\alpha_{CLB} = \alpha_a N_{4\_input\_LUT} + \alpha_b N_{3\_input\_LUT} + \alpha_c N_{DFF} \quad \textbf{(6.6)}$$

Where N is the number of LUTs or DFF used in the design and α_i are constant in mA/MHz. In this section we only show the power consumed by LUTs and, in section 6.3, we will present the power consumption of DFFs. Table 6.3 shows the results obtained for both devices:

Table 6.3. Look-Up Tables

Device	Element	P (mW/MHz)	I (mA/MHz)
FLEX10K	4-input LUT	0,15	0,03
XC4000E	4-input LUT	0,1	0,02
XC4000E	3-input LUT	0,075	0,015

6.2 Outputs and Inputs

From equation (4.3), we can isolate the Power consumption of I/O cells configured as input or as outputs. It can be represented as follows:

$$\varepsilon_{I/O} = \varepsilon_a N_{inputs} + \varepsilon_b N_{outputs} \quad \textbf{(6.7)}$$

Where N is the number of Inputs or Outputs used in the design and ε_a, ε_b are constants in mA/MHz. The following tables show the results obtained for both Altera and Xilinx devices:

Table 6.4. Outputs

Device	P (mW/MHz*pF)	I (mA/MHz*pF)
FLEX10K	0,065	0,013
XC4000E	0,028	0,0056

Table 6.5. Inputs

Device	P (mW/MHz)	I (mA/MHz)
FLEX10K	0,456	0,0912
XC4000E	0,225	0,045

6.3Clock Tree and DFFs

The equation that corresponds to the Clock tree contains basically the clock buffer and the interconnect that serve to distribute the clock signal. This equation for a Flex 10K can be expressed as follows:

$$\delta_{CLK} = \delta_a N_{CLK\_buffer} + \delta_b N_{mux\_LAB} + \delta_c N_{MUx\_LE} \tag{6.8}$$

The following equation corresponds to the power contribution of the clock tree in a XC4003E:

$$\delta_{CLK} = \delta_a N_{CLK\_buffer} + \delta_b N_{CLB\_mux} + \delta_c N_{interconnect} \tag{6.9}$$

The following tables show the numerical values of these sub-elements:

Table 6.6. Flex 10K100

Feature	Element	P (mW/MHz)
CLOCK	Clk_buffer	12
	Mux_LAB	0,05
	Mux_LCELL	0,005

Table 6.7. XC4003E

Feature	Element	P (mW/MHz)
CLOCK	Clk_buffer	4
	CLB_mux	0,1

Equation (6.5) contains the power contribution of the DFFs. The following table shows the results obtained using both Altera and Xilinx devices:

Table 6.8. DFFs

Device	P (mW/MHz)	I (mA/MHz)
FLEX10K	0,12	0,024
XC4000E	0,21	0,042

6.4 Memory

The following equation represents the power contribution of the Altera's EABs and Xilinx's CLB configured as memory:

$$\varphi_{Memory} = \varphi_a N_{EAB} \tag{6.10}$$

$$\varphi_{Memory} = \varphi_a N_{CLB\_configured\_as\_Memory} \tag{6.11}$$

Where N is the number of used EABs or CLB configured as memory, and φ is a constant in mA/MHz. An Altera's EAB can be used to build a 2K*1 memory, on the other hand, a CLB configured as memory can serve to build a 16*1 memory. In order to compare both resources, we implemented a 2k*1 memory block. Table 6.9 shows our results:

Table 6.9. 2k*1memory in Altera and Xilinx

Device	Memory size	Resources	P (mW/MHz)
FLEX10K	2K*1	1 EAB	1,44
XC4000E	2K*1	64 CLBs	12,16

Memory implemented in XC4000 devices represents some disadvantages compared with the use of EABs in Flex10K devices. The use of embedded memory cells is useful to reduce power. On the other hand, The implementation of memory using CLBs increases power consumption and it sacrifices also logic resources.

6.5Global Results

The following graphics contains the five elements of a FPGA that consumes and its percentage. This representation of the power distribution inside a FPGA will be useful to understand the comportment of this kind of device and to identify the elements that have to be optimized. The following example represents a hypothetical case in both Flex10K100 and XC4003E with internal activity rate average equals to 30%, CL equals to 12 pF, and using the 80 percent of all internal resources; the power consumption is distributed as follows:

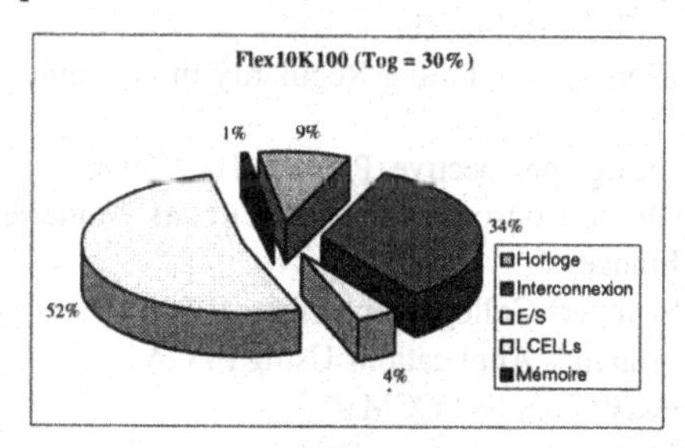

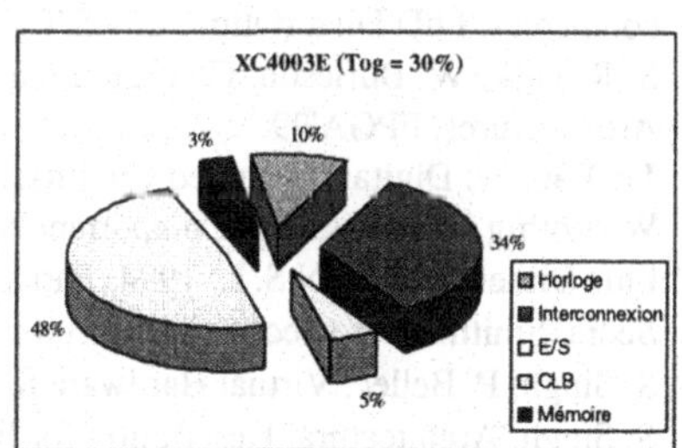

Fig. 6.1. Power distribution with a = 30%

In the XC4003E case, the 25% of CLBs were configured as RAM blocks (320*1). We can observe that most part of the total power consumption comes from the LCELL or the CLB (LUT + DFF). The second most important element in power consumption of these devices is the interconnect resources. The power consumed by a CLB configured as Memory is higher than an EAB from Flex10K. Using the model to estimate the power consumption of a design with a fixed (or known) toggling rate, we can obtain results closer to the measurement results. It means that the results obtained from this study could be useful to design techniques of power reduction.

7 Conclusions

Measurement results have shown that the clock buffer represents less than 10 % of the global power consumption. They shown also that most part of power consumption comes from the Logic Cells (DFFs and LUTs) and the interconnect. The use of 4-input LUT to generate memory structures increases power consumption. Embedded memory represents a good solution to reduce power since it allows to build low power memory blocks and big glue logic blocks also. Another solution is the use of smart place a route tools that allow a more effective place and route process, minimizing critical paths and optimizing surface. Recently, other authors like Vaughn Betz from the University of Toronto have developed some methods and tools to optimize the place and route. This work could be useful to decrease power consumption. Finally, partial reconfiguration could be a good solution to reduce power consumption in applications based on FPGAs.

References

1. Altera. Flex10K Data Sheet and Max+PlusII. Altera Corporation. 1998.
2. E. Boemo, G. Gonzalez de Rivera, S. Lopez-Buedo, J. Meneses. Some Notes on Power Management on FPGA-based Systems. Proceedings of FPL'95.
3. S. D. Brown, R. J. Francis, J. Rose, Z. G. Vranesic. Field Programmable Gate Arrays. Kluwer Academic Publishers. 1992.
4. T. Callahan, P. Chong, A. DeHon, J. Wawrzynek. Fast Module Mapping and Placement for Datapaths in FPGAs. Proceedings FPGA'98.
5. G. Kélemen. Conception des circuits intégrés pour la basse consommation : Méthodes comparées. PhD Dissertation. E.N.S.T.. 1997. Paris, France.
6. S. R. Park, W. Burleson. Configuration Cloning: Exploiting Regularity in Dynamic DSP Architectures. FPGA'99.
7. Jan Rabaey. Digital Integrated Circuits; a design perspective. Prentice Hall. 1996.
8. W. Röthig. Modélisation Comportamentale de Consommation des Circuits Numériques. PhD Dissertation. E.N.S.T.. 1994. Paris, France.
9. Sedra, Smith. Microelectronics Circuits. Saunders College Publishing. 1989.
10. S. Singh, P. Bellec. Virtual Hardware for graphics Applications Using FPGAs.
11. S. Singh. Architectural Description for FPGA Circuits. FCCM'95.
12. K Skahill. VHDL for programmable Logic. Addison Wesley. 1996.
13. C. C. Wang, C. P. Kwan. Low Power Technology Mapping by Hiding High-Transition Paths in Invisible Edges for LUT-based FPGAs. ISCAS'97.
14. Vaughn Betz, J. Rose. VPR: A New Packing, Placement and Routing Tool for FPGA research. Proceedings FPL'97.
15. N. Weste, K. Eshraghian Principles of CMOS VLSI Design. A system perspective. Mc Graw Hill. 1995.
16. Xilinx. XC4000E/EX/XL Data Sheet and Xilinx Foundation. Xilinx Corporation. 1998.

NEBULA: A Partially and Dynamically Reconfigurable Architecture

Dinesh Bhatia, Kuldeep S. Simha, PariVallal Kannan

Design Automation Laboratory
Department of ECECS
University of Cincinnati
Cincinnati, OH 45221–0030
dinesh@ececs.uc.edu

Abstract. Field programmable gate array based reconfigurable computing machines have successfully demonstrated their performance advantages for a large class of compute intense problems. In this paper, we present a partially and dynamically reconfigurable architecture, NEBULA, which reduces the inherent overheads of reconfigurable computing. NEBULA is built around Xilinx 6200 family of devices. NEBULA is supported by a collection of drivers and CAE tools that allow rapid implementation of algorithms in the reconfigurable core. We also present experimental results that highlight the performance capabilities of NEBULA. With very small reconfiguration and communication overhead, we are able to use NEBULA to speed up a large class of applications.

1 Introduction

General purpose fixed instruction set processors are universally used for day-to-day computing applications. Most of the high performance microprocessors implement a mix of RISC and CISC concepts [4] and are usually clocked at very high frequencies. In spite of the vast improvement in their performance in recent years, there is a class of very important applications which do not show adequate speedup when implemented on these general purpose processors. Such applications can be targeted by custom ASICs, fine-tuned for the specific problem. Reconfigurable computers provide the option of implementing these applications in hardware yielding desired performance levels.

Reconfigurable computing systems can be found in a variety of configurations – stand-alone systems as in logic emulators, application specific computing systems, and as co-processors to a host system. SPLASH[1] was the earliest example of FPGA based custom processor. Some of the well known co-processing units developed so far are the RACE [5] from University of Cincinnati, Hotworks [3] from Virtual Computers Corp., and ACE [6] from TSI-Telsys. These computing machines use a standard bus interface to communicate with the host processor.

In this paper we present NEBULA, a reconfigurable co-processor architecture developed at the University of Cincinnati. The architecture attempts to provide a stable reconfigurable co-processor system with a high bandwidth interface to the host, low reconfiguration times and support for partial reconfiguration. It reduces the configuration overheads by using partially reconfigurable XC6200 family [2] FPGAs from Xilinx.

The architecture also reduces the data communication overhead by using the high speed PCI bus, operating at 33MHz for the host interface. In the following sections, we describe the architecture and design of NEBULA. Also, we compare the performance of several general purpose computers with the performance of NEBULA for the execution of an image processing application.

2 NEBULA: Hardware Architecture

NEBULA is a partially reconfigurable co-processor fabricated as a PCI card. The card can be attached to any PC or a workstation featuring the PCI bus and a spare PCI slot. In our architecture, the NEBULA card is attached to an Intel based PC running Linux. The choice of Linux was influenced by the fact that Linux is an open source operating system which makes device driver development very easy. The architecture of NEBULA is shown in Figure 1. The main components of the board are

- The Reconfigurable Units : Two Xilinx XC6264 FPGAs R1 & R2
- On board Memory : 2MB of 32bit wide fast SRAM divided into two blocks M1 & M2
- Host Interface : PCI Controller implemented on a Xilinx 4020 FPGA using Xilinx LogiCORE PCI.
- Switch Logic : Switches that configure the board in one of 4 memory access modes, described later.
- Clock Generator : 139kHz-100MHz programmable clock generator, which can be used to clock user design on the FPGAs.

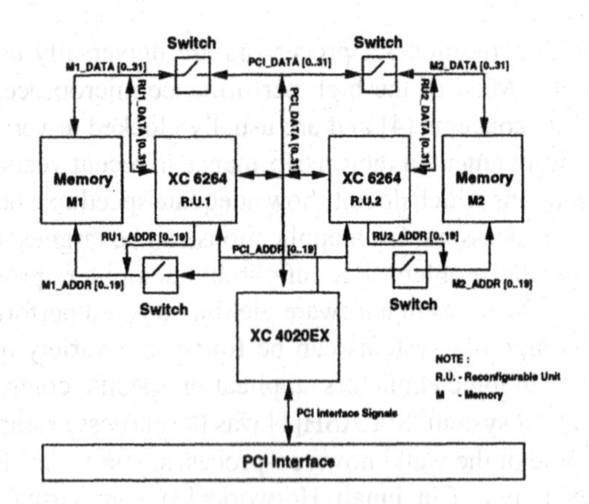

Fig. 1. *NEBULA Architecture*

2.1 NEBULA: Reconfigurable Units

The NEBULA board has a large reconfigurable logic area equivalent to 128k-200k gates, provided by the two XC6264 devices from Xilinx. The two XC6264 FPGAs

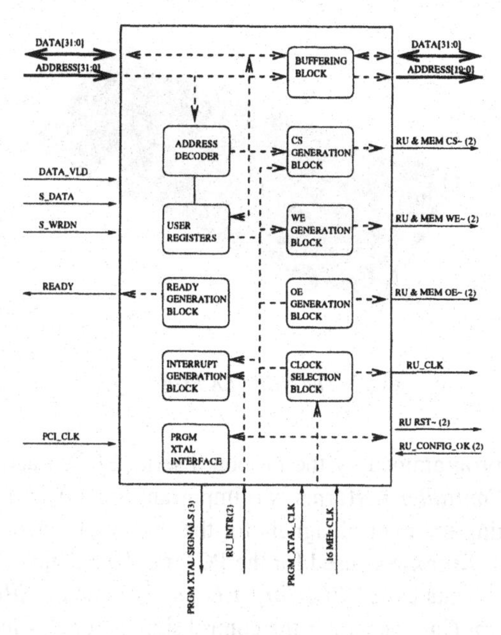

Fig. 2. *Main Controller*

are indicated by R.U.1 and R.U.2. The XC6264 devices have many unique features like partial reconfiguration, memory mapped configuration & state space and a very fine grained architecture. The cells in the XC6200 devices are configured by writing into a portion of on-chip memory called the configuration space. This configuration memory space is memory-mapped to the host and appears to a host program as a contiguous memory buffer. The memory mapping is performed by the device driver and is transparent to the user code. Similar to the configuration space, any registers in the user's design are also available to the host program through the device driver and is called the State space. Through this interface, the host program can read or write to the registers, flip flops, and latches in the user's design, configured on the FPGA.

2.2 NEBULA: On Board Memory

The memory block comprises of 2 MB of fast static RAM, divided into two banks of 1MB each (M1 and M2) which act as local memory units for the two reconfigurable units(RUs). The memory units can be accessed by either the host processor through the PCI interface or the RUs. This is done by enabling or disabling the switches shown in the Figure 1. The control switches are accessible to the host through an interface on the control FPGA. The NEBULA library provides an abstraction to these switches. The memory blocks can be accessed as 8/16/32 bit wide words from the FPGAs and from the host.

2.3 NEBULA: Host Interface

The board has a PCI bus interface that operates at 33 MHz. The PCI interface logic along with the NEBULA control logic resides in a XC4020EX FPGA. The XC4020

Fig. 3. *The NEBULA PCB board*

FPGA after being programmed by the on-board serial PROM acts as the *Main Controller*. The *Main Controller* performs two important functions: handling the PCI interface and generating the control signals for the on-board functional units. The PCI LogiCORE [7] from Xilinx was used for the PCI interface logic. The Main Controller has two functional blocks called *PCI Target Logic Core* and *NEBULA Control Core*. The *NEBULA Control Core* generates the control signals for all functional units on the board.

Figure 2 shows the different functional blocks within this *NEBULA Control Core*. A set of user programmable registers is used to control various events on the board. It includes the *Chip Select*, *Write Enable* and *Output Enable* signals for the RUs and memory units. It also has a *Clock Selection* block which selects one of PCI clock, 16 MHz clock, and the programmable clock to be used as one of the global clock of the RUs. An *Interrupt Generation* block generates an interrupt to the host processor when a interrupt-generating event occurs on the board. The *Ready Generation* block generates wait states for memory and RU read operations. Figure 3 illustrates the board design of the NEBULA PCB.

3 Software Support

The NEBULA board is supported by a powerful and complete software and hardware suite. The software include a fully tested and portable Linux device driver and NEBULA_library, which contains the various routines required for the host program to control the NEBULA board. Notable features of the library include routines to open and close the board, configure the 6200 devices, program the clock generator, access the on board memory, access the user (state space) registers on the 6200 devices, control the access mode and interrupt servicing. The hardware library includes schematic components in Viewlogic for the memory interface, clock interface and a set of commonly used operators not available with the Xact6000 library like multipliers dividers, etc. Presently the physical design is through the Xact6000 tool supplied by Xilinx, however a replacement tool suite is being developed.

The library based approach can be visualized as a library of functions or objects provided in a high level language with various levels of abstraction for the user. The

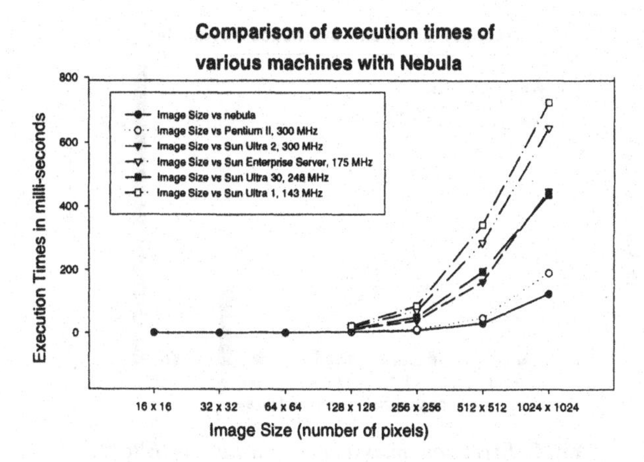

Fig. 4. *Execution times for the Smoothing filter on different machines*

user can use the objects provided at the higher levels of abstraction or can start building new objects from the lower levels of abstraction according to specified requirements. Our approach is to provide a complete prototyping environment to the user thereby allowing customization to meet the requirements.

4 Case Studies and Performance

Time-consuming complex software operations could be implemented on NEBULA to get a better performance. However all complex applications cannot be mapped onto the board. *Only operations in which multiple computational units could be parallely executed on hardware benefit from hardware execution.* Rich benefits in performance can be achieved by using bit-slice designs since such structures are readily mappable on the XC6200 family of FPGAs. Keeping these considerations in mind, we designed a bit-sliced implementation of a *Image Smoothing Filter*. The function of this application is to create a blurring effect on a given grey-scale image. A *Smoothing* or *Blurring filter* is also implemented by a neighbor averaging methodology. In this methodology, the average value of grey scales of all the neighboring pixels is used to replace the grey scale of a given pixel. The area required for processing a basic averaging step is very small. This allows us large scale parallelism. Thus, when the reconfigurable unit is programmed, it can carry out filtering operation on a 20×20 image in a single clock cycle.

We executed the hardware implementation of the *Smoothing Filter* application on NEBULA for different image sizes. We then executed the software implementation of the same application on different computing platforms. The execution on NEBULA has the configuration overheads and the device driver overheads. In spite of these overheads, the performance on NEBULA was found to be orders of magnitude better than the execution on other high performance processor based systems. The Figure 4 shows the

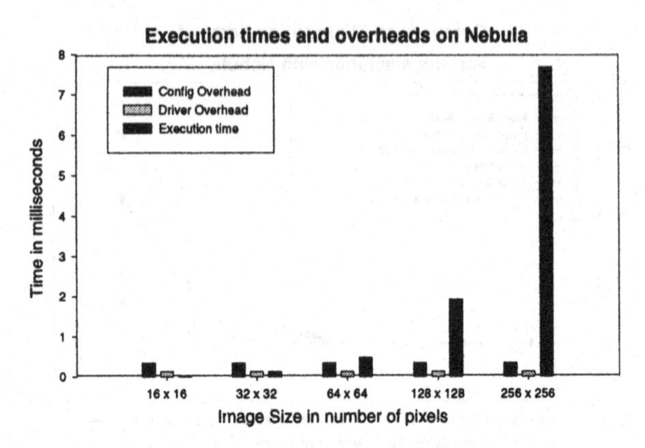

Fig. 5. *Execution times and Overheads on NEBULA*

performance comparison of the execution of the image filtering application on different computing platforms. Figure 5 shows the configuration and device driver overheads in relation to the actual hardware execution time for different image sizes. It can be seen that the configuration overhead and device driver overheads remain almost constant for different image sizes. This indicates that for larger images, the performance on NEBULA would perform even better than the other computing platforms discussed earlier.

5 Concluding Remarks

We have developed a partially and dynamically reconfigurable PCI bus based system, NEBULA, which can be utilized to gain high performance in the execution of a certain class of applications. A robust Linux device driver supports the reconfigurable core. A set of utility functions help a user to rapidly integrate hardware functions with a software envelope and subsequently use them as library modules for different applications. We have made a study on the classes of applications which would perform well on NEBULA.

References

1. Jeffrey M. Arnold, D.A.Buell, and E.G.Davis. *Splash* 2. Proceedings of 4th Annual ACM Symp. on Parallel Algorithms and Architectures, 1992.
2. XC6200 Field Programmable Gate Arrays. 1997.
3. Steven Casselman. *H.O.T. Works User's Guide.*
4. John L. Hennessy and David A. Patterson. *Computer Architecture: A Quantitative Approach.* Morgan Kaufman Publishers, Inc., 1990.
5. Doug Smith and Dinesh Bhatia. *RACE: Reconfigurable and Adaptive Computing Environment.* 6th International Workshop on Field-Programmable Logic and Applications, 1996.
6. TSI-Telsys Inc. *ACE Card, User's manual, Version 1.0*, 1998.
7. Xilinx. *LogiCORE PCI Master and Slave Interface User's Guide.* July 1997.

High Bandwith Dynamically Reconfigurable Architectures Using Optical Interconnects

Keith J. Symington[1], John F. Snowdon[1] and Heiko Schroeder[2]

[1] Department of Physics, Heriot-Watt University, Edinburgh U.K.
kjsymington@iee.org and J.F.Snowdon@hw.ac.uk
[2] Department of Computer Science, Loughborough University, Loughborough, UK.

Abstract. Optoelectronic interconnects are one means of alleviating the ever growing communications bottlenecks associated with silicon electronics. In chip-to-chip and board-to-board interconnection, the bandwidths presently (if experimentally) available far outstrip what is predicted possible in electronics until into the next decade. Such high bandwidth possibilities demand a rethink of conventional computer architectures where bandwidth is always at a premium. The combination of dynamic reconfiguration in electronics with this new technology may enable a new generation of architectures.

1 Introduction

Dynamically reconfigurable field programmable gate arrays (FPGAs) combine, in principle, the speed of dedicated hardware with the flexibility of software. As with all VLSI systems, the increasing density and speed of silicon circuits frequently transfers the performance bottleneck of a system to its communications. Optoelectronic interconnects are widely considered as a potential solution to this problem and are already being deployed (at a crude level) in commercial systems. Aside from the potential of optics for raw data throughput (unattainable in conventional systems), what is perhaps more exciting is the enabling of new architectural concepts by a combination of this throughput with the potential to reconfigure at high speed. This paper begins with a discussion of the potential for free space optoelectronic interconnects and a description of the types of hardware currently available. Three optoelectronic systems are then described; a dedicated digital sorter, a neural network switch controller and a generic parallel processing interconnect harness with a role for FPGAs suggested in each case.

2 Optoelectronic Interconnects

As the length or density of electrical interconnection increases, it suffers from increased wire resistance, residual wire capacitance, fringing fields and cross-talk. The maximum bandwidth of electrical systems has been independently estimated [1] as being around $500A/L^2$THz (where A/L^2 is known as the aspect ratio), which for a 10cm off chip connection is around 150GHz. Free space optics can deliver far higher bandwidths. The free space optical systems considered here are constructed by flip-chipping optical detectors and emitters (or modulators) onto silicon circuitry (see section 3) so the whole ensemble can be viewed as "conventional" silicon equipped with a large number of "optical pins" normal to the chip surface. The advantages of such a system are as follows

Off Chip Data Rates: The number of optical pins that can be driven depends primarily on thermal and real estate considerations. Currently we can drive 4,096 channels from $1cm^2$ and see no real obstacle to reaching >10,000 channels. These channels can be driven at speeds much greater than that of the CMOS and take considerably less power to drive than pads and wire bonds. Of course one may still connect conventionally to the chip as well. The sorting demonstrator system described in section 4 has an off chip data rate of 200 Gbs^{-1} which is what will be required by the semiconductor industry (according to the SIA roadmap) in 2007.

Bandwidth in "Busses": The bandwidth sustainable in a free space optical relay (see section 4) is far higher than that in an electrical bus. A $1cm^2$ relay can carry >100,000 channels. Currently we are driving at 200MHz (CMOS limited) giving a bandwidth of approximately 20 Tbs^{-1}. Devices may routinely be driven at 10 Gbs^{-1} so the relay can handle 1,000 Tbs^{-1} if we are not CMOS limited (the theoretical limit is actually much higher).

Distance: Optical signals (both guided and free space) are attenuated far less than electrical. Optical lengths of the order of meters are attainable without a significant increase in driving power.

Non-Local Interconnections: The non-interacting nature of free space optical channels means that they can pass through each other to form any desired interconnection topology without cross-talk. Interconnects such as the perfect shuffle and the hypercube thus become relatively simple to implement. This also reduces skew as very large variations in wire length can be avoided.

Data Acquisition: The naturally parallel nature of the connections implies high parallelism around the machine: e.g. to and from memory and peripherals.

Reconfiguration: It is also possible to reconfigure optical relays in the optical domain. This is not considered here.

3 Integrating FPGAs with Optoelectronics

The novel components may be thought of as consisting of three basic stages. The first stage, the input stage, consists of a detector array that is capable of receiving digital optical input. The second stage is the processing stage and consists of a dynamically reconfigurable FPGA system that could be considered as one or more configurable logic blocks (CLBs) corresponding to a single detector input. The final stage is the output stage and consists of an optical emitter or modulator, again corresponding to one or more CLBs. This combination will be referred to as an Optical FPGA (OFPGA).

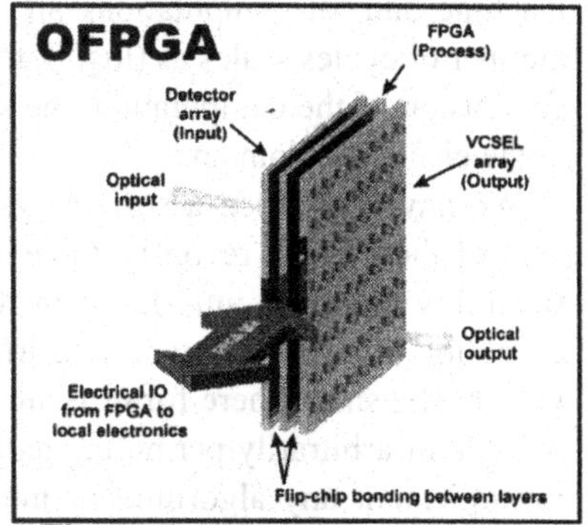

Figure 1: Combination of detector, FPGA and VCSEL arrays.

The OFPGA (figure 1) is capable of communicating internally electronically, or with any other local electronics for that matter, and can be viewed as a standard dynamically reconfigurable FPGA with an extra optical input and output available to a specific CLB or set of CLBs. Any optical input has the potential to reprogram a CLB by interlacing configuration information using a predefined protocol, or to reprogram another CLB in another system by interlacing the same configuration information onto its optical output stream.

In general the optoelectronic interface is constructed of hybrid processing chip technologies, which employ GaAs optical chips hosting detectors and emitters, flip-chipped [2] on top of the Si FPGA. The combination of input, processing and output elements is generally known as a smart pixel. Figure 1 shows a three layer chip which makes visualisation of circuits simpler (actual components at present are two layer with input and output on the same surface).

4 Optoelectronic Dynamically Reconfigurable Systems

The Optoelectronic Sorting Demonstrator [3]: The architecture of the demonstrator utilises optoelectronics as described above and exploits a non-local interconnect, in this case the perfect shuffle. Figure 2 shows a schematic of the sorting demonstrator. Each OFPGA array can convert electrical or optical inputs into electrical or optical outputs. The data to be sorted are entered sequentially into the processing loop through electrical I/O as shown. Sixteen bit planes of 32x32 bits (the number of optical communication channels) may be entered in this version. At run

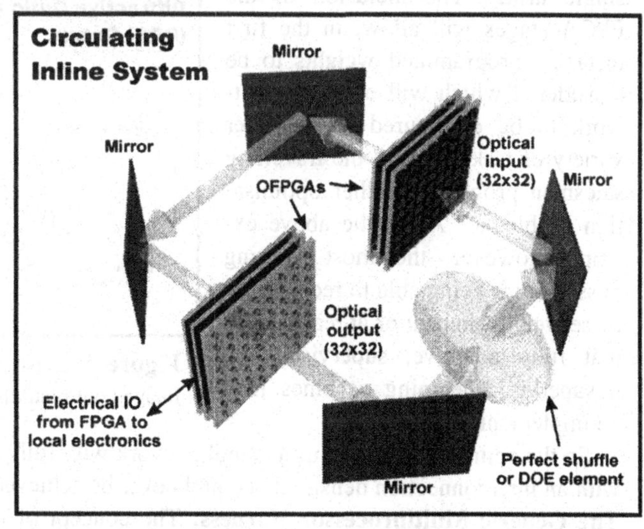

Figure 2: Optoelectronic sorting demonstrator modified to include inline components.

time, the 2D perfect shuffle is performed by a lens operation during each cycle of the machine and all computations are performed in parallel by the FPGA. The total number of cycles scales as $(\log N)^2$ for a Batcher's bitonic sort of N data points. At the conclusion of the computation, the sorted set of data can be sequentially downloaded to the electronic domain.

We have described a sorting system built using two OFPGAs; however it is the uses of the FPGAs reconfigurable aspects that are of interest here. A first level of flexibility could be introduced in that the CLBs could be reconfigured to optimise execution time for different data array sizes [4]. As a further level, the iterative perfect shuffle shown here forms (with suitable node switching) an omega class network capable of arbitrarily permuting data. Bearing this in mind, the set up could be used to implement any algorithm or multistage switching function given the right logic configuration. Indeed it is known that many algorithms map exceedingly efficiently onto this topology (the FFT being the classic example) and performance results at this level of parallelism would be of some interest to the engineering community. Therefore the combination of the FPGA and the high bandwidth optical interconnect gives us a general purpose fine grained massively parallel processor.

The Neural Network Packet Switch Controller [5]: Although the physical layout of the neural switch controller is somewhat different to the sorting module described above, from an architectural perspective the neural network may be formed from the module merely by replacing the optics that are used to generate the perfect shuffle with a diffractive optical element (DOE) [6] such as that shown in figure 3. The effect of the element is to fan-out in a space invariant manner every input channel into the pattern shown. This may readily be seen to effectively amount to an analogue sum over rows and columns of the outputs of the previous in-line device. The system as it stands at present has a fixed set of weights and is designed to perform this single task. The inclusion of the FPGA stages will allow, in the first instance, programmed weights to be considered which will allow the network to be configured for a wider variety of tasks such as the travelling salesman problem or other optimisation problems. As in the above example however, the most exciting possibility is being able to reconfigure these weights in near or at real time so that fully adaptive, supervised and unsupervised learning schemes may be implemented.

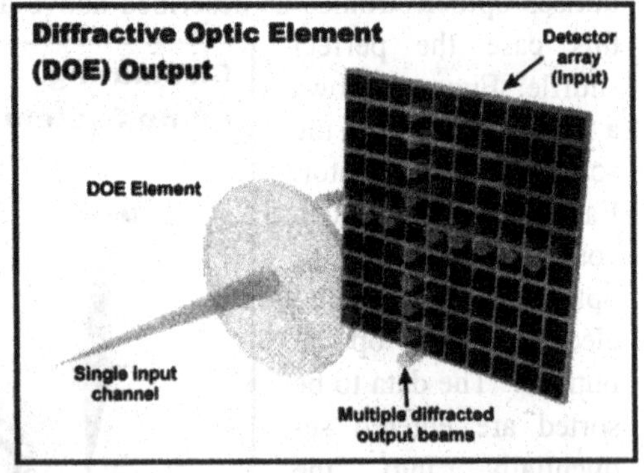

Figure 3: Illustration of a single input beam imaged onto multiple detectors using a DOE.

So the combination gives us a neural network with fully real time adaptive weights with an interconnection density that could never be achieved electronically.

The Generic Multiprocessor Harness: The concept of optical highways [7] is of perhaps the most interest in that a general purpose multiprocessor solution can be envisaged. Figure 4 schematically shows a highway used to connect nodes in an arbitrary topology, with each node having access to >1,000 channels. The interconnect is point to point and hard wired, with several thousand channels being passed from each node via a smart pixel interface into a free space optical relay system which can hold (under reasonable assumptions regarding aberrations) several hundred thousand channels. Polarising optics are used to determine the position of every individual interconnection link thus defining the computational topology.

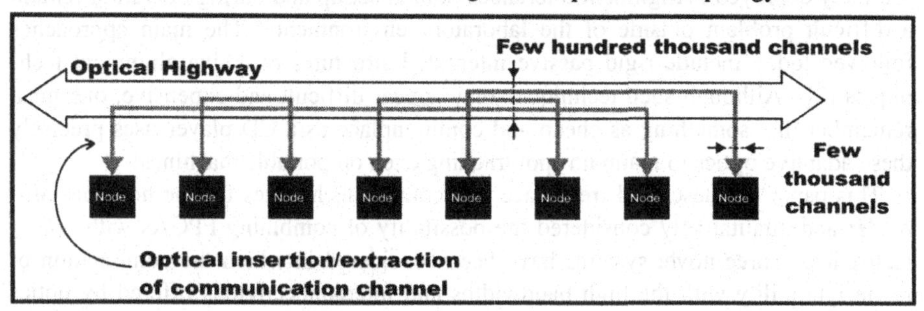

Figure 4: Optical highway for connection of multiple processing nodes.

The data used in this design is extrapolated directly from the components used in the sorting demonstrator (number of optical channels off chip is 2,500, clock speed 250MHz, data width is 64 bits) which shows the highway to support bisection bandwidths in excess of 10,000 Tbs^{-1} connecting up to 1 million nodes (depending on topology) [7]. Each node might consist of a microprocessor, memory and cache, and one or more OFPGAs to handle communications.

The role of OFPGAs in this architecture is to optimise the communications and processing in real time during execution of a range of algorithms. The bandwidth of the communications in this system is sufficiently high to implement a flat memory model, but superior performance on particular algorithms may be obtained by changing topology in the interconnect harness. In essence, the destination of any data channel output into the optical domain is determined by the spatial location of the emitter output upon the OFPGA. Thus by re-routing signals within the OFPGAs, a particular global topology may be established. The machine could of course be configured arbitrarily into several differently connected regions if this was desirable. In addition, the functionality of the interface may be changed, for example, OFGAs allow us to configure the interface as a router (necessary if we wished to utilise a hypercube in the optical domain) or as a simple switch of high throughput (necessary if we wished to use a large crossbar in the optical domain). For nodes of sufficient complexity, the maximum throughput of the system may often be attained by changing the width of data words as well as the topology so as to keep all communication channels busy. OFPGAs could support variable width multiplexing as well as routing at the interfaces.

For the generic interconnect the combination of FPGAs with high bandwidth optoelectronics enables an intelligent communications interface to be constructed which allows the maximisation of the ultra high throughput available. In turn this enables real time optimisation and load balancing of the whole machine over a range of computational models.

5. Discussion

Free space optical interconnects appear to offer a tremendous enhancement in all fields of computation and raise interesting computer science and architecture issues. There are many engineering issues to be confronted before such interconnects can be routinely deployed. Alignment tolerances both at set up and during operation remain a difficult problem outside of the laboratory environment. The main approaches followed today include rigid passive integrated structures or active alignment techniques [8]. Although such techniques may appear difficult and expensive, one must remember that something as cheap and commonplace as a CD player uses precisely these adaptive optics to maintain spot tracking even on portable machines!

This paper has described free space interconnect techniques for the non-specialist reader and qualitatively considered the possibility of combining FPGAs with optoelectronics. Three novel systems have been considered in which the combination of reconfigurability with the high bandwidths and interconnectivities offered by optics enable new architectures.

References

1. Miller, D. A. B., Ozaktas, H. M.: Limit to the Bit-Rate Capacity of Electrical Interconnects from the Aspect Ratio of the System Architecture, Journal Of Parallel And Distributed Computing, 41, No. 1, pp. 42–52, (1997).
2. Makiuchi, M., Hamaguchi, H., Kumai, T., Aoki, O., Oikawa, Y., Wada, O.: GaInAs pin Photodiode/GaAs Preamplifier Photoreceiver for Gigabit-rate Communications Systems using Flip-Chip Bonding Techniques, Electronics Letters, volume 24, number 16, pages 995-996, (August 4th 1988).
3. Gourlay, J., Yang, T., Dines, J. A. B, Snowdon, J. F., Walker, A. C.: Development of Free-Space Digital Optics in Computing, Computer, 31, 2, pp. 38- 44, (1998).
4. Akl, S.G.: Parallel sorting algorithms, Academic Press, (1985).
5. Webb, R.P., Waddie, A.J., Symington, K.J., Taghizadeh, M.R., Snowdon, J.F.: An Optoelectronic Neural Network Scheduler for Packet Switches, Submitted to Applied Optics, (May 1999).
6. Taghizadeh, M., Turunen, J.: Synthetic Diffractive Elements for Optical Interconnection, Optical Computing and Processing, 2 (4), pp. 221-242, (1992).
7. Dines, J.A.B; Snowdon, J.F; Desmulliez, M.P.Y; Barsky, D.B; Shafarenko, A.V., Jesshope, C.R.: Optical interconnectivity in a scalable data-parallel system, Journal of Parallel and Distributed Computing, 41, pp. 120-130, (1997).
8. Yang, T. Y., Gourlay, J., Walker, A. C.: Adaptive Alignment with 6-Degrees of Freedom in Free-Space Optoelectronic Interconnects, OSA Technical Digest, Optics in Computing, pp. 8-10, Snowmass, (April 1999).

AHA–GRAPE: Adaptive Hydrodynamic Architecture – GRAvity PipE

T. Kuberka[1], A. Kugel[1], R. Männer[1], H. Singpiel[1]
, R. Spurzem[2], and R. Klessen[3]

[1] Dept. for Computer Science V, University of Mannheim, B6-26,
D-68131 Mannheim, Germany,
kugel@ti.uni-mannheim.de

[2] Astronomisches Rechen-Institut, Mönchhofstraße 12-14,
D-69120 Heidelberg, Germany,
spurzem@ari.uni-heidelberg.de

[3] Sterrewacht Leiden, Postbus 9513, NL-2300 RA Leiden, The Netherlands,
klessen@strw.strw.LeidenUniv.nl

Abstract. [1] In astrophysics numerical star cluster simulations and hydrodynamical methods like SPH require computational performance in the petaflop range. The GRAPE[2]-family of ASIC-based accelerators improves the cost-performance ratio compared pared to general purpose parallel computers, however with limited flexibility. The AHA-GRAPE architecture adds a reconfigurable figurable FPGA[3]-processor to accelerate the SPH computation. The basic equations of the algorithm consist of three parts each scaling with the order of $\mathcal{O}(N)$, $\mathcal{O}(N \cdot N_n)$ and $\mathcal{O}(N^2)$, respectively, where N is in the range of 10^4 to 10^7 and $N_n \approx 50$. These equations can profitably be distributed across a host workstation, an FPGA processor and a GRAPE-subsystem. With the new ATLANTIS FPGA-processor we expect a scalable SPH-performance of 1.5Gflops per board. The first prototype AHA-GRAPE system will be available in mid 2000. This 3-layered system will deliver an increase in performance by a factor of 10 as compared to a pure GRAPE solution.

1 Introduction

1.1 Multi-particle interactions

The gravitating N-body-problem is one of the grand challenges of theoretical physics and astrophysics. Its accurate solution for very large particle numbers cannot generally be obtained by mathematical considerations (series evaluations) as it was possible for the historical treatment ment of the classical two- and three-body problems. Only computer modeling on the fastest available hardware using specialized mathematical-numerical algorithms can be used as an appropriate tool here. Hydrodynamical problems fall into the same category: Analytic solutions exist only for a limited number of

[1] This paper is also published in the 1999 PDPTA conference proceedings, CSREA press.

[2] GRAPE: Gravity Pipe: An ASIC for parallel calculation of the gravitational force [18].

[3] FPGA: Field Programmable Gate Array. FPGAs are the core elements of reconfigurable custom computing machines.

highly simplified cases. Hence, understanding the time evolution of a gaseous system in most cases involves sophisticated cated numerical modeling. The questions related with astrophysical systems typically involve addressing both sets of problems simultaneously. For instance, clusters of stars form from collapse and fragmentation mentation of self-gravitating gas clouds and grow in mass by accretion of the available gas reservoir. In their later dynamical evolution stellar clusters dissolve due to a combination of close and distant encounters between the stars. Eventually they blend into the overall stellar distribution of the Milky Way. In addition to the fundamental theoretical interest of large gravitating N-body systems such models are essential for our understanding of the structure and evolution of many astrophysically relevant objects, as there are our planetary system, our own and other galaxies and the entire universe seen as an object forming structure via gravitational interaction between particles. Such numerical modeling is also important for the interpretation of a wealth of new observational data from space based instruments, as e.g. of the dense centers of galactic nuclei observed with the Hubble Space Telescope (HST). They have recently improved evidence for the existence of supermassive black holes in their centers [19]. Surrounding them is a very dense star cluster, rotating, axisymmetric if not triaxial. Despite recent attempts to tackle this problem the physical interplay between relaxation, star accretion and black hole growth in such situations remains an unsolved and challenging theoretical and numerical problem for the astrophysical N-body simulators. For this physical situation a particular class of 'high-accuracy' numerical models following the orbit of each particle due to the 'exact' gravitational forces of all the other particles in a many-body system has to be used. We use a fourth order Hermite predictor-corrector scheme with hierarchically blocked individual time steps, Ahmad-Cohen (AC) neighbor scheme, and regularization of close encounters [20] and hierarchical subsystems [21, 22]. This method is widely known as Aarseth scheme [1, 2, 26][4]. Furthermore, particle based methods can also be applied to solve the equations of hydrodynamics. A widely used scheme is SPH (smoothed particle hydrodynamics). In this approach the fluid is represented by an ensemble of particles each carrying mass and momentum (analog to the N-body problem) and additional properties like temperature, pressure, entropy, and so forth. Thermodynamic observables ables are determined in a local averaging process ess over a given set of neighboring particles [4, 23]. This method is fully Lagrangian and is successfully applied invarious fields of numerical astrophysics. It is especially useful when dealing with the interaction between stellar and self-gravitating gaseous systems, as it elegantly unites the hydrodynamical and the gravitational N-body approach within one numerical scheme.

1.2 Implementations

The algorithms require most of their computational tional time to accumulate the mutual pairwise gravitational forces between the particles (N^2 problem!) and to compute a list of neighbors. Until recently this limited the maximum particle number for SPH calculations to $\approx 10^5$ even on large super computers [27]. The constraint is even more

[4] Such codes are denoted by NBODYx, where x is an integer denoting various degrees of complexity in the algorithm taken into account.

severe in the case of collision dominated N-body calculations. For example, three years ago the record particle number used to follow a globular cluster into core collapse was only 10^4 particles [25]. This situation improved considerably with the advent of the special purpose pose computers of the GRAPE series which were developed in Japan [18] and are also used in Germany and many other countries in the world. In the context of globular cluster simulations this pushed the record to 32k particles and proved the existence of gravothermal oscillations [15]. Despite being constructed to solve the gravitational N-body problem with high speed, GRAPE devices are useful for a large variety of other astrophysical applications as well, ranging from cosmological problems [3, 23] down to studies of the dynamical friction of a binary black hole in galactic nuclei [17, 16]. In particular, the combination with the particle based SPH method has opened the door to also study hydrodynamical problems with GRAPE. For example, it has been used to investigate the properties of X-ray halos around galaxy clusters [1], or the properties of interstellar turbulence [3, 23]. GRAPE has proven to be especially useful for hydrodynamical collapse calculation in the context of star and planet formation [11, 12, 3]. However, the special purpose machines of the GRAPE series reach their highest efficiency only for problems, which can be tackled with pure and clean N-body algorithms such as NBODY4 or KIRA. For SPH or standard N-body simulations using an AC neighbor scheme or a very large number of close (so-called primordial) binaries, or even worse for molecular dynamics simulations with potentials other than the Coulomb potential (e.g. van der Waals) they are not the optimal choice. One commonly used solution is to use general purpose massively parallel machines as the CRAY T3E, for which a competitive imple implementation of NBODY6++ exists using MPI and SHMEM [26]. While its performance compares well with one of the single GRAPE-4 boards, a larger scale GRAPE machine or the coming GRAPE-6 are still much more efficient for the pure N-body case. There is still work in progress, however, to improve the implementation on the general purpose parallel computers. The new solution presented here is to build a hybrid machine, which uses for the intermediate range forces a reconfigurable custom computing machine: an FPGA processor. This new system will profit from both the extremely high performance of the GRAPEs for the $\mathcal{O}(N^2)$ gravitational force computation and the high degree of flexibility of the FPGA processor which lets it adapt to the needs of the various hydrodynamic (SPH) oriented computations in the $\mathcal{O}(N \cdot N_n)$ region.

1.3 FPGA processors

The family of FPGA devices was introduced in 1984 by Xilinx. FPGAs feature a large number of relatively simple elements with configurable interconnects and an indefinite number of reconfiguration cycles with short configuration times. All configuration information is stored in SRAM cells. The basic processing element [5] (PE) of all current mainstream FPGAs is a 4-input/1-output look-up-table (LUT) with an optional output register. The functionality of the FPGA is thus determined by the contents of the look-up-tables within the PE's and the *wiring* between these elements. Over the last few years FPGA performance has increased tremendously as it profits from both:

[5] The currently largest devices offer approx. 10.000 of these PEs and more than 400 I/O pins.

- Increased density by a factor of 24 from 1993 through 1998 (Xilinx XC4000: 400 to 18400 elements)
- Increased speed by a factor of 3 from 1994 through 1998 (Xilinx XC4000: 133 to 400MHz internal toggle rate).

Eight years of experience at the University of Mannheim with FPGA based computing machines shows that this new class of computers is an ideal concept for constructing special-purpose processors combining both the speed of a hardware and the flexibility of a software solution [13, 14]. The so called FPGA processors consist of a matrix of FPGAs and memory forming the computational core. In addition there are a (programmable) I/O unit and an internal (configurable) bus system. As processing unit, I/O unit and bus system are implemented in separate modules, this kind of system provides scalability in computing power as well as I/O bandwidth. FPGA processors have shown to provide superior performance in a broad range of fields, like encryption, DNA sequencing, image processing, rapid prototyping etc. Very good surveys can be found in [6] and [5]. The hybrid microprocessor/FPGA systems developed at the University of Mannheim are in particular suitable for:

- acceleration of computing intensive pattern recognition tasks in High Energy Physics (HEP) and Heavy Ion Physics,
- subsystems for high-speed and high-frequency I/O in HEP,
- 2-dimensional industrial image processing,
- 3-dimensional medical image visualization [8] and
- acceleration of multi-particle interaction calculations in astronomy.

A well-tried means to adjust a hybrid system to different applications is modularity. ATLANTIS implements modularity on different levels. First of all there are the main entities host CPU and FPGA processor which allow to partition an application into modules tailored for either target. Next the architecture of the FPGA processor uses one board-type to implement mainly computing tasks and another board-type to implement mainly I/O oriented tasks. A backplane based interconnect system provides scalability and supports an arbitrary mix of the two board-types. Finally modularity is used on the sub-board level by allowing different memory types or different I/O interfaces per board type. The most important parameters of the system are listed in Table 1.

2 Architecture

2.1 FPGA performance assessment

Using FPGAs to accelerate complex computations tions using floating-point algorithms has not been considered a promising enterprise in the past few years. The reason is that general floating-point as well as particular N-Body implementation have shown only poor performance ance on FPGAs[6]. Usually N-Body calculations and particle based hydrodynamical simulations need a computing performance in at least teraflop range

[6] In 1995 approx. 10 Mflops [24] per Xilinx chip were reported for 18 bit precision, and 40 Mflops [10] with 32 bit precision on an 8 chip Altera board.

Table 1: FPGA Processor Specs

Number of FPGAs per computing board	4
Memory size per computing board	40MB
Computing board memory bandwidth	4 GB/s @ 50MHz
Max. number of FPGA boards per PCI bus	7
Max. PCI bandwidth	125MB/s @ 33MHz, 32Bit PCI
PCI bandwidth @4k blocks	75MB/s @ 33MHz, 32Bit PCI
Private bus bandwidth	800MB/s @ 50MHz per board pair
Expected SPH floating-point performance per computing board	1.5Gflops @ 50MHz
I/O board external bandwidth	4*200MB/s @ 50MHz
Number of supported host CPUs per PCI	1
Supported OS	Win NT, Linux

and are accelerated with the help of ASIC-based co-processors like the GRAPE-series. Nonetheless we have recently investigated the performance of a certain sub-task of the SPH algorithm on the Enable++ system [9]. The results indicate that FPGAs can indeed provide even in this area a significant performance increase. The piece of code shown in Figure 1 was implemented on 15 out of 16 core FPGAs[7] of the Enable++ system making heavy use of the configurable interconnect structure, as shown in Figure 1. For the implementation a 28bit floating-point format was used: 1 sign-bit, 7 bits exponent, 20 bits mantissa. The maximum pipeline depth is 6 stages and a result is produced at every clock cycle. The total performance for the code in the loop is therefore 16*13MHz = 208Mflops with the XC4013-5 chips and 16*32MHz = 512Mflops with the XC4028-2 chip respectively. If the XC4036-3 implementation will allow – as we expect – that 2 instances can run in parallel, the performance will increase to 1.024 Gflops. Parallel I/O is also done with 52 or 128MB/s on the input side plus a few MB/s on the output side. ATLANTIS will support two instances of this code to run in parallel on one computing board.

2.2 AHA-GRAPE

For astrophysical particle simulations including self-gravity, the determination of the gravitational potential at each particles position is usually the most expensive step in terms of computational time required. This step shall be done by the special hardware GRAPE for force computation in N-body simulations, which proved highly efficient in the case of a pure point-mass simple algorithm (NBODY1) case. For many more realistic applications however, some parts of the code become important bottlenecks if the gravitational force calculation is done very fast. They are usually of order $\mathcal{O}(N \cdot N_n)$ – where N_n is a neighbor particle number and $N_n \ll N$ – and comprise

1. Computation of the neighbor force when using a more complicated, but more efficient N-body algorithm (about 20 flops per pairwise force, of which order N_n per particle per time-step have to be computed).

[7] The present system uses Xilinx XC4013 FPGAs. A new system is currently being assembled equipped with XC4036 FPGAs.

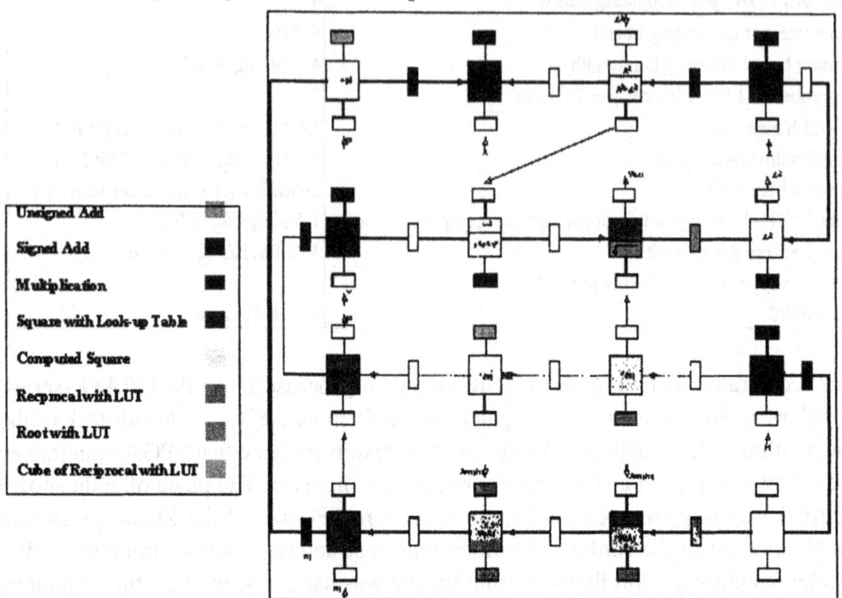

Figure 1: Implementaion of SPH-Fragment on the FPGA-Processor Enable++

Figure 2: SPH Code Fragment

```
Do i = 1, N
    Do j = 1, Nn
        r_ij = r_i - r_j      /*(3-d vectors)*/
        r_ij = | r_ij |
        h_ij = (h_i+h_j)/2
        1/h_ij
        W(r_ij, h_ij)    /*(table look-up)*/
        ρ_i = ρ_i+m_j W(r_ij, h_ij)/ h_ij^3
    Enddo
Enddo
```

2. Computation of the kernel function, its derivatives, and terms related to gas dynamical quantities[8] in the SPH algorithm (about 100 flops per pairwise particle interaction, of which again $\mathcal{O}(N_n)$ per particle per time-step have to be computed[9]).

3. Integration of binary motions in regularized coordinates as a function of near perturbers (order N_n); this is a very sophisticated algorithm and cannot easily be estimated now in its complexity [20].

4. Integration of SCF force (self-consistent field [7] to compute approximately the gravitational potential of distant particles) for hybrid N-body models[10].

[8] Like density, pressure, viscosity, energy fluxes, etc., with summation over N_n.

[9] Calculation of 1) can be done as a subset of operations within 2) for a combined high-precision gravity SPH-code.

[10] The complexity of this algorithm differs from the standard scalings.

Figure 3: Expected SPH-Performance

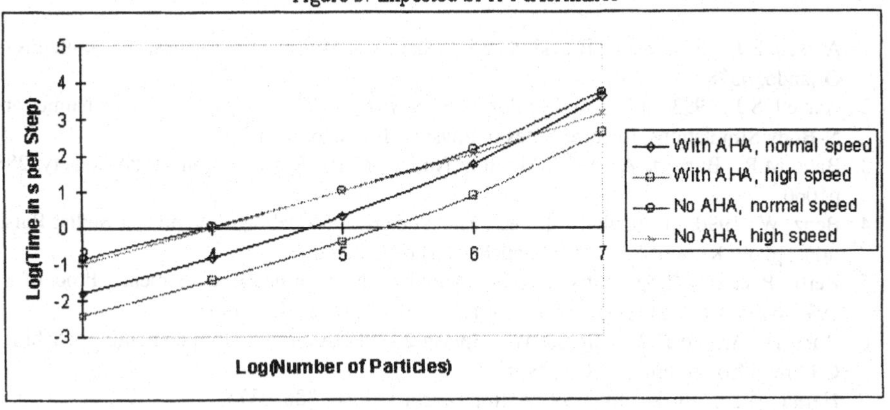

The floating point operations related to 1) and 2) are in principle straightforward to map onto an FPGA processor, however critical for the performance is the word length which is sufficient for each component of the sum of pairwise forces and other SPH expressions. Test calculations have to be performed to clarify this. The two subsystems – GRAPE cluster and FPGA processor – will be connected to the host workstation by the PCI bus, either directly or via an interface. Within the subsystems the respective local buses will be used to broadcast sample data and intermediate results. In a second step and by close cooperation with the Tokyo group a hierarchical coupling of the FPGA device with GRAPE, including memory and control of the GRAPE, could be envisaged. This will further improve performance by parallelization of force computations and data communication. Figure 3 displays the performance estimates for various systems with and without FPGA processor.

3 Status and plans

At present (June 1999) a test implementation of the SPH-loop/step1 on ENABLE++ is carried out to verify the estimated performance. By mid 99 the new ATLANTIS system will be available where the full SPH-code has to be implemented. A communication library for LINUX must be developed, supporting simultaneous transfers between host/GRAPE and host/FPGA respectively. We expect the first prototype AHA-GRAPE system to be available in mid 2000. The key figures for this prototype are 50Mflops for the host workstation, 5Gflops for the FPGA processor and 500Gflops for the GRAPE subsystem. The presence of the FPGA processor will lead to an increase in performance ance by a factor of 10 and will allow us to handle up to approx. 10^6 particles[11] in collision dominated N-body simulations and a few 10^7 particles in SPH.

[11] At very large particle numbers the N^2 term becomes dominant which is the domain of the GRAPE subsystem.

References

1. Aarseth S.J., 1985, in J.U. Brackbill, B.I. Cohen, eds, Multiple time scales, Academic Press, Orlando, p378
2. Aarseth S.J., 1993, in G. Contopoulos, N.K. Spyrou, and L. Vlahos, Galactic Dynamics and N-Body Simulations, Lecture Notes in Physics, Thessaloniki, p365
3. Bate, M.R., Burkert, A.: 1997, Monthly Notices of the Royal Astronomical Society, 288, p1060
4. Benz, W.: 1990, in Buchler, J. R., ed., The Numerical Modeling of Nonlinear Stellar Pulsations, p269, Kluwer Academic Publishers, The Netherlands
5. Bertin P. et al., "Programmable active memories: a performance assessment", Proc. of the 1993 symposium on research on integrated systems, pp.88-102, 1993
6. Buell D., Arnold J., Kleinfelder W., "Splash-2 – FPGAs in a custom computing machine", CS Press, Los Alami- tos, CA, 1996
7. Hernquist L., Ostriker J.P., 1992 Astrophysical Journal, 386, p375
8. Hesser J., Vettermann B., "Solving the Hazard Problem for Algorithmically Optimized Real-Time Volume Rendering", Submitted for publication to "International Workshop on Volume Graphics", 24–25 March 1999, Swansea, United Kingdom
9. Hoegl H. et al., "Enable++: A second generation FPGA processor", Proc IEEE Symposium on FPGAs for custom computing machines, pp. 45–53, 1995
10. Kim H.-R.., Cook T. A.., Louca L., "Hardware Acceleration of NBody Simulations for Galactic Dynamics". In SPIE Photonics East Conferences on Field Programmable Gate Arrays (FPGAs) for Fast Board Development and Reconfigurable Computing, pp. 115–126, 1995.
11. Klessen R., 1997, Monthly Notices of the Royal Astronomical Society, 292, p11
12. Klessen, R.S., Burkert, A., Bate, M.R.: 1998, Astrophysical Journal, L205
13. Kugel A. et al.: ATLAS level-2 trigger demonstrator-A activity report, ATLAS DAQ-note 85, March 26, 1998
14. Kugel A. et al., "50kHz pattern recognition on the large FPGA processor Enable++", Proc. IEEE Symp. on FPGAs for custom computing machines, CS Press, Los Alamitos, CA, 1998, pp.1262-263
15. Makino J., Astrophysical Journal, 1996, 471, p796
16. Makino J., Astrophysical Journal, 1997, 478, p58
17. Makino J., Ebisuzaki T., 1996, Astrophysical Journal, 465, p527
18. Makino J., Taiji N., Ebisuzaki T., Sugimoto D., 1997, Astrophysical Journal, 480, p432
19. van der Marel R.P., Cretion N., de Zeeuw P.T., Rix H.W., 1998, Astrophysical Journal, 493, p613
20. Mikkola S., 1997, Celestial Mechanics and Dynamical Astronomy, 68, p87
21. Mikkola S., Aarseth S.J., 1996, Celestial Mechanics and Dynamical Astronomy, 64, p197
22. Mikkola S., Aarseth S.J., 1998, New Astronomy, 3, p309
23. Monaghan, J.J.: 1992. Annual Review of Astronomy and Astrophysics, 30, p543
24. Shirazi N. et al., "Quantitaive analysis of floating-point arithmetic on FPGA-based computing machines", Proc. IEEE Symp. on FPGAs for custom computing machines, CS Press, Los Alamitos, CA, 1995, pp.152-162
25. Spurzem R., Aarseth S.J., 1996, Monthly Notices of the Royal Astronomical Society 282, p19
26. Spurzem R., Baumgardt H., 1999, Monthly Notices of the Royal Astronomical Society, subm., preprint at ftp://ftp.ari.uni-heidelberg.de/pub/spurzem/edinpaper.ps.gz
27. Whitworth, A.P., Chapman, S.J., Bhattal, A.S., Disney, M.J., Pongracic, H., Turner, J.A.: 1995 Monthly Notices of the Royal Astronomical Society, 277, p727

DIME - The First Module Standard for FPGA Based High Performance Computing

Malachy Devlin[1] and Allan J. Cantle[1]

[1] Nallatech Ltd, 10-14 Market Street, Kilsyth, Glasgow, G65 0BD, Scotland
m.devlin@nallatech.com, a.cantle@nallatech.com

Abstract. FPGAs are offering a novel mechanism for the development of new architectures in high performance computing. Constructing a system with a single FPGA node is a simple exercise, however to construct a scalable system requires a sound foundation. A modular system, DIME, for constructing multiple FPGA systems is introduced which is based on the experiences of traditional distributed parallel processing approaches.

1 Introduction

With the introduction of one million gate FPGAs, novel computing platforms are becoming more practical and a serious contender to conventional high speed processors. Conventionally FPGAs have been used for absorbing glue logic, performing control operations and some simpler processing operations. These larger devices are now capability of carrying out more complex operations and with the proliferation of system on a chip design methodologies, IP is becoming more readily available. This enables more substantial designs to be performed in a single FPGA and thus allowing new approaches to application implementation.

However the development of an FPGA design itself is one aspect of a complete system design. Electronic systems must take into account factors such as I/O, product availability, environment, etc.

In this paper we will address the issues involved in constructing very high performance computing systems which utilise FPGAs as a key component and draw on lessons in traditional computing architectures while leveraging the benefits in new technology developments.

We will focus on creating a foundation FPGA platform that can be used to achieve the need for large-scale system construction using off the shelf components that offer flexibility and scalability.

2 Scalable High Performance Computing Systems

Although new generation FPGAs can be used to offer outstanding performance for data processing applications, a constant demand exists for pushing the performance barriers of technology by applications such as image processing, radar, sonar or communications. A number of processor architectures have been developed to offer maximum performance to try and address the needs of these applications. However when very large scale computing is required the distributed parallel processing architecture has shown to be scalable and practical for a range of applications. These processors used the concept of Communicating Sequential Processes, CSP, [1], with one of the early processors to implement this concept being the Transputer[2]. This has been followed by a number of successful processors such as the TMS320C4x family from Texas Instruments [3] and the SHARC processor from Analog Devices [4]. Systems based on these processors have been scaled to several hundred processors and have proven successful in numerous applications.

3 Modular System Construction

3.1 Processor Based

Undoubtedly the key to successful parallel system building using multiple processors has been the use of a fundamental modular approach. This enables complex and diverse systems to be constructed from off the shelf units to create application tailored solutions

To match the flexibility of these parallel processors and the practical construction of large scale distributed parallel processing systems several board level module systems have been developed which are associated with the processors previously described, these include

- TRAM - First Generation **TRA**nsputer **M**odule
- HTRAM - Second Generation **H**igh Speed **TRA**nsputer **M**odule
- TIM40 - **T**exas **I**nstruments **M**odule Standard for the **C40** Family of DSPs
- SHARCPAC - Analog Devices Module Standard for the SHARC family of DSPs

These module standards have always been tightly coupled to the DSP that they are associated with and they provide the ability to easily scale a parallel processing system with the number of processors to perform any given task. These standards have also been extended to support many different I/O capabilities in such systems that have made it possible to easily construct systems that are suitable for a wide variety of applications.

3.2 I/O Based

In addition to processor based modules a number of I/O module standards also exist, these include

- PMC - PCI Mezzanine Card
- PC-MIP - PCI based Module with additional I/O

These modules are based on the PCI interface for communication on and off the module and offer the ability customised the I/O facilities of a carrier motherboard.

4 FPGAs - A New Era in Computing

FPGAs are now available with over 1 Million gates capacity and have become a true contender to the traditional processor architectures. This is further reflected in the proliferation of conferences such as FPL focusing on the use of FPGAs for carrying out intensive computational operations. In practice it is straightforward to construct a system which harnesses a single FPGA and has pre-defined I/O requirements. Construction of a scalable FPGA based computing platform both in terms of computational performance and I/O requirements need much more attention. The modular approach has proved to be the most practical method of building scalable and flexible systems.

However the current range of processor and I/O module standards detailed above are becoming out-dated or inappropriate in FPGA based systems. These standards do not cater for current and next generation silicon technologies which require operating voltages down to as little as 1.8V and ever increasing bandwidths demand very high pin count, miniature connectors that interface the motherboard to the module at every greater speeds. Furthermore, these modules are not appropriate for FPGA systems.

The I/O modules are PCI based modules, this interface standard does not allow deterministic data transfers to be carried out between modules with low latency. Each module must also include PCI circuitry which in itself takes up only a small percentage of the new large FPGAs, however from a system perspective the application must be designed to interact with the PCI interface and hence constraints are always imposed on the system from the outset. The application design will also need to contain additional protocol overheads to provide for the communication over PCI.

Looking at the processor based modules, these have all been defined to be tightly coupled to the specific processors. Therefore to use these for FPGA systems would involved recreating the particular processor interface in the FPGA to interface to the module interface which again introduces a predefined constraint on the system design.

5 DIME – DSP and Image Processing Module for Enhanced FPGAs

The DIME (**D**sp and **I**mage processing **M**odule for **E**nhanced fpgas) Module standard [5][6] was developed to address this change in the technology arena. The main focus of the DIME standard has been that it is a general purpose module standard based around the FPGA rather than any specific Microprocessor or DSP. The DIME module is based on a multi-level standard. At the lowest level, Level 0, it defines the physical format and the essential pin connections. The higher level, level 1, defines the system level. There can be several different system Level 1 standards for different applications offering flexibility to the system designer.

5.1 DIME Level 0 - The Physical Level

The Physical Level specifies the basic mechanical and electrical properties of the module. It also defines two JTAG Boundary Scan chains for the modules. The first is used for configuring the modules FPGA's and the second is in place for DSP and Microprocessor "In Circuit Emulator" support, if these devices are designed onto the module. FPGA configuration in this manner ensures that the module standard is 100% compatible with a wide variety of FPGAs from different vendors. Further to this, there are over 200 unspecified I/O lines between the module and it's motherboard.

5.2 DIME Level 1 - The Virtual Level

The Virtual Level then provides a rigid definition of the Unspecified I/O Pins. For example common busses can be defined between modules that include specific communication protocols. The standard can support many different Level 1 standards. Ideally all of the I/O on a module will be routed into its FPGA. Therefore all modules designed to the standard will be compatible with any Level 1 standard simply by a firmware update to the FPGA. Only DIME carrying motherboards will need to be designed to specifically support the different level 1 standards.

5.3 Features

The physical size of the DIME module measures 59.5mm by 97mm offering a compromise in small size which allows compact systems to be developed but large enough to enable substantial functionality on a single module. The diagram below illustrates the organisation of components on a module and how they interrelate with other DIME modules in the system.

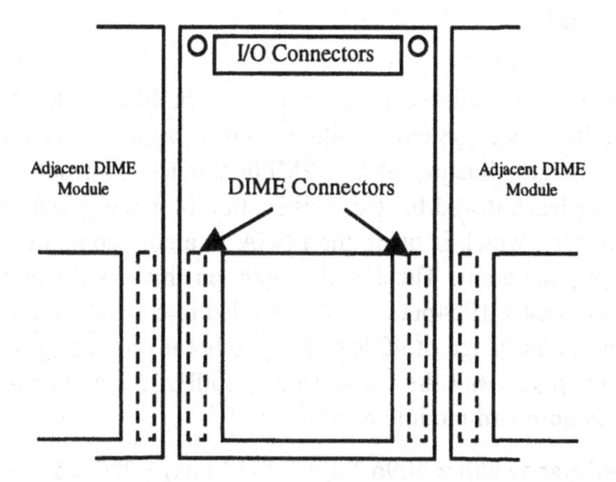

Fig 1. Illustration of physical layout of components on a DIME module and the typical relationship between several DIME modules.

DIME has been designed to take advantage of advanced packaging technology such as ball grid arrays and as illustrated in fig 1 all current packages can be located between the connectors of the module for maximum use of space.

Another critical element is getting 'services' on and off the module and in particular with higher processing rates required, the movement of data is very critical. Data throughput is sometimes an underestimated parameter and it is this factor that can choke the performance of modern day processors. The DIME module is capable of achieving bandwidths of over 2.5Gigabytes/sec which ensures that the FPGA is capable of being full utilised.

5.4 Dynamic Configuration

When building large systems of FPGAs it is desirable to have the capability to reconfigure individually FPGAs both partially and completely while allowing the rest of the system to continue operating. The DIME standard incorporates plug and play FPGA configuration which enables intelligent and reliable configuration of the complete system by carrying out diagnosis and integrity operations. DIME provides JTAG boundary scan services, which maximises the compatibility of this open and accepted standard to allow for the control of all available FPGAs on the market. In this way heterogeneous systems can be constructed which can incorporate FPGA and non-FPGA devices from a wide range of sources and more critically maintains the methodology of design for test, DFT.

6 Application

DIME has already been established itself as a viable building block for constructing real-time high performance systems. With modules capable of capturing and displaying real-time images, hosting over 2.2Million gates, external transfer of high speed data, etc. This has allowed for the construction of a new generation of 3D real-time simulation system, which is using the FPGA as a core computing node for performing the image generation. The I/O demands on this system alone required an image resolution of 1024 x 1024pixels which needed to be over sampled by 4 in both directions resulting in an image of 4096x4096 pixels for processing. Giving also that the image has 14bit pixels and the frame rate is 200Hz, a simple calculation shows that this requires an on-board module bandwidth of : -

On Board Bandwidth = 4096 * 4096 * (14 bits) * 200 = 5.87 GBytes/sec

The corresponding bandwidth for taking the data onto and from the module of

Off Board Bandwidth = 1024 * 1024 * (14bits) * 200 * 2 = 732Mbytes/sec

Note that these simple calculations do not take into account associated control signals and other concurrent communications which are also be carried out to perform addition system level functions which increases further the required bandwidths.

The new generation of FPGAs are capable handling these on board bandwidth and the DIME module offers a corresponding performance to fuel the FPGAs.

7 Conclusion

We have highlighted the traditional framework of constructing scalable distributed systems based on parallel processing technologies and re-stated the novel capabilities of FPGAs as a viable computing platform. From this knowledge and experience a novel module standard, DIME, has been introduced which focuses on maximising the potential of FPGAs and how it eases the construction of semi-custom systems for practical applications.

References

1. C.A.R. Hoare, "Communicating Sequential Processes", Prentice-Hall, 1985, ISBN 0-131-53271-5
2. "Transputer Reference Manual", Inmos, Prentice Hall, 1988, ISBN 0-13-929001-X
3. "TMS320C4x User manual", Texas Instruments, 1991
4. "SHARC User Manual", Analog Devices, 2nd Edition, 1996
5. DIME Module, Physical Level 0 Specification, Nallatech Ltd, NT301-0001
6. Video processing, implementation Level 1 of the DIME Module, Nallatech Ltd, NT301-0002

The Proteus Processor — A Conventional CPU with Reconfigurable Functionality

Michael Dales

Department of Computing Science, University of Glasgow,
17 Lilybank Gardens, Glasgow, G12 8RZ, Scotland.

Abstract. This paper describes the starting position for research beginning at the Department of Computing Science, University of Glasgow. The research will investigate a novel microprocessor design incorporating reconfigurable logic in its ALU, allowing the processor's function units to be customised to suit the currently running application. We argue that this architecture will provide a performance gain for a wide range of applications without additional programmer effort. The paper gives some background information into similar research, argues the benefits of this idea compared to current research, and lists some of the questions that need to be answered to make this idea a reality. We end by specifying the initial work plan for the project.

1 Motivation

There has been a lot of research into combining microprocessor technology with Field Programmable Logic (FPL) (typically using a Field Programmable Gate Array (FPGA)). Whilst these solutions claim some performance over a conventional hard–code CPU, they typically make significant alteration to the CPU, or make the reconfigurable logic detached, available over a bus. In both cases, significant work is required by the programmer to make general purpose applications take advantage of the hardware.

Another recent trend is the appearance of specialised function units in some microprocessors. As silicon technology is progressing to smaller die sizes, Integrated Circuit (IC) designers have more and more gates at their disposal. In CPU designs, they have more gates on their die than they need to build just a CPU, so the designers are looking for extra ways to utilise these gates. Typically designers use two methods to utilise this extra resource. The first is to add more memory onto the CPU, e.g. bigger caches. The second is to add more Functional Units (FUs) to the processor, such as the new MMX instructions added by Intel to its Pentium processors [8] and the 3D acceleration built into the new Cyrix MediaGX processors [3]. These designs add specialised instructions to the CPU in order to accelerate certain types of application (typically graphics based).

With more specialised function units being added to CPUs we begin to see a lower utilisation of the silicon. The new 3D graphics FUs added are fine for CAD programs or games, but are not required for databases or word processing. Another problem is that the functionality in these FUs must either be generalised versions of a problem, in which case it may not solve a programmer's specific problem [1], or tackle a specific problem and become out of date (e.g. MMX adds graphics functions working at a colour depth of 8 bits per pixel (bpp), but most programs now are looking at greater colour depths, such as 16 and 24 bpp).

In this paper we present a novel architecture that we argue will provide a performance increase for a wide range of applications without significant overhead for the programmer. Instead of using the extra available gates to add new problem specific functions to the ALU of a conventional microprocessor, our architecture places reconfigurable logic inside the ALU, from where its functionality can be accessed in the same way as any other ALU instruction. Thus, we have a single processing device that utilises the flexibility of reconfigurable logic with a modern high speed hard–coded CPU design.

2 Current Work

Several attempts are being made to utilise the advantages of FPL in the field of microprocessor design. Here we look at each of the different categories in turn, followed by a discussion of these efforts as a whole.

2.1 Coupling a CPU and an FPGA over a Bus

In this category, a conventional microprocessor is connected to an FPGA over a bus. Efforts in this area can be split into those which couple the two parts loosely and those that use tight coupling.

In loosely coupled designs, an FPGA is placed on a separate piece of silicon and attached across an external bus (e.g. PCI bus), just as a 3D acceleration card might be today [10]. Typically such a device will have a certain amount of dedicated memory attached to it (like with a video card) to reduce the number of individual bus transfers. Circuits are loaded onto the FPGA, data fed across the bus, processed and passed back. As the FPGA is on its own piece of silicon it can have a high cell count, allowing for complex circuits to be instantiated on it. Typically, no low–level support is provided for the programmer — they have to explicitly load a circuit onto the FPGA and transfer any data between it and the CPU. Some run–time system support methodologies for such devices have been investigated [2], but have failed to gain widespread acceptance.

The second method moves the FPGA onto the same die as the processor [7, 12]. This reduces the space available for the FPGA, as it has to share the available gate space with the CPU. However, this means that data transfers between the CPU and the FPGA are quicker as the bus between the two parts is much shorter and can be clocked at a higher rate. Note, however, that the FPGA is usually connected via a separate bus from the main internal data–paths. With the closer coupling comes lower–level support for accessing the FPGA. The CPU can have instructions built in for interacting with the FPGA, such as "load design" and "run circuit for *x* clock ticks".

2.2 Building a CPU on an FPGA

The second method involves replacing the conventional processor with an FPGA and applying some form of CPU framework on top [14, 13, 4]. Different groups have tackled this idea in different ways, but they share a common principle. In such a scenario, only the actual parts of the CPU required for a given application are instantiated; this allows middling complex circuits to be instantiated on the CPU. Because the routing of the CPU can be reconfigured, these designs typically make use of an optimised data–path for their circuit layout. Solutions in this category implicitly have low–level support for the programmer. Typically they use new programming models — they work in a fundamentally different way from current CPU designs.

2.3 Static Framework with a Reconfigurable Core

The third category describes systems that have taken a microprocessor shell and placed FPL inside this to allow its functionality to be tailored during execution. Two examples of this approach can be seen in [6] and [9]. The former is based on a data–procedural architecture, called an Xputer. This design utilises an area of FPL in which it can place multiple FUs. The second is closer to a conventional microprocessor, placing a block of reconfigurable logic inside its equivalent of the ALU, allowing the contents of the FPL to be accessed as if it were a conventional instruction. Both of these processor designs provide low–level support for the reconfigurable functionality.

2.4 Discussion

The FPGA based CPU solution has a problem as a general processor. By moving everything onto the FPGA, the entire CPU can be reconfigured, however this is not typically required. Consider operations such as 2s complement arithmetic, boolean logic, and effective address calculation, along with techniques such as pipelining and branch prediction. These standard components are always used and do not need to be reconfigurable. Thus by placing these components on an FPGA they will suffer a performance loss and fail to utilise the flexible nature of the platform. By adding a level of indirection (the reconfigurable logic) between the *entire* CPU and the silicon, it is likely that the general performance of such a CPU will not match a generic hard–coded CPU.

Now consider the problems of latency and synchronisation. The bus technique suffers from latency problems as the data being processed has to move outside the normal CPU data–paths. This technique also suffers from synchronisation problems: how is the processing in the FPGA interleaved with the normal processing of the CPU? This becomes more apparent if being used in a multiprogramming environment.

Finally, if we contrast the the approaches described above from the point of view of the programmer we see another problem. Although the tighter coupling gives programmers greater low–level support, most of the designs use a non–standard programming model. This is a drawback as it inhibits the wide–spread acceptance of such a technology — very little existing software can easily be ported to the new platform. The importance of supporting legacy systems can be seen clearly in the Intel x86 family, which still supports programs written back in the early 80's.

Thus, from a general perfromance point of view, we see a preference for the static framework and reconfigurable core technique, and from a programming view a preference for the use of a traditional Von Neumann programming model.

3 The Proteus Architecture

The idea behind our Proteus[1] processor is to place a smaller amount of reconfigurable logic inside the ALU of a conventional microprocessor where it would be tightly integrated into the existing architecture. In this model, along with the normal ALU functions, such as add and subtract, we propose to deploy a bank of Reconfigurable Function Units (RFUs) whose functionality can be determined at run–time, with applications specifying the functions that they require. Functions can be loaded into the RFUs as and when they are needed, suggesting a much better utilisation of the silicon. When functions are no longer needed they can be removed to make way for new circuits. This turns the ALU into a Reconfigurable ALU (RALU)[2]. An application could use multiple RFUs at once, loading in the set of circuits it needs. The circuits loaded into the RALU could change over time. If an application needs to load another circuit but there are no free RFUs then the contents of one could be swapped out (cf paging in a virtual memory system). This results in a CPU that can be customised to suit the application currently in use, be it database, spreadsheet, word processor or game, allowing each one to take advantage of custom FUs. These FUs can be tailored to the exact needs of a given problem, matching the programmer's requirements.

By placing this logic inside the ALU it will have access to the processor's internal data–paths, removing the latency problem described above. At a higher–level, it will appear as if the CPU has a dynamic instruction set, one that meets the needs of each program. The RFUs have the same interface as any other FU in the CPU. This approach allows the CPU to behave normally in terms of pipelining and caching — the new instructions are just like normal ALU instructions in that respect. Programs can be written using a normal programming model. The programs simply need to load their FU designs into the RALU before they are required and from then on invoking a custom function can be made syntactically the same as issuing any other instruction to a traditional ALU: all it requires is an opcode and a list of standard operands.

This technique is different from altering the microcode in the processor's control unit. Some modern processors, such as Intel's Pentium Pro and Pentium II processors, allow the microcode inside the processor to be updated [5]. Changing microcode allows machine instructions to carry out a different sequence of operations, but only operations which already exist on the processor. It does not allow any new low-level functionality to be added. Our proposal does allow for new low-level constructs to be added to the processor.

3.1 The Reconfigurable ALU

The external interface of the RALU is similar to a normal ALU. An ALU contains a set of operations of which any one can be chosen per instruction. The RALU is similar, containing a set of RFUs, any of which can be used in an individual instruction. The difference is that in the RALU the actual operations contained within each RFU can change (independently) over time to suit the currently executing application(s).

[1] Proteus - A sea-god, the son of Oceanus and Tethys, fabled to assume various shapes. O.E.D.

[2] Note that this is different from the *rALU* described in [6] — see Section 3.1 for a more detailed description of our RALU.

A conventional ALU takes in two operands *a* and *b*, and returns a result *x*. The function the ALU carries out on the operands is determined by control lines. The RALU presents the same interface to the CPU for conventional operation. It also has two additional inputs: lines which take in a circuit bit–stream to configure the individual RFUs and additional control lines needed during reconfiguration. Thus, apart from when being configured, the RALU can be used in the same way as a traditional ALU.

Inside each RFU is an area of FPL. This logic can take inputs from the operands *a* and *b* and, when selected by the *op* lines, place a result on *x*. The FPL inside each RFU is independent of that in other RFUs, thus reprogramming one will not affect the others. The amount and type of the FPL inside an RFU is an area of research (see Section 4.1).

4 Areas of Possible Research

4.1 Hardware Design

Obviously, there are the low–level aspects to consider. At the design stage we need to look at the relationship between the RALU and the rest of the CPU. It may be that the traditional ALU and RALU can be put together into a single unit, or should be kept separate. The CPU layout will be easier if the two parts are combined, but there could be good reasons for not doing so. Many modern CPU designs only place single cycle instructions in the ALU, moving multi–cycle instructions into external FUs which can run in parallel. It is likely that we would like to allow RFUs to run for multiple clock cycles, suggesting we adapt such an architecture instead.

An important question is how much FPL should each RFU contain, and how many RFUs to place in the RALU. These factors determine how complex a circuit (and thus how much functionality) can be encoded in each RFU, and how many RFUs an application can use at a given time. If the RFUs are too few or too small, applications may be constantly swapping circuits in and out, wasting time (cf thrashing in virtual memory). Too many or too big RFUs could lead to space and routing problems.

We also need to consider the necessary changes required to the control algorithm for a CPU to support the new functionality. New instructions will be required so that the run time environment for a program can set up the circuits it needs, and new instructions for invoking the functions will be needed. A related question is whether there should be caches for RFU circuits, similar to the caches for instructions and data which are already found inside CPUs?

If we are to allow the Operating System (OS) to virtualise the RALU resource (see Section 4.2), then the application can not make assumptions as to where a specific circuit will reside in the RALU (cf to virtual/physical address translation in a virtual memory system). A layer of indirection between an instruction the application issues and the RFU called will be required. The translation operation will need to be very fast — even a tiny overhead can add up to a large one if an application relies heavily on the RALU.

4.2 Operating System Issues

On the software level, we have the question of resource management of the RFUs. Just as an OS in a multiprogramming environment must manage resources such as memory and CPU time in a secure and efficient manner, there needs to be both a mechanism and policy for management of the RFUs. Potentially the RFUs can be split amongst multiple programs, virtualising the resource. If this is done, then we need to consider how many RFUs should be dedicated to a single program, and what sort of replacement policy should be used. This could possibly also have an adverse effect on context switch times it we need to swap FUs in and out along with everything else.

Another issue concerns maintaining the pretence of a Virtual Machine (VM) to application programs. The program's functionality, when encoded as a circuit in a RFU has access to the underlying hardware, by-passing all the OS security checks. This means that malicious or badly written programs can pose security threats to other programs, or possibly hog resources (e.g. a circuit in an infinite loop). This poses the same questions that have been addressed in research into extendible OSs [11].

4.3 Programmer Support

For the new CPU design to be accessible, there are high-level programming issues to be considered too. If these issues are not tackled, then it is unlikely that the technology will be widely accepted, despite the anticipated performance increase.

One obvious question is whether software designed to utilise this new architecture will show an improvement compared to that written for a hard–coded CPU (although at this stage it is assumed that there will be some improvement). Applications will incur new overheads in terms of managing their circuits, which might detract from the performance of the system. Questions such as how many instructions need to be optimised into a single new FU to be efficient need to be examined and answered.

It is unreasonable to assume that most software engineers want to design circuits to go with their program, and software companies will not like the added expense of hiring hardware engineers. Thus, the hurdle of generating circuits needs to be removed, or at least made considerably smaller. One way to do this is to have pre–built libraries of circuits, which the programmer can invoke through stubs compiled into their program. To the programmer it will look like a normal function call that is eventually mapped onto hardware.

This technique could be improved, however, by having a post–compilation stage which analyses the output of the compiler for sections of the program that are often called and can be synthesised automatically into RFU designs. This approach is more flexible, releasing software engineers from a fixed set of circuits. This raises the important question of what functionality can we extract from programs to place into circuits? A simple way is to look for anything which takes two inputs and one output as a traditional FU does, but more complex algorithms, especially those based on iteration, may not be so easy to find. Another issue is proving that the synthesised circuit is functionally equivalent to the software instructions it replaces.

5 Current Position

Research on this project will begin in October 1999, funded by EPSRC and Xilinx Edinburgh. In the initial phase we will address the low–level design issues of a microprocessor with a RALU. The viability of such a platform will be examined, investigating such issues as size and number of RFUs, and the type of logic that should go inside them. If this stage is successful then we plan to take an existing conventional CPU design and modify it to include a RALU. This will be simulated, and possibly prototyped, to prove the concept works.

This fundamental low–level work can then be used as a starting point for further work. This includes full analysis of the performance of the proposed architecture compared with both conventional microprocessors and the approaches taken by others (as described in Section 2), along with research into the topics discussed in Section 4.

6 Conclusion

In this paper we have presented a novel architecture for combining traditional CPU design with reconfigurable logic based on the idea of a Reconfigurable ALU, as well as highlighting many research areas that need to be investigated before such a design becomes practical. We argue that this new architecture, along with proper support in the OS and programming tools, will provide a noticeable performance gain for a wide range of applications without a significant overhead for the programmer.

References

1. M. Abrash. Ramblings In Real Time. *Dr. Dobbs Source Book*, November/December 1996.
2. J. Burns, A. Donlin, J. Hogg, S. Singh, and M. de Wit. A Dynamic Reconfiguration Run-Time System. In *IEEE Workshop on FPGAs for Custom Computing Machines*, April 1997.
3. Cyrix Corporation. Cyrix MediaGX Processor Frequently Asked Questions, January 1999.
4. A. Donlin. Self Modifying Circuitry — A Platform for Tractable Virtual Circuitry. In *8th International Workshop on Field Programmable Logic and Applications*, September 1998.
5. E.E. Times. Intel preps plan to bust bugs in Pentium MPUs, June 1997. 30th June.
6. R. W. Hartenstein, K. Schmidt, H. Reinig, and M. Weber. A Novel Compilation Technique for a Machine Paradigm Based on Field-Programmable Logic. In *International Workshop on Field Programmable Logic and Applications*, September 1991.
7. J. R. Hauser and J. Wawrzynek. Garp: A MIPS Processor with a Reconfigurable Coprocessor. In *IEEE Workshop on FPGAs for Custom Computing Machines*, September 1997.
8. Intel Corporation. *Intel Architecture Software Developer's Manual, Volume 1: Basic Architecture*. Intel Corporation, January 1998.
9. S. Sawitzki, A. Gratz, and R. G. Spallek. CoMPARE: A Simple Reconfigurable Processor Architecture Exploiting Instruction Level Parallelism. In *Proceedings of the 5th Australasian Conference on Parallel and Real-Time Systems*, 1998.
10. S. Singh, J. Paterson, J. Burns, and M. Dales. PostScript rendering in Virtual Hardware. In *7th International Workshop on Field Programmable Logic and Applications*, September 1997.
11. C. Small and M. Seltzer. A Comparison of OS Extension Technologies. In *Proceedings of the USENIX 1996 annual technical conference*, January 1996.
12. Triscend Corporation. Triscend corporation web page, January 1999.
13. M. J. Wirthlin and B. L. Hutchings. A Dynamic Instruction Set Computer. In *IEEE Workshop on FPGAs for Custom Computing Machines*, April 1995.
14. M. J. Wirthlin, B. L. Hutchins, and K. L. Gilson. The Nano Processor: a Low Resource Reconfigurable Processor. In *IEEE Workshop on FPGAs for Custom Computing Machines*, April 1994.

Logic Circuit Speeding up through Multiplexing

Valeri F. Tomashau

Institute of Engineering Cybernetics,
Logical Design Laboratory (http://www.bas-net.by/~logic),
220012 Minsk, Belarus
toma@bas-net.by

Abstract. Discrete devices speeding up is the actual aspect of logic design. The combinational circuit delay essentially depends on the number of levels involved. To reduce the number of circuit levels as well as its delay a multiplexing-based technique is proposed. At first, a given Boolean function (more precisely, its SOP) is represented in partially orthogonalized form: $SOP=k_1(SOP_1)+k_2(SOP_2)+...+k_r(SOP_r)$, where the factor-products $k_1,...,k_r$ are orthogonal to each other, and $k_1+k_2+...+k_r\equiv 1$. Then this Boolean function is implemented through multiplexing of the subcircuits corresponding to SOP_i , $i=1,...,r$. Switching of these subcircuits carries out by three-state gates, controlled by mutually orthogonal factor-products k_i , $i=1,...,r$. The decrease of delay is achieved by calculating of the SOP_i and k_i in parallel and thanks to an ordinary wired joint of the subcircuit outputs instead of using the multi-input OR. The method suggested is focused on the speeding up of some critical circuit path in conditions then wired OR is impermissible.

1 Introduction

Timing optimization of a digital circuit is an important problem in logic synthesis. The combinatorial circuit delay essentially depends on the number of levels involved. Each extra level increases the delay and reduces the circuit speed.

However, a multi-level circuit appears to spare more area. To find some compromise between area and speed different resynthesis techniques are usually proposed for multi-level circuit timing optimization.

In this paper another approach concerning the improvement of the critical part of the circuit, specified by some Boolean function, is presented. This approach targets timing optimization through lowering the number of circuit levels. The key to improve the delay by circuit levels lowering lies with the three-state gates (transmission gates), which are used here for dynamic switching from one subcircuit to another.

The main idea of the method is to break down the hole Boolean space ($\{0,1\}^n$ - cube) into some non-intersecting regions in order to replace an initial Boolean function by a set of the more simple ones.

Each of them should be equal to the initial Boolean function within the boundaries of the corresponding region but can be arbitrary redefined beyond it. This opens up wide possibilities for simplification of these Boolean subfunctions, in particular, thanks to the reducing the number of variables on which every subfunction depends. The originally specified behaviour of the circuit is achieved by the appropriate switching of the simple Boolean subfunctions corresponding to the regions mentioned above.

2 The Main Idea

Let a Boolean function $f(x_1,...,x_n)$ be presented in the form of Sum of Products (SOP). Say that Boolean function f is *simple* if it can be represented by the SOP_f wherein the number n of products and the product ranks r_i, $i=1,...,n,$ no greater then number of inputs of the available logic gates (OR, AND, NOR, NAND, ...). A *simple* Boolean function usually can be implemented by the two-level circuit if both variables and their complements are available.

If a given Boolean function is not a *simple* one, only a multilevel SOP-based implementation is possible. The delay of such circuit is essentially defined by its depth measured with the number of circuit levels.

The better way to reduce the circuit delay is to lower the number of circuit levels. For this purposes the transformation of the SOP representing the initial Boolean function f into the form (1) is proposed.

$$SOP_f = k_1(SOP_1) + k_2(SOP_2) + ... + k_r(SOP_r). \tag{1}$$

Here the factor-products $k_1,...,k_r$ are orthogonal to each other, and $k_1+k_2+...+k_r \equiv 1$. To be implemented by no more then two levels, all the factor-products must have a rank no greater then p^2, where p is the number of inputs of a basic logic gate. Each SOP_i must be *simple* in above-mentioned sense to be implemented by two-level circuit. Otherwise the SOP_i itself calls for a further expansion in a similar manner, if its implementation as a multi-level circuit is undesirable.

The transformation (1) will be named the *partial orthogonalization*. The Boolean function represented by its partially orthogonalized form can be implemented through multiplexing, because one, and only one, factor-product k_i $(i=1,...,r)$ is equal to 1 at every instant. It is proposed to use *transmission gates* (tree-state gates) for the purpose of switching from one subcircuit (that implements some SOP_i) to another (that implements some another SOP_j). In fact, some specially adapted multiplexers, distributed among other logic, are built of transmission gates. The given low-level and thereby more flexible method is beneficial to the approach based on the classical multiplexer which calls for the Shannon's expansion under a *fixed subset* of input variables. Note that the expansion of this kind is a special case of more general one (1). In case of the Shannon's expansion under a *fixed subset* of variables each factor-product has the same rank and involves all variables (or their complements) of that *fixed subset*.

Generally the proposed implementation of Boolean function requires a greater number of gates in compare to the corresponding multilevel circuit. However, this approach may be beneficial to re-design (in the goal of speeding up) some critical paths of the circuit in spite of consuming of larger area.

The decrease of delay is achieved by calculating of the SOP_i and k_i in parallel and thanks to the ordinary wired joint of the subcircuit outputs instead of using the multi-input OR. Note that a *Wired-OR* does not take place here because the factor-products, control the transmission gates, are orthogonal to each other, hence, one and only one transmission gate is active at each instant of time.

3 Partial Orthogonalization

To represent a given Boolean function f in the form (1) the *Partial Orthogonalization* should be performed. This procedure applies the Shannon's expansion recursively.

The Shannon's expansion of some current SOP by variable x (2) produces two another SOPs, namely, SOP_x and $SOP_{\bar{x}}$, which should be subjected to the Shannon's expansion at the next step.

$$SOP = x(SOP_x) + \bar{x}(SOP_{\bar{x}}). \qquad (2)$$

The important point is that a variable for the next expansion is chosen independently of the previous choices in accordance with some performance considerations.

It is obvious that this problem has a many solutions similar to other combinatorial problems [1]. How to choose a variable for the next Shannon's expansion more correctly taking into account that the increasing the number of products leads to a rise in the circuit area. Some considerations on this topic are given below.

- It is desirable to choose such variable x, which is present as x or $\bar{x}$ into the greatest number of products of the current SOP. If some product t does not involve x or $\bar{x}$ it should be represented by the sum $t = x\,t + \bar{x}\,t$. Thus the overall number of products is increased and as a result additional gates would be needed.
- The choice of variable x, if an empty (or near-empty) SOP_x (or $SOP_{\bar{x}}$) results from the current Shannon's expansion (2), is not desirable, because the corresponding subcircuit will be also empty (or near-empty). The uniform distribution of products over SOPs is favored.

The basic combinatorial objects associated with the partial orthogonalization procedure are rather economy presented by the *Binary* and *Ternary* matrices. Many efficient logic design algorithms using these matrices may be founded in [1], [2].

4 Example

In the following, an efficiency of proposed technique is illustrated by the multiplexing-based implementation of some Boolean function. The example shows that the delay can be decreased by no less then 25 percent in comparison with its SOP-based implementation.

Let us consider a Boolean function F [3], represented by the next SOP:

$$F = x_1x_2x_3x_4 + x_1x_2x_3x_5 + x_1x_2x_4x_5 . \quad (3)$$

Fig.1 shows two possible implementations of this Boolean function based on the simple gates (OR, AND), as given in [3]. The left-hand circuit is built immediately by the SOP (3), except of the repeated using of the result of the product x_1x_2. The right-hand one is built by an expression (4), obtained from (3) by factorization.

$$F = x_1x_2(x_3x_4 + x_3x_5 + x_4x_5) . \quad (4)$$

The circuit, shown in Fig.1b, uses the smaller area but is not faster, then the circuit, shown in Fig.1a. Its delay is equal to 4τ (τ denotes the delay of 2-input gate). Note that there is 3-input gate in the circuit. To represent it through 2-input gates an additional level is required.

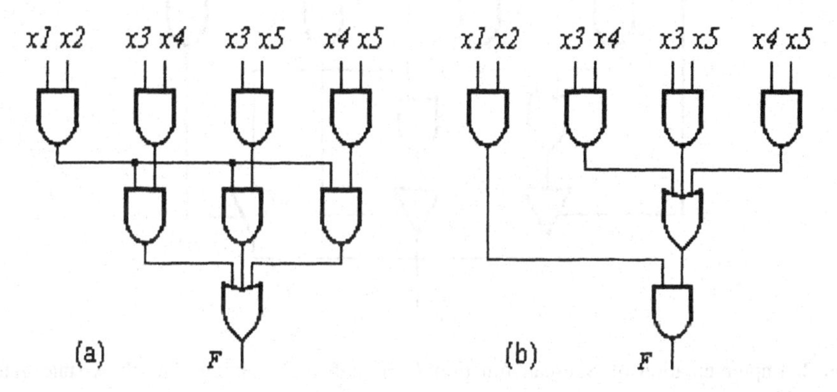

Fig. 1. (a) The circuit built by the SOP (3). (b) The circuit built by the expression (4) resulting from factorizing of (3).

Seemingly, there is no way to speed up this circuit in the framework of 2-input gate without resorting to *wired OR*. However, the suggested technique enables to decrease the delay to 3•. (Note, if *wired OR* is permissible the circuits in Fig. 1 can be transformed in two-level form with the delay of 2•.)

In the beginning, the partial orthogonalization of the SOP (3) of the given Boolean function F should be performed. Let us choose variable x_3 for an initial iteration of the Shannon's expansion. Since variable x_3 does not present in the product $x_1x_2x_4x_5$, the last should be replaced with the sum $x_1x_2x_3x_4x_5 + x_1x_2\bar{x}_3x_4x_5$ where x_3 appears explicitly. The Shannon's expansion by x_3 gives the next result:

$$F = x_3(x_1x_2x_4 + x_1x_2x_5 + x_1x_2x_4x_5) + \bar{x}_3(x_1x_2x_4x_5). \quad (5)$$

Furthermore, the next Shannon's expansion by x_5, applied to an expression enclosed in the left-hand parentheses, brings the expression (5) to the form (6).

$$F = x_3x_5(x_1x_2x_4+x_1x_2+x_1x_2x_4) + x_3\,\bar{x}_5(x_1x_2x_4) + \bar{x}_3(x_1x_2x_4x_5). \quad (6)$$

The expression in the left parentheses (6) may be simplified. First, one of the identical product $x_1x_2x_4$ can be eliminated. Second, the product $x_1x_2x_4$ itself can be eliminated because a product x_1x_2, absorbing the product $x_1x_2x_4$, exists (the cube $x_1x_2x_4$ is covered by the cube x_1x_2).

Finally, Boolean function F is represented by the form (7).

$$F = x_3x_5(x_1x_2) + x_3\,\bar{x}_5(x_1x_2x_4) + \bar{x}_3(x_1x_2x_4x_5). \quad (7)$$

In accordance with suggested technique the partially orthogonalized form (7) can be implemented by the appropriate switching of the subcircuits corresponding to the SOPs, enclosed in the parentheses (Fig. 2).

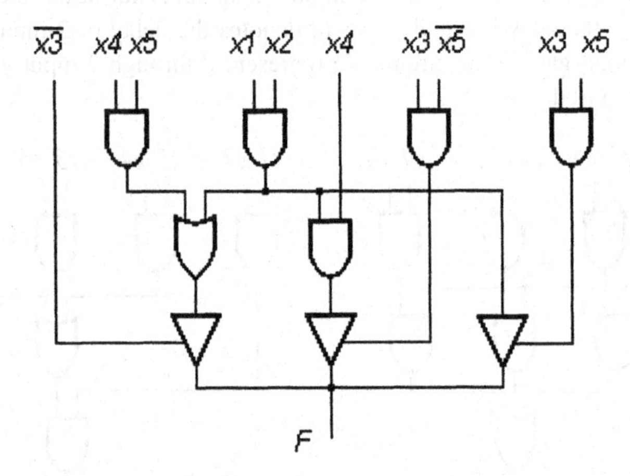

Fig. 2. Implementation of Boolean function F through multiplexing. The three-state gate is denoted by a triangle.

The delay of this circuit equals to 3τ. Thus, the technique makes possible the 25 percent decrease of the circuit delay. However the circuit shown in Fig. 1(b) is 30 percent more effective by area then the circuit in Fig. 2.

5 Results and Further Work

1. The multiplexing-based technique for logic circuit speeding is proposed. It makes possible the 25 percent decrease of the circuit delay when *wired OR is* impermissible.
2. The problem of partial orthogonalization of Boolean function is formulated.

The nearest goal is to develop the efficient algorithm to solve the partial orthogonalization problem.

Further the partial orthogonalization of Boolean function can be generalized to the case of arbitrary regions specified by some Boolean functions, instead of Boolean cubes as it has been considered here.

6 Area of Application

The approach suggested here can be applied in the design of the fast ASICs (Application-Specific Integrated Circuits) or, more precisely, for speeding up their critical parts. The method can be also applied to the ASIC's librarian cells design. Another area of application is the timing optimization of the small scale integrated circuits and the PLD-based circuits.

Note that the suggested technique owes its origin to the search of a non-trivial using [4] of the three-state control tools provided in the PAL-type devices.

Acknowledgments

Author thanks to colleagues from Logic Design Laboratory (Institute Engineering Cybernetics, NAS, Belarus), especially to prof. Arkadij Zakrevskij and prof. Peter Bibilo for useful discussions and help. Thanks to Albina Zakrevskaya for improvement of English whereby this paper initially was written.

References

1. Zakrevskij, A.: Combinatorial Problems over Logical Matrices in Logic Design and Artificial Intelligence. Revista do DETUA, Janeiro, Vol. 2, No. 2 (1998) 261-268
2. Zakrevskij, A.D.: Logic Design of Cascade Circuits. "Science", Moscow (1981) 416p. (In Russian)
3. Badulin, S.S. (ed.): Computer-Aided Design of Digital Devices. "Radio & Communications", Moscow (1981) 240p. (In Russian)
4. Tomachev (Tomashau), V.: The PLD-Implementation of Boolean Function Characterized by Minimum Delay. In: Hartenstein, R.W., Keevallik, A. (eds.): Field Programmable Logic and Applications. Lecture Notes in Computer Science, Vol. 1482. Springer-Verlag, Berlin Heidelberg New York (1998) 481-484

A Wildcarding Mechanism for Acceleration of Partial Configurations

Philip James-Roxby, Elena Cerro-Prada

School of Electronic and Electrical Engineering
University of Birmingham, Edgbaston, Birmingham
United Kingdom, B15 2TT
P.B.James-Roxby@bham.ac.uk

Abstract. Wildcarding is a hardware feature specifically intended to accelerate the configuration process by reducing the number of configuration items required, and programming cells which share the same configuration simultaneously. In order to ascertain which cells can be programmed together, wildcarding has to be built into the address decoding hardware. In the Xilinx XC6200 series, wildcarding operates prior to decoding on bit positions of the target address. In this paper, we consider an alternative scheme, where wildcarding operates after address decoding, allowing individual cells to be addressed. This is shown to be beneficial for accelerating partial configurations where constant propagation is used to give new configurations at run-time, whilst supporting more configuration geometries than the current wildcarding mechanism.

Introduction

Run-time reconfigurable systems operate by modifying the configuration of a programmable device (usually an FPGA) during application execution. However, the configuration process is not without cost : the programmable device is often not available for computation during configuration limiting the possible overall speedup over a software implementation. In order to assist reconfiguration, modern devices are often equipped with hardware features, such as wide parallel configuration paths.

In addition to increasing the width of configuration paths, devices may also offer other hardware features. The Xilinx XC6200 series contains two wildcard registers, which allow multiple cells to be loaded with the same configuration. Wildcarding is useful in configurations offering true random access requiring address-memory pairs. In frame-based configuration, a form of wildcarding could possibly be used to load the same frame at multiple places within a device. Given the large granularity of change of devices currently using frame based configuration, it is difficult to envision a wide range of design examples which could use this wildcarding. For example, the smallest frame size in the Xilinx Virtex family is 384 bits in the XCV50.

This paper considers an alternative to the wildcard register in the XC6200 series. Initially, the existing wildcarding scheme is described with examples of its use at design compile time, and at application run-time. The alternative scheme is defined, and its performance compared to the existing scheme. Finally, some pointers for future work are given.

Wildcarding

The XC6200 is a family of fine-grained sea-of-gates devices, offering partial reconfiguration through a microprocessor interface called the FastMap interface. In order to perform configuration, address and data pairs are presented to the FastMap interface in a similar manner to interfacing to a standard memory device. The address format for the XC6216 is shown below in table 1.

Table 1. Address format for the XC6216

Mode(1:0)	Column(5:0)	Column Offset (1:0)	Row(5:0)
15:14	13:8	7:6	5:0

The operation of the microprocessor interface is described in [1], including the operation of the wildcard registers. It is important to realize that the wildcard registers only affect the column and the row fields of the address. In some cases, three writes to column offsets 0, 1 and 2 are required to reconfigure a cell or a number of cells if wildcarding is used. Also, the number of cells that can be configured simultaneously using wildcarding is limited in the XC6200 series. As this is an implementation issue, we ignore this limitation for this study.

Wildcarding operates before the address decoder, and determines which bits of the row or column address are to be considered for decoding, and which are don't care. Consider the row wildcard RW[5:0] operating with the row address A[5:0]. If the row wildcard is set to 000001, then the value of A[0] does not affect the decoding process. With the wildcard set, when the entire address shown in table 1 is constructed, and presented at the microprocessor interface, the configuration corresponding to the particular mode and column offset will be applied to 2 cells whose location is given by A[5:1]. For example, a write to 00 000000 00 000000 will set the neighbor routing multiplexers for cells <0,0> and <0,1>.

This wildcarding mechanism supports square blocks with sizes which are powers of two at certain locations within the device, often limited by powers of two. An m bit wildcard containing n 1's can configure cells at 2^{m-n} locations, i.e. for the 6 bit row wildcard in the XC6216, blocks of 32 cells require a wildcard containing 5 1's, and can be placed in two locations determined by A[5], in this case <col,0> and <col,32>.

Due to the limited number of locations that can be configured simultaneously, in order to use wildcarding effectively, provision has to be built in at design time. For example, a 32-bit ripple carry adder contained within the bounding box <0,0>, <2,31> can be configured using the wildcard 00011111, but a similar adder at <0,4>, <2,35> would require a series of wildcarded configurations, since there is no unique wildcard register value that covers the bit patterns from 00000100 to 00100011.

Wildcarding can assist in both compile-time reconfiguration and run-time reconfiguration. An example of wildcarding at compile-time is presented in [2]. Wildcarding is used to allow the host processor to compress configuration bitstreams prior to configuring the device without requiring any decoding hardware at the device end. An algorithm is presented which exploits overlay-based reconfiguration that is that a cells configuration can be written many times with different values, but will only implement the last value sent. Therefore, large regular areas can be set using a small number of wildcarded writes, before smaller areas overwrite this configuration. Hauck reports typical compression ratios of around 4:1. In [3], this compression method is compared to more mainstream lossless compression methods including run-length encoding and Lempel-Ziv encoding : Hauck's wildcarding compression method produces similar compression ratios, without requiring additional hardware.

An example of the use of wildcarding at run-time is shown in [4]. The technique of constant propagation is used to construct a custom comparator based on a slowly varying threshold. In order to change the threshold without wildcarding, 8 configuration writes are required. On average with wildcarding, this drops to 7 writes. It is interesting to see how these transfers arise: initially, 2 transfers are required to set the wildcard register to cover all 8 bits of the comparator, and then set the byte of configuration data to configure the slice where the relevant bit in the constant is zero. The wildcard register is set to all zero, which requires a single trans-fer. Then of the 8 slices, it is assumed that on average 4 of them will need to be changed to the case where the relevant bit is one. This requires a further 4 transfers, since the slices for 0 and 1 share a common configuration byte. The comparator can only be placed at certain row positions in the device : since the wildcard is 00000111 containing three 1's, from the expression given above, the number is $2^{(6-3)} = 8$.

Alternative wildcarding method

The row and column wildcards are identical: consider initially the row wildcard. The current wildcarding method in the XC6200 works on bit positions within the row address, and operates before the address decoder. An alternative wildcarding method would work after the address decoder. The wildcard register WR would contain flags for each individual row within the device. A 1 at bit position i within WR means that that row will be written to no matter what the configuration address is. A 0 at bit position i within WR means that that row will only be written to when directly addressed.

Immediately, it can be seen that the wildcard registers increase in size. For a device containing n cells, rather than a row wildcard register requiring $\log_2 n$ bits, the new method requires a row wildcard register with n bits. For example, the 6 (usable) bit wildcard register for the XC6216 would be replaced by a 64 bit register under the new scheme. The new wildcard register effectively acts as an OR on each address line following decoding. An address line can either be selected by explicitly being addressed or by being included in the wildcard register.

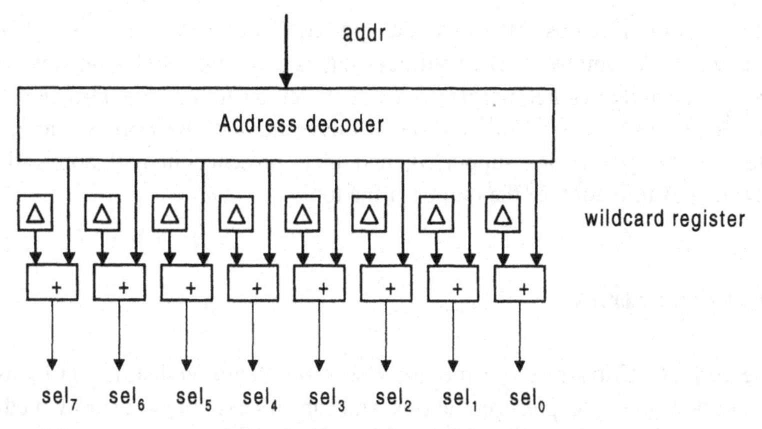

Fig. 1. Alternative wildcarding scheme

Consider initially the case where the column wildcard register WC is all zero : all column addresses must be directly addressed in order to write. The new scheme offers all the supported patterns of the original wildcarding scheme. In order to instantiate a block of four identical cells at <0,col>, the old row wildcard would be set to 00000011, followed by a write to <0,col> with the required configuration data. Under the new scheme, the 8 least significant bits of the row wildcard register would be set to 00001111 followed by a write to <0,col>.

One specific advantage of the old wildcarding scheme is that it can be used to instantiate same sized blocks at different offsets without requiring a rewrite of the wildcard register. Thus blocks which cannot be written using a single wildcard write due to their location could be written by a series of smaller but fixed, wildcard writes. For example, 8 identical cells could be set up vertically from <4,col> by two 4 bit wildcarded writes. Under the new scheme, same sized blocks at different offsets would require a rewrite of the wildcard register. If they were in the same column, they could be written by 1 wildcarded write, no matter how they were separated.

Software support

A tool has been written to assist in the search for useful wildcarding sequences. Initially, the configuration file is parsed, and each address is split into its four constituent fields : mode, column offset, , row and column. Wildcards only affect the areas of configuration memory dealing with actual cell configuration, namely modes 00, 01 and 10. They do not affect writes to configuration registers, mode 11 : these writes are simply maintained and are not considered as candidates for wildcarding. The tool sorts the configuration file into ascending data items order, and sorts each of these into mode and column offset order, since the same data item with different modes or column offsets would require separate writes. Currently, the tool outputs a list of candidates for wildcarding as comma separated values for manual analysis in a spreadsheet package.

We believe it would be possible to use the approach described by Hauck in [2] to determine an efficient compression of configuration data. Unlike Hauck however, we are working on a wholly synthetic problem, and any demonstrated compression could not be realised in practice. For this reason, we have limited our work so far to an investigation of the geometries supported, and their presence in real applications based on manual examination of the output of the tool.

Supported geometries

As with the current wildcarding scheme, the row and column wildcards can be used together to support quite complex geometries, and can still exploit the overlay method of reconfiguration discussed previously. An example geometry supported by the new mechanism are given below. Due to the location of the configuration, this would require 21 wildcarded writes under the existing XC6200 wildcarding scheme, compared to 11 under the new scheme.

Supporting partial configurations for constant propagation

In order to effectively use constant propagation for performance gains, it is important to have a mapping between a constant and the logic circuit which embodies the constant which can be calculated in a short time. Otherwise, the gains made from the faster, specialized hardware will be offset by the time to calculate the new circuit configuration plus the time to download the new configuration data.

Often, constant propagation gives rise to a circuit comprising of a number of slices, whose configuration depends on whether a bit at a certain position within the constant was 1 or 0. For example, in the case described in [4], constant propagation is applied to a comparator giving rise to slices consisting of either an AND gate or an OR gate. The new wildcarding scheme is very efficient at mapping such circuits. The wildcard is initially set to the complement of the constant (or to cover the whole constant), and the slice configuration corresponding to zero is written. The wildcard is then set to

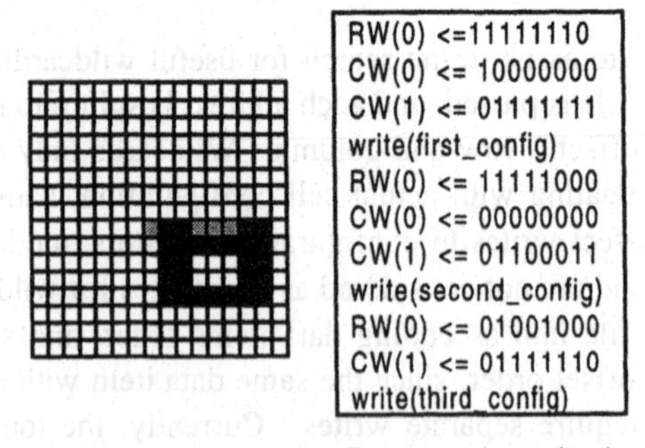

Fig. 2. Example supported geometry

the constant value, and the slice configuration corresponding to one is written, as shown below for the value 74.

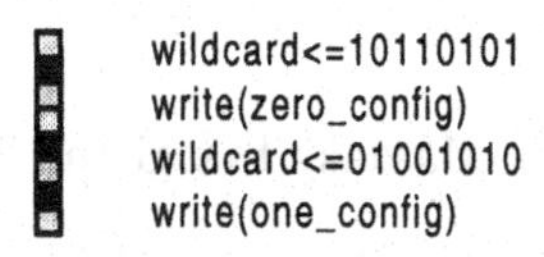
wildcard<=10110101
write(zero_config)
wildcard<=01001010
write(one_config)

Fig. 3. Accelerating constant propagation

Conclusions and future work

Wildcarding is an important technique for reducing the configuration overhead which may limit the speed-up possible from reconfigurable computing applications compared to software or fixed hardware implementations. We have presented an alternative wildcarding scheme to the address bit based scheme implemented in the XC6200 family, and have shown that the alternative scheme supports a number of configuration geometries commonly found in designs.

Future work includes automating the search for efficient wildcards, to allow a full comparison between this method, and the work of Hauck with the existing wildcard mechanism. An automated method would also help in the search for common configuration geometries. Also this study has not considered the silicon overhead required for each decoding scheme which we intend to address in future work.

Finally, we wish to investigate using 32 bit transfers rather than the 8 bit transfers considered here. Whilst increasing the width to 32 bits would allow more of the wildcard to be written at one time, effectively the configuration granularity is increasing possibly limiting the amount of wildcarding that can be performed, and we believe this needs more investigation.

References

1. Churcher, S., Kean, T., Wilkie, W. : The XC6200 FastMap™ processor interface. In: Moore, W., Luk, W. (eds.): Field Programmable Logic and Applications. Lecture Notes in Computer Science, Vol. 975. Springer-Verlag Heidelberg New York (1995) 36-43
2. Hauck S., Zhiyuan, L., Schwabe, E. : Configuration compression for the Xilinx XC6200 FPGA. Proc. FCCM98 (1998) 138-146
3. Hauck S., Wilson, W. : Runlength compression techniques for FPGA configurations. Preliminary Proc. FCCM99 (1999)
4. Cerro-Prada, E. Charlwood, S., James-Roxby, P : Constant propagation by partial reconfiguration in the Xilinx XC6200 FPGA. Proc. PREP99 (1999) 311-321

Hardware Implementation Techniques for Recursive Calls and Loops

Tsutomu Maruyama, Masaaki Takagi and Tsutomu Hoshino

Institute of Engineering Mechanics, University of Tsukuba
1-1-1 Ten-ou-dai Tsukuba Ibaraki 305 JAPAN
maruyama@darwin.esys.tsukuba.ac.jp

Abstract. Field Programable Gate Arrays (FPGAs) begin to show better performance than microprocessors in many application areas because of drastic improvement of the size and speed. In the near future, FPGAs will be directly attached to or involved in microprocessors as accelerators that execute algorithms written in programming languages. In this paper, we show hardware implementation techniques (multi-thread execution and speculative execution) for recursive calls and loops, which are the most time exhaustive parts in many application programs written in programming languages. These techniques can be employed with very little overheads in clock cycle speed and circuit size. Experiments on simple combinatorial problems show 4.1 - 6.7 times of speedup compared with a workstation (Ultra-Sparc 200MHz).

1 Introduction

Many approaches toward accelerators based on reconfigurable hardwares have been proposed to date. The size and the speed of field programmable gate arrays (FPGAs) have been drastically improved recently, and even the small systems begin to show much better performance than microprocessors in mary application areas. In the near future, FPGAs will be directly attached to or involved in microprocessors as accelerators[1–3], and execute algorithms written in programming languages.

Recursive calls and loops are the most time exhaustive parts in many algorithms written in programming languages, and the high speed computation by accelerators is expected. For example, tree search (combinatorial) problems which are very important problems in many application areas, require a lot of computation time and the most of the time is spent in recursive calls and loops.

In this paper, we show hardware implementation techniques for recursive calls and loops in reconfigurable hardwares. The techniques (multi-thread execution and speculative execution) which are well-known in computer architecture designs show drastic performance improvement in reconfigurable hardwares. These techniques can be employed with very little overhead in clock cycle speed and circuit size. Experiments on combinatorial problems show very good results (6.7 times speedup in the multi-thread execution and 4.1 times speedup in the speculative execution compared with Ultra-Sparc 200MHz).

2 Multi-thread Execution

2.1 Multi-thread Execution of Tree Search Problems

In this section, we show implementation techniques for problems that require the best solutions searching all search spaces. Figure 1 shows a program that solves knapsack problem[4]. The last recursive call in figure 1 can be translated to loop instruction (tail recursion optimization). Therefore, the structure of the program becomes as shown in Figure 2.

```
try(int i, int tw, int av, int obj_set)
{
    int avl;

    /* part-A */
    if (tw + obj[i].w <= limw) {
        if (i < N-1)
            try(i+1, tw+obj[i].w, av, obj_set|(1<<i));
        else {
            if (av > maxv) {
               maxv = av;  max_obj_set = obj_set|(1<<i);
            }
        }
    }
    /* part-B */
    avl = av - obj[i].v;
    if (avl > maxv) {
        if (i < N-1) {
            try(i+1, tw, avl, obj_set);
        } else {
            maxv = avl;  max_obj_set = obj_set;
        }
    }
}
```

Fig. 1. Program 2: Knapsack Problem

```
loop:
    cal-A:
    if (cond-A)
        recursive-call();
    cal-B;
    if (cond-B)
        goto loop;
```

Fig. 2. The Structure of the Program 1

Figure 3 shows a block diagram for sequential execution of figure 2 and how the stages become active (suppose that `cal-A` and `cal-B` in figure 2 require two clock cycles respectively). In figure 3, when `cond-A` becomes true in stage S2, the current environment variables are pushed on the stack and the new environment variables for `recursive_call()` are forwarded to S0 in S3. When `cond-B` becomes true in stage S6, new environment variables are just forwarded to S0 in S7. When `cond-B` is false, environment variables on the stack are popped, and the computation for the environment is resumed from stage S4. In this sequential execution, only one of the eight stages becomes active at a time.

Figure 4 shows a block diagram for multi-thread execution. In figure 4, when `cond-A` in S2 becomes true and `recursive-call()` in S3 is called, stages S4 - S7 for the current environment variables are continuously executed. The new environment for `recursive-call` is forwarded to stage S0, if S7 is idle (S0 is not occupied by the thread on S7 in the next cycle), or pushed on the stack if not. The environment variables on the stack is popped and forwarded to stage S0, when S3 is idle (no push operation) and S7 is idle.

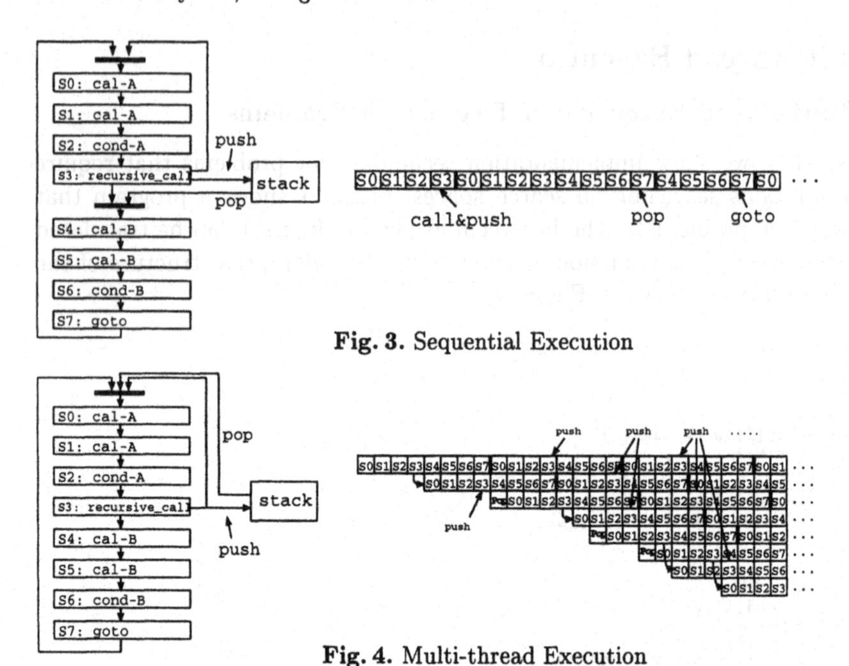

Fig. 3. Sequential Execution

Fig. 4. Multi-thread Execution

In this multi-thread execution, all stages becomes active at the same time as shown in figure 4. The maximum speedup by this methods is the depth of the pipeline. Therefore, by processing more complicated programs which require more pipeline stages, we can expect higher performance, though more amounts of hardware resources become necessary. The difference from the sequential execution is very small, and there are almost no overheads in hardware size and clock speed.

2.2 Evaluation

Figure 5 shows a multi-thread pipeline for the knapsack problem implemented on a FPGA chip (ALTERA EPF10K100). The structure of figure 5 is exactly same with figure 4.

All memories including a stack are implemented using internal RAMs of a FPGA chip because of the two reasons. First, three memory accesses may be executed at the same time (array accesses in S0 and S4, and stack access in S3). And second, the width of the stack is very wide (69 bits: 5 bits for 32 objects (`i`), 16 bits for weights (`tw`), 16 bits for values (`av`) and 32 bits for current status (`obj_set`: object is selected or not selected)).

Table 1. Comparison of Computation Speed of Knapsack Problem (32 objects, average of 100 runs)

Ultra-Sparc 200MHz	1.00	(0.84 sec)
sequential execution (8 stages) 35MHz	0.74	(1.13 sec)
sequential execution (4 stages) 35MHz	1.55	(0.55 sec)
multi-thread execution (8 stages) 35MHz	6.68	(0.13 sec)
multi-thread execution (4 stages) 35MHz	6.70	(0.13 sec)

```
S0:
   w = obj[i].w;
S1:
   nw = tw + w;  iMax = N - 1;
S2:
   c1 = (nw<=limw) && (i<iMax);  c2 = (nw<=limw) && (i>=iMax);
   n_i = i + 1;   n_obj_set = obj_set | (1 << i);
S3:
   if (c1) {
      if (S7!=valid||(S7==valid&&!c3_in_S7)) forward_to_S0(n_i,nw,av,n_obj_set);
      if (S7==valid&&c3_in_S7) push(n_i,nw,av,n_obj_set);
   }
   if (c2 && av>maxv && av>=av1_of_S7) {maxv = av;max_obj_set=n_obj_set;}
S4:
   v = obj[i].v;
S5:
   av1 = av - v;
S6:
   c3 = (av1>maxv) && (i<iMax); c4 = (av1>maxv) && (i>=iMax);
   n_i = i + 1;
S7:
   if (c3) {i = n_i; av = av1; goto S0;}
   if (c4 && av1>maxv && av1>av_of_S3) {maxv = av1;max_obj_set=obj_set;}
--------------------------------------------
POP: (this stage is always active)
   if ((S3!=valid||(S3==valid&&!c1_in_S3))&&(S7!=valid||(S7==valid&&!c3_in_S7)))
       pop & goto S0;
```

Fig. 5. A Multi-thread Pipeline for Knapsack Problem

Table 1 compares the computation speed of each method. In figure 5, the pipeline can be reduced to 4 stages by executing S0 and S4, S1 and S5, S2 and S6, S3 and S7 at the same time respectively. The results in the table are the average of 100 runs with different objects sets (values and weights are decided at random) because the amounts of the computation by each method are different owing to the branch-and-bound method used in the program. In table 1, multi-thread execution shows very good results regardless of the pipeline depth. Maximum operation clock speed of the circuit is 35 MHz, and 17% of logic cells of ALTERA EPF10K100 is used.

3 Speculative Execution

3.1 Pipeline Evaluation of Branches of Search Trees

In this section, we show implementation techniques for problems that require only one solution as fast as possible. Figure 6 shows a program for knight's tour problem. The multi-thread execution method described in the previous section can improve the total search time to find all solutions, but are not effective for finding one solution faster.

Figure 7 shows the structure of the knight's tour program, and figure 8 shows how the stages become active in speculative execution. In figure 8, loop index variable `i` is incremented each clock cycle, and stages S0 - S2 for different value of `i` are executed continuously. If `cond-A` for `i` becomes true in stage S2, computations for `i+1,i+2` and `i+3` are cancelled, and `recursive_call()` for `i` is executed at stage S3. In this stage, current environment variables are pushed on the stack, and the new environment variables for `recursive_call()` are forwarded to S0. The computation for the new environment variables is started

```
try(int x, int y, int array[N][N], int n_move)
{
    int i, nx, ny;

    for (i = 0; i < 8; i++) {
        if ((nx = x + move[i][0]) < 0 || nx >= N) continue;
        if ((ny = y + move[i][1]) < 0 || ny >= N) continue;
        if (array[nx][ny] == 0) {
            array[nx][ny] = n_move;
            if (n_move == N*N) {
                print_array(array); exit(0);
            } else
                try(nx, ny, array, n_move+1);
            array[nx][ny] = 0;
        }
    }
}
```

Fig. 6. Program 3: Knight's Tour

```
for (i = 0; i < N; i++) {
    cal-A;
    if (cond-A) {
        recursive-call();
        cal-B;
    }
}
```

Fig. 7. The Structure of Knight's Tour Program

from i=0 in stage S0. When cond-A becomes false for all value of i, environment variables on the stack are popped at stage S3 and the computation for the environment is resumed from stage S4.

3.2 Evaluation

Figure 9 shows a pipeline implemented on the FPGA, and table 2 compares the computation speed of each method. The computation speed by sequential execution is faster than the workstation because many operations are executed in one stage as shown in figure 9 (stage S2 and S3). The speedup by speculative execution is 2.36 (4.06/1.72 = 2.36).

Table 2. Comparison of Computation speed of Knight's Tour Problem (size=8×8)

Ultra-Sparc 200MHz	1.00 (15.1 sec)
Sequential Execution 31MHz	1.72 (8.78 sec)
Speculative Execution 31MHz	4.06 (3.72 sec)

4 Conclusions

We showed that FPGAs can achieve high performance in the computation of recursive calls and loops which are the most time exhaustive parts in many

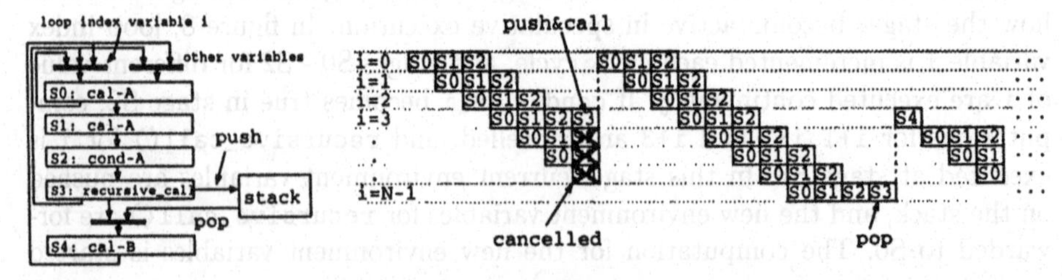

Fig. 8. Speculative Execution

```
S0:
   dx = move[i][0]; dy = move[i][1];
   {
     sc = (i < 8);  // stage control
     i = i + 1;     // loop control var.
   }
S1:
   nx = x + dx;  ny = y + dy;
S2:
   c1 = ((nx>=0) && (nx<N)) && ((ny>=0) && (ny<N)) && (array[nx][ny]==0);
   c2 = (n_move == N_x_N);  c3 = (i == 7);
   n_n_move = n_move + 1;   ni = i + 1;
S3:
   if (c1) array[nx][ny] = n_move;
   if (c1&c2) goto print_array & exit;
   if (c1&!c2) {
      push (ni,x,y,n_move,nx,ny,c3);                    // push
      i = 0; x = nx; y = ny; n_move = n_n_move;         // forward
      cancel S0-S2 & goto S0;
   }
   if (!c1 && c3 && !stack_empty) pop;
S4:
   array[nx][ny] = 0;
   if (c3 && !stack_empty) goto POP;
   if (!c3) {i = ni;  goto S0; }
```

Fig. 9. A Pipeline Implemented on a FPGA

tree search (combinatorial) problems. Our current implementations on FPGA (ALTERA EPF10K series) run more than 30 MHz and the speed up over Ultra-Sparc 200 MHz is 6.7 times in the knapsack problem, and 4.1 times in the knight's tour problem. The speedup is limited by the speed of wide bit-width adders and selectors (more than 16 bit). And, the problem sizes are limited (in order to maintain the performance) by the speed of external memory accesses (speed of I/O pins). With a FPGA that has special supports for these functions, we will be able to achieve more speedup for larger problems.

The implementation techniques (pipeline processing by multi-thread and speculative execution) can be applied other tree search (combinatorial) problems by changing only computation parts which can be easily extracted from programs written in programming languages, without changing pipeline control circuits. The speedup by the techniques is almost proportional to the depth of the pipeline (the amount of the hardware resources used for the computation), and we can expect more speedup for more complex problems. We are now developing a software system that generates circuits for multi-thread and speculative execution from programming languages.

References

1. Timothy J.Callahan and John Wawrzynek, "Instruction-Level Parallelism for Reconfigurable Computing", Field-Programmable Logic and Applications, 1998, pp.248-257.
2. John R. Hauser and John Wawrzynek, "Garp: A MIPS Processor with a Reconfigurable Coprocessor", FPGAs for Custom Computing Machines, 1997 pp.12-21.
3. Ralph D. Witting and Paul Chow, "OneChip: An FPGA processor with reconfigurable logic", FPGAs for Custom Computing Machines, 1996 pp.126-135.
4. Niklaus Wirth, "Algorithms and data structures", Prentice-Hall, 1986

A HW/SW Codesign-Based Reconfigurable Environment for Telecommunication Network Simulation

Juanjo Noguera[1], Rosa M. Badia[2], Jordi Domingo[2], Josep Solé-Pareta[2]

[1]Dept. Computer Science, Universitat Autònoma de Barcelona
Edifici C, Campus UAB, E08193 Bellaterra, SPAIN
juanjo@cnm.es

[2]Dept. Computer Architecture, Universitat Politècnica de Catalunya
Campus Nord - Mòdul D6, C/ Jordi Girona, 1-3 E08034 – Barcelona, SPAIN
{rosab, jordid, pareta}@ac.upc.es

Abstract. Sequential network simulation is a high time-consuming application, and with the emergence of global multihop networks and gigabit-per-second links is becoming a challenging problem. A new approach to this open problem is presented, based on HW/SW co-design. A complete modular and scalable reconfigurable system architecture is explained. Most important features of this simulation framework are: (1) efficient and flexible network simulation, and (2) transparent use of the reconfigurable system by telecommunication networks engineers because of the use of a high level network modeling language.

1 Introduction

Network engineers and researchers routinely use simulations in their daily network design, analysis and evaluation tasks. With the emergence of global multihop packet networks and gigabit-per-second links, the network simulation community is faced with new and significant challenges. First, actual packet traffic is dominated by long-range correlations, which means that realistic models have to be simulated for very long time-scales. Second, network configurations of really large size have to be simulated to study issues such as scalable routing or packet loss probability.

The capabilities of sequential simulation techniques, are inefficient to address such simulation requirements, due to the several days-long simulation execution time. Parallel computing architectures [3] and reconfigurable computing techniques [1, 2] can be used for simulation execution time improvement. However, the lack of established easy-to-use modeling methodologies suitable for parallel execution or reconfigurable hardware mapping, the absence of mature software environments and comprehensive feasibility demonstrations have prevented the widespread use of such techniques in network research and industrial practice.

This paper presents a complete HW/SW codesign-based reconfigurable network simulation framework devoted to new network simulation challenges. The rest of this paper is structured as follows: in section 2, the basics of our new approach are explained. The reconfigurable system architecture is presented in section 3. Finally, section 4 summarises the conclusions of our research.

2 Our New Approach to Telecommunication Network Simulation

Our approach to telecommunication network simulation tries to overcome problems presented by existing simulation frameworks, which have been summarized in [4]. Desired features in our new simulation framework are:

- Efficient simulation. It must be able to simulate long time-scales and network configurations of really large size in an acceptable execution time.
- General purpose simulation framework. It must be capable to simulate actual, as well as, future telecommunication networks.
- Easy use and a well defined working methodology must be desired. That is, the acceleration technique must be transparent to the telecommunication user.
- Low cost. The cost of our system must be affordable by all telecommunication research community, and lower than the cost of a parallel computer.

A network modeling language (i.e TeD [3]) will be used to obtain the feature of easy use, so networks engineers only have to be concentrated in network modeling. To achieve the features of low cost and efficient simulation, a reconfigurable computing system will be used. Finally, a HW/SW codesign tool will be the responsible for mapping the network model to the hardware platform, achieving the desired transparency in the use of the reconfigurable system.

2.1 Simulation Algorithm

Our simulation algorithm is based on event-driven simulation paradigm. Sequential simulators utilize three data structures: (1) the state variables that describe the state of the system, (2) an event list containing all pending events that have been scheduled, and (3) a global clock variable to denote how far the simulation has progressed.

Basically, an event-driven simulator works on a producer-consumer manner. There are events to process which are saved in an ordered list (event memory), and simulation processes (P_i) which are responsible for the event processing. A simulation controller is required for assigning events to processes, save new events generated by processes and ordering the event list. See fig. 1.

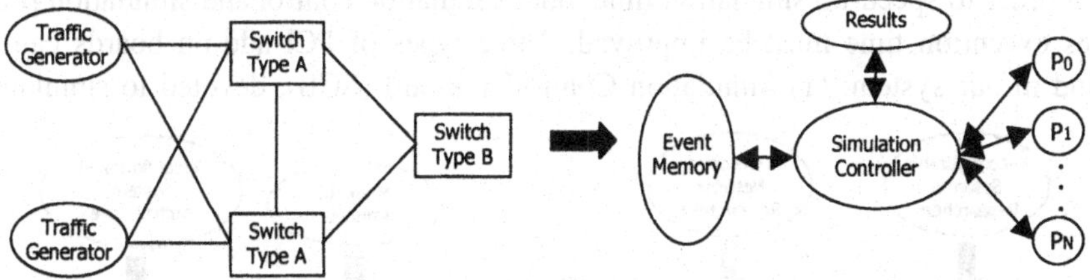

Fig. 1. Network Elements to Simulation Processes mapping

The key point when planning to design and develop a network simulation framework is how to implement the mapping between network elements and simulation processes. Mainly, two approaches are possible: (1) One network element is mapped on a simulation process, and (2) Several network elements with the same behavior (or the same type) are mapped on a single simulation process.

2.2 HW/SW Codesign-Based Approach

A network event-driven simulation acceleration, as explained in the previous subsection, is a perfect application for HW/SW codesign methodology. The main idea is to implement using reconfigurable devices, as many simulation processes as possible, with the final objective to speed-up simulation execution time.

As stated in the introduction, a new challenge in network simulation is the simulation of very large scale networks, during long time-scales. We plan to use the multiple-to-one mapping strategy between network elements and simulation processes, in order to simulate large scale networks. It is obvious that the number of reconfigurable devices is limited, and as the number of different types of network elements increases, a selection of simulation processes to be implemented using reconfigurable devices must be done: the most time-consuming simulation processes will be implemented using reconfigurable devices. Our new network simulation framework will be based on a HW/SW codesign tool. The main tasks or functions of this HW/SW codesign tool are:

- Obtain a list of all simulation processes from the network model which is specified using a high-level network modeling language (i. e. TeD).
- Perform the partition between processes that have to be mapped into the reconfigurable hardware and the processes that have to be mapped to software.
- VHDL code generation of simulation processes that have to be implemented in hardware, which could be synthesized using commercial tools.
- C code generation for the simulation processes that have to be implemented in software, which could be executed in any commercial microprocessor.

3 Reconfigurable System Architecture

Our main goals, when planning to design the reconfigurable system architecture, are to implement a low-cost, efficient, modular and scalable reconfigurable system. For such purposes a PCI-based system architecture is proposed; see fig. 3.

In order to speed-up simulation time, both simulation control and simulation processes execution time must be improved. Three types of PCI add-on boards can be found in our system: (1) Simulation Controller Board (SCB), devoted to simulation

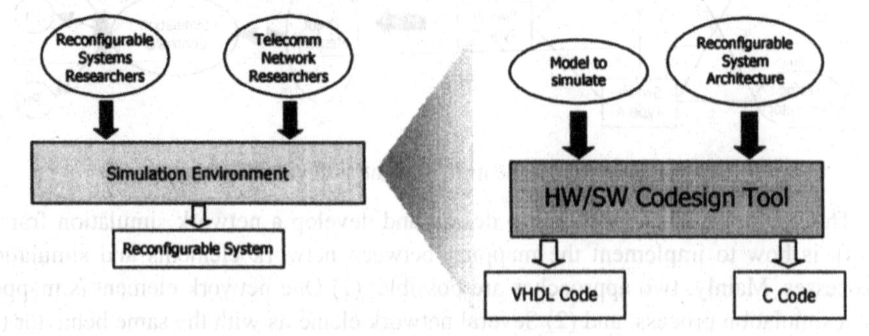

Fig. 2. HW/SW Codesign based approach

control execution time improving. (2) Simulation Compile-time-reconfiguration Processing Board (SCPB) and (3) Simulation Run-time-reconfiguration Processing Board (SRPB), devoted to simulation processes execution time improving. The SCPB will be used to implement coarse grain simulation processes, while the SRPB will implement low granularity simulation processes. PCI add-on boards are interconnected through an internal dual-bus in a daisy-chain manner, for fast event transfer. Simulation processes implemented using SW will be mapped to the host CPU.

3.1 Simulation Controller Board Architecture

The SCB architecture can be found in fig. 4. The main function of this board is to control the event list processing, basically inserting and deleting events. The time-stamp ordered event list is stored in the event memory, which is composed of several memory blocks to obtain a higher flexibility (event size and events of different type). Each event will be composed of a fixed data structure and an application-dependant data structure. The basic fields of the fixed data structure are: (1) the timestamp, (2) the processing element identifier, that is which CPLD will execute the event, and (3) the network element identifier, as we are using a multiple-to-one mapping strategy.

The first event in the list will be sent to the processing boards chain through the events-to-process bus, until the event arrives to the right processing element. If the processing of such event generates a new event, this event will be transported through the event-to-save bus until it arrives to the SCB which will store it in the event list.

The SCB will be responsible for the SRPB reconfiguration control through the reconfiguration control bus, based on the next event to process (event list) and the state of the SRPB (which simulation processes are "active").

3.2 Simulation Compile-Time-Reconfiguration Processing Board Architecture

The SCPB board architecture (see fig. 5) is composed of a linear array of Compile-time-reconfiguration Processing Elements (CPE) obtaining a superscalar architecture. Several CPEs can be mapped into a single CPLD. Each CPE is composed by a Processing Element (PE) and a SRAM memory block. A PE corresponds to a simulation process type. The SRAM memory block is used for saving the different states of the different network elements which are mapped into the single CPE. Processing an event by a PE means: (1) load the state of the network event to which the event is

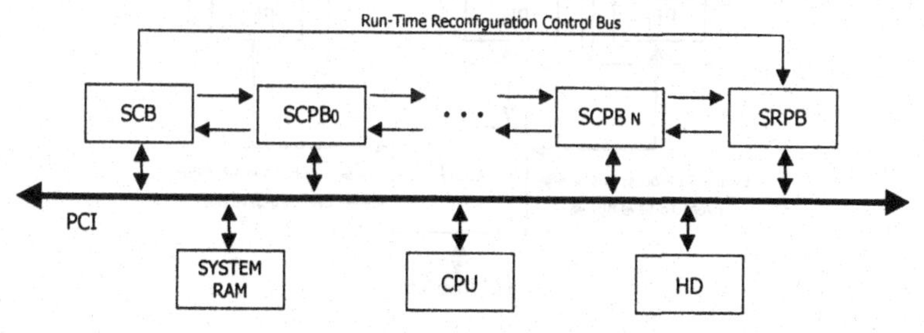

Fig. 3. PCI-based System Architecture

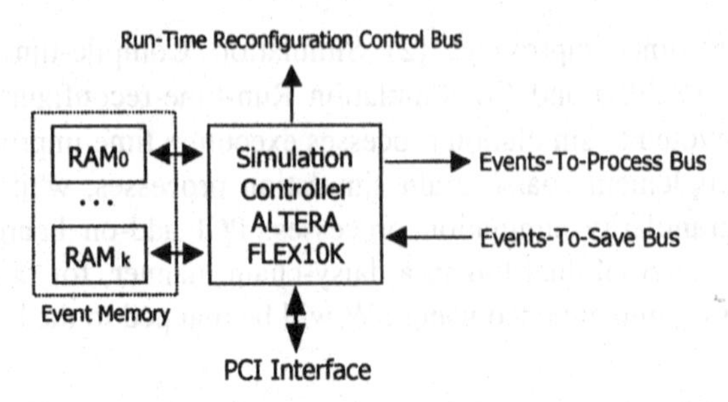

Fig. 4. SCB (Simulation Controller Board) Architecture

addressed, (2) process the event and (3) generate a new event to the SCB, if it is necessary. Another element present in the SCPB is the Control Board Unit (CBU). When an event reaches a concrete SCPB, the CBU checks the event processing elements identifier. If it corresponds to a PE present in the board, the event is processed, otherwise, the event is sent to the next processing board (SCPB or SRPB) present in the daisy-chain. The CBU is also responsible of the communication between CPEs and the PCI interface.

3.3 Simulation Run-Time-Reconfiguration Processing Board Architecture

The SRPB board architecture (see fig. 6) is composed of an array of global Run-time-reconfiguration Processing Elements (RPE) connected to the CBU which mainly works as in the SCPB. Each RPE is composed by a CPLD and a Reconfiguration-RAM (R-RAM) memory block. Each CPLD will be used to implement a single low-granularity simulation process type, at the same time. In the R-RAM memory block, different CPLD configurations are stored, corresponding to different network element types. A concrete CPLD configuration is loaded upon a run-time reconfiguration message arrival to the CBU through the reconfiguration control bus. The reconfiguration message is sent by the SCB, based on the events scheduled in the event list. A look-ahead technique is used to minimize reconfiguration time penalties, with the final objective that the CPLD is configured before the event arrival. The state of network elements is stored in a shared memory block, connected to the CBU.

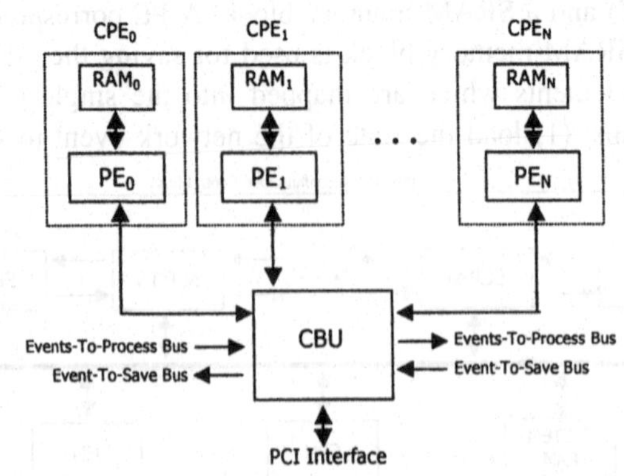

Fig. 5. SCPB (Simulation Compile-time-reconfiguration Processing Board) Architecture

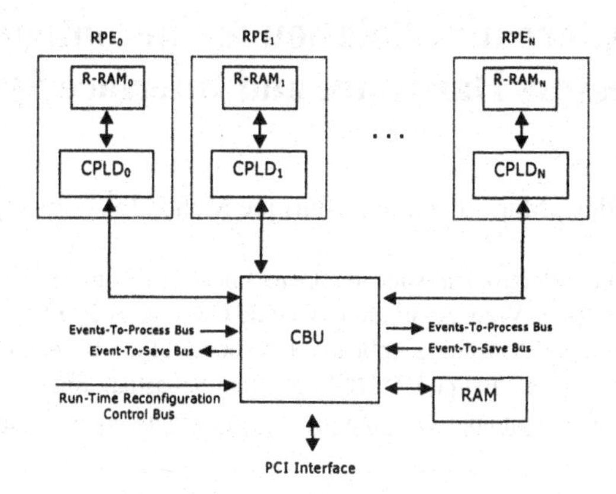

Fig. 6. SRPB (Simulation Run-time-reconfiguration Processing Board) Architecture

4 Conclusions

A new simulation framework has been proposed, which is focused on network simulation challenges which are not solved nowadays by existing simulation environments. A HW/SW codesign approach based on reconfigurable devices will be used, achieving an efficient simulation framework where reconfigurable hardware is transparent to telecommunications engineers.

The PCI-based reconfigurable system architecture has been explained. Special features of the proposed system are its modularity and scalability. The internal archtecture of each PCI board type (SCB, SCPB and SRPB) has been explained. A prototype of the proposed system will be based on Altera ARC-PCI platform. A more accurate study of the codesign tool is subject of actual work, with emphasis in the partitioning.

Acknowledgements

This work has been founded by CICYT projects TIC98-0410-C02-01 & TEL97-1054-C03-03, and by the Altera Programmable Hardware Development Program.

References

1. A.Touhafi, W.F.Brissinck, E.F.Dirkx, "Simulation of ATM switches using Dynamically Reconfigurable FPGA's" FPL'98. Lecture Notes in Computer Sciences 1482.
2. D. McConnell, P. Lysaght, "Queue Simulation Using Dynamically Reconfigurable FPGAs", UK Teletraffic Symposium, Glasgow, March 1996.
3. S. Bhatt, R. Fujimoto, A. Ogielski, K. Perumalla, "Parallel Simulation Techniques for Large-Scale Networks". IEEE Communications Magazine, pp. 42-47. August 1998.
4. J. Noguera, R. Badia, J. Domingo, J. Sole, "Reconfigurable Computing: an Innovative Solution for Multimedia and Telecommunication Network Simulation". Proceedings of the 25th EUROMICRO Conference. Milan, September 1999. IEEE Computer Society.

An Alternative Solution for Reconfigurable Coprocessors Hardware and Interface Synthesis

María D. Valdés[1], María J. Moure[1], Enrique Mandado[1], and Angel Salaverría[2]

[1]Instituto de Electrónica Aplicada. Department of Electronic Technology.
University of Vigo. Apartado Oficial de Correos. 36200 Vigo. Spain
[2]Instituto de Electrónica Aplicada. University of the Basque Country.
Avda. Felipe IV, 1B. 20011 San Sebastian. Spain
Mvaldes@uvigo.es, mjmoure@uvigo.es, emandado@uvigo.es, jtpsagaa@sp.ehu.es

Abstract. The design of co-processing and interface functions required in processor based control systems (PBCS) is subject to a hardware/software (Hw/Sw) co-design process that results in a tedious task demanding a lot of time and work. This paper proposes an alternative solution for re-configurable co-processors rapid prototyping using a library containing over 200 reusable functions that makes the hardware design and synthesis process easier. The library has been developed to be implemented in a Xilinx FPGA but it can be easily portable to other FPGAs families.

1 Introduction

A typical digital processing system comprises a set of communicating components some of them implemented in software and others in hardware. Generally, software is used for features and flexibility and hardware is used for performance. Hardware implementation allows tasks being able to meet hard real time constraints and reduces the system execution time. In this way, the hardware/software co-operation allows a higher system performance [1].

In the majority of the cases digital processors (microprocessors, microcontrollers or DSPs) execute software tasks while hardware tasks are implemented using glue-logic that can be considered as a very simple co-processor with limited capabilities.

Common functions that can be mapped using hardware are timing processing and interrupt control functions. On the other hand, there are many microcontroller applications where the internal interfacing resources of the are not enough and external interfaces is required. From this perspective the co-processor can be used to support not only co-processing functions but interfacing digital circuits as for example RAM or ROM memories, input/output ports, or address decoders [2].

The design of co-processing or interface logic has many different constrains, including timing, reusability, cost, power consumption, or configuration flexibility. Therefore, generally, the development of such a kind of systems becomes a tedious task requiring a lot of time and work.

The aim of this paper is to present a software tool oriented to co-processor systems rapid prototyping using reconfigurable logic devices (FPGAs). The developed tool makes the design, verification and implementation of some timing processing functions, interrupt control functions and interfacing logic easier.

2 Hardware/Software Design Methodology

Prototyping of Hw/Sw systems encloses many phases requiring different CAD tools. Figure 1 shows a summarized diagram of a Hw/Sw co-design methodology:

- *System behavior:* Comprises the general description of the system behavior specifying if it is a general or a specific purpose system and the definition of the different performing tasks.
- *Mapping:* Comprises the definition of the software and the hardware tasks.
- *Mapped specification:* Comprises the definition of the design constraints associated to each part of the system (task) including timming, power consumption, reconfiguration, etc.
- *Software synthesis:* Comprises the edition, compilation and debugg of the software tasks to be executed by the processor. Depending on the processor the software synthesis use assembly or high-level languages and specific or general purposes tools.
- *Hardware synthesis:* The co-processor implementation differs from one application to another. It can be an embedded or an external system using fixed or reconfigurable architecture and depending on hardware characteristics the designer will use a synthesis process or another.
- *Interface synthesis:* Interface between software and hardware tasks is generally made of a data transference protocol and a hardware support. The first one responds to the communication protocol of the processor and the second one corresponds to the signals intervening in the data transference process. In most cases the hardware is supported by the co-processor, so the synthesis process is the same as the one used for hardware synthesis. Nevertheless the design constraints of the Hw/Sw interface are more severe than the hardware tasks due to the processor timing characteristics.
- *Co-simulation:* Comprises the system behavior simulation combining hardware and software tasks.

At this point an optimized software and hardware is obtained implementing a final system or a prototype.

Our work is focused on Hw/Sw digital processing systems rapid prototyping oriented to control application (processor based control system [PBCS]). Using FPGA based co-processors results in PBCS rapid prototyping, but software tools are necessary [3]. In this way a library (called INTERFACE) containing common timing, interrupt control and interface functions used in PBCS has been developed. This library can be used for FPGA based co-processors design and implementation reducing the hardware and the interface synthesis phases and making them easier [4].

The library INTERFACE is oriented to a Hw/Sw system made of a processor connected trough buses to a reconfigurable co-processor and a RAM memory (Figure 1). The memory contains a set of different configuration files corresponding to the different hardware tasks that will be supported by the co-processor and the processor decides the task been executed each time.

To obtain a useful library the following considerations have been taken into account:

- The design constraints of the co-processing and interface functions.
- The selection of the suitable FPGA.
- The implementation constraints due to the characteristics of the reconfigurable device.
- The characteristics of the interface between the control device and the FPGA, that is, the Hw/Sw interface.

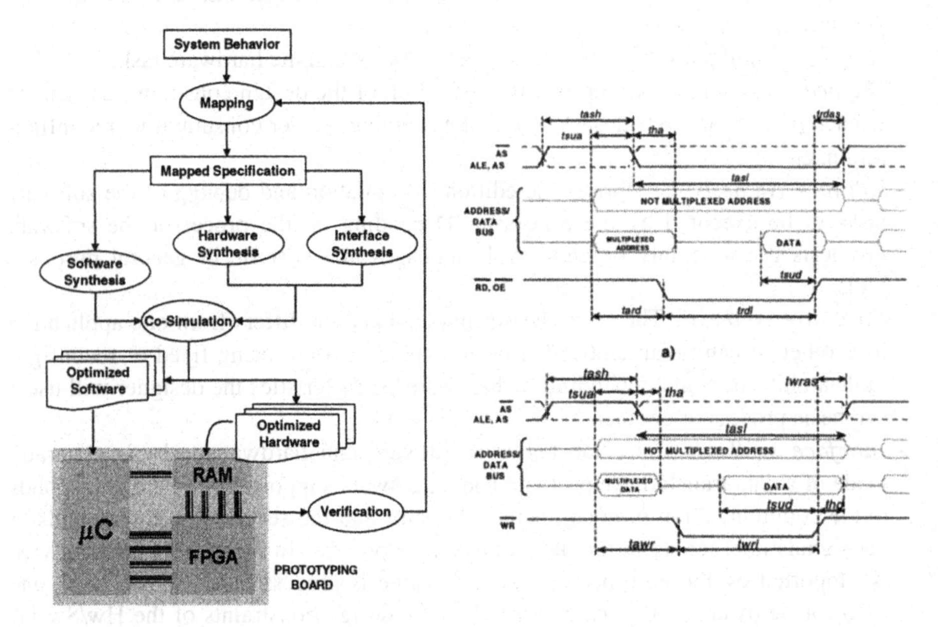

Fig. 1. Hw/Sw co-design methodology.

Fig. 2. Data transference protocols. a) Read operation. b) Write operation.

3 Hardware/Software Interfaces

As we said before the Hw/Sw interface depends on the way processor sends and receives data from the co-processor. Although there are different kinds of interface architectures this work focuses on parallel ones where the main processor is connected to the co-processor (FPGA) through data and address buses [3]. Nevertheless not all the devices have the same kind of buses nor use the same data transference protocol and control signals nor the same operation rates. In this way, different interfaces are required for microprocessors, microcontrollers or DSPs of different manufacturers.

The latter points out the interest in obtaining Hw/Sw interfaces with the following characteristics:

1. Compatible with microprocessors, microcontrollers or DSPs from different manufacturers.
2. Easily fitted to the characteristics of the co-processing functions.
3. Reusable in different applications.

To ensure the co-processor compatibility (designed using our library) with a wide range of commercial processors, a set of 86 microprocessors, microcontrollers and DSPs of different manufacturers were studied (Echelon NEURON, HITACHI, INTEL, MICROCHIP, MOTOROLA, PHILIPS, SGS-THOMSON, SIEMENS, TEXAS INSTRUMENTS, TOSHIBA and ZILOG). The characteristics of the address and data buses, the data transference process, and the timing constraints were defined in each case. As a result 8 different data transference protocols were established. Figure 2 illustrates one of them.

Timing constraints associated to each protocol ensure its compatibility with all the analyzed processors. Besides each protocol has several versions differing in the size of the bus and the names of the transference signals. As a result about 50 different functions were designed fitting the specifications of the analyzed microprocessors, microcontrollers and DSPs.

4 FPGA Features

The design constraints of the functions included in the library define the hardware support required for its implementation, that is, the FPGA features. In the same way the selected FPGA defines the way in which the library is created and used. This situation is due to the fact that FPGAs from different manufacturers use different and proprietary CAD tools. On the other hand, although there are some CAD tools compatible with different FPGAs, the best design performance is obtained using their own CAD tools.

In our case the characteristics of the required FPGA are: SRAM technology, embedded RAM blocks or CLBs configured as RAM, configurable logic blocks (CLBs) with function generators based on look-up tables (LUTs), CLBs with edge-triggered D-type flip-flops or latches (clock enable inputs), dedicated high-speed carry logic, internal tri-state bus capability, wide edge decoders, global low-skew clock or signal distribution networks, and IOBs with:

- Input and output registers configured as an edge-triggered flip-flop or latches with independent RESET.
- Independent output enable signals.
- Direct input.
- Individually programmable output slew-rate.
- Inputs and outputs globally configured for either TTL or CMOS thresholds.
- Programmable input pull-up or pull-down resistors.

The selected FPGA is the Xilinx's XC4000E Serie [5] because its architecture and its logic resources fit the design specifications associated to the logic systems which are to be implemented. Nevertheless, there are also other FPGAs that can be used, as for example the FLEX10K from Altera [6].

The library INTERFACE has been designed using the Xilinx's XACT CAD tools.

5 Library Functions

The library INTERFACE includes the following functions [4]:

Co-processing functions.

- Timing processing functions: Include events counting, waveforms and pulse generation, time capture, pulse width modulation, period detection, etc. To obtain an optimum implementation of any of these functions the use of dedicated carry logic is recommended [7][8].
- Interrupt control functions: Simple and complex controllers consist of a set of registers and additional logic detecting external interruptions, decoding its priority, and generating the request signals. In every case the registers can be implemented using flip-flops with asynchronous reset and enable inputs. Additional logic differs from simple to complex controllers but in both cases they are functions with more inputs than outputs. Therefore the use of LUTs is a suitable solution.
  According to the possible configurations, three different interrupt controllers were included in the library: interrupt controllers with program identification, interrupt controllers with fixed priority and programmable clear, and interrupt controllers with fixed priority and automatic clear. In these three cases several options with different number of interrupt sources and programmable active level/edge of the interrupt signal are available.

Hw/Sw interfaces: To avoid interconnection delays address and data buses must be implemented using the FPGA input/output blocks (IOBs).
  The bi-directional data bus uses tri-state buffers controlled by a transference signal. Both the address and data buses can be multiplexed or not multiplexed. As multiplexed address/data buses use the same pins for address and data signals, an input flip-flop is required to latch the address signal and a direct input to receive the data. The address latch register is controlled by an external signal that in some cases is a positive pulse and in others a negative one.

Interface circuits:

- Address decoder: Usually the address decoder of an interface device is a logic function with more inputs (the size of the address bus) than outputs. For this reason there are three possibilities for its implementation: using edge decoders, using LUTs function generators, or combining both resources.
  The first solution does not use internal logic resources of the FPGA. This is the suitable solution for wide decoders with a large number of inputs (20 or more) and outputs (12 or more).
  The second solution uses of CLBs. Each CLB of the XC4000E device can be configured as a logic function with a maximum of 9 inputs and two outputs. To implement a complex decoder a lot of CLBs will be necessary and the delay time associated to the interconnections is considerable. For this reason the address decoders

implementation using CLBs can be a good approach for a simple decoder (not many inputs and outputs) but not for a complex one.
On the other hand the combination of edge decoders and CLBs allows the implementation of designs with a good performance using FPGA's internal logic resources to their greatest advantage. In this case the MSBs (common to all addresses) are decoded using the edge decoders, and the rest of the bits are decoded using function generators (CLBs). This is the best solution for the implementation of fairly complex decoders.

- I/O ports: The input/output ports must offer the following features: bi-directional pins, individually configurable input/output pins (single bit addressing), programmable on the fly, input and output latches and high operation speed.
These features demand two characteristics from the FPGA's IOBs: independent output enable control signal associated to each tri-state buffer to accomplish the single bit addressing and latches/flip-flops to store the configuration bits and the output signals.

6 Hardware and Interface Synthesis

The library INTERFACE makes PBCS's Hw/Sw co-design easy and short improving the hardware and interface synthesis process. A diagram describing this process (used also to generate the library) is shown in figure 3. Using the Xilinx's XACT Tools the following phases are executed:

- *Design Entry:* According to the characteristics of the main processor and the application itself, the designer selects the required functions from the library INTERFACE and interconnects them. If necessary the XACT library can be used. To achieve the best performance the library functions have been designed considering some special partitioning attributes.

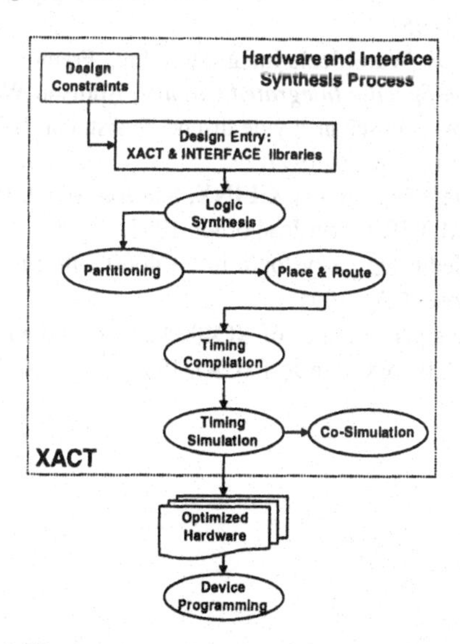

Fig. 3. Hardware and Interface synthesis process.

- *Logic Synthesis, Partitioning, Place and Route and Timing Compilation:* All these are automatic processes that generate the final configuration file of the FPGA taking into account the design constraints defined in the previous phase.
- *Timing Simulation:* Allows the simulation of the co-processor considering the interconnection delays.
- *Co-simulation:* Finally the designer can use the timing simulation tool to simulate the behavior of the Hw/Sw PBCS introducing partial results of the software tasks as stimuli of the simulation environment.

This process must be used to implement each hardware function of the PBCS. The resulting configuration files can be stored in any memory device connected to the FPGA (according to the architecture of the system). The main processor reconfigures the co-processor to execute the appropriate hardware task when necessary.

7 Conclusions

Using the library INTERFACE many reconfigurable PBCS oriented co-processors have been designed successfully. Therefore this library constitutes a suitable software tool for rapid prototyping and implementation of this kind of Hw/Sw digital processing systems. Although the library has been designed for FPGAs of the Xilinx's XC4000E Serie [5], it can be easily portable to other FPGAs families as for example the FLEX10K of Altera [6].

References

1. *The Codesign of Embedded Systems : A Unified Hardware Software Representation*, Kluwer Academic Publishers, 1995.
2. *PSD Products Data Book,* WaferScale Integration, Inc., Fremont (CA), 1996.
3. M.J. Smith, *Application-Specific Integrated Circuits*, Addison-Wesley, 1997.
4. M.D. Valdés, "Métodos de enseñanza y de diseño de sistemas basados en FPGAs", *Doctoral Thesis*, University of Vigo, Spain. 1997.
5. *The Programmable Logic Data Book*, XILINX, San José (CA), 1995.
6. *ALTERA Data Book*, ALTERA, San José (CA), 1995.
7. B. New, "Using the dedicated carry logic in XC4000", in *The Programmable Logic Data Book*, XILINX, San José (CA), 1995.
8. B. New, "Estimating the performance of XC4000 adders and counters", in *The Programmable Logic Data Book*, XILINX, San José (CA), 1995.

Reconfigurable Programming in the Large on Extendable Uniform Reconfigurable Computing Array's: An Integrated Approach Based on Reconfigurable Virtual Architectures

A. Touhafi[1,2], W. Brissinck [1]E.F. Dirkx[1]

(1)Vrije Universiteit Brussel, Pleinlaan 2, B1050 Brussel, Belgium.
(2)Erasmus Hogeschool Brussel, Nijverheidskaai 170,
B1070 Brussels, Belgium.
email: atouhafi@info.vub.ac.be
phone:+32.(0)2.629.29.85

Abstract. In this paper we focus on the implementation problem of large reconfigurable computing programs on extendable multi-node reconfigurable computing systems. We will discuss a methodology and an integrated graphical design environment for the development of large reconfigurable computing programs.

1 Introduction

Reconfigurable Programming in the Large (RPiL) is a term used to stress the ability to implement program's with a high complexity in terms of the need for a high amount of reconfigurable computing resources and Hugh design time. For this kind of programs it is expectable that their will be always a need for multi node reconfigurable computing systems. Such systems must be able to exploit parallelism on different levels of the application and must be able to deal with programs of which the full spatial implementation exceeds the available active resources (i.e. FPGA's, DPGA's or other reconfigurable computing processors). Many systems which can be classified as a multi node Reconfigurable Computing (RC) system have been built but none of them was really able to deal with RPiL in a convenient way. This due to the complexity of the RC system and the lack of a programming methodology and supporting tools.

In this paper we will first outline a programming methodology which can deal with RPiL. The programming methodology is based on Algorithmic Objects and Virtual Reconfigurable Architectures (VRA) which are mapped on a physical multi node RC system. By the end the developed programming tools and an integrated programming and implementation environment are discussed.

2 Reconfigurable Programming in the Large

Dealing with RPiL requires a programming methodology, the definition of a suitable reconfigurable computing (RC) system and a set of tools supporting the methodology. The methodology developed to solve RPiL organizes computation around the flow of data instead of the flow of control. Programs are represented as sets of algorithmic objects (see next section) which can intercommunicate trough predefined unidirectional asynchronous communication channels. The sets of interconnected algorithmic objects are mapped on a virtual reconfigurable architecture (VRA) which is on its turn mapped on a convenient RC system. This implicates that the developed tools should solve or give the application writer the ability to solve the following problems:

1:Definition of hierarchical algorithmic objects and their interconnections.
2:The definition of a VRA which implements the algorithm or intercommunicating algorithms.
3: Partitioning of the program into concurrent computing nodes of the virtual machine.
4: Mapping the virtual machine on a given RC-system.

The RC system has to deal with:

1:Being able to emulate a VRA on it's available active resources.
2:Being able to start the execution of a program and to detect the completion of it.

2.1 Atomic Algorithmic Objects

Definition 1. *An atomic algorithmic object is a circuit implementation of a process with zero or more (a)synchronous input communication channels and zero or more output (a)synchronous communication channels. The process implements a datapath and a microcoded sequencer or a naked datapath executing an atomic amount of computation.*

Definition 2. *An hierarchical algorithmic object or shortly algorithmic object is obtained by making a valid combination of atomic algorithmic objects. Where the objects are not necessarily interconnected.*

Definition 3. *An algorithm is defined as a valid interconnected combination of algorithmic objects. An algorithm has zero or more (a)synchronous input channels and zero or more (a)synchronous output channels.*

Definition 4. *A set of atomic algorithmic objects is named complete towards a class of algorithms if it is possible to implement all the necessary hierarchical algorithmic objects needed by the class of algorithms, using the basic algorithmic objects.*

2.2 Virtual Reconfigurable Architectures

Definition 5. *A virtual reconfigurable architecture (VRA) is a combination of interconnected virtual computing nodes. On each virtual computing node it is possible to map one or more algorithms.*

The full spatial implementation of a Virtual Reconfigurable Architecture can require much more active (i.e. dynamically reconfigurable) resources than a RC system provides. Further it is possible to classify VRA's as being static or dynamic:

Definition 6. *A static VRA is characterized by an unchanging number of virtual nodes during the execution of the program.*

Definition 7. *A dynamic VRA is characterized by a changing number of virtual nodes during runtime.*

2.3 Execution Methods for VRA's on Multi-node RC systems

A VRA is represented as a hierarchical directed graph. The nodes of the hierarchical directed graph represent the computing resources and the directed edges represent communication channels. The hierarchical nodes have three levels of hierarchy: on the lowest level we find a directed graph of interconnected atomic algorithmic objects (i.e. Atomic level) which are grouped into algorithmic objects (i.e. Algorithmic level) which are then further grouped into Virtual computing nodes (i.e. VRA level).

The developed execution model supposes that the VRA is composed of three parts: the first part is a set of sources (processes that create data or data buffered in memory), the second part is an unidirectional non cycled computing part and the third part is a set of sinks (memory for buffering the results of the computation). First the graph is partitioned based on the changing granularity of the dataflow graph. The partitioning algorithm minimize the difference between the granularity of the multi node RC system and the granularity of the dataflow graph. Ones the dataflow graph is partitioned it can be mapped in space and time on the multi node RC-system. The result is a scheduling and a mapping such that the VRA is correctly executed. From a performance point of view it is not needed to have a full spatial implementation to achieve the required performance. For a given parallel reconfigurable computing system it is possible to implement a part of the graph during a certain amount of clock cycles and then to reconfigure it's resources partially or completely to compute the next part of the directed graph. The number of computing-cycles during which a computation can be done without the need for reconfiguration is specified by the Lookahead (i.e. When does a certain input value of the computing node affect new inputs in the future) of the implemented part of the directed graph. In the case where the directed graph is unidirectional the Lookahead is infinite which means that as long as there is input data and enough memory resources to save intermediate results for the next stage, the system does not need to be reconfigured.

3 The RPiL Design Environment

In section 2 we gave some basic definitions and explained the programming methodology to support RPiL. We will now give some details on the design tools developed to support the design for - and execution of RPiL applications on the EUReCA system.

The programming methodology is based on Virtual Reconfigurable Architectures. Practically this imply's that the programmer thinks in terms of virtual computing nodes (rather than available computing nodes) which communicate by an asynchronous communication interface. The computing nodes are connected in a dataflow fashion and perform a certain amount of computation on incoming data. The result of the computation is then send to other computing nodes or to the memory. We've tried to relieve the programmer as much as possible from any details at the hardware- or circuit level. The design flow to complete an application for EUReCA is given in fig.1.

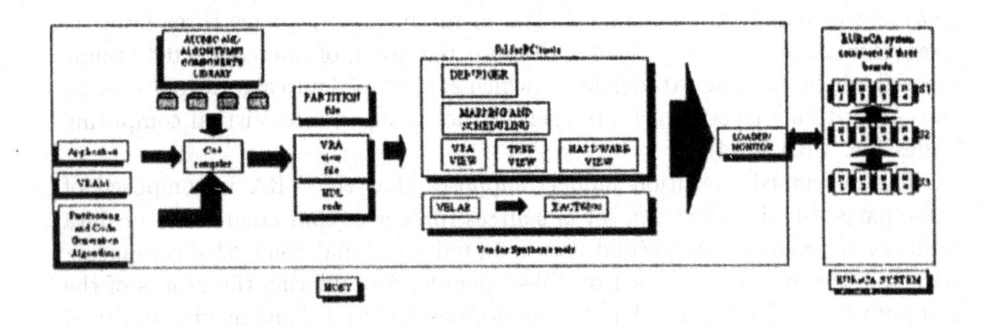

Fig. 1 RPiL design flow

The steps in the design flow are:

- Atomic Object Specification: Starting an application from scratch requires the user to implement a library of Atomic Objects and if necessary algorithmic objects. The first step to implement an Atomic Object is to describe its behavior in HDL code. The second step is to make a *name.h* file in which the input and output ports of the Atomic Object and their type is specified. This can be done by some VRAlib functions. The third step is to specify the computation time and communication bandwidth of each input and output port, the estimated area for implementation is also required. The second and third step are important and necessary for the partitioning mapping and scheduling algorithms. The code generator will implement automatically an asynchronous interface for the Atomic Object.
- Application: The application can be written using the Atomic Objects and the VRAlib functions. The VRAlib functions developed make it possible to specify an hierarchical dataflow graph.

- compilation of the application: The compilation of the Application can be done using any C++ compiler. The application includes VRAlib.h, CodeGen.h and Atomic Object libraries in it's code. CodeGen contains the graph-partitioning algorithm and a HDL code generation algorithm. The compiled program can then be run and gives as output a set of generated HDL files, a file required for the graphical visualization of the hierarchical dataflow graph and partitioning information. In case there are errors or warnings an error report is generated.
- Mapping and Scheduling: The output of the compiler is then used in the PiLforRC tools. The tools enable the mapping and scheduling of the HDL code. The mapping can be done automatically or manually using a graphical placement tool. Ones the placement is done the mapping tool generates a mapping file and an address list of the used communication registers. The next step is to generate a schedule list for the computing nodes. For that we have developed a scheduling algorithm that supports the scheduling-synchronization of the different computing nodes.
- HDL Synthesis: The generated HDL file must be synthesized in valid bit-streams for the DPGA's. Vendor specific tools are required for that [2].
- Downloading and running: A loader program which runs on the host, sets-up EUReCA using the schedule file, the placement files and the DPGA bit-streams.

The PiLforRC environment combines a set of interacting tools designed to facilitate RPiL on a EUReCA like computing platform. New in the approach of the PiLforRC environment is that it is closer to a parallel programming environment than an ASIC- or FPGA design environment. The PiLforRC tools are integrated in one graphical design and debug environment written in Microsoft Visual C++ and runs on the windows 95/98 and NT operating system.

The developed tools which are available can be classified in four groups: The visualization tools, The mapping and scheduling tools,the EUReCA setup tools and the debugging tools.

3.1 The Visualization Tools

Three kinds of visualization tools are developed. Visualization tools are very important for checking on correctness and during debugging. A program which is written can be visualized in a graphical dataflow graph browser. The user has also direct access to all the hardware resources in the hardware-viewer. The third view is a tree-view of the EUReCA hardware resources and the contents. The Hardware viewer is an hierarchical viewer allowing the user to see, on the top level how different EUReCA boards are interconnected and to find out how each computing node is configured until circuit level.

3.2 The Mapping and Scheduling Tools

For the mapping and scheduling of the VRA program we have developed a graphical mapping and scheduling tool allowing the user to experiment with manual placement and scheduling. Further it is also possible to use automatic mapping and scheduling. The user has to specify choose which execution model he wants to use, then he has to choose an existing mapping and scheduling algorithm.

4 Conclusion and Future Work

In this paper we have proposed a methodology for dealing with large reconfigurable programs on multi node reconfigurable computing systems. The methodology is based on virtual reconfigurable architectures. Such a virtual reconfigurable architecture is composed of virtual computing nodes rather than real computing nodes. We have also introduced the EUReCA system. This RC system is able to deal with emulating large reconfigurable programs and supports debugging facilities. A practical implementation of the proposed programming methodology is realized in the form of VRAlib, a HDL code generation algorithm, a mapping and scheduling algorithm and a graphical design environment. The graphical design environment distinguishes itself from other environments by it's approach which is based on parallel computing design environments.

References

1. Xilinx inc., Xilinx Programmable Gate Array Data book, release 1998
2. Xilinx inc., Velab, VHDL elaborator for XC6200, http://www.xilinx.com/apps/velabrel.htm (1998)
3. Xilinx inc, XC6000 series databook, www.xilinx.com,1996
4. Kay Hwang and Faye A. Briggs, Computer Architecture and Parallel Processing, section 10.3.2: Mapping algorithms into VLSI Array's. Mc Graw Hill International Edition
5. A.Touhafi et al, Applying Field Programmable Logic for Generic Network Adaptor Design and Implementation, IEEE Lan/Man workshop on Local and metropolitan Area Networks, 1996 Potsdam Germany

A Concept for an Evaluation Framework for Reconfigurable Systems

Sergej Sawitzki and Rainer G. Spallek

Institute of Computer Engineering
Dresden University of Technology
D-01062 Dresden, Germany
{sawitzki, rgs}@ite.inf.tu-dresden.de

Abstract. The design of reconfigurable systems is a hard task due to a huge amount of optimization trade-offs and constraints. Deficiencies at the higher level of conceptual decisions may result in critical bottlenecks of the system implementation. This paper explores the design space and performance evaluation criteria of reconfigurable systems and introduces a concept for an evaluation framework based on these explorations helping to avoid such deficiencies.

1 Motivation

The developments in reconfigurable logic over the last decade result in fascinating achievements both in device architecture and software environment. High-density million-gates devices, partially reconfigurable logic and multiple-context FPGAs raise new challenges to system developers and researchers. The success of a particular project always depends on the experience and skills of the design team which has to find a suitable solution driven by optimization trade-offs and design space constraints. This process usually results in the time-consuming development of dedicated simulation and prototyping environments, especially if the design of a novel architecture is considered. A general purpose framework which allows systematical exploration of the design space supported by the performance evaluation steps may reduce the complexity of this problem. The following sections describe the models required to develop such a framework as well as a first conceptual approach.

2 Design Space and Performance Evaluation

The modelling of the design space is the most important task while developing a computing system. A suitable model helps to classify the existing approaches and shows niches for novel architectures and paradigms. The analysis of the known classification schemes like 4- and 3-class-model [1, 2] or RP-space [3] and Olymp taxonomy [4] draws some basic parameters of the design space. A review of the newest logic devices and CCM implementations helps to establish a more general view of the problem resulting in the following criteria for a generalized model.

Hardware structure HS. This parameter describes different grades of reconfigurability, which can be either restricted to functional units fu (like within the Spyder processor [5]) or interconnection network ic (like within the KressArray [6]) or affect both components (like in most SRAM-based FPGA and CPLD families). The formal notation for this parameter is $HS = \wp(\{fu, ic\})$, whereby $\wp$ is the power set of the corresponding set.

Flexibility F. To allow the consideration of hybrid systems, the programmability in common sense is seen as an additional grade of flexibility. Thus, the computing system can be either programmable p or reconfigurable r or a combination of both: $F = \wp(\{p, r\})/\emptyset$.

Granularity G. The functional units within the reconfigurable device are either coarse-grained cg (e.g. complete ALU), medium-grained mg (e.g. Lucent ORCA Series 3 FPUs) or fine-grained fg (e.g. Xilinx XC6200 FUs), thus $G = \wp(\{cg, mg, fg\})/\emptyset$.

Host coupling HC. Most reconfigurable systems are implemented as a combination of hardwired and programmable logic. The coupling between these components is either loose l (e.g. a host workstation with an FPGA extension board), middle m (e.g. different ICs on the same printed circuit board) or tight t (single die). If a stand-alone reconfigurable device like a single FPGA or CPLD is considered, this parameter can be omitted, thus $HC = \wp(\{l, m, t\})$.

Memory coupling MC. Two basic approaches are known from the existing CCMs: reconfigurable processing units may have their own local memory lm or share the memory resources sm with the hardwired host (see [1] for a detailed discussion). For a single "memoryless" device, the MC value can be omitted, $MC = \wp(\{lm, sm\})$.

Reconfiguration time RT. The value of this parameter depends on the host coupling paradigm. Systems and devices which can be reconfigured during runtime (RT of several milliseconds and below) are called dynamically reconfigurable (dy). The reconfiguration time of statically reconfigurable systems (st) reaches from several milliseconds to several seconds or minutes. In general terms $RT = \wp(\{dy, st\})/\emptyset$.

Programming model PM. The reconfigurable resources are accessed by the software either via special instructions si or in a memory-mapped fashion mm. The synchronization of these accesses within the global control flow uses either the global system clock sc or interrupts in. Addressing these criteria separately results in $PM = (\wp(\{si, mm\})/\emptyset) \times (\wp(\{sc, in\})/\emptyset)$.

The design space DS is described as a cartesian product of corresponding power sets: $DS = HS \times F \times G \times HC \times MC \times RT \times PM$. If some attributes should be omitted, different projections can be defined. Considering a reconfigurable system $a = (\{fu, ic\}, r, fg, l, \{lm, sm\}, st, (mm, in))$, $\pi_{G,HC,RT}(a) = (fg, l, st)$ is a projection which describes only the granularity, the host coupling and the reconfiguration time of the system. Additional metrics can be considered for each parameter to improve the accuracy of the model, e.g. detailed values for

Table 1. DSP algorithm run times (rt, in milliseconds) and speedups (sp)

	FFT (512pt)		LPC Filter		Viterbi Decoder	
	rt	sp	rt	sp	rt	sp
AMD K6 200	1.2	8.33	0.00100	10.00	6.95	1.00
Pentium II 350	0.5	20.00	0.00059	16.94	3.20	2.17
ARM stand-alone	10.0	1.00	0.01000	1.00	4.80	1.45
ARM with FPGA	1.4	7.14	0.00060	16.67	0.27	25.74

the reconfiguration time, the access cycle time over the host bus, the amount of LUTs or logic gates per functional block etc.

The performance of a computing system is usually measured in terms of run-time and throughput. A benchmark set is used to determine these and other related system attributes like latency, response time, bandwidth etc. An evaluation framework must be able to provide the designer with corresponding values which can be extracted through simulation, timing backannotation or requests to knowledge bases. An important criterion in context of reconfigurability is the functional density FD, which expresses the efficiency of the system in terms of both area and time and is defined as the number of operations n executed within the time slice T_n on a unit of silicon of size A: $FD = \frac{n}{AT_n}$ [7]. An additional metric for the efficiency is the performance/cost ratio. The following example illustrates the relevancy of the efficiency-oriented performance evaluation (for a detailed description of the experiment discussed below see [8]).

A benchmark set consisting of three digital signal processing algorithms was run on a variety of different computing platforms (including standard PCs and an FPGA extension board with 2×XC4013 devices and an ARM AT91 micro-controller). Table 1 summarizes the run times and the speedups achieved. Albeit these values suggest that the FPGA extension board is not the most powerful solution in every case, the analysis of the performance/cost ratio and the functional density (based on the transistor count for each chip) encourages the choice in favor of FPGAs (see Table 2) since the functional density improvement resulting from replacing the Pentium PC by the FPGA board is about 635%.

3 Framework Concept

Some basic ideas within the framework described in this section are adapted from the DCS concept [9]. DCS, however, concentrates on the dynamic reconfiguration and involves the implementation details for particular devices more tightly. In contrast to the DCS, the concept proposed here is tailored to evaluation aspects and uses the design space and performance evaluation models discussed above, hence it should be seen as a system-level approach. The design principles of the framework are discussed below.

Modularity. The global system view introduces three basic framework modules, namely front-end, kernel and back-end. The front-end unifies all tools required

Table 2. Cost/Performance Ratio and Functional Density

Circuit	Price [$]	Performance/$ [arb. units]	⌀FD $\left[\frac{OP}{Trans \cdot ms} \cdot 10^{-9}\right]$
AMD K6 200	65	99	42
Pentium II	250	52	108
XC4013XL-3	60	280	794

for the description of system components. The kernel controls the evaluation flow and provides the communication layer for other framework modules. The back-end accumulates the evaluation results and provides the associated statistics. These global modules are split further into smaller submodules, which implement particular tools and functions. The communication between the modules is realized via standardized interfaces, which use both industrial standards (EDIF, Verilog or VHDL for netlists or STF and SDF for timing backannotation) and dedicated description formalisms. The interfaces further include format converters to achieve more transparency.

Open architecture. The concept of interfaces allows the integration of any additional commercial and industrial tools which provide a description of their data interchange formats. Therefore, the proposed framework should rather be seen as an extension to than a replacement of the existing design flows. It depends on the decision of the system designer whether a dedicated or an existing commercial tool is used.

Scalability. The framework does not set any constraints concerning the complexity of the system to be evaluated or the amount of tools which can be involved in the particular evaluation flow. Furthermore, the evaluation proceeds at different levels of detailization depending on the design space projection, description formalisms and tools used. E.g. if the impact of configuration context caching on the functional density shall be explored, one can first create the projection $\pi_{G,HC,MC,RT}$ into the design space and then activate the memory subsystem simulator to create the required statistics.

The main part of the evaluation flow will be implemented within four kernel submodules. The run-time processor controls the evaluation flow and provides the data logging and debugging functionality. The design space evaluator reads the system description with related projections according to the model discussed in the previous section and extracts the required parameters for the run-time processor. The memory subsystem simulator is used to simulate all memory requests including caching effects and configuration context control. The performance evaluator extracts the performance data according to the performance evaluation criteria described in Section 2 and creates the evaluation statistics.

The first design flow to be evaluated includes the SIS logic synthesizer [10] and the SUIF [11] compiler as front-end and the VPR placer and router [12] as back-end. This configuration provides enough capabilities to evaluate both problem- and architecture-oriented approaches. In the first case, the optimal system ar-

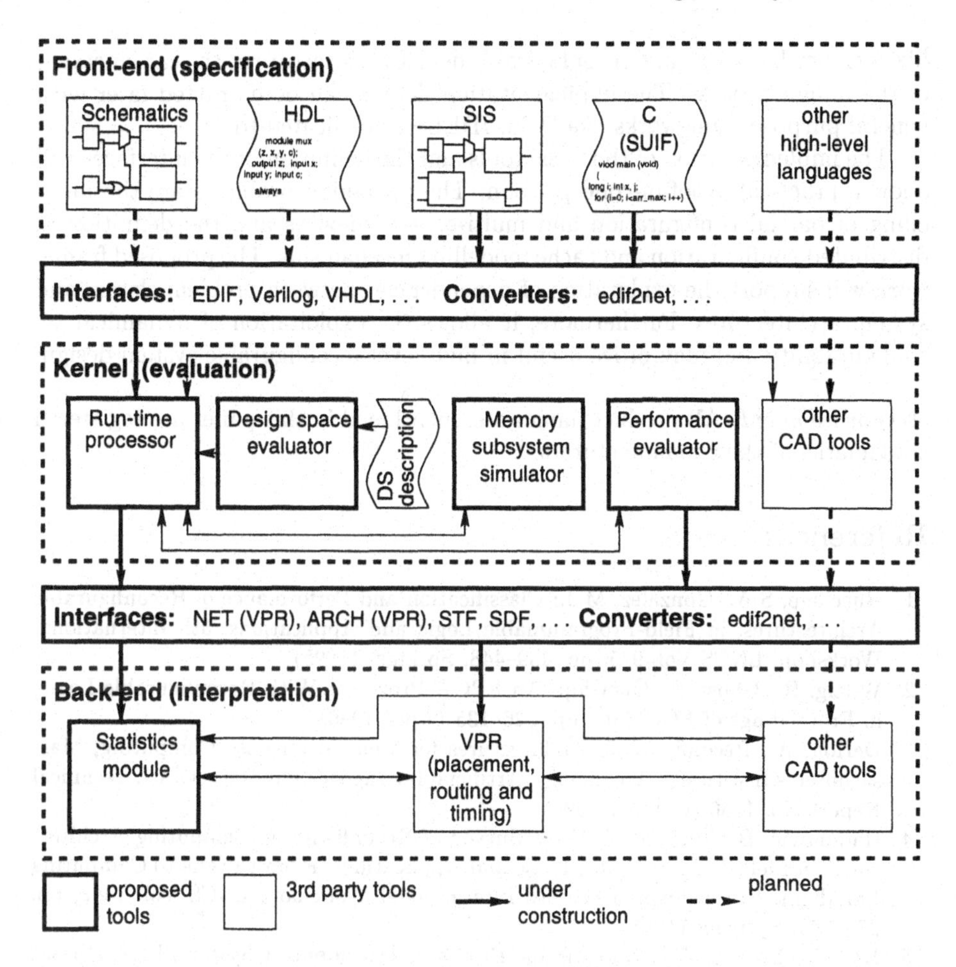

Fig. 1. Framework structure and interfaces

chitecture for a given problem class (like the DSP algorithms mentioned above) is to be determined, in the second case the best hardware/software partitioning is in demand for a given system architecture and application. The integration of commercial CAD tools and of graphic user environment follows. The structure and interfaces of the framework are shown in Fig. 1.

4 Conclusions and Further Work

This paper has introduced an evaluation framework for reconfigurable systems based on the generalized view of the design space and efficiency-oriented performance evaluation criteria. It provides a high-level extension to (instead of replacement for) the existing tools. Due to the modular and open concept this framework can easily interact within established design flows. The main advantage of this approach is the specification support for global system parameters like memory hierarchy and reconfigurable device architecture at the earlier stages of the design process. The implementation details can be extracted later with general purpose frameworks like DCS, IDELS or dedicated tools.

The implementation of the basic tools and the definition of the interfaces will allow a practical proof-of-concept soon. The upcoming problems are the modelling of partial configuration and multi-context devices and the definition of distributed configuration and cache modelling mechanisms. The proposed framework will support the exploration of dependencies between problem classes and system architectures. Furthermore, it allows the exploitation of dynamical reconfigurability and may prove useful in high-level reconfigurable system design.

Acknowledgment. All product names etc. mentioned in this paper are registered trademarks of their manufacturers.

References

1. Guccione, S.A., Gonzalez, M.J.: Classification and Performance of Reconfigurable Architectures, in Field-Programmable Logic and Applications: 5th International Workshop, LNCS Vol. 975, pp. 439–448, Springer (1995)
2. Wittig, R., Chow, P.: OneChip: An FPGA Processor With Reconfigurable Logic, in Proceedings of FCCM'96, pp. 126–135 (April 1996)
3. DeHon, A.: Reconfigurable Architectures for General-Purpose Computing, Massachusetts Institute of Technology, Artificial Intelligence Laboratory, A.I. Technical Report No. 1586 (October 1996)
4. Radunović, B., Multinović, V.: A Survey of Reconfigurable Computing Architectures, in Field-Programmable Logic and Applications: From FPGAs to Computing Paradigm. Proceedings of the 8th International Workshop, LNCS Vol. 1482, pp. 376–385, Springer (1998)
5. Iseli, C.: Spyder: A Reconfigurable Processor Development System, Ph.D. thesis, Département d'Informatique, École Polytechnique de Lausanne (1996)
6. Kress, R.: A Fast Reconfigurable ALU for Xputers, Ph.D. thesis, University of Kaiserslautern (1996)
7. Wirthlin, M.J.: Improving Functional Density Through Run-Time Circuit Reconfiguration, Ph.D. thesis, Brigham Young University (1997)
8. Köhler, S., Sawitzki, S., Gratz, A., Spallek, R.G.: Digital Signal Processing with General Purpose Microprocessors, DSP and Reconfigurable Logic, in Proceedings of IPPS/SDPS'99, Reconfigurable Architectures Workshop (April 1999)
9. Robinson, D., McGregor, G., Lysaght, P.: New CAD Framework Extends Simulation of Dynamically Reconfigurable Logic, in Field-Programmable Logic and Applications: From FPGAs to Computing Paradigm. Proceedings of the 8th International Workshop, LNCS Vol. 1482, pp. 1–8, Springer (1998)
10. Sentovich, E.M., Singh, K.J. et.al.: SIS: A System for Sequential Circuit Synthesis, Technical Report No. UCB/ERL M92/41, University of California, Berkeley (1992)
11. Wilson, C.: The SUIF Guide, Stanford University (1998)
12. Betz, V., Rose, J.: VPR: A New Packing, Placement and Routing Tool for FPGA Research, in Field-Programmable Logic and Applications. Proceedings of the 7th International Workshop, LNCS Vol. 1304, pp. 213–222, Springer (1997)

Debugging Application-Specific Programmable Products

Rainer Kress and Andreas Pyttel

Infineon Technologies AG i. Gr., Corporate Research, CPR 6
D-81730 Munich, Germany
{Rainer.Kress|Andreas.Pyttel}@infineon.com

Abstract. This paper describes different approaches to debug application specific programmable products (ASSPs). These products combine a core based design (e.g., a microcontroller) with embedded programmable logic. The various debugging levels of the microcontroller as well as the debugging possibilities of the programmable hardware are explained. Furthermore, methods of interaction between software and hardware are shown for different debugging levels.

1 Introduction

Application-specific programmable products (ASPPs) represent an emerging category of system-on-a-chip products. These products combine intellectual property (IP) core-based ASICs with on-board embedded programmable logic. Jordan Selburn, principal analyst for ASICs and system-level integration at Dataquest Inc.´s Semiconductors Group, sees ASPPs as the predominant system-on-a-chip product in the next five years [3]. As the manufacturing capacity becomes more expensive and system-level integration (SLI) advances to be a feasible solution for volume production, programmability is becoming a source of added value.

Today's ASSPs are dominated by the field-programmable gate array (FPGA) part, which is enhanced by standard hardware such as microcontroller cores or specific interfaces. Lucent [7] sells a device that comprises the ORCA OR3T series FPGA and Embedded Master/Target PCI Bus Interface. This device, called ORCA OR3TP12, combines the high bandwidth of an industry-standard PCI interface with the programmability of FPGAs. SISDA [5] offers the FIPSOC field-programmable system-on-chip based on an 8051 microcontroller, a configurable analog block, and a RAM-based on-chip FPGA. Infineon Technologies (the former Siemens Semiconductors' Group) is planning to combine the 32-bit microcontroller TriCore with a Gatefield´s flash-based FPGA [1]. The National Napa1000 Reconfigurable Processor is a single-chip implementation for signal processing that provides both a 32 bit RISC core and a 50k gate reconfigurable logic part, called Adaptive Logic Processor [6].

ASPPs can be customized by software as well as via the programmable hardware. Therefore, debugging is needed for software and hardware. Modern microcontrollers provide an on-chip debugging (OCD) subsystem. The term OCD describes a variety of approaches used to allocate on-chip resources for the purpose of debugging. OCD provides two modes: *run-mode* and *debug-mode*. When the target processor is running an application normally, the OCD engine is in a passive state. This mode, called run-mode, is the preferred one to debug real-time processes, where the processor need not be stopped, but where the ongoing interactions between processes and system blocks should be monitored. In debug mode, the target processor stops running and the OCD engine becomes fully active for debugging operations.

The availability of programmable hardware enables the annotation of the circuit with specific hardware elements that provide debugging functionality such as breakpoint detection, breakpoint setting, register reading etc. In the final design of the circuit, these specific hardware functions can be removed. Debugging programmable hardware in this manner can be compared to monitoring of software programs which run on a microcontroller.

Koch et. al. [2] proposed a method called source level emulation (SLE) that allows for symbolic debugging of a hardware design. It runs on an FPGA-based hardware emulator. The emulation can be controlled at the source code level specified as behavioral VHDL. During run-time, the values of registers are read from the emulator and annotated back to the VHDL source code. The correlation between the emulated hardware and the source code is obtained by logging the synthesis steps.

This paper concentrates on the combined debugging of hardware and software in ASPPs. The various debugging levels are explained for microcontrollers (in section 3) and their corresponding counterparts in the programmable hardware are shown in section 4. Methods of interaction between hardware and software are addressed for these debugging levels. The paper closes by presenting results of our work.

2 Problem description

ASPPs (figure 1) offer the possibility to implement an application partially in software that runs on the microcontroller core, and partially on the embedded programmable logic (embedded FPGA). The microcontroller's development environment supports compilation and debugging of the software. Furthermore, a simulator allows us to validate the software in an early design stage. In section 3, we discuss several software and hardware methods that support debugging of a microcontroller's application software.

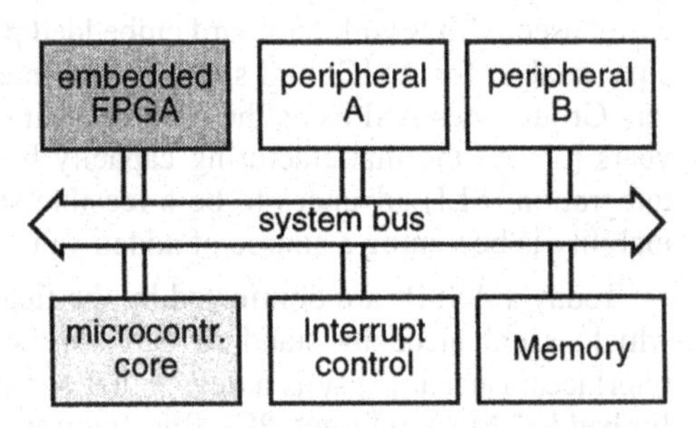

Fig. 1. Microcontroller with embedded FPGA as an example of an ASPP

The development environment of the embedded FPGA part comprises tools to simulate and to synthesize the design (which is described by a hardware description language) either at the behavioral level or at the register-transfer level. Debugging of the design on the FPGA platform is supported by additional logic which is inserted into the design during synthesis. This logic facilitates the observation of signals and registers and controls the debugging process. We discuss this in section 4. The only way to debug the complete application of the ASPP is to co-debug the microcontroller's software together with the hardware design on the embedded FPGA.

3 Debugging support for microcontrollers

Debugging embedded systems has always been a challenging task. Early microcontrollers did not offer any debugging support by hardware features. Traditional microcontrollers use monitor programs to control the application. The monitor program is located in the same address area as the application. The advantage of a monitor program is its low cost. The disadvantage is the partial sharing of the serial interface by the debugger. Therefore, the serial interface cannot be fully used by the application. Additionally, the debugger may crash if an error occurs in the application.

A better solution for debugging is to use an in-circuit emulator (ICE) that works with a bond-out version of the microcontroller. The ICE gives an optimal inside view of the internal workings of the microcontroller. However, this good visibility requires a high cost investment.

On-chip debug modes (OCD) provide a way to stop a microcontroller when required. This stopping can be controlled by comparators inside the chip. This means that an OCD has two modes: run mode and debug mode. In run mode, the OCD is generally in a passive state, especially when a real-time application is running. Nevertheless, some microcontrollers support a real-time trace. If a real-time trace support is available and enabled, some real-time information can be read via the so-called helper outputs. These outputs give information about the internal operations. Helper inputs are available for control by external logic. In debug mode, the CPU of the microcontroller stops and the OCD subsystem becomes fully active for debugging operations. The OCD is actively communicating with the debug host computer. In general, three approaches can be distinguished: a resident monitor program, a microcoded OCD engine, and instruction stuffing [4]. A *resident monitor program* is implemented in the firmware of the controller. It uses a serial link to communicate with the host computer. The monitor program is the least flexible in terms of non-intrusive run mode debugging. A *microcoded OCD engine* can be triggered by OCD-specific microcode in the target CPU. This monitor program is independent from the instruction set of the target CPU. It permits the use of OCD-specialized instructions to compute complex operations in less time. *Instruction stuffing* suspends the normal instruction fetches and obtains instructions via the OCD subsystem.

The on-chip debug support system (OCDS), which is supported by the TriCore 32-bit microcontroller from Infineon Technologies, is targeted for real-time, non-intrusive emulation. The OCDS uses a modular concept of three levels [8]. OCDS level 1 and level 2 have been created to meet 80% of all emulation and debugging requirements. OCDS level 3 is available for high end support, such as ICEs.

OCDS level 1 provides real-time run-control and internal visibility of resources for data and memories. Level 1 is an extremely low cost solution. It is targeted to a software development environment. It can be combined with a standard debugger and simulator. Level 1 is realized by a peripheral connected to the on-chip system bus with emulation hooks to the CPU core. Communication with the debugger is performed via an expanded JTAG interface. The heart of the OCDS is the debug event generation unit (DEG) and the debug event processing unit (DEP), which are shown in figure 2. The DEG unit contains various comparators on the address, data, and control buses. The DEP unit can respond to other events, such as the status of a dedicated break pin, execution of debug instructions, and access to a general-purpose register.

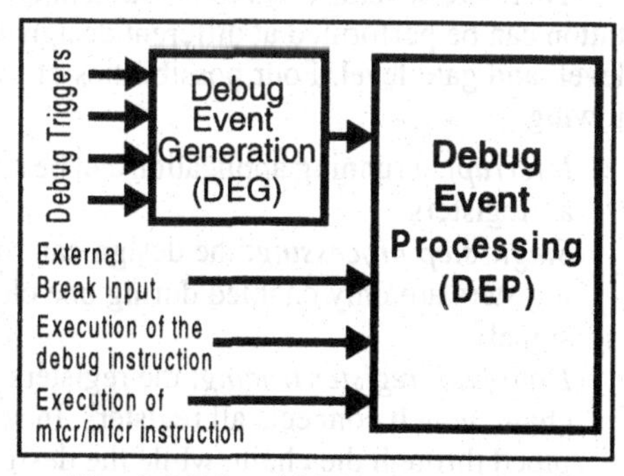

Fig. 2. DEG and DEP units of OCDS

OCDS level 2 provides run control and, in addition, a trace expansion. The tracing of the instruction pointer is performed through extra signals. Thus, level 2 realizes the internal visibility of the program flow and therefore allows for performance analysis. The precise breakpoint capability combined with a real-time trace of the instruction flow provide building blocks for all advanced statistic and real-time debugging tools. In addition to instruction flow information, level 2 also provides pipeline status information. OCDS level 3 is a bondout oriented solution that requires a complex chip with all internal address and data information busses driven externally and supported by a specifically designed emulator.

4 Debugging support for programmable hardware

Programmable hardware enables the addition of specific logic to the original application for debugging support. The specific logic can be removed after the circuit has been debugged. Source-level debugging facilitates symbolic debugging of running hardware. This includes the examination of variables, the setting of breakpoints, and performing single steps. Source code is either used at behavioral or register-transfer (RT) level. In the case of behavioral level, source-level debugging is closely related to high-level synthesis (HLS). HLS is comprised of three steps: allocation, scheduling, and assignment. *Allocation* determines a set of components available for the implementation of a design. *Scheduling* assigns a component type and a control step to each operation of the design. *Assignment* maps each operation to a physical component of the type it was assigned to by scheduling.

Koch et. al. [2] use HLS to write correlation information into a map file during the synthesis of the design. In addition, the circuit is annotated with special hardware, allowing for breakpoint detection, breakpoint setting, register reading, etc. The debugger controls the circuit operation on the emulator via the special hardware and displays register values in the context of the description. Thus, the designer deals only with the source code, but not with the lower levels of abstraction which are generated by the tools involved in synthesis. The additional hardware includes a debugging controller as well as annotations of the application´s data path and finite-state-machine (FSM).

There are several ways to instrument code for debugging purposes. The instrumentation can be performed at different design levels: the behavioral level, register-transfer level, and gate level. Four possibilities of the instrumentation are explained in the following:

- *Interrupt*: a running application can be stopped by disabling the system clock for all registers.
- *Single step processing*: the design can run in single-step mode, where the registers are only enabled during one clock cycle. (e.g., via the clock enable signal).
- *Data path register tracing*: the registers may be traced by using a scan-path chain, which connects all registers. In scan-path mode, the register values are piped through the chain, while the design that is debugged halts.
- *Breakpoint*: a breakpoint stops the design in a state corresponding to a specific instruction of the source code. Breakpoints are identified by comparing the controller state with a given breakpoint identifier. The encoding of the breakpoint identifier is determined during synthesis.

On the one hand, the interface to the outside world of the embedded programmable logic may be implemented by helper inputs and outputs. These helpers are configured especially for debugging purposes. On the other hand, Boundary Scan/JTAG ports of the programmable logic can be used as an interface. Of course, those interfaces may also connect to the system bus. They can be read or written by the software program of the microcontroller.

The Boundary Scan/JTAG standard was formally known as IEEE/ANSI Std 1149.1. It was a set of design rules, which facilitated testing, device programming and debugging at the chip, board and systems level. The standard came about as a result of the efforts of the Joint Test Action Group (JTAG) formed by several North American and European companies. IEEE Std 1149.1 was originally developed as an on-chip test infrastructure capable of extending the lifetime of available automatic test equipment (ATE). This methodology of incorporating design-for-test allows complete control and access to the boundary pins of a device without the need for a bed-of-nails or other expensive test equipment. Each JTAG compliant device includes a boundary-scan cell on each input, output, or bi-directional device pin that under normal conditions is transparent and inactive, allowing signals to pass normally. When the device is placed in the test mode, input signals can be captured for later analysis and output signals can be set to affect other devices on the board [10]

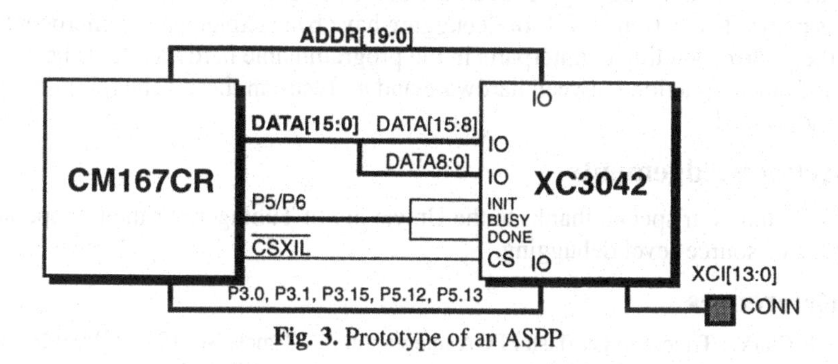

Fig. 3. Prototype of an ASPP

5 Results

The prototype of an ASPP is shown in figure 3. It consists of a CM176CR 16-bit microcontroller connected to a Xilinx® XC3042 FPGA. The FPGA is configured by the microcontroller in peripheral mode [9]. After configuration, the 16-bit data bus, the 20-bit address bus, and 5 pins of the microcontroller's ports are available for the design that is implemented on the FPGA. Additional 14 pins of the FPGA can be used as flexible I/O. Several applications were implemented to assess the debugging possibilities of the FPGA.

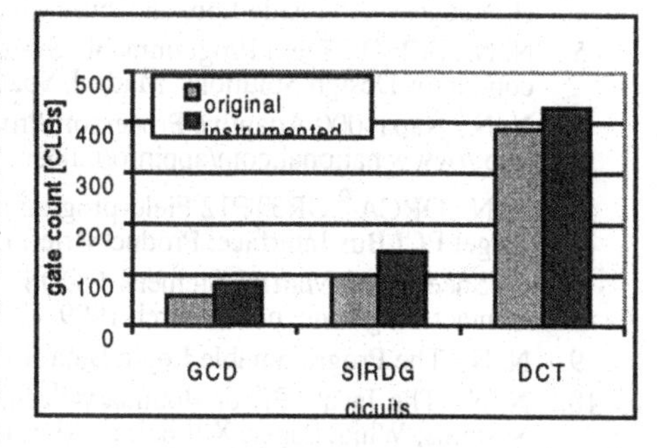

Fig. 4. Area comparison

The instrumentation of code increases the required gates count. Furthermore it decreases the speed of the circuit. The following applications were all synthesized, downloaded, and run on a prototyping board [2]: GCD, SIRDG, and DCT. Greatest common division (GCD) is a standard example for high level synthesis. SIRDG computes "single image random dot stereograms". The digital cosine transform (DCT) computes a two-dimensional 8x8-DCT.

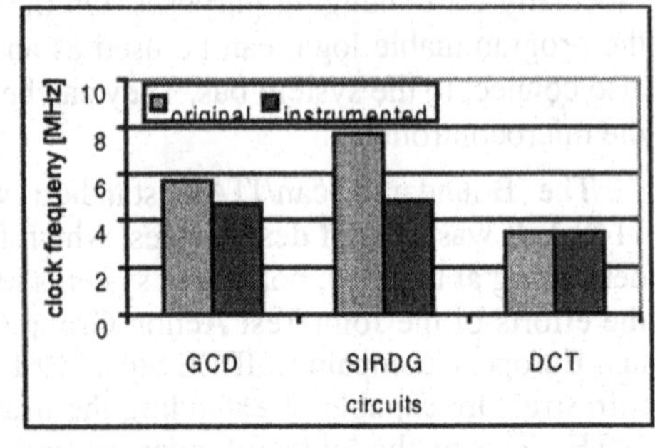

Fig. 5. Speed comparison

As shown in figure 4, the instrumentation of the code does not significantly increase the required area. Furthermore, the maximal clock frequency does not decrease considerably (figure 5).

6 Summary

The combined debugging of hardware and software in ASPPs has been presented in this paper. The various levels of debugging have been explained for microcontrollers and their corresponding counterparts in the programmable hardware have been shown. Methods of interaction between hardware and software in these debugging levels have been addressed.

Acknowledgements

We extend our special thanks to the University of Tübingen for their cooperation in the area of source-level debugging.

References

1. P. Clarke: Tricore to get flash FPGA integration. EE Times, No. 1000; CMP Media, 1998
2. G. Koch, U. Kebschull, W. Rosenstiel: Breakpoints and Breakpoint Detection in Source Level Emulation; ACM transactions on Design Automation of Electronic Systems, vol. 3, no. 2, 1998
3. N. Mokhoff: Analyst sees ASPPs becoming volume products; EE Times, Oct. 21, 1998
4. H. Neugass: An inside Look at Chip Debugging; Insight, vol. 3, issue 1, 1998
5. N. N.: FIPSOC Field Programmable System-on-Chip; FIPSOC Data Sheet, SIDSA Semiconductor Design Solutions, Madrid, Spain, 1998
6. N. N.: Napa1000 Adaptive Processor; Product Brief, National Semiconductor, http://www.national.com/appinfo/milaero/napa1000/, 1998
7. N. N.: ORCA® OR3TP12 Field-programmable System Chip (FPSC) - Embedded Master/Target PCI Bus Interface; Product Brief, Lucent Technologies, May 1998
8. G. Sheedy, G. Martin: Siemens OCDS: The Next Generation On-chip Debug Support; Contact Magazine, no.4, March 1999
9. N. N.: The Programmable Logic Data Book; Xilinx Inc., San José, CA, 1998
10. N. N.: The Java API for Boundary-Scan: Enabling Technology for Internet-Driven PLD Systems; White Paper, Xilinx Inc., May 1999

IP Validation for FPGAs Using Hardware Object Technology^{TM}

Steve Casselman, John Schewel and Christophe Beaumont

Virtual Computer Corporation, 6925 Canby Ave #103
Reseda, California, USA 91335
Email: {sc|jas|cb}@vcc.com

Abstract. Although verification and simulation tools are always improving, the results they provide remain hard to analyze and interpret. On one hand, verification sticks to the functional description of the circuit, with no timing consideration. On the other hand, simulation runs mainly on subsets of the entire input domain. Furthermore, these tools provide results in a format (e.g. state graphs, bit vectors or signal waves) that remain disconnected from the real output of the application.

We introduce in this paper the process of validation applied to digital designs in FPGAs. It allows the designer the ability to test his/her implementation using the real data of the application and providing real results. With such real data, it becomes easier to identify where the error occurs and then to understand it

1. Introduction

In the standard design flow used for implementing IP into FPGAs, designers use mainly verification and simulation tools to check their circuits. These tools are always getting more powerful, but they remain generic: the designer has to translate the real input data to a particular format and then interpret results which are not directly in the real output format (e.g. video). Such translations often make the understanding of errors more difficult. One easier way is to feed an operator with the real data it should receive and check if the resulting output matches the expected values. If not, wrong results may be directly interpreted without translation and thus without loosing important debugging information. The designer validates his design and implements it with real data from the circuit. Having the real results helps in verifying the design and, once verified, in fine-tuning the implemented circuit.

In the first section, we present the basic design flow used for designs to be implemented into FPGAs. We then introduce validation and the role it takes in this basic flow. In this paper, we will use an implementation of the Mandelbrot set generator to illustrate the introduced validation concept and its use in error tracking and performance tuning. Section 3 first describes the fractal operator and the choices we made for its implementation in hardware. Then two uses of validation are exposed: help for verification and a tool for precise performance measurement. We then concluded with a summary of these first applications and further directions to be investigated.

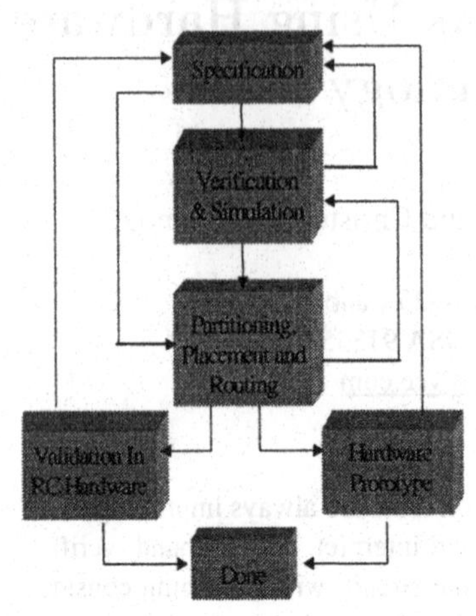

Figure-1 - Design Flow

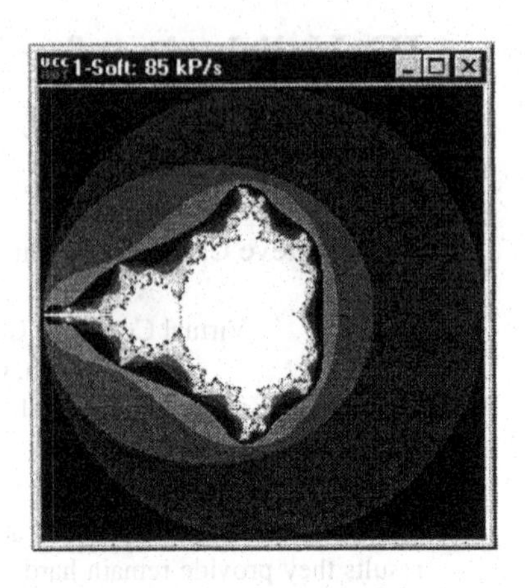

Figure 2. Graphical output of the Mandelbrot set fractal

2. Basic FPGA Design Flow

The most costly pitfalls encountered are the ones that have not been covered by simulation or not detected when analyzing simulation outputs. The principle we introduce is Validation. By this, we mean using real application data to validate the synthesized design. This will provide results in a more understandable form (i.e., the same as the real output provided by the operator). With such a representation of the data, which sticks to the currently developed circuit, analyzing output is much easier than with standard tools. Given real input data, the corresponding real result is calculated with the actual circuit. There is no need to convert input and output to particular formats.

Reconfigurable technology allows the designer to rapidly generate designs at very low cost. These designs may easily be downloaded to the component and exercised to check their behavior. Having a predefined interface helps the designer in focusing only on the operational part of the circuit. Once described and synthesized with the standard design flow, the real circuit is tested with some (or all, depending of the covered data space) of the input data and the output may be then compared to known values.

Let us consider a very basic example to illustrate this. If we want to check if an adder meets some timing constraints, using a timed simulation will require many processor cycles (even with a high level model) for each set of inputs. Validating the design into an FPGA will simply require some tens to a few hundreds of nanoseconds (depending on the circuit frequency). Based on such a simple example, validation provides a speed-up that may range from 10s to 1,000.

3. Uses of Validation

We will use in this section the fractal function computing the well-known Mandelbrot set. This section will briefly introduce some implementation choices we made. Two main uses we have identified for validation in the digital circuit design process are then presented: the first one illustrates the ease in debugging and the other the accurate information it provides for timing and optimization.

3.1. Mandelbrot Set Definition

Named after Benoit Mandelbrot, the Mandelbrot set is one of the most famous fractals in existence. It was born when Mandelbrot was playing with the simple quadratic equation $z=z^2+c$. In this equation, both z and c are complex numbers. In other words, the Mandelbrot set is the set of all complex c such that $z=z^2+c$ does not diverge.

To generate the Mandelbrot set graphically, the computer screen becomes the complex plane. Each point on the plain is tested into the $z=z^2+c$. If the iterated z stayed within a given boundary forever, convergence, the point is inside the set and the point is plotted black. If the iteration went of control, divergence, the point was plotted in a color with respect to how quickly it escaped. When testing a point in a plane to see if it is part of the set, the initial value of z is always, zero. This is so because zero is the critical point of the equation used to generate the set. For the most part, critical points are a subject better left to a mathematical course. However, they have an application in chaos, and fractals in particular. Critical points are used as the starting value for specific variables in a function while calculating fractal sets. For example, in the Mandelbrot set, the initial value of z in the function $z=z^2+c$, is always zero because zero is the critical point of the function.

3.2. Implementation Choices

For each point of the complex plan, the function involves 3 products and 4 additions for each iteration of the function. The calculated function is defined for each point using the C-source code fragment in Figure 3. Integration issues led us to implement these operators on 24 bits with a fixed-point representation:

- 1 bit for the overflow indicator,
- 1 for the sign,
- 2 bits for the integer part of the coordinate,
- the remaining 20 bits for the decimal part (leading to a $2^{-20}=10^{-6}$ precision. Pixelization appears when zooming in, reflecting the lack of accuracy).

```
for (i=0;
      (i<=m_iterationCount)&&
      (distance<m_bailout); i++)
{
   rePart=Zre+x;
   imPart=Zim+y;
   xSquare=rePart*rePart;
   ySquare=imPart*imPart;
   Zre=xSquare-ySquare;
   Zim=2*rePart*imPart;
   distance=xSquare+ySquare;
}
return (byte)(i-1);
```

Figure 3 - C Code for Mandelbrot

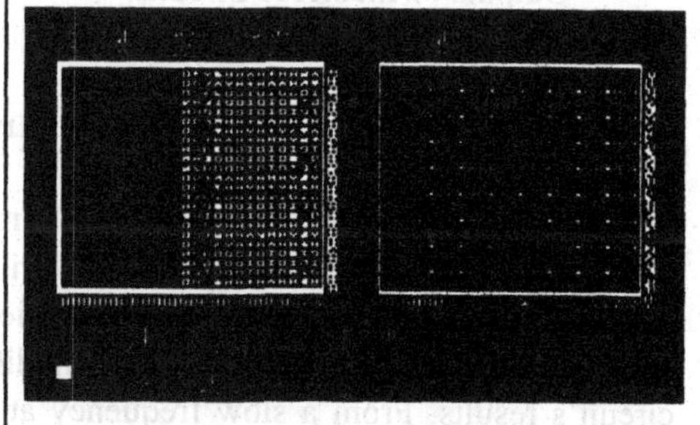

Figure 4 - PCI Core in FPGA

The result of the function evaluation is an 8-bit unsigned integer for each point. It represents the drawing color for the corresponding point on the screen. See Figure 2.

With our current implementation of the function into the programmable logic, a single operator fits into the Xilinx Spartan XCS40 and up to 4 operators into the larger Xilinx XC4062 FPGA.

3.3. Validation Reducing Verification

The verification process remains quite easy in the basic flow when the circuit completely crashes and does not give the right result for any value. Nevertheless, it becomes very hard to track every single error if they only appear for inputs with a particular property. Then it may be difficult to define what is responsible for the error (and thus to correct it).

Offering to the designer a way of seeing its design crash with the set of real input data will help him understand what is really happening in the designed circuit.

Figure 4 gives outputs, which expose different errors: in the first one (Fig. 4.a), the output shows an almost working circuit with erroneous values following a particular shape. Having such a pattern led us to consider an error affecting inputs and results with some particular bit configuration. This error was generated by the wrong management of the overflow bit. In the figure 3.b, one notices that the output has vertical stripes, but a "globally correct" output. The design is replicated into the circuit synthesized and 2 output points are computed at once. One can conclude the design seems to be right but that one of the operators is not operating or is not properly connected to the input/output registers.

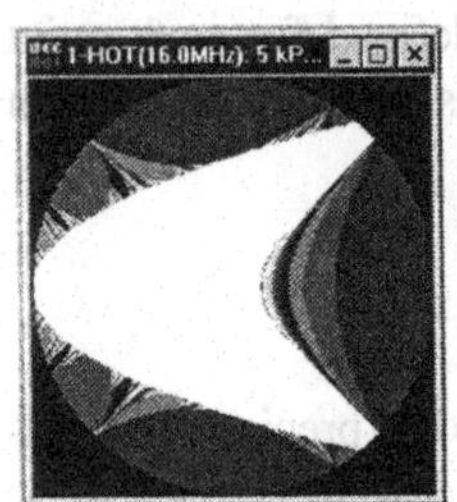
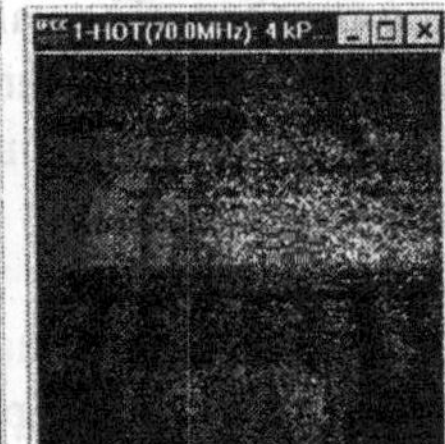
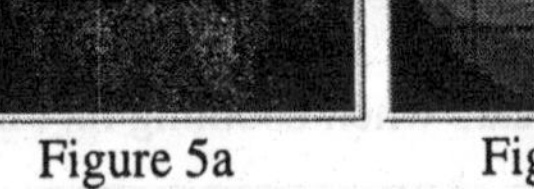
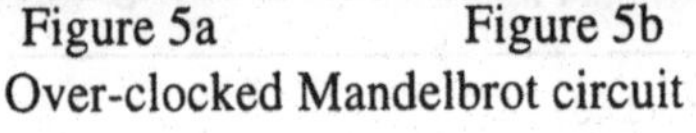

Figure 4a Figure 4b
Outputs of Incorrect Circuits

Figure 5a Figure 5b
Over-clocked Mandelbrot circuit

3.4. Validation for Performance Tuning

An information difficult to extract from simulation results is the actual and accurate timing of a design. Some delays may be critical in all cases, but some may have only little influence on the overall implementation. Validation will give the user a unique way of timing its circuit: by using it, modifying the clock frequency and seeing the circuit's results. From a slow frequency at which the circuit operates normally, the frequency is increased until the operator no longer gives the right output. Viewing the operator crash at a certain clock speed and for a certain values gives much more information than the concatenated list of all signal delays. It then becomes easy to optimize and fine tune a design as one can focus on the slow signal.

With the Mandelbrot example used, we could tune the clock frequency to more than 70MHz after 3 successive optimizations (first timings only allowed about 20MHz). More improvement was possible at the cost of an increasing number of errors (compared to software or a slower hardware implementation). If the timing deadline is met before reaching maximum possible frequency, there is then no more need to focus on placement and/or routing. Further optimization may aim to improve the spatial aspect of partitioning (i.e., to reduce the number of used logic elements or to get a more compact design). See Figure 5a @54MHz & 5b @70MHz.

A 70MHz implementation with 2 operators outperforms by 10 to 20% a mid-sized Pentium PC. As no data dependency exists between the computation of any two points, the implementation of 4 operators will at least double these performances (some overhead is then saved as writing a word is less expensive than writing 4 bytes).

If more improved performances are required, one may consider an implementation where the circuit will store all the results in the board memory and transfer them back in burst mode to the main memory of the system. The implementation is totally different and such an optimization, which is beyond the scope of the paper, has to be done at the description level (i.e., the beginning of the entire design flow from figure 1).

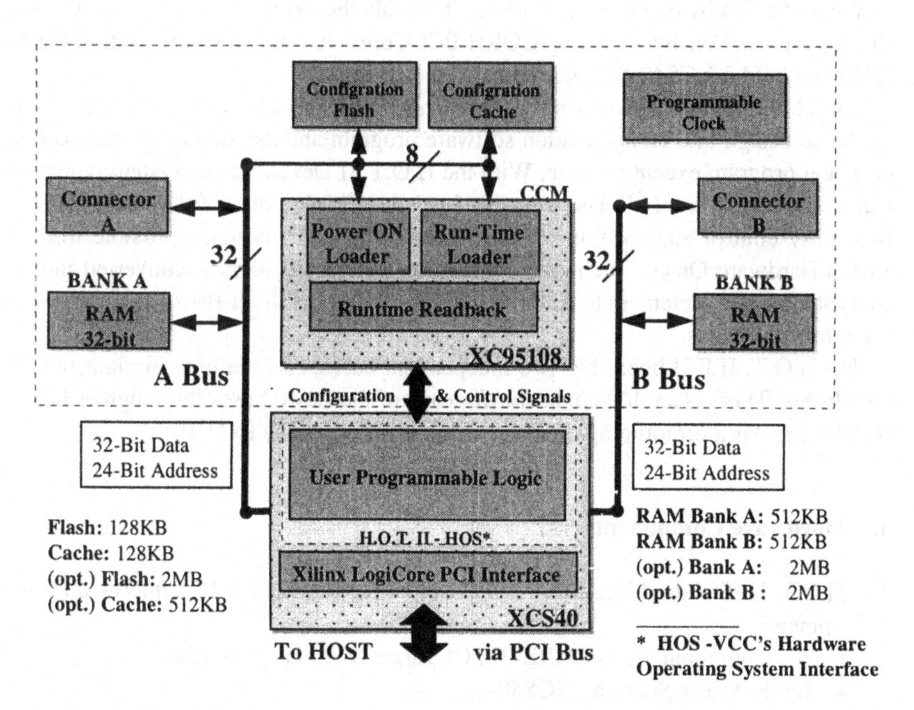

Figure 6 -- HOT II -- Block Diagram

4. HOT II - IP Validation Platform

The H.O.T. II Development System is a powerful tool for evaluation, customizing and prototyping designs using the Xilinx LogiCORE PCI Interface. It is ideal for validating IP products. The PCI based board includes a bus controller, a compute element (FPGA), on board programmable clock and other system features. The combination of these features enables System Level Validation of both hardware and software components of your design. VCC's unique Run-Time Reconfiguration Programming Method to configure the FPGA on the fly (without re-booting), enables running your designs with real data. See Figure 6.

4.1. Functional Description

On power up the FPGA boots up with the H.O.T. II Interface. The H.O.T. II Interface contains the Xilinx PCI LogiCORE Interface Macro and a VCC custom backed that lets users communicate with two fully independent 32-bit banks of RAM and the Configuration Cache Manager (CCM). The CCM controls the Run-Time Reconfiguration (RTR) and reload behavior of the H.O.T. II System. The CCM can configure the FPGA from two on board sources, the Configuration Flash and the Configuration RAM. During reconfiguration, access to the board is disabled by the driver.

When the FPGA comes back on-line, it signals the driver, which reloads the PCI Header information into the LogiCORE PCI Core. A 128KB Configuration Cache RAM can hold 3 XCS40 configurations.

The H.O.T. II Configuration Cache Manager (CCM) allows you to place your hardware design into an application software program and dynamically download the design at program execution time. With the H.O.T. II Development System, you have *Hardware on Demand* ™. The user can load the Configuration RAM over the PCI Bus. Easy control and loading of the configurations (IP) is made possible through VCC's Hardware Object Technology-API. The design bitstream is converted into an encrypted program element to be downloaded into the H.O.T. II Board via application program commands. .

The H.O.T. II PCI board has two independent buses, each with 32-bit data and 24-bit address. There is an I/O connector for each of these two buses for daughter boards (HOT2_PDC is a prototyping daughter card available for the HOT II).

4.2. Hardware Components

1. H.O.T. II Standard Version {HOT2} Ideal for LogiCORE PCI Interface Development
   - PCI Compliant - LogiCORE ™ PCI Target & Initiator Interface
   - Single Xilinx Spartan XCS40-4
   - Single Xilinx XC95108-15
   - 1MB of fast SRAM organized as 2 Independent Banks of 32-bit RAM

- Configuration Flash 128KB
- Configuration RAM Cache 128KB
- Programmable Clock Generator Module (360KHz to 100 MHz)
- Mezzanine Connectors for daughter cards
- Security Jumper
- LED's for DONE, 5v and 3.3v
- Universal 3.3v or 5v PCI board
- 3 Split Power Planes & 1 Ground Plane
- 4 Signal Layers
- Download/Cable98 Module

2. H.O.T. II Expanded Version {HOT2-XL} for CORE Development
   - Single Xilinx XC4062XLT-1 (replaces XCS40-4)
   - 4 MB of fast SRAM organized as 2 Independent Banks of 32-bit RAM
   - Configuration Flash 2MB
   - Configuration RAM Cache 512KB
   - Other features same as Standard Version

4.3. Software Components

With the use of a High Level Hardware Description Languages (HDLs), Hardware Object Technology (H.O.T.) and the HOT II PCI board, the engineer can begin to use Run-Time-Reconfiguration techniques for Validation. The integration of drivers, API and C++ functions in the HOT II DS makes testing designs in Real-Time using Real Data by configuring hardware from executable programs possible.

I. Design Entry -- Use Standard VHDL/Schematic Tools.

II. Design Implementation -- Use Xilinx Foundation or Alliance Software.

III. Make Bitstream -- Use Xilinx Foundation or Alliance Software.

IV. Convert Bitstream File into Run Time Program Mode. -- The Development System supports two methods for Run-Time Reconfiguration. Both require C++ routines for control of the Board. The first method loads the FPGA from a file on the hard disk via program control. The other method compiles the Hardware Object directly into the executable program, loading the FPGA during application run-time. The necessary API and programming routines are included on the Development System CD.

5. The Application Interface

The H.O.T. II Interface includes the PCI/Target as well as the logic for managing the HOT2 board features. The user connects his/her backend to the HOT II Macro (see Figure 7). All handshaking with the host is automatically handled. The example below shows a simple data register circuit.

6. Conclusions

We have shown in this paper a new supplemental way of verifying the correctness and effectiveness of hardware design. Validation allows the designer to check his design with real input data. As a first advantage, we exposed the resulting improvement over timing simulation. Furthermore, as the operator gets real data as inputs, it will provide result in the same format as the output. Instead of analyzing bit vectors or state graphs, the design can compare directly the output to some expected known results. This comparison is much easier and will help to focus on the exact problem.

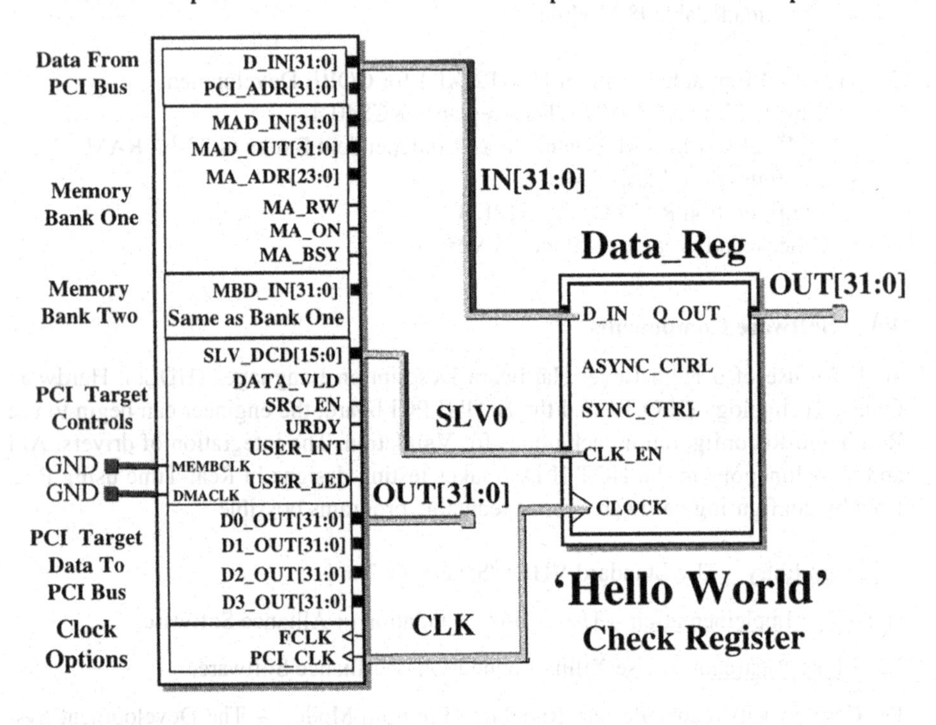

Figure 7 - HOT II Interface

This method makes fine-tuning and timing of the circuit extremely easy and straightforward: increase the frequency until the circuit no longer operates correctly. More applications of validation are still to be investigated. We are currently working on a solution to automate as much as possible of the concept (which still requires a lot of user interaction). Another interesting issue is that applications may be less graphical than the exposed example. In such a situation, an immediate visualization of the errors is then more difficult to obtain. A solution may consist in providing different pre-defined graphs with a simple interface (an operator with one input and one output can be represented by different kinds of planar curves,).

A Processor for Artificial Life Simulation

Matthias Böge and Andreas Koch

Technical University of Braunschweig, Dept. E.I.S., Gaußstr. 11, D-38106 Braunschweig, Germany {boege,koch}@eis.cs.tu-bs.de

Abstract. We present a processor architecture and initial implementation specialized for the simulation of biological evolutionary processes. The CPU simulates a MIMD shared memory computer and executes a set of instructions whose operation is subject to random mutations. Furthermore, it transparently provides memory management and thread control at the assembly level. The current implementation relies on a complex controller using two-level micro-code to generate the required control sequences.

1 Introduction

Computer simulation of digital "organisms" can be employed to experimentally examine natural evolutionary processes that might take millions of years in biological systems. Among the aspects observed are competitive exclusion and coexistence, host/parasite density dependent population regulation, the effect of parasites to enhance the diversity in a community of organisms, and the "evolutionary arms race" (hosts, parasites, hyper-parasites, ...) [1].

TIERRA [2], which forms the basis for this work, is one of several "artificial life" simulation systems [3] [4] that have been developed in the past. It provides a software model of a virtual MIMD shared memory computer and operating system, with the organisms being programs that are executed in parallel. In contrast to conventional processors, the TIERRA CPU is designed to allow the evolution of the organisms by mutation (changing the program code on the fly) and recombination (exchanging code segments between programs). Organisms that still remain functional (contain valid instructions) after such modifications continue to live (run), while those that are damaged (contain invalid instructions) die (are marked for removal from execution and memory).

We present our first attempt to actually implement such a processor for evolvable computation in hardware.

2 TIERRA Virtual CPU

While TIERRA can be configured for multiple instruction sets and in terms of the number of registers and the stack depth, we will concentrate on just a single architecture with the following characteristics:

It has four general purpose registers named AX, BX, CX and DX, all of them 16 bits wide. The architecture also contains a 16-entry stack with an associated stack pointer SP as well as an instruction pointer IP. The instruction set executed encompasses the

32 operations shown in Table 1. A dedicated fault flag register FL is set when an invalid instruction would be executed next (which will be ignored instead). The program memory holding the organisms, called the *soup*, has a size of 64KB.

While this instruction set seems to be rather conventional, it is made unique by the following aspects: In order to simulate random mutations ("cosmic rays"), the execution of each instruction is randomly modified with a user-defined probability. One such modification is the alteration of a register specification. E.g., a destination of DX might become AX or CX for a specific instruction execution. Another mutation introduces +/- 1 inaccuracies into arithmetic and logic (e.g., shift distance) operations. Furthermore, organisms procreate by copying themselves, an operation that is also subject to random changes and thus creates mutated offspring. This approach might result in invalid instructions, which will be detected and skipped during the execution.

Code	Description
nop0	no-op, code 0x00
nop1	no-op, code 0x01
pushA	push AX on stack
pushB	push BX on stack
pushC	push CX on stack
pushD	push DX on stack
popA	pop AX from stack
popB	pop BX from stack
popC	pop CX from stack
popD	pop DX from stack
movDC	DX $\leftarrow$ CX
movBA	BX $\leftarrow$ AX
movii	[AX] $\leftarrow$ [BX]
subCAB	CX $\leftarrow$ AX $-$ BX
subAAC	AX $\leftarrow$ AX $-$ CX
incA	AX $\leftarrow$ AX +1
incB	BX $\leftarrow$ BX +1
incC	BX $\leftarrow$ BX +1
decC	CX $\leftarrow$ CX $-$1
not0	CX $\leftarrow$ CX *xor* 1
zero	CX $\leftarrow$ 0
shl	CX $\leftarrow$ 2$\cdot$CX
ifz	skip next instr if CX $\neq$ 0
jmpo	jump to nearest label
jmpb	jump backwards to label
call	call to nearest label
ret	return from call
adro	AX $\leftarrow$ find nearest label
adrf	AX $\leftarrow$ forward find label
adrb	AX $\leftarrow$ backward find label
mal	allocate memory
div	create new thread (cell division)

Table 1. TIERRA Instruction Set 0

Furthermore, control flow instructions do not operate on absolute or relative addresses, but instead search for sequences of nop0 and nop1 instructions that label a specific location within the code. In this manner, organisms may recognize and jump to certain signatures within themselves or other organisms. Note that this approach requires the relevant instructions to actually search forward and/or backward through the soup for an arbitrarily long sequence of label nops, making them very slow to implement.

Similarly, the mal and div instructions are quite complex: mal implements a first-fit memory allocation scheme which also performs garbage-collection (see below) if required, while div spawns another thread of execution and thus simulates the separation of a child from a parent organism.

In addition to instruction execution, the TIERRA virtual CPU also updates various data structures for memory management and scheduling. The most crucial of these are the reaper and slicer queues. The first is used to determine the "fitness" of an organism: Each time a program attempts to execute an invalid instruction, it is moved upwards in the reaper queue. Each time it successfully procreates using a mal/div combination, it is moved downwards. When the time comes, and a mal request requires more space than currently available, organisms die (are garbage collected) in the order of ascending fitness until the request can be satisfied. Conversely, when a new organism appears, it is allowed to execute immediately after separation from its parent by inserting it appropriately into the slicer queue, which allocates time slices to the organisms.

The software realization also updates various statistical functions that track the diversity and life cycles of organisms in the soup. In this first attempt, these operations were not considered for hardware implementation.

3 Hardware Realization

Even in this simplified version, our attempts at realizing the TIERRA processor push the limits of established FPGA technology. Only the very latest devices [5] are able to accommodate the synthesized circuits efficiently.

The initial target for TIERRA was the SPARXIL architecture [6] (three XC4010 FPGAs, two memory banks). Space limitations forced us to to abandon the planned RISC-like design with fast execution units dedicated to specific TIERRA functions (e.g., a label search engine), instead settling on a realization named TIERRA/1 with a higher degree of operation sharing, but heavily dependent on multiple levels of hard-wired micro-code.

Compared to most conventional CPUs, TIERRA/1 manages a diverse set of data structures. This includes the current data context for each cell (registers, stack, and offspring reference), a map of free memory locations, and the soup holding the program code for the cells itself. Additional locations are used to implement cell scheduling and garbage collection.

Figure 1. TIERRA/1 Hardware Architecture

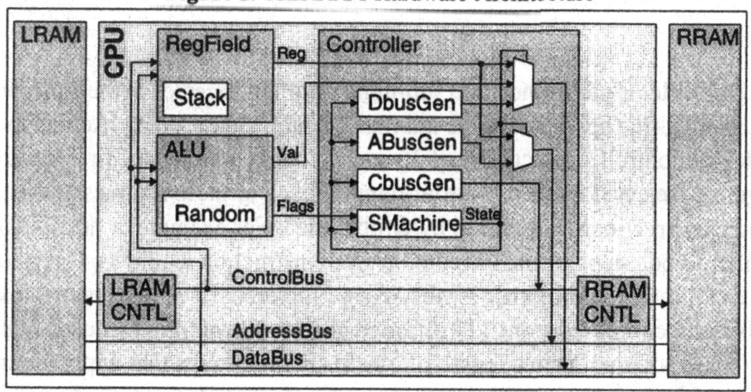

The hardware architecture is sketched in Figure 1. In addition to the registers and internal stack required by the base architecture, the ALU has been extended with a pseudo-random number generator based on a linear-feedback shift-register. All of these units together require only a small fraction of the total device area, the far larger remainder is occupied by the central controller.

4 Hierarchical Micro-Code

Small implementation size was the main design goal for the controller. To this end, it employs a hierarchical micro-code scheme with a considerable sharing of resources between steps. In addition to the instruction semantics defined by the TIERRA specification, the controller also performs the context switches required to emulate a MIMD architecture on a single processor.

Figure 2. Hierarchical micro-coded controller

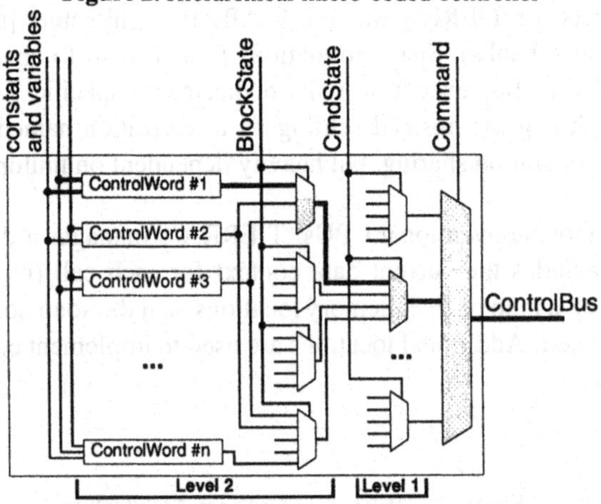

The controller is partitioned into sub-units handling accesses to the data and address busses as well as the operation of the internal execution units. After fetching the instruction code, the controller selects one of 25 micro-code routines at the first hierarchy level (see Figure 2) selected by the Command input. Each of these routines consists of multiple micro-instructions (sequenced by CmdState). For example, a nop0 is processed in two micro-code steps, while the complex mal instruction requires 64 steps (also containing conditionals and loops). At the second hierarchy level, the micro-instructions themselves are composed from 129 different nano-code routines (sequenced by BlockState). These routines have a length of one to six nano-steps that apply control words to the various control signals. The control words contain both hardwired as well as data-dependent control bits. Each of the nano-routines is re-used an average of 2.6 times in the implementation of the micro-instructions, with the highest degree of sharing being 21 times. One nano-step currently requires three clock cycles (two memory accesses plus bus turn-around time).

Figure 3. Control flow for the SHL instruction

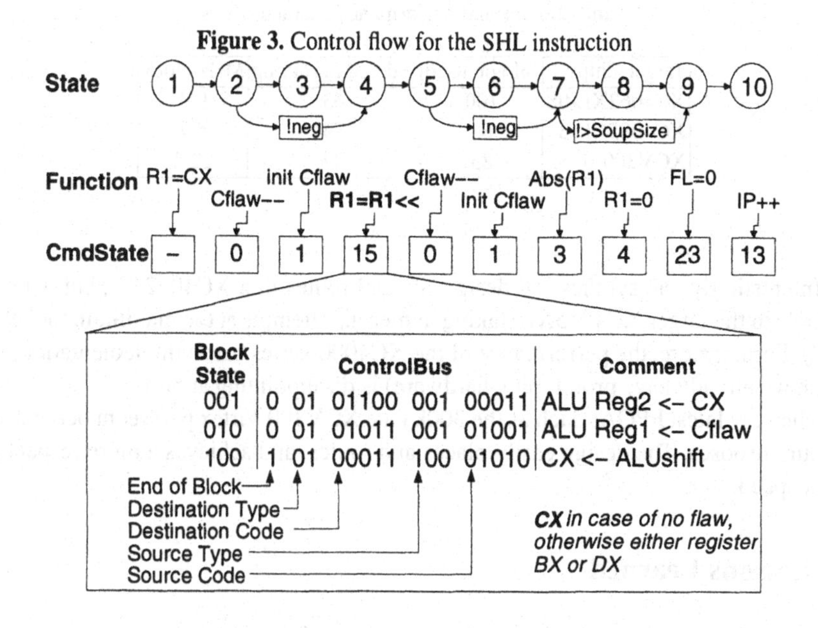

Figure 3 is an example the for the operation of the central controller. It sketches the execution of the "shift left" instruction, which ideally (assuming no mutation) shifts the CX register one bit to the left. The micro-code statements are sequenced by CmdState. In the State 1, we set the target register as CX, mutating it into BX or DX if Cflaw is 0. Afterwards, we decrement Cflaw and re-initialize it to a positive random value. In State 4, we perform the actual shift. If Cflaw is zero at this point, we randomly offset the shift distance by plus or minus 1 (leading to an effective shift by 2 or 0 bit positions). After decrementing and optionally re-initializing Cflaw, we constrain the shifted value to lie within SoupSize (since it might be used for addressing within the soup) and mark the instruction just executed as valid before advancing to the next one.

To illustrate the functionality of the nano-routines, we look at the internal implementation of the `R1=R1<<` micro-instruction, which consists of three nano-instructions generating 16-bit control words. These encode data sources and destination types (b01 and b001 indicate a register, b100 is an ALU output) as well as specifics (b01100 and b01101 are the two ALU input registers, b00011 is CX, b01001 the mutation flaw, and b01010 is the ALU shifter). The BlockState is one-hot encoded, and the nano-routine signals its completion by setting an end-of-block indicator.

5 Performance

Due to the strong control emphasis, the resulting circuit is extremely irregular, and has a large number of high-fanout (> 100) nets. It quickly became apparent that SPARXIL would not be large enough to accommodate this area-optimized design. But even larger chips are hard pressed (Table 2).

Table 2. Area and performance characteristics

Target Chip	%logic used	%delay in routing	MHz clock
XC4062XL-08	100	63	9.3
XC4085XL-3	68	74	5.4
XCV300-6	83	87	22.2

Interestingly, the synthesized design fits and routes in a XC4062XL, but is barely routable in the larger XC4085XL (taking numerous attempts at the maximum tool effort level). Furthermore, the performance of the XC4000-series based implementations (for which we already have prototyping hardware) is disappointing at best.

The very latest FPGAs such as the 300k gate XCV300 Virtex part seem better suited for our purposes. The design easily places and routes, and achieves a more respectable clock speed.

6 Lessons Learned

While our first attempt at implementing a processor for artificial life simulations must be considered less than successful from a performance view point, we learned several important lessons. The first is, that all but the latest devices are simply too small to implement even an area-optimized micro-coded design efficiently. Furthermore, even when state-of-the-art chips are used, the performance is far below those of pure software implementations running on today's CPUs. However, with FPGA capacities growing into the millions of gates, the initially envisioned version using dedicated (hardwired) execution units instead of multi-cycle micro-code might become feasible at last. We plan to follow these developments closely when making another attempt at tackling this fascinating problem.

References

1. Ray, T., "What Tierra Is", http://www.hip.atr.co.jp/~ray/tierra/whatis.html, 03/1999
2. Ray, T., "Tierra Home Page", http://www.hip.atr.co.jp/~ray/tierra/tierra.html, 03/1999
3. Adami, C., Brown, T., "Evolutionary Learning in the 2D Artificial Life System Avida", *Proc. of Artificial Life IV*, MIT Press, 1994
4. Menczer, F., Belew, R., "Latent Energy Environments", http://dollar.biz.uiowa.edu/~fil/LEE/, 03/1999
5. Xilinx Inc., "Virtex 2.5V Field Programmable Gate Arrays", *Advance Product Specification*, 01/1999
6. Koch, A., Golze, U., "Practical Experiences with the SPARXIL Co-Processor", *Proc. Asilomar Conference on Signals, Systems, and Computers*, 11/1997

A Distributed, Scalable, Multi-layered Approach to Evolvable Systems Design Using FPGA's

Craig Slorach, Steve Fulton, and Ken Sharman

Department of Electronics and Electrical Engineering, University of Glasgow
Glasgow, Scotland, G12 8LT, United Kingdom
{c.slorach, s.fulton, k.sharman}@elec.gla.ac.uk
http://www.elec.gla.ac.uk/~craigs/ehw.html

Abstract. In this paper a distributed framework for evolvable hardware system design is presented. This framework allows the automatic evolution of digital circuits through the use of evolvable hardware and offers distribution of the evolutionary algorithm and fitness evaluation resources. The system is scalable and permits evolution of circuits over a wide range of complexities, with the use of a layered approach offering architecture independence. An illustration of the system is given where a circuit is automatically evolved for a classical signal processing problem.

1 Introduction

In recent years, there has been considerable interest in the concept of evolvable hardware (EHW). Indeed, there have also been efforts to produce single chip implementations of evolvable systems [1]. However, investigation to date has centred around finding solutions to specific problems and as such there has been little effort into investigating general frameworks suitable for the evolution of such systems. This paper offers such a general framework, offering distribution, scalability and reconfigurable architecture independence.

There is further motivation behind this research. Firstly, there currently exist freely available and stable evolutionary algorithm packages, such as GA-Lib [2] which provides various forms of Genetic Algorithms (GA's). Such packages offer off-the-shelf evolutionary capabilities using a simple interface and are often already utilised in evolvable hardware research. Furthermore, similar packages also exist which provide abstracted access to reconfigurable resources- for example the JERC package [4] written in Java for the Xilinx XC6000 series [5]. Combining these two resources permits the possibility of facilitating evolution of digital circuits with the minimum of coding effort.

It is also desirable to investigate different evolutionary parameters when working with evolvable hardware, such as the best *genotype* to *phenotype* mapping, the *population size*, and the number of *generations*. Such investigation requires a loosely coupled approach to system integration, ideally with minimal changes being made when adjusting the evolutionary parameters. Current

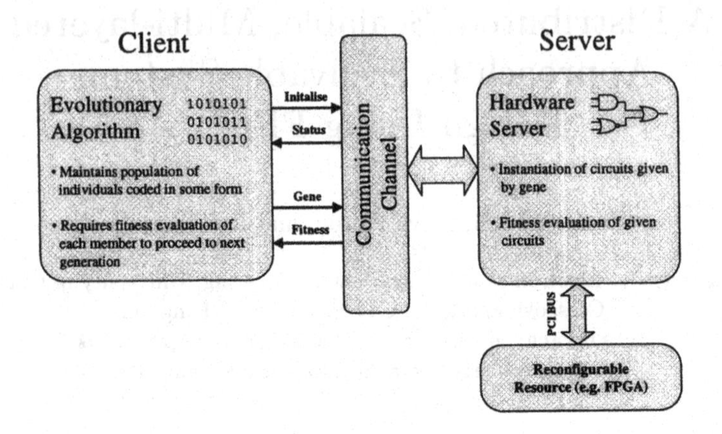

Fig. 1. System architecture, showing a single server.

systems are tightly coupled and require significant coding effort to change the evolution parameters. We also envisage that hierarchical evolution is the way to proceed, with the evolution firstly of primitive building blocks and then evolution involving the placement and routing of these blocks. An example of this is outlined later in this paper.

Considering the above, we provide an architecture which permits both evolution of circuits and also allows changes to be made to the evolutionary parameters with minimal effort. The system also offers distribution allowing non-local reconfigurable resources to be utilised. Architecture independence is offered through the use of a layered approach to resource management.

2 System Architecture

2.1 Client Overview

The client's function (as shown in Fig. 1) is to run an evolutionary algorithm (such as a GA) which is used to maintain a population of individuals which represent some particular circuit configuration. For each individual *gene* in the population there is a corresponding *fitness* value- i.e. how well the particular circuit matches the target specification which is used in constructing a new *generation*. Measurement of this fitness requires the circuit to be instantiated on some reconfigurable device (typically an FPGA). Coding of the client is in C++, as evolutionary algorithms are typically coded in this language.

2.2 Server Overview

The function of the server is to provide some form of reconfigurable resource to instantiate the circuit configuration described by a particular gene in the population, i.e. it performs *genotype* to *phenotype* mapping. Typically, this resource is in the form of a plug-in card containing one or more programmable logic devices. After instantiation, the server then performs fitness evaluation of the given circuit by applying test data and comparing the output with the desired result.

2.3 Client-Server Framework

Given the client and server specifications discussed above, some method of client and server communication is necessary. Here, there are several issues to consider. Firstly, the data passing between the client and server is likely to change as the form of gene coding changes- so any communication method should be able to handle this with minimal changes. Secondly, the communication method must not have a significant overhead in complexity and data transmission requirements, as it is undesirable to have to consider all the underlying communications issues such as network protocols etc. Furthermore, it is desirable to have a single piece of code that can deal with models involving the client and server operating on a single machine, or distributed over a network.

Considering the above, we have chosen Microsoft's Common Object Model (COM) as the framework for client-server interaction. This approach offers maximum flexibility and maximum decoupling of the client and server components. Indeed, it offers a framework for the communication of different programming languages also (in the case of this implementation, Java and C++) in a simple manner. In this implementation, the client simply creates instances of server objects at run-time. Communication is then in the form functions which take standard data type representation of their arguments and return values. The use of the COM model also permits distributed access to objects with little additional effort using Microsoft's Distributed Common Object Model (DCOM) which allows creation of server objects over a network.

2.4 Client and Server Interaction

There are two functions used in the client-server interaction, viz.:

```
int initSystem( <args> )
```

This first function performs server initialisation, requesting access to a set of server based resources and passing configuration settings. In the current implementation, the function takes an argument of an experimental parameters file located somewhere on the network. The parameters file contains details of the experiment including the circuit layout on the FPGA, evolutionary parameters and also the data set(s) used in fitness evaluation. A flag is returned indicating whether or not initialisation has been successful. If not, then the client may opt to try to connect to another server. This process may then be repeated for a number of servers so that fitness evaluation may be undertaken on multiple servers to exploit parallelism.

```
float evaluateGene( <genein> )
```

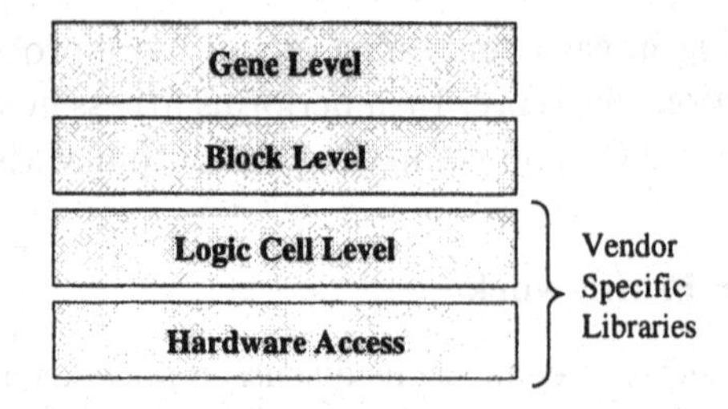

Fig. 2. Layered approach to system resource management.

Assuming access has been granted and the server is ready to provide resources, this function submits a gene for evaluation. Here, the gene is generated by the evolutionary algorithm running on the client. It is then sent to the server and a single result, in the form of a floating point representation of the fitness of the gene is returned. This result is then passed back to the client evolutionary algorithm and evolution proceeds.

3 Abstraction of Hardware Resources

In order to map circuits to reconfigurable resources, some level of abstraction is required. That is, some management of how the gene is instantiated on the FPGA (with the gene being independent of the underlying architecture if evolution is proceeding at a hierarchical level where blocks are being placed on the device). Our choice of abstraction is to use a multi-layered approach which builds upon existing interfaces to FPGA resources as shown in Fig. 2. The hardware access and logic cell layers are based on existing vendor function libraries [4] which are typically provided for access to reconfigurable processor boards. The block level captures logic cells of similar functionality, for example a block of product terms with cells characterised by complemented inputs or pass through cells. Such blocks have fixed inputs and outputs in a dataflow arrangement, with a layout manager controlling their placement. Typically a number of different function blocks are connected together to form a complete system, with the characteristics of the blocks being determined by the gene. Above the block level is the gene management level which controls how a given gene maps to given blocks on the FPGA and how the gene is interpreted (i.e. the genotype to phenotype mapping).

Fig. 3 shows an example of a small scale system, where 2 blocks of AND and OR terms are connected together to form a complete system. Additionally, input registers are instantiated to write the fitness evaluation data (in the final configuration these would take the form of input registers). A counter is also present to count the number of logic 0's and 1's at the output of the system (when used in classification problems)- though this could alternatively be replaced by an output register for other types of system.

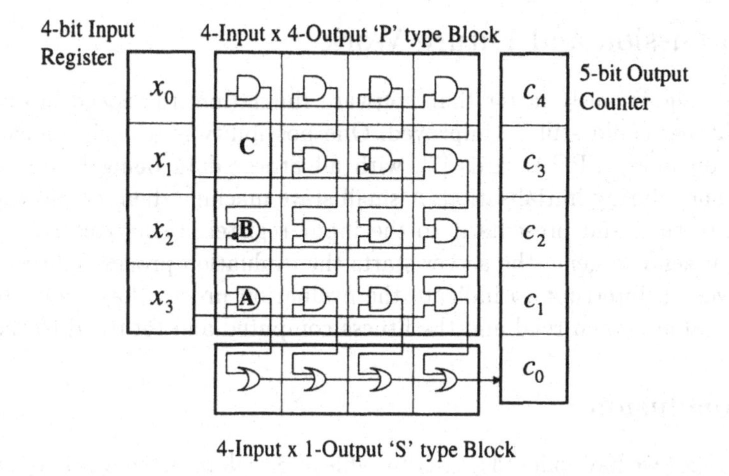

Fig. 3. Example System. Cell 'A' configured as a normal product term, Cell 'B' configured as inverted data input, Cell 'C' configured as a pass through cell.

4 Example System

To illustrate the use of our framework, we have chosen channel equalisation as a target application. This problem has already been investigate extensively, typically involving the use of large computation blocks such as adders and multipliers. The target of the experiment is to evolve a system which can classify input data into one of two sets (denoted by a logic '0' or '1' at the output).

To achieve this, a product and sum architecture (Fig. 3) was chosen. In this experiment, the characteristics of the gene dictated the logic function of each cell within the product or sum block (normal function, complemented input, pass through). The experiment made use of the GA-Lib library and targeted the Xilinx XC6200 series. Results are presented in Table 1. Further results are presented in [6].

Table 1. Experimental Results

Input data width	16 bits	Generations to evolve	10
Number of product columns	2	Population size	500
Total number logic cells	34	Resulting fitness	1.0

5 Discussion and Future Work

Although significantly faster than software simulation, the speed of evaluation of the fitness could still be improved. One possibility is to utilise local SRAM present on many FPGA cards [7], with all fitness data being loaded into the SRAM once during initialisation. A small state machine then fetches each piece of data in turn and presents it to the input register. Using this configuration the client sends a gene, the server starts the evaluation process and waits until it receives an interrupt to indicate the evaluation process has completed. The counter value is then read and the fitness computed and returned to the client.

6 Conclusion

In this paper we have offered a distributable and scalable framework for evolution of digital logic circuits. The framework offers the basis for evolution of a wide range of systems and is architecture independent. We have also demonstrated the use of such a system to evolve a solution which performs channel equalisation.

Acknowledgements

This work uses the GA-Lib genetic algorithm package, written by Matthew Hall at the Massachusetts Institute of Technology. We also wish to thank Xilinx Inc. for their provision and assistance regarding various aspects of the JERC6K software. This research is funded through an EPSRC studentship.

References

1. C. G. Slorach and K. C. Sharman. A novel single chip evolutionary hardware design using FPGAs. In John Schewel, editor, Configurable Computing: Technology and Applications, Proc. SPIE 3526, pages 114-123, Bellingham, WA, November 1998. SPIE – The International Society for Optical Engineering.
2. GAlib: A C++ Library of Genetic Algorithm Components, August 1996. http://lancet.mit.edu/ga
3. Wayne Luk, Peter Y. K. Cheung, and Manfred Glesner, editors, Field-Programmable Logic and Applications. Springer-Verlag, Berlin, September 1997. Proceedings of the 7th International Workshop on Field-Programmable Logic and Applications, FPL 1997. Lecture Notes in Computer Science 1304.
4. Eric Lechner and Steven A. Guccione. The Java environment for reconfigurable computing. In [3], pages 284-293.
5. Xilinx Inc., XC6200 Field Programmable Gate Arrays Datasheet V1.10, San Jose, CA, 1997
6. Craig Slorach, Steve Fulton, and Ken Sharman. A Distributed Approach to Evolvable System Design using FPGA's. In Proceedings of the 3rd World Multiconference on Systemics, Cybernetics and Informatics (SCI'99) and 5th International Conference on Information Systems Analysis and Synthesis (ISAS'99), August 1999.
7. Stuart Nisbet and Steven A. Guccione. The XC6200DS development system. In [3], pages 61-68.

Dynamcially Reconfigurable Reduced Crossbar: A Novel Approach to Large Scale Switching

Tri Caohuu[1]
Department of Electrical Engineering
San Jose State University
San Jose, California 95192-0084
caohuut@email.sjsu.edu

Thuy Trong Le
High Performance Computing Group
Fujitsu America, Inc.
San Jose, California 95134
thuy@fujitsu.com

Manfred Glesner, Jürgen Becker
Institute of Microelectronic System
Darmstadt University of Technology
64283 Darmstadt, Germany
glesner@mes.tu-darmstadt.de, becker@mes.tu-darmstadt.de

Abstract: The cross-bar is the fastest switching architecture for multiprocessor system, yet the most expensive in terms of hardware cost. The hardware complexity is of order $O(n^2 w)$ where n is the number of processors and w is width of the data path. In this paper we present a novel reconfigurable architecture where the hardware cost is *reduced to* $O(w(n/k)^2)$ while maintaining the same operating speed *most* of the time, k is the reduction factor. The approach bases on a cache-like *connectivity table*. This table controls the reconfigurable data path of the cross bar. A hit in the connectivity table results in a direct connection of unit delay while a miss result in a miss penalty for updating the connectivity table. We assume that the local and spatial locality principles are applicable for this class of switching networks.

I. Introduction

The interconnection networks allow high-end multiprocessor system to fully exploit parallelism for higher performance. The ideal interconnection network connects instantaneously any output port of a processor to the input port of any other processor in the system. In practise, the physical location, the connection topology, and the contention of resources prevent this idea situation from happening. Typically, the interconnection can be a bus-based system, a multistage network, or a cross-bar and the design trade of is speed vs cost.

The bus-based network allows only one processor to communicate at anytime puts a significant constraint on the bus bandwidth particularly when a large number of processors is involved. This method is however the cheapest in term of hardware cost. The multistage network is a more popular choice for lager number of processors

[1] Currently a Visiting Professor at TU Darmstadt

which allow several transactions to take place simultaneously. However the packets must travel through $log_p(n)$ where p is the switching factor and n is the number of processors. A major drawback of this scheme is that the latency increases as packets encounter delays due to contention for common paths. The multistage network therefore represents a middle ground between the performance speed and the hardware cost. The cross-bar switch connects each processor to any other processor in a system in one time unit delay and allows several transfers to occur simultaneously. Provided that no two processors trying to access the same destination at the same time, the cross-bar facilitates the connection of $n!$ permutations without blocking . The cost of the hardware is however inhibiting for large size systems since it is of the order $O(n^2 w)$ where n is the number of processor and w is the bus width [1,2,3].

In reality, the applications of "one-to-all broadcast", "single-node accumulation", "all-to-all broadcast", etc... are very limited to the cases such as dense matrix-vector operations and database related operations. As a matter of fact, most of parallel scientific applications strongly have spatial and temporal communication localities [7]. Scientific computing problems generally are represented by systems of algebraic equations that are discretized in forms of sparse matrix and vector operations with some specific patterns. For example, the most popular sparse matrix patterns are the block-tridiagonal and the banded unstructure which are mostly originated by finite difference and finite element discretization classes of system of partial differential equations, respectively [8,9]. Parallel implementation of these methods will lead to the communications of only neighbouring processors. The number of neighbouring processors in fact depends on the decomposition of the physical domain and mostly is fixed for each parallel implementation.

In this paper we investigate the implementation of new type of reconfigurable cross-bar, coined *reduced-crossbar*, which offers the performance speed approximating that of a cross-bar at a fraction of the hardware cost. The novel approach is based on cache-based principle, which allows unit delay access for a hit and a longer penalty in case of a miss, will be presents in the next section. As in case of cache, if the hit rate is high, the reduced cross-bar is virtually a cross-bar at a fraction of the hardware cost.

II. Approach

In the past two decades, cache memory is a very important concept that had been successfully implemented to improve the processing speed of a computer system. Loosely speaking, if the data is the cache memory, the processor can access to it at a much faster speed that that of the main memory (hit). If the data is not in the cache (miss) then addition delay will be incurred to read in new line from memory. For a hit rate of 95% or higher, the processor is mostly operate from the cache with much faster access time.

Similar to the cache concept, the reduced-crossbar design is based on the assumption that the spatial and temporal locality properties also exist in large scale switching application. For instance, in telephony, these properties are translated as: a number called will be called again (temporal locality) and more numbers in the same city will be more likely be called by a particular user than a line outside the city (spatial locality). In our approach, we use a connectivity table in conjunction with scale-down multiplexers to implement the reduced cross-bar. If all the destinations for a particular permutation is in the table then the latency of the crossbar is a unit time delay. In case of a miss, a new line will be read in or generated to be stored in a look-up table.

In this paper we present the design of a 16 x 16 reduced cross-bar with the data path of 16 bits which connects 16 processing elements (PE) to 16 memory modules (ME). Fig. 1 shows the top level block diagram. In this implementation, reduced multiplexer size of 1-to-8 is used, i.e. only 8 destinations are allowed for each PE. The accessibility of each ME is kept in a look-up table which is a cache-like memory. This table is called the connectivity table which control the configuration of the data paths. Each bit in the word represent the access code for a destination memory. If a miss occurs, a new word must be fetched (or generated) and loaded into the table. The algorithm to decide which destination addresses to be loaded into the table is application dependent and hence is not described as a part of the design.

The proposed approach could be easily extended to dynamically adapt the connectivity strategy to actual application requirements. The purpose is to avoid communication bottleneck as well as to exploit free switching resources [5]. Appropriate algorithm for such dynamically adaptation will be incorporated in the future version.

III. Algorithm

The flow chart shown in Fig. 2 represents the implementation of the reduced crossbar. In every cycle, each PE generates an address indicating which destination ME it desires to access. The first step in the flowchart is to determine if the particular PE can gain access to the desired ME (a *hit*). This is done by looking up the connectivity table. Corresponding to each PE address, there is an access word. The access word is 16 bit long with each bit representing accessibility to a ME. An "1" for allowed, an "0" for not-allowed. If the PE and ME address mapping is not in the connectivity table, then a *miss* signal is generated, a new access word will be load and the address is to be regenerated in the next cycle.

In the current cycle, if it is a *hit*, the destination address is decoded, routed through the arbitration logic. In case a single memory module is request by more than one PE, the arbitration logic checks and gives the access to the request with the highest priority and generate the appropriate multiplexer select signals.

A simulation program written in C is used for testing the correctness of the algorithm while the entire crossbar is designed and simulated using Opus and CSIM. Since our interest in the scope of this paper is the correctness of the design and algorithm, we use a random number generator instead of using a test bench in which the generation of the addresses is based heavily on spatial and temporal locality.

IV. Implementation Details

The main building block of the reduced crossbar are the multiplexer logic, the arbitration logic and the miss generation logic as shown in Fig. 3.

The miss generation logic block diagram is shown in Fig. 4 which includes an address decoder block, the connectivity table and the comparison logic. The arbitration logic has similar structure as that of the miss generation unit. The PE address is used as the key to find out which ME it is allowed to access. If the particular PE is allowed access to an ME then multiplexer select signals are generated for giving access to the ME. The generation of the multiplexer select signals is done on a pre-determined priority basis (for simplicity, in our design the PE with lower address has higher priority). Since we have a 16 x 16 reduced cross-bar, 16 set of multiplexer selects signals are generated. Each set of multiplexer select signals are 8-bit wide and are mutually exclusive. The final block of the reduced cross-bar is 8-1 mux block. Although the concept of this block is simple, the implementation is however non-trivial.

The address decoder block is fairly straight forward with a decoder for each PE. Each decoder converts a 4-bit address to a 16-bit access vector. Nand gates are used to improve rise time and the availability of parallel p-stacks [4].

The arbitration unit decides which of the PEs can gain access to a ME in case of contention. A part of the arbitration logic for PE0, for instance, is shown in Fig. 7. Similar circuit is used for PE1 to PE15. PE addresses are decoded in the decoder block and a set of decoded addresses are generated. A total of 16 arbitration blocks are used to generate the multiplexer selects for all 16 PEs.

The Miss Generation Logic generates a miss signal when a particular destination address is not found in the connectivity table. The miss signal will treated as a cache miss signal which causes update of the connectivity table in the next cycle.

The 8-1 multiplexers provide data paths for selected PE and ME pairs. We use pass transistors for the implementation of the multiplexers for speed and area considerations. For each of the ME we need one 8-1 multiplexer controlled by the me#muxel<7:0> signals. Fig. 8 shows typical design of the multiplexer using pass transistors.

The connectivity table is designed so that rows represent PEs and columns represent MEs . *Access allowed* is represented by a "1" (Vcc), and *access inhibited* is represented by a "0" (Gnd).

The connectivity table is to be updated by the table controller to assure the highest hit rate possible. We use a hard-wired controller based on the least recently used (LRU) algorithm for simplicity of implementation. This however can be easily replaced with any new application dependent algorithm to dynamically reconfigure the crossbar for maximum hit rate.

V. Hardware Cost Analysis

The following table summarize the total number of gates used in a 16 x16 reduced cross-bar with data path of 16 bits.

Block Name	*# of gates*
Decoder logic	384
Arbitration block	576
Look-up table	304
Miss generation block	1072
Multiplexers	5888
Latches	64
Total	8288

Comparing this hardware cost with a similar design for full cross-bar we have the following gate counts, for a *n x n* cross-bar with *w*-bit data path. For compatibility with the reduced crossbar design we use *n* equals to 16 and *w* equals to 16.

Block Name	*2-input AND*	*2-input OR*	*# of gates*
MUX Block	$2n^2w$	$2(n^2-n)w$	16,128
Arbiter Block	n(n-1)		230
Total			16,358

Comparing the hardware cost we draw the conclusion that the hardware is significantly reduced by a factor of approximately 50%. In this design example the reduction factor k equals 2 is applicable to a single dimension (i.e. 8-1 MUX is used for 16 x 16 cross-bar) . Similar concept can be extended for two dimension reductions (i.e., *n/k x n/k* true cross-bar is used to implement reduced *n x n* cross-bar). The key factor to a successful reduction is the ability to generate new access word for the connectivity table so that the spatial and temporal localities of the application is maximized.

VI. Conclusion

We have thus represent the concept and design of reduced cross-bar. In the best case, all the destination of all active processors are in the "cache-like" connectivity table, we have the performance latency of a cross-bar. In the worst case, if the destination is not in the connectivity table we need to read from main memory or to generate a new "cache" line and this requires and additional cycle.. The algorithm to generate of the new connectivity word and the strategy to select which entries in the table to be discarded are application dependent. The crossbar is reconfigured dynamically as the new "cache" word is loaded. Given a high hit rate is achievable, the reduced crossbar operates at the speed of crossbar most of the time at a fraction of the hardware cost. Perhaps it is worthwhile point out that the circuit to generate the new cache word can be designed to run in parallel with the execution delay of the cross-bar so that no additional time will incur to the total reduced cross-bar latency. Further work is required to enhance the effectiveness of the connectivity table controller.

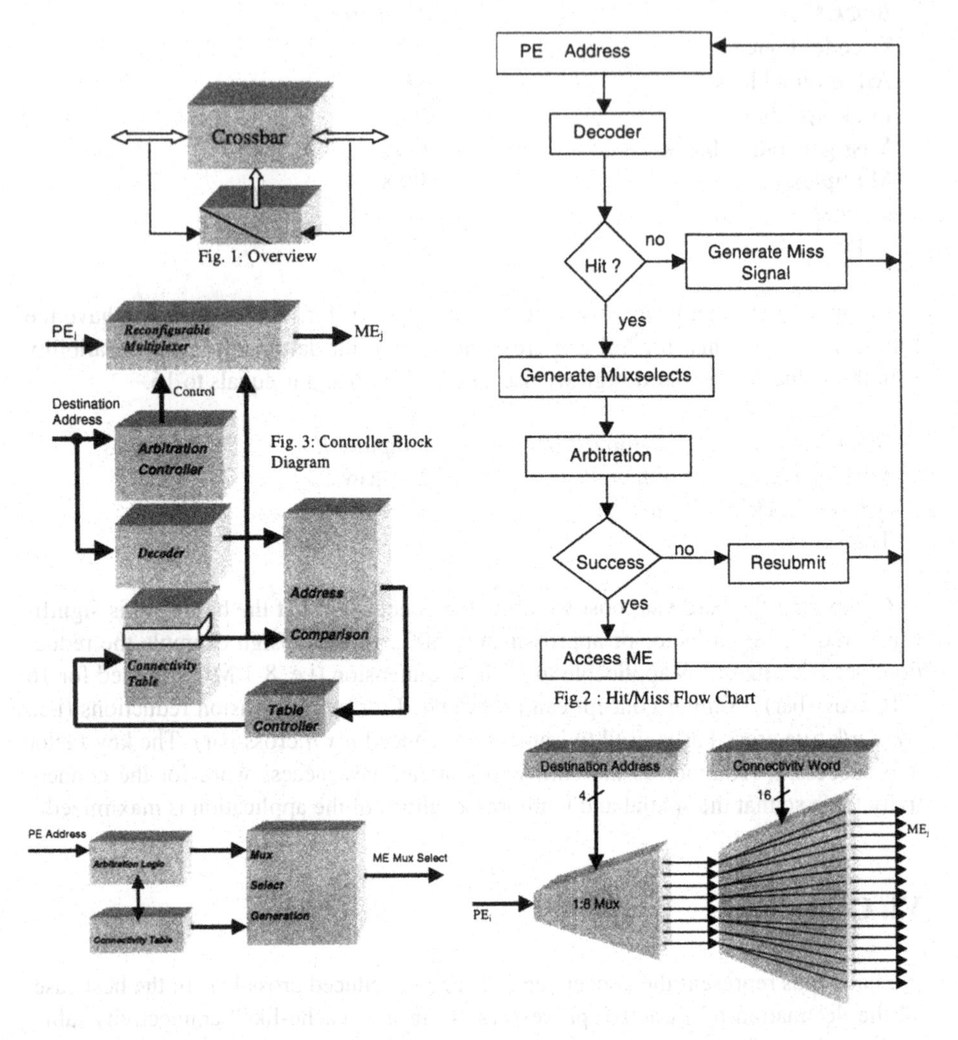

Fig. 4: Miss Generation Block Diagram

Fig. 5: MuxSelect Block Diagram

References

1. M M. Denneau, "The Yorktown Simulation Engine," Proc. 19th Design Automation Conference, pp. 55-59, 1982.
2. G. Broomell and Jr. Heath, "An Integrated-Circuit Crossbar Switch System Design", Int. Conf. on Distributed Computing, pp. 278-287, Apr., 1984
3. Frank E. Barber, et al., "A 64 x17 Non-blocking Crosspoint Design", Int. Solid State Circ. Conf, pp. 116-117, Feb. 1988.
4. S. E. Butner, et al., " A Fault-Tolerant GaAs/CMOS Interconnection Network for Scaleable Multiprocessors", IEEE Journal of Solid State Circuit, Vol. 26, No. 5, pp. 692-704, May 91.

5. D. C. Chen et al., " A Reconfigurable Multiprocessor IC for Rapid Prototyping of Algorithmic-Specific High-Speed DSP Data Paths", IEEE Jou. of Solid State Circuits, Vol. 27, No. 12, pp. 1895-1992, Dec. 1992.
6. B. Quatember, " Concept of a Crossbar Switch for Large-Scale Multiple Processor Systems in the Field of Process Control," Proc. 18th Parallel Computing, pp. 1415-1431, 1992
7. L. Adams, "Iterative Algorithms for Large Sparse Linear Systems on Parallel Computers," Ph.D. Thesis, University of Virgina, 1982
8. Arvind and R. Bryant, "Parallel Computers for Partial Differential Equations Simulation," Proc. Scientific Computer Information Exchange Meeting, Livermore, CA, 1979 pp. 94-102
9. D. Evans, "On the Numerical Solution of Sparse Systems of Finite Element Equations," Mathematics of Finite Elements and Applications III, Mafelap 1978 Conference Proceedings, J.R. Whiteman, ed., Academic Press, New York, pp. 448-458

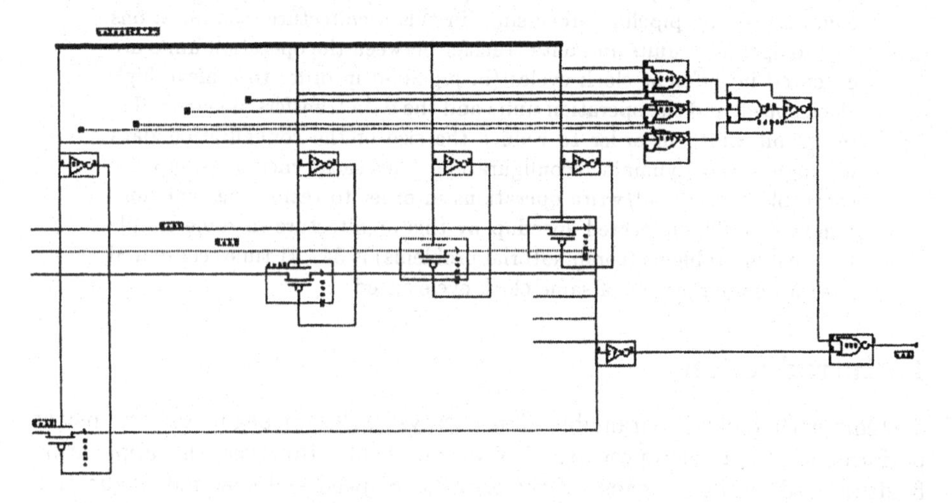

Fig. 7: Reconfigurable Multiplexer Circuit

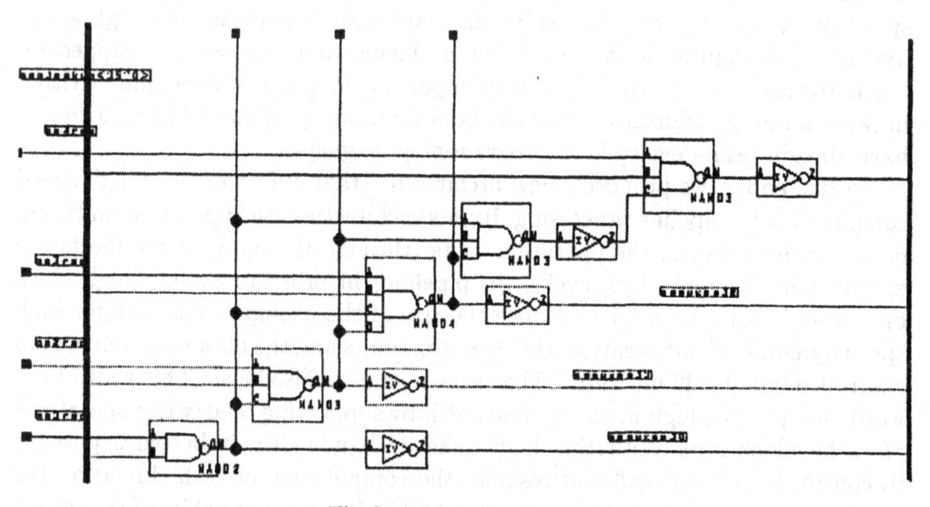

Fig 8: Arbitration Control

A Reconfigurable Architecture for High Speed Computation by Pipeline Processing

Tsutomu Maruyama and Tsutomu Hoshino

Institute of Engineering Mechanics, University of Tsukuba
1-1-1 Ten-ou-dai Tsukuba Ibaraki 305 JAPAN
maruyama@darwin.esys.tsukuba.ac.jp

Abstract. We propose a new reconfigurable architecture for high speed computation by pipeline processing. In this architecture, computations by the operation units and data transfer between the operation units are executed in different clock cycles (in pipeline) in order to achieve high clock speed, and the operation units and the data transfers are controlled by 32 bit width in order to reduce the size of the configuration data for high speed dynamic reconfiguration. The cache memories support cache block size read/write operations in order to realize high memory bandwidth. The expected speedup by this architecture in some simple tree search problems (combinatorial problems) is 34 - 41 times compared with a microprocessor of same clock cycle speed.

1 Introduction

Systems with Field Programable Gate Arrays (FPGAs) begin to show better performance than microprocessors in various fields. However, the application fields are still limited because of the slow clock speed and slow reconfiguration speed of FPGAs, and several new FPGA architectures have been proposed[1–4] to date. In order to achieve high performance in more application fields, high speed operations for wide bit-width data are very important, and high speed dynamic reconfiguration is a must for problems that can not be mapped on the hardware at once. And, it is also important to process algorithms written in programming languages efficiently, because in many applications, algorithms have already been developed in programming languages.

In this paper, we propose a new architecture that aims to realize high speed computation by pipeline processing. In this architecture, the operation units and data transfer between the operation units (through the data interconnect) are executed in different clock cycles (in pipeline) in order to achieve high clock cycle speed. And, in order to reduce the size of the configuration data for high speed dynamic reconfiguration, the operation units and the data interconnect are controlled only by 32 bit width. The cache memories also have 32 bit width for a word, and provide high memory bandwidth by supporting read/write operations of cache block size. With this high memory bandwidth, it becomes possible to control (start, suspend and resume) the computation of each thread on the pipeline in one clock cycle by loading/storing environment variables of the thread at the same time.

2 High Speed Computation by Reconfigurable Hardware

Reconfigurable hardwares can achieve higher performance than microprocessors for the algorithms written in programming languages because of little overhead for controlling data-flow, instruction level parallelism, SIMD processing, and pipeline processing.

In reconfigurable hardware, there is no overhead for controlling data flow such as instruction fetch and decode, and branch instructions, because programs (instructions) are already mapped onto hardware. Speedup by this factor is, however, very small because of pipeline processing of instructions and very high hit percentage of branch prediction in microprocessors.

As for the instruction level parallelism, reconfigurable hardware can execute as many instruction as possible in parallel, while the number instructions executed in parallel by super-scalar microprocessors are limited (usually 2 - 4). However, in many application, the number of instructions that can be executed in parallel is not so large, and speedup by this factor is not drastic.

In many applications where reconfigurable hardwares have showed very high performance, parallel processing (SIMD) and pipeline processing are used very effectively. Parallel processing achieves N times of speedup with N times of hardware resources and N times of memory bandwidth. In parallel processing, it is very important to implement more processing units and more memories in a single chip. Thus, hardwares need to be configured bit by bit. However, this high flexibility reduces the operation speed, and increases the size of the reconfiguration data, which consequently limits the application fields for the reconfigurable hardwares.

Pipeline processing can achieve almost L times of speedup (L is the number of stages) with a single processing unit (a unit for pipeline processing is a bit larger than units for parallel processing because of registers required between the stages) when the number of data processed by the pipeline is large enough. In many application problems that require a lot of computation time, same sequences of operations (by loops and recursive calls) are repeatedly applied to long sequences of data. With reconfigurable hardwares, we can provide pipelines for these computations while supporting the maximum instruction level parallelism.

3 A New Reconfigurable Architecture

3.1 Overview

Figure 1 shows a conceptual block diagram of the architecture (details are given in figure 2) and its processing timing. The architecture consists of operation units (ALUs with selectors and registers), a data interconnect, memories (cache memories and local memories) and a control logic block (PLD) that controls the selectors and registers (enable signals) in the operation units, and read/write operations of memories. Data width on the architecture is 32-bit, and the operation units and the data interconnect is also controlled only by 32-bit width.

In this architecture, data transfer through the data interconnect requires one clock cycle as shown in figure 1-(1). Figure 1-(2) shows the processing timing of multi-thread execution on the pipeline. Cache memories provide read and write operations of cache block size in order to load/store environment variables of each thread in one clock cycle. Direct map cache memories are used in order to achieve fast access time, and send out data before finishing tag comparison.

With this architecture, high clock cycle speed and small size of configuration data (for fast dynamic reconfiguration) are achieved by executing operations and the data transfer in different clocks, and by configuring the operation units and the data interconnect by 32 bit width.

3.2 Details of the Architecture

Figure 2 shows how each block is connected by the data interconnect. The data interconnect consists of horizontal and vertical switches as shown in figure 2. The control logic block (PLD) consists of AND/OR logic plains and registers. The PLD controls the selectors and enable signals of registers in the operation units, and read/write signals to memories.

Figure 3 shows a block diagram of a pair of the operation units. The operation unit has four inputs from the data interconnect (two from a horizontal switch and two from a vertical switch), a self feedback loop and a direct path from a paired operation unit. The self feedback loop is used for mainly adding/comparing the data in its register with its input data sequences every clock cycle. The direct path is used for operations between a variable in a thread and another variable in the next thread.

As shown in figure 2, four cache memories (direct map) which support block size read/write operations are provided. Direct map cache memories are used to reduce cache accesses time, and to send out the data before finishing tag comparison. By strictly managing which data are read from or written into which cache memories throughout the computation, we can achieve high hit ratio and high memory bandwidth using these direct map cache memories.

When a cache miss (for *data-1*) happens, all enable signals of registers are de-asserted, and whole the pipeline is stopped. However, before a cache miss

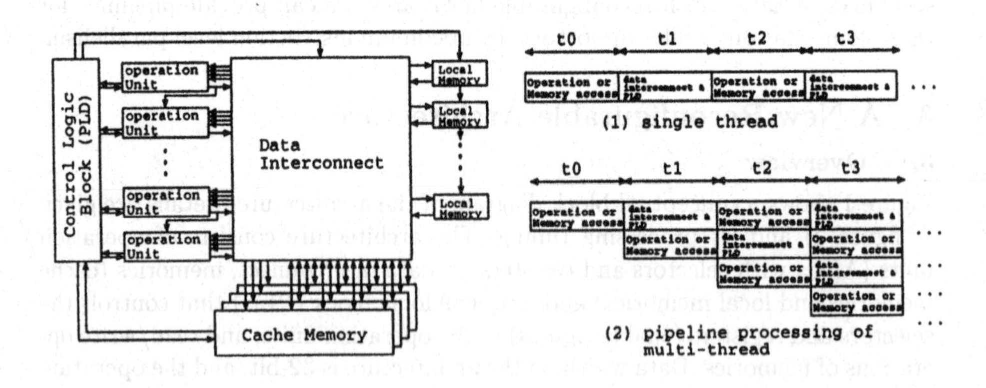

Fig. 1. The Concept of the Architecture and Its Processing Timing

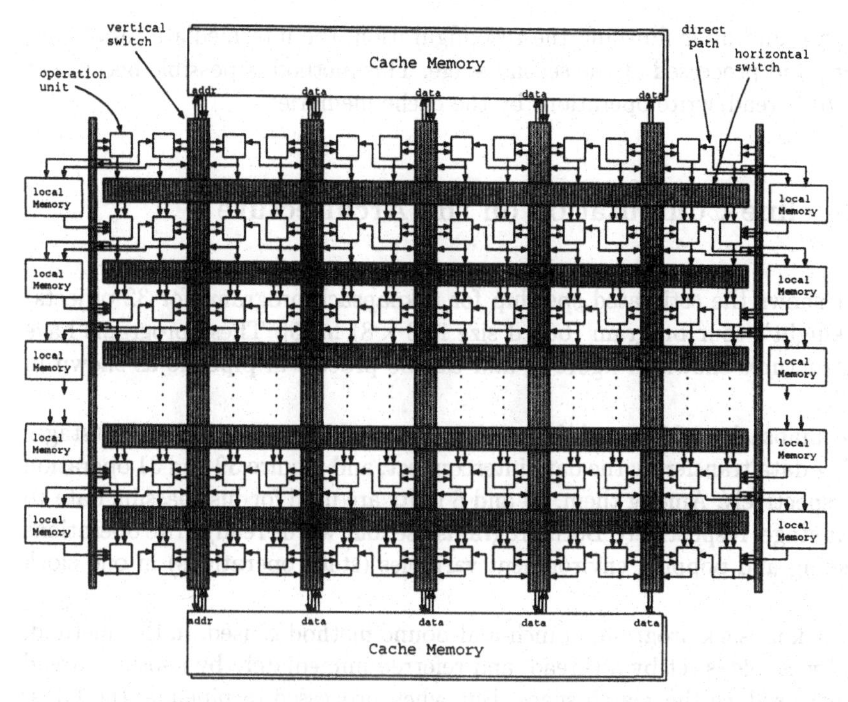

Fig. 2. Detail of the Connection by the Data Interconnect

is detected, one cycle of computation has already finished. Therefore, memory access in the clock cycle (for *data-2*) has already finished, and *data-2* is in the data-output registers of the cache memory when the pipeline is stopped. In order to continue the computation correctly, the four data-output registers in each cache memory have shadow registers, and when *data-1* are loaded from external memories, *data-2* in the data-output registers is pushed into the shadow register. After delivering *data-1* to the operation unit that requires *data-1* through the data interconnect, *data-2* in the shadow register is popped, and the computation is resumed. In this procedure, we need to make only the input registers of operation unit for *data-1* active. Therefore, two enable signals are necessary for cache miss recovery.

In order to process large programs that can not be mapped on the architecture at once, several banks of configuration memories are necessary. With several banks of configuration memory, we can realize very quick dynamic reconfiguration by exchanging an active bank with another bank, and we can update one of the banks while using another bank. The exchanging and the updating of the banks are controlled by some of the operation units and the PLD configured for the purpose. The operation units read configuration data directly from external memories and update one of the bank, and the bank is switched on by the PLD using extra registers that hold the signals beyond the dynamic reconfiguration.

Figure 4 shows how a loop that requires dynamic reconfiguration is processed. In figure 4, the loop computation is divided into three stages. At the end of the first stage, the intermediate results are stored in memory (cache block size boundary), and after finishing the reconfiguration, the intermediate results are read out, and processed at the second stage. This method is possible because of multi-words read/write operations by the cache memories.

4 Pipeline Computation on the Architecture

Table 1 shows the estimated speedup for a knapsack program (for 32 objects) and a knight's tour program (board size is 8 × 8) in [5]. These programs have same structures shown in figure 5, and can be process in pipeline as shown in figure 6.

The knapsack problem and the knight's tour problem require 12 and 10 stages including data transfer by the data interconnect, and require 39 and 31 operation units respectively. Among them, 14 and 6 units are used for just passing data to the next stage, respectively. Both programs use four words read/write operations for pushing and popping environment variables (stack operations) in one clock cycle.

In the knapsack program, branch-and-bound method is used. In this method, a global variable is set by a thread, and referred immediately by another thread in order to reduce the search space. But when processed in pipeline, the latest value of the global variable can not be sent to the next thread immediately. Therefore, the search space by pipeline processing may be a bit larger than the sequential execution (there is no significant difference in the evaluation of the knapsack problems using 100 different set of data).

Table 1. Estimated Speedup

operation frequency (microprocessor=f)	f×0.75	f×1.0	number of operation units used
Knapsack Problem	31.1	41.4	39
Knight's Tour	25.6	34.1	31

Performance of Ultra-Sparc 200MHz = 1.0

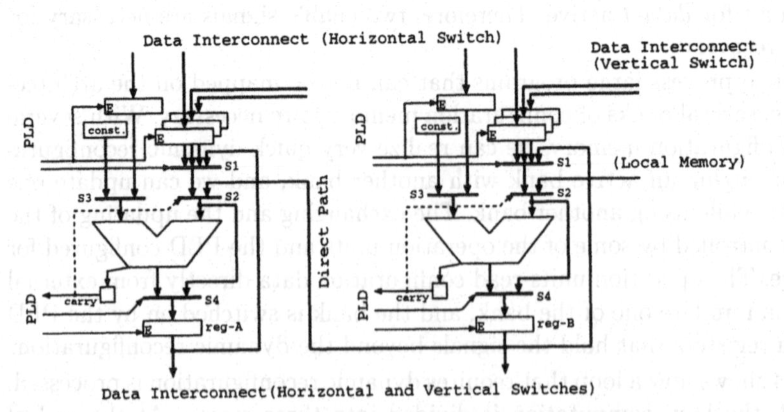

Fig. 3. A Pair of Operation Units

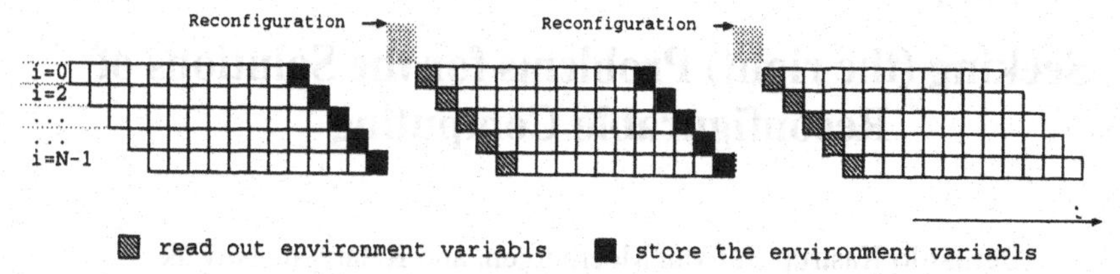

Fig. 4. Pipeline Processing of a loop that requires dynamic reconfiguration

```
for (i = 0; i < NumberOfBranches; i++) {
    cal;                     /* computation for the condition */
    if (cond) {              /* test the condition */
        setup;               /* set up variables for the call */
        recursive_call();    /* recursive-call */
        undo;                /* undo the update by the setup */
    }
}
```

Fig. 5. The Structure of the Programs

5 Conclusions

This paper describes a new architecture for high speed computation by pipeline processing. The speedup by this architecture in some tree search (combinatorial) problems is 34 - 41 times compared with a workstation of the same clock speed. This architecture requires a lot of hardware resources, but the size of LSIs has been steadily improved compared with the improvement of the operation speed. With this architecture, we can achieve more speedup using more amounts of hardware resources, though the performance of microprocessors depends on the improvement of the clock cycle speed. The architecture still has many parameters to be fixed, and redundant circuits to be removed off through the evaluation of real application problems. We are now developing a software system for mapping algorithms written in programming languages onto the architecture.

References

1. John R. Hauser and John Wawrzynek, "Garp: A MIPS Processor with a Reconfigurable Coprocessor", FPGAs for Custom Computing Machines, 1997, pp.12-21.
2. Ralph D. Witting and Paul Chow, "OneChip: An FPGA processor with reconfigurable logic", FPGAs for Custom Computing Machines, 1996, pp.126-135.
3. M. J. Wirthlin and B. L. Huchings, "A Dynamic Instruction Set Computer", FPGAs for Custom Computing Machines, 1995, pp.99-107.
4. E. Waingold, et al., "Baring it all to Software: Raw Machines", IEEE Computer, Sep. 1997, pp86-93.
5. Niklaus Wirth, "Algorithms and data structures", Prentice-Hall, 1986

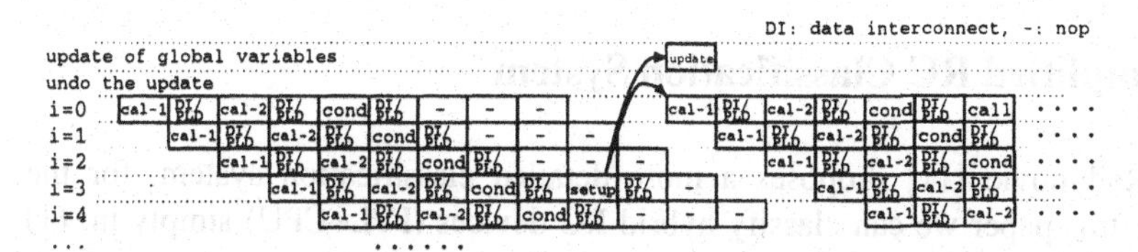

Fig. 6. Pipeline Processing of The Problems

Seeking (the right) Problems for the Solutions of Reconfigurable Computing

Bernardo Kastrup, Jef van Meerbergen, and Katarzyna Nowak

Philips Research Laboratories, Prof. Holstlaan 4 (WL11), 5656 AA Eindhoven, The Netherlands. Tel.: +31 40 274 4421. FAX: +31 40 274 4004.
{kastrup, meerberg, knowak}@natlab.research.philips.com

Abstract. After a decade of active research, Reconfigurable Computing (RC) has yet to prove itself competitive enough to establish a commercial presence. Why? This paper looks for reasons from a pragmatic perspective based on cost-effectiveness. It illustrates a think-model of four dimensions of reasoning that can help evaluate the efficiency of RC approaches. It contributes to augmenting an RC classification system with a choice criteria for each category. RC is found cost-effective for some embedded applications, if certain guidelines are observed. We try to point out practical ways to identify those cases.

1 Introduction

This paper discusses issues facing Reconfigurable Computing (RC) as a promising paradigm for *high volume* embedded systems. RC may fill the gap between programmable cores and ASIC cores. Programmable cores are limited for high throughput applications like video and often dissipate 2 orders of magnitude more power than ASIC cores. ASICs, however, lack flexibility and require large design effort and time. For most functions in embedded applications the implementation choice is often black and white, *i.e.* a function is either mapped on the programmable core or in hardware. It is expected that RC offers different levels of granularity so that a better balanced trade-off becomes possible. In spite of amazing progress in the last 10 years, however, RC has not yet found its way into the computing market. The lack of a clear commercial application is acknowledged in the community [1]. On the eve of the new millennium, we ask ourselves why the approach has not yet proven competitive enough to become a viable commercial option in the high-volume computing main stream. While not pretending to come up with answers, this paper tries to contribute to the discussion from a pragmatic and applications-oriented perspective. We try to survey well-known notions regarding the utilisation of RC and organise them into a structured framework.

2 A Simplified RC Classification System

Although Radunovic [11] proposes a more detailed classification system, for the purposes of this paper we can classify hybrid RC devices (FPL+CPU) simply in: (1)

Loosely coupled. Devices featuring FPL resources as a co-processor. Typically, co-processor and host-processor are connected via a bus and can operate concurrently, in an asynchronous fashion. To exploit concurrent computation, typically large segments of the application are mapped onto the co-processor, which must have a relatively large amount of FPL resources available; (2) *Tightly coupled.* The FPL is integrated within the datapath of the host-processor (compile-time scheduling). The tight integration eliminates the problem of synchronisation and communication latency between host and FPL unit. There are two main lines of usage: (2.1) Using the FPL unit as a *Reconfigurable Functional Unit* (RFU) (e.g. [8]), in a way analogous to the use of an ALU or a multiplier. The execution latency is typically low. RFUs are considerably smaller than reconfigurable co-processors; (2.2) Using the FPL unit for building *deep custom pipelines* (e.g. [10]), with relatively long execution latency.

3 The Need for Hardware Reconfigurability

The area overhead of FPL is largely inherent to its flexibility. The counter-argument is that FPL is re-usable, in the context of reconfiguration. The same piece of silicon, re-used repeatedly for different circuit implementations, can justify the area penalty it implies for a single implementation. *Re-usability through reconfiguration is the only justification for the silicon overhead of FPL implementations, to the extent that the FPL resources are used to implement a number of different digital circuits throughout the device's lifetime.* A typical example can be found in the video domain. Functions like noise reduction and zooming are characterised by a high speed data path and a small controller which is responsible for supplying new settings at frame rate (50 or 100 Hz). Using a programmable processor for this is usually an overkill. It better fits a reconfigurable architecture.

For high-volume electronics, however, hardware reconfigurability is not an issue when: (1) The algorithms a computing device must run during its operating lifetime are known at design-time. In this case, ASICs are more cost-effective (see the discussion in [3]); (2) Standard programmable architectures can fulfil the performance requirements. These architectures can be customised before fabrication [6], which is claimed to be viable even for low-volume production [7]. The stability of the hardware platform makes application programming tools much easier to develop for standard programmable platforms. Reconfigurable Computing platforms are way behind programmable architectures (and their well-developed compiler technology counterpart) in terms of programming friendliness. *The closer an RC platform is to the standard programmable architectures, the greater are the possibilities of adapting standard compiler technology to make application programming less of a problem. Tightly-coupled platforms are suggested as good candidates (Section 2) in this aspect.*

The less a device is targeted at a specific application, the more unknown-at-design-time algorithms it must run in operation, and the better it can benefit from hardware reconfigurability. This is represented in Figure 1. However, the performance deficiency aspect of FPL with respect to ASICs must also be taken into account, as discussed in the next Section.

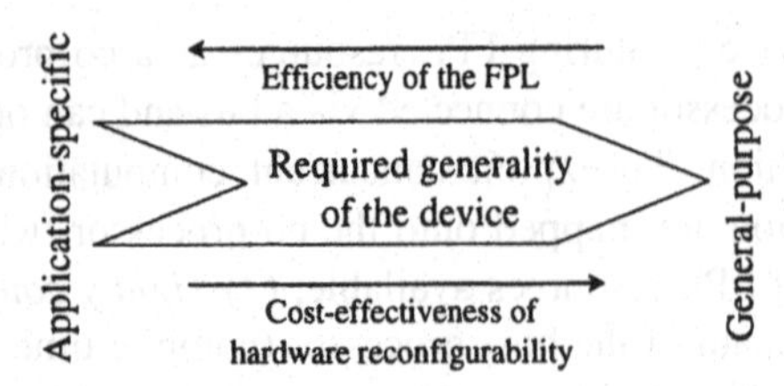

Fig. 1. Cost-effectiveness of the RC paradigm, and efficiency of the FPL, in perspective.

4 The Right FPL for the Right Problem

Once the usefulness of reconfigurability has been verified, there are yet other issues to look at. As the enabling technology of the RC paradigm, FPL is a promising solution for a wide range of problems. A solution, however, that comes in different flavours. The choice of the right flavour for the right problem is not always obvious.

Mapping a multiply-rich application segment onto a general-purpose FPL architecture is like using a hammer to tighten a screw. If the target applications are known to be biased towards a certain kind of computation, a suitable FPGA architecture can be chosen that performs best (and with the least silicon overhead) for that particular kind of computation. For instance, "island-style" FPGAs [4], like the Xilinx XC4000 family, have arbitrary long-distance communication lines suitable for complex, irregular random logic. In contrast, fine-grained "cellular-style" FPGAs [4], like the Atmel's AT6K family, are better suited for highly local, pipelined circuits such as systolic arrays. Hauck [5] discusses these issues thoroughly.

Architectural optimisations that improve FPL performance for regular DSP arithmetic have been developed more extensively in the Academia, in the form of coarse-grained FPGAs [9][10] (or "chunky functional units", as in [1]). Hard-wired computing cores as ALUs or multipliers are embedded into the framework of a reconfigurable interconnect matrix. This allows for a boost in performance and a reduction in the area overhead for the target applications. The loss in flexibility, in turn, renders chunky units inefficient for irregular bit-wise computations. Another limitation is that reduction of order is no longer possible (for instance, a multiply by a constant 2^n can no longer be order-reduced to a mere left logic shift of n bits).

Generally speaking, the FPL architecture can be fine-tuned towards a specific set of applications (a domain) by varying the degree of flexibility of the interconnect and the logic blocks, and by specific performance-enhancing features. This fine-tuning, in turn, usually renders the FPL inefficient for other application domains.

Returning to our point regarding reconfigurable general-purpose processors (Section 3), it is likely that any such device would be required to run as much DSP-like computing kernels as anything else, due to its broad application nature. General-purpose computing, however, is the extreme of a spectrum that goes all the way down to ASIC devices in the opposite extreme. A fundamental dilemma in RC then becomes clear in Figure 1. *In our view, the essential design challenge is to find an application domain wide enough to justify hardware reconfigurability, while specific enough to allow for proper fine-tuning of the FPL resources. Different domains may require different FPL flavours and different integration methods* (see Section 2).

5 Trading Off Time and Space

Impressive speed-ups achieved with the use of FPL in different computing applications have been reported (for instance, [10]). However, as with any other promising technology out there, the use of FPL for computations involves a trade-off. *FPL is cost-effective for computations only if the benefits it offers are necessary or desired in the context of the trade-off it implies.*

Brebner [2] has discussed issues involving the use of control-flow ("computing in time") and data-flow ("computing in space") approaches in the framework of RC. He notes that RC platforms support both control and data-flow programming. The speed-ups FPL allows for, when compared to programmable processors, are related to the fact that the intrinsic parallelism of an application can be fully exploited in hardware, in a data-flow computing fashion. Computing units (at whatever level, from logic gates to full multipliers) can be replicated as necessary for parallel data manipulation (limited only by the amount of programmable logic available in the device). In this context, coprocessor platforms (see Section 2) are suggested as the best candidates, due to the large amounts of FPL resources they deploy. Small RFUs typically cannot achieve the same level of speed-up, but represent only a modest investment in silicon. On the other hand, standard programmable architectures use the control-flow computing paradigm, and process data in a sequential way. A defined and limited number of computing units (functional units), executing pre-defined instructions, is utilised for all data manipulations. Hardware is not spatially replicated, but cyclically re-used over a period of time (without reconfiguration). This typically leads to a more compact and cheaper hardware implementation. Therefore, orders of magnitude speed-ups are the consequence of resources replication.

There is no panacea in here. High levels of parallelism and hardware replication (i.e., FPL utilisation) are only justifiable if the trade-off with performance is cost-effective. In spite of all the real benefits reconfigurable computing allows for, it is then not surprising that, many times, the trade-off renders FPL not competitive at all.

6 Wrap-Up

A cost-effective implementation of the RC paradigm will depend upon four main dimensions of reasoning, which define a think-model:

Dimension 1: Is there a real need for RC, or can either ASICs or programmable cores do the job? (Section 3)

Dimension 2: If there is a need, then what is the right FPL architecture? (Section 4)

Dimension 3: If we have the right architecture, what is the right trade-off between computing in time and computing in space? (Section 5)

Dimension 4. Specifics. Particular requirements must be taken into account, like the need for low-power, user-friendly programming tools, device testability, etc.

Table 1 below is derived from the application of this think-model. The rows represent different levels of granularity. The columns represent the classification criteria of Section 2. The table summarises application requirements that map effectively onto each group and gives examples of applications.

Table 1. Mapping application requirements to the classification categories of RC platforms.

	Loosely-coupled	Deep pipelines	RFUs
General-purpose, fine-grained FPL	High-throughput concurrent processing; bit-level computations; real-time; dynamic rate. (e.g. networking)		Modest investment in FPL; multi-threaded platforms; programmers have no hardware background; gradual transition to RC; alternative to bigger caches and faster clock. (e.g. embedded crypto)
Domain-specific, fine-grained, cellular-style FPL	Order-reducibility; high-throughput concurrent processing; systolic computations; real-time; dynamic rate. (e.g. radar apps)	Order-reducibility; high-throughput and low-power; mix of word and bit-level computations; fixed-rate. (e.g. finite-field computations)	
Domain-specific, coarse-grained FPL	High-throughput concurrent processing; word-level computations; real-time; dynamic rate. (e.g. video I/O processing)	High-throughput; alternative to superscalarity; fixed-rate. (e.g. filter sections)	Modest investment in FPL; multi-threaded platforms; programmers have no hardware background; gradual transition to RC; data-parallel processing; alternative to increased superscalarity. (e.g. multimedia instruction set extensions).

7 An Example: Philips ConCISe

Philips ConCISe [8] is a tightly-coupled single-RFU approach based on a fine-grained CPLD architecture. The CPLD is placed in parallel with the ALU in the execution stage of a RISC pipeline, and can execute Application-Specific Instructions (ASIs).

Standard processors are known to be very inefficient for bit-level operations due to their fixed word-size. The ConCISe RFU can be of benefit for a broad set of applications for easing that limitation. Multiple, small application segments, different for each application it runs, can be mapped onto the RFU (dimension 1). The CPLD architecture is optimised for bit-level manipulations, and the compiler makes sure the RFU is never used for any other sort of operation (dimension 2). This is possible due to the tightly-coupled RFU integration approach used (see Section 3). The CPLD core occupies an estimated 4 mm^2 of silicon surface (yet to be confirmed) in a 0,35μm process. This is a modest investment when compared to the 22mm^2 of a MIPS PR3930, in the same process. We expect ConCISe to allow for more than 50% speedup in critical applications (namely in the cryptography domain), based on simulations currently been carried out. The small investment ConCISe represents makes it cost-effective (dimension 3). A special compilation chain has been developed, which automatically translates application segments into hardware descriptions for the RFU. The device is as easy to program as any microprocessor, therefore representing no extra costs in the programming flow (dimension 4).

8 Conclusions

Reconfigurable Computing (RC) implies a number of trade-offs in terms of: (1) the demand for hardware reconfigurability, (2) the choice of the correct FPL architecture for the kind of computations at hand, (3) the silicon overhead of a "computing in space" approach, and (4) special requirements related to the target applications. These trade-offs spawn a 4-dimensional think-model that can help evaluate the cost-effectiveness of RC approaches. The RC paradigm is not a cost-effective solution where the trade-offs do not lead to a commercial edge.

References

1. W. H. Mangione *et al.* "Seeking Solutions in Configurable Computing", *Computer*, 30(12), pp. 38-43, December 1997.
2. G. Brebner. "Field-Programmable Logic: Catalyst for New Computing Paradigms", *Proc. Of Field-Programmable Logic and Applications*, pp. 49-58, Estonia, 1998.
3. R. Wilson. "Large PLDs face big uncertainties", *EE Times*, Issue 1050, March 1st, 1999. http://www.techweb.com/se/directlink.cgi?EET19990301S0002
4. S. Trimberger. "Field-Programmable Gate Array Technology", Kluwer, MA, 1994.
5. S. Hauck. "The Roles of FPGA's in Reprogrammable Systems", Proc. of the IEEE, Volume 86, Issue 4, April 1998.
6. D. Bursky. "Tool Suite Enables Designers to Craft Customized Embedded Processors", *Electronic Design*, pp. 33-38, February 8, 1999.
7. Wolfe. "HP lays foundation for embedded's future", *EDTN Network*, March 1st, 1999. http://www.edtn.com/story/tech/OEG19990226S0010-R
8. B. Kastrup *et al.* "ConCISe: A Compiler-Driven CPLD-Based Instruction Set Accelerator", *Proc. IEEE Symp. on Field-Programmable Custom Computing Machines*, Napa Valley, April 1999.
9. R. W. Hartenstein *et al.* "Using the KressArray for Configurable Computing", *Proc. of SPIE*, Vol. 3526, Boston, MA, November 2-3, 1998.
10. C. Ebeling *et al.* "RaPiD – Reconfigurable Pipelined Datapath", *Proc. Of Field-Programmable Logic and Applications*, 1996.
11. Radunovic *et al.* "A Survey of Reconfigurable Computing Architectures", *Proc. Of Field-Programmable Logic and Applications*, pp. 376-385, Estonia, 1998.

A Runtime Reconfigurable Implementation of the GSAT Algorithm

Wong Hiu Yung, Yuen Wing Seung, Kin Hong Lee, and Philip Heng Wai Leong

Department of Computer Science and Engineering
The Chinese University of Hong Kong, Shatin, N.T. Hong Kong
{hywong2,wsyuen,khlee,phwl}@cse.cuhk.edu.hk

Abstract. Boolean satisfiability (SAT) problems are an important subset of constraint satisfaction problems (CSPs) which have application in such areas as computer aided design, computer vision, planning, resource allocation and temporal reasoning. In this paper we describe an implementation of an incomplete heuristic search algorithm called GSAT to solve 3–SAT problems. In contrast to other approaches, our design is runtime configurable. The input to this system is a 3–SAT problem from which a software program directly generates a problem–specific configuration which can be directly downloaded to a Xilinx XC6216, avoiding the need for resynthesis, placement and routing for different constraints. We envisage that such systems could be used in hardware based real time constraint solving systems.

1 Introduction

A constraint satisfaction problem (CSP) is a problem with a finite set of variables. These variables can take values within a certain finite domain subject to a set of constraints which restricts them. The solution of a CSP involves finding an assignment of the variables which violates no constraints. Many real life problems such as scheduling, graph coloring, circuit test pattern generation, circuit synthesis and scene labeling can be formulated as constraint satisfaction problems. These are mostly NP hard problems and algorithms to efficiently solve them have been the field of active research in artificial intelligence (AI).

Algorithms for solving CSPs can be complete or incomplete. Complete algorithms, by definition, can find all of the solutions to a problem and typically involve pruned tree searches. However, since CSPs are NP hard, in practice the long execution times involved make it difficult to find any solution. In many applications, completeness is unnecessary and by removing this restriction, much faster heuristic algorithms can be employed to quickly scan the search space. Such incomplete algorithms are able to solve otherwise intractable CSPs. GSAT [1] is an incomplete algorithm for solving the boolean SAT problem. The boolean SAT problem is a CSP in which the variables are binary and the constraints are represented by a boolean equation in a product of sums form. Each sum term is called a clause and is the sum of single literals, where a literal is a variable or its negation. If there are n literals in each clause, the problem is called an n–SAT problem.

Several hardware designs which solve CSPs have been proposed [2–5] and all shown significant speed improvement over software implementations. An important limitation of all of the complete and incomplete implementations described above is that they generated a high level description of a circuit which was customised for the constraint problem. Thus a complete iteration of the synthesis, place and route cycle was required for each new set of constraints. These steps are time consuming (it can take several hours to synthesize, place and route a large design) and precludes its use in real time systems.

In this paper we describe an FPGA architecture and its implementation which enables the direct generation of a problem specific configuration for solving the GSAT problem. Our design exploits the open architecture of the Xilinx XC6200 series FPGAs which is used to directly configure the circuit from the constraints. The remainder of the circuit is fixed. Using this technique, the need for synthesis, placement and routing of a new circuit for different constraints is eliminated, resulting in a large savings in compilation time. It should be easy to adapt this reconfiguration technique to other constraint solving algorithms.

2 Implementation

The GSAT algorithm [1] is a simple greedy search based method for solving SAT problems. Basically, each iteration of the algorithm is performed by flipping the bit of the variable which would result in the largest improvement in the number of satisfied clauses. This process is repeated Maxflips times. If no solution is found, a new random variable assignment is made and the process repeated (up to Maxtries times). For a boolean constraint equation F, the algorithm can be described by the following pseudocode

```
GSAT(int Maxtries, int Maxflips) {
   for (i=1 to Maxtries) {
       V = a random instantiation of the variables;
       for (i = 1 to Maxflips) {
           p = variable whose negation yields largest
               increase in number of satisfied clauses;
           V = V with p flipped;
       }
       if (F(V) is true)
           return V;
   }
   return the best instantiation found;
}
```

Our implementation of the GSAT algorithm uses reconfigurable hardware to implement the inner loop of the algorithm described in the previous section, i.e. the calculation of p. The rest of the GSAT algorithm is implemented in software which maximises the flexibility of our hardware. The software downloads a variable assignment to the board. The hardware flips each variable in turn and computes the number of satisfied clauses and after each variable has been flipped, the variable whose negation yielded largest increase in number of satisfied clauses is returned to the software program. The software then computes the next variable assignment by flipping that variable.

A block diagram of the hardware architecture is shown in Figure 1. The *variable memory* is a register that is used to hold the current variable configuration is called the variable memory. The host computer writes a new variable configuration every iteration of the algorithm. The *flip–bit vector* is a shift register used to cycle through each variable of the variable memory in sequence. Only one bit of the flip–bit vector is asserted at any time so an exclusive–OR of the flip–bit vector and the variable memory will negate that bit (see Figure 2). The *clause checkers* are used to implement the sum terms of the constraint equation. Inputs to the clause checkers are variable assignments. For a problem with n clauses, n parallel clause checkers are used to implement all of the clauses and since our hardware is restricted to 3–SAT problems, each clause checker has 3 inputs. Thus each clause checker is simply a 3 input OR gate with active low or active high inputs depending on the equation. The *sum register* is used to store the largest number of satisfied clauses. The *M–bit adder* is used to calculate the number of satisfied clauses. For a problem with M clauses, M 1 bit numbers must be summed to find the number of satisfied clauses. Since the adder is in the critical path of the design, a tree adder was used. As a result, there will be $\log_2 M$ levels of delay. Moreover, each level towards the root of the tree has an additional bit of precision. Thus in the kth level, all inputs are k–bits and they are added together in a pairwise fashion to generate a k+1 bit result. The *comparator* is used to compare the current number of the satisfied clauses with the value in the sum register to see if a better result had been found. In the event that the current number of satisfied clauses is equal to the best value stored in sum register, the *random bit generator* is used to decide which solution to keep. This prevents the algorithm from being captured in a local minima. The Random bit generator is implemented as a linear feedback shift register and the equation used is bit(0) = bit(3) xor $\overline{\text{bit(4)}}$ xor bit(5) xor bit(7).

The target device for our implementation was a Xilinx XC6216 reconfigurable processing unit (RPU) [6]. This device was selected mainly because the open architecture of the RPU documents how the low level configuration bits relate to the hardware of the device. Other FPGA architectures do not provide this information making it extremely difficult to generate configurations without going through the vendor's CAD software. The Virtual Computer Company (VCC) H.O.T. Works development system was used to test the design. This PCI board includes a Xilinx XC6216 reconfigurable processing unit (RPU) together with SRAM memory and a PCI interface.

With the exception of the clause checkers, the circuits were synthesised from VHDL descriptions using the Velab VHDL elaborator version 0.52. They were then placed and routed using the Xilinx XACTstep 6000 tools. The clause checkers are problem dependent and are customised by a C program. The implementation of a clause is illustrated in Figure 3. All of the variables are routed in horizontal lines and the logic to implement a particular clause are distributed in a vertical direction. In order to implement the routing and logic, 5 different logic configurations are required as detailed in the figure. Compared with the general problem of placement, routing or logic synthesis, this task is trivial and is performed in linear time. The software customises the logic equation of each clause checker and writes the new configuration into the address mapped configuration of the XC6200 memory [6]. In the XC6200 devices, this can be done without affecting the nonconfigurable parts of the circuit.

Ultimately, the compactness of the circuit implementation determines the size of the SAT problem that can be solved using the system. For v variables and n clauses, the amount of logic required for our design is variable memory (v); flip–bit ($2v$); clause checkers (nv); sum register ($3\log_2 v$); adder ($3n\sum_{k=1}^{\log_2 n}(k/2)^k + \log_2 n$); comparator ($8\log_2 n$); and random number generator (12). For the XC6200 device, we estimated that the total resources required (i.e. logic+routing) were approximately 4× that of the logic alone. The clause checkers dominate the size of the circuit. A method of overcoming this problem is to use a non–runtime reconfigurable clause checker which utilises exactly the same design except that the clause checker circuitry is synthesised from VHDL. This customisation of the clause checker reduces its logic from nv to $2n$ since a specific routing pattern is much more compact than the general case. With a non–runtime reconfigurable clause checker, the 50 variable 80 clause problem to fit on the XC6216 which contains 4096 cells. A smaller 8 variable 16 clause problem was used to test the runtime reconfigurable clause checker.

3 Results

The design was tested on the aim-50-1_6-yes1-1.cnf (aim) benchmark problem from the Second DIMACS Implementation challenge on NP Hard Problems: Maximum Clique, Graph Coloring, and Satisfiability [7]. This problem has 50 variables and 80 clauses. A smaller problem (small) with 8 variables and 16 clauses was also tested. This problem was generated using the *mwff* program from DIMACS [7, 8].

Table 1 shows the total VHDL compile time, VHDL compile time of just the clause checker and the execution time of our runtime configuration approach using a Pentium II 300MHz 64MB Ram system running Winows 98. It can be seen that a four orders of magnitude improvement in the time required to translate the problem into the clause checking circuit is seen. The same table also compares the execution time of a software implementation of the GSAT program version 41 written by Selman and Kautz [1] running on a Sun Ultra–5 machine with the hardware execution times. The final column shows the maximum clock frequency given by the XACTstep software.

Our design is currently faster than the software version for the small problem but slower for the aim problem. Although a software implementation on a typical workstation is roughly two times faster for the aim problem, it must be

Problem	Total (s)	Clause Checker (s)	Runtime Config (ms)	Software (ms)	Hardware (ms)	Frequency (MHz)
small	52	16	1.2	0.59	0.032	9.12
aim	900	90	7	5.14	13.5	4.17

Table 1. Table comparing the compilation and execution times of various GSAT implementations.

remembered that it is achieved using very few logic gates. For embedded systems where circuit size, cost, power dissipation and reliability are concerned the RPU implementation offers significant advantages.

4 Conclusion

Using a problem specific architecture and runtime configuration, a four order of magnitude speedup in the reconfiguration time of the GSAT algorithm over the conventional approach involving resynthesis, placement and routing was demonstrated. The design was tested on hardware and achieved approximately the same performance as that of a modern workstation but at greatly reduced hardware cost, power consumption and memory requirements. We envisage that such systems could be used in hardware based real time constraint solving systems and may have applications in signal processing, robotics and control.

References

1. B. Selman et. al., "A new method for solving hard satisfiability problems," in *Proc. of AAAI-92*, pp. 440–446, 1992.
2. M. Yokoo et. al., "Solving satisfiability problems using field programmable gate arrays: First results," in *Proc. of the 2nd Inter. Conf. on Principles and Practice of Constraint programming*, pp. 497–509, 1996.
3. P. Zhong et. al., "Accelerating boolean satisfiability with configurable hardware," in *IEEE Symposium on Field-Programmable Custom Computing Machines*, pp. 186–195, 1998.
4. Y. Hamadi and D. Merceron, "Reconfigurable architectures: A new vision for optimization problems," in *Principles and Practice of Constraint Programming CP97*, (Austria), pp. 209–215, 1997.
5. T. Suyama et. al., "Solving satisfiability problems using logic synthesis and reconfigurable hardware," in *Proc. 31st Annual Hawaii Internation Conf. on System Sciences*.
6. X. Inc., *XC6200 Programmable Gate Arrays (data sheet)*. 1997.
7. *Dimacs challenge benchmarks*. ftp://dimacs.rutgers.edu/pub/challenge.
8. D. Mitchell et. al., "Hard and easy distributions of sat problems," in *Proc. AAAI-92*, pp. 459–465, 1992.

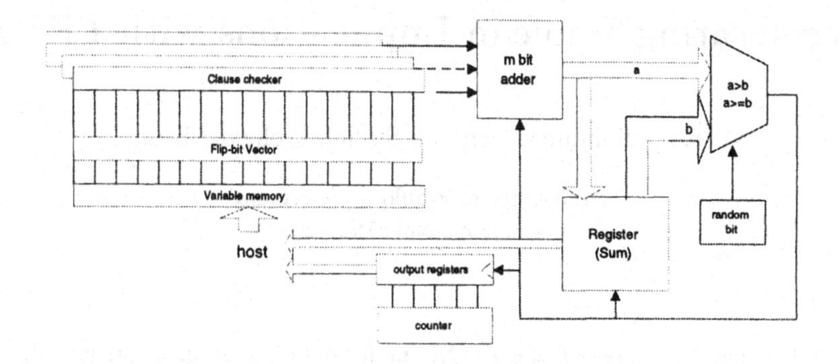

Fig. 1. Block diagram of the GSAT hardware.

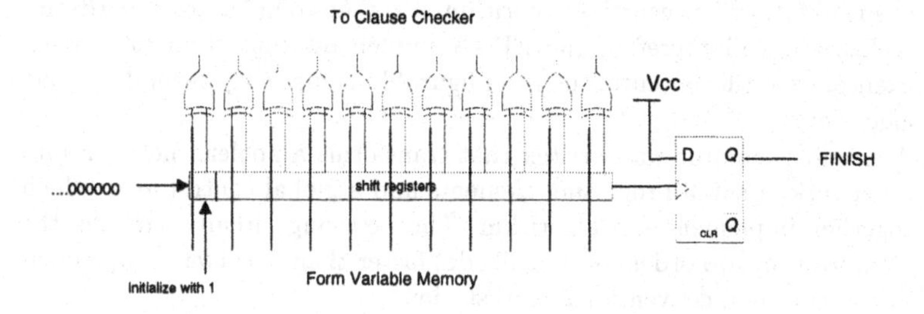

Fig. 2. Flip bit vector detail.

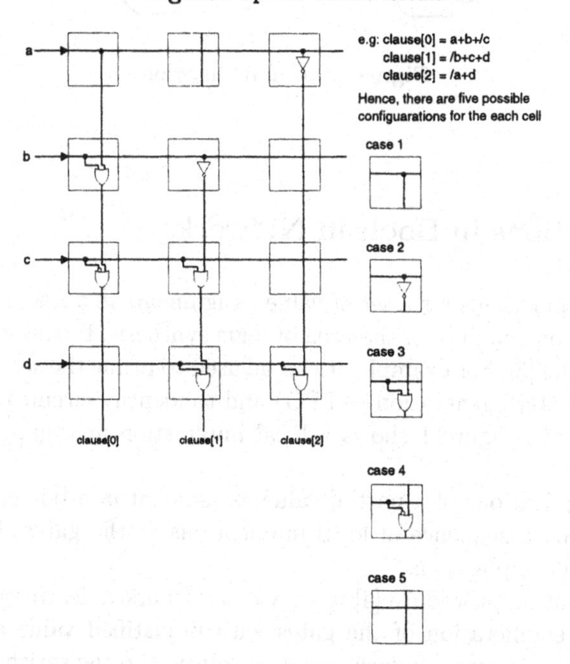

Fig. 3. Implementation of the reconfigurable clause checker.

Accelerating Boolean Implications with FPGAs

Kolja Sulimma, Dominik Stoffel, Wolfgang Kunz

University of Frankfurt, Germany
kolja@prowokulta.org

Abstract We present the FPGA implementation of an algorithm [4] that computes implications between signal values in a boolean network. The research was performed as a master´s thesis [5] at the University of Frankfurt. The recursive algorithm is rather complex for a hardware realization and therefore the FPGA implementation is an interesting example for the potential of reconfigurable computing beyond systolic algorithms.
A circuit generator was written that transforms a boolean network into a network of small processing elements and a global control logic which together implement the algorithm. The resulting circuit performs the computation two orders of magnitudes faster than a software implementation run by a conventional workstation.

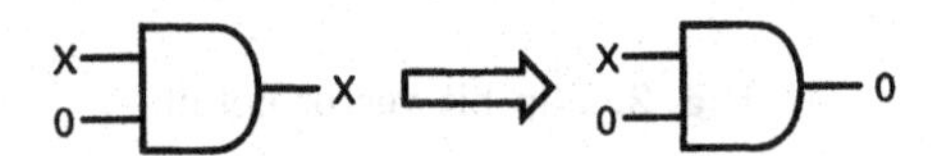

Figure1. Local implication

1 Implications in Boolean Networks

Computing implications for a set of value assignments in a boolean network has many applications, mainly in the area of logic synthesis, testing and verification of digital circuits [3]. For example, it can be used to prune the search space in automatic test pattern generation (ATPG) and to identify circuit transformations in logic synthesis. Figure 1 shows a local implication at a gate with a partial value assignment.

Not all implications of a partial value assignment in a boolean network can be computed by a sequence of local implications at the gates. Figure 2 shows such an indirect implication.

Our implication procedure [4] is known as "recursive learning" and relies on an AND/OR enumeration of the gates with unjustified value assignments. It computes all implications and represents a solution to the satisfiability problem which is key to many CAD problems.

Recursive learning finds indirect implications by identifying gates for which there are multiple possible value assignments and then recursively computing the implications for each assignment. If there are implications that are common for all assignments they must be valid indirect implications that can be learned.

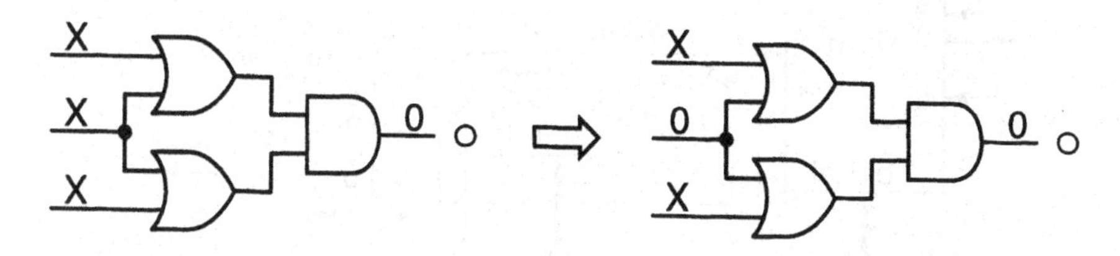

Figure2. Indirect implication

2 FPGA Implementation

The recursive algorithm is rather complex for a hardware realization and therefore the FPGA implementation is an interesting example for the potential of reconfigurable computing beyond systolic algorithms.

The control flow of the recursive AND/OR enumeration is hard to parallelize. Furthermore the control logic is too large to replicate it several times. Furthermore the control flow may be crucial for the success of the algorithm, but most of the runtime is spent in the computation of implications locally at the gates.

Therefore, it was decided to implement a single, sequential control flow and to parallelize the operations on the data structure that represents the boolean network.

To take full advantage of the possible fine grain parallelism of FPGAs, a circuit generator was written using the BOOM Package of UC Berkeley [1] that transforms a boolean network into a network of small processing elements (PEs). One element is required for each gate and each variable, respectively, as shown in figure 3. The network of processing elements inherits the structure of the circuit under examination. The PEs communicate exclusively along the edges of the boolean network. A central FSM controller connects to all elements via a few global control signals.

The connections between the PEs are two bits wide because the algorithm operates on four-valued logic. The same encoding as in [6] was used. With this encoding all bits have a monotonic behaviour during the central implication procedure which allows asynchronous operation and simplifies the PE circuits.

There are two types of processing elements, one type representing gates and the other representing variables. Each PE contains the logic necessary to perform the operations required by the algorithm and a stack that allows to save the state of the element up to an recursion depth of 16.

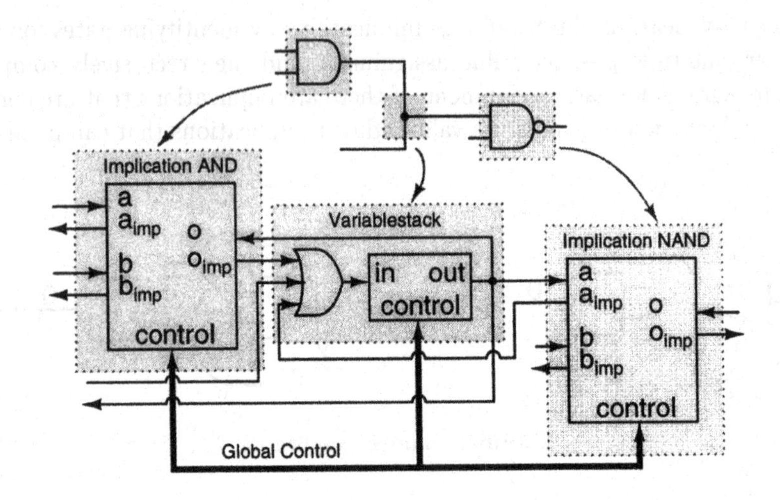

Figure3. Mapping of gates to processing elements.

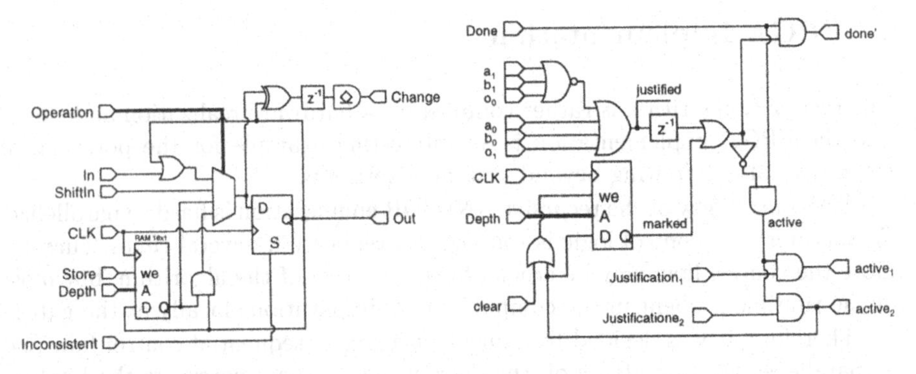

Figure4. PE for a boolean variable (left) and the control portion of a gate PE (right).

Figure 4 shows half of the schematic of a processing element for a boolean variable on the left. It stores the current value of one bit of the encoding of the variable in a flipflop as well as a sixteen entry recursion stack for this bit. The variable type PE can perform five operations on this bit. During initialization the bit can be shifted in from a daisy chain. It can be pushed to or popped from the recursion stack and a logical AND of the current value and the value on the stack can be computed to determine the assignments common to the last recursion call and the one stored on the stack. During the computation of the direct implications a logical OR of the current value and local implications computed by the neighbouring gate type PEs is stored into the flip flop.

The gate type PEs consist of two parts. One controls the AND/OR enumeration in a distributed way. Its schematic is shown in figure 4 on the right. It calculates whether the gate needs to be part of the enumeration (we call this unjustified) and remembers in a recursion stack whether this gate has allready been part of the enumeration during the current recursion call. This information is used to determine whether this gate should be justified next.

The other part of the gate type PEs compute local implications from the value assignments of the variables connected to the gate. It also assigns justifications to these variables if necessary.

Each processing element requires four Xilinx XC4K logic blocks. This means that with today's FPGAs implications can be performed in circuits with up to 1000 Gates in a single chip. Arrays of multiple FPGAs can be used with only a minor performance degradation, as the implementation still works correctly even if the delay and bandwidth vary from connection to connection.

3 Operation

The FPGA configuration depends on the input network and therefore must be synthesized individually for each boolean network under examination. The circuit can be reused without changes for different initial value assignments.

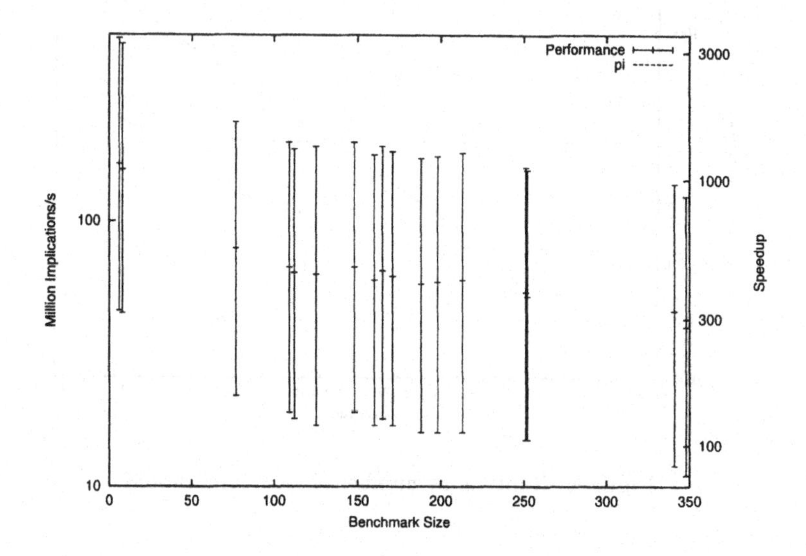

Figure5. Performance of the accelerator by benchmark size

The performance of this approach has been evaluated for ISCAS benchmarks with up to 600 gates. For each benchmark a configuration for XC4000XV FPGAs was generated using the BOOM generator and the Xilinx Foundation synthesis tools. The performance of the various operations required by the algorithm was then analyzed using the Foundation static timing analyzer.

Compared to a 220 MHz Ultrasparc workstation a speedup of two orders of magnitudes was achieved as shown in figure 5. The diagram shows quite large error bars because the parallelism of the computation and hence the achievable speedup depends highly on the initial value assignment. The center values are typical values for satisfiability problems. The lower values are valid for worst case assumptions that have never been observed during our experiments. The upper bounds are results for assignments close to high fanout nodes.

These results show, that the computation of indirect implications can be sped up by a factor of several hundreds using reconfigurable hardware if the synthesis times can be neglected. This shows that there is a clear potential of reconfigurable computing beyond systolic algorithms. On the other hand, with currently available synthesis tools for FPGAs, resynthesis of a full circuit for each problem instance is a large overhead. In our experiments the place and route times for the larger examples exceeds the typical run time of the software version of the algorithm.

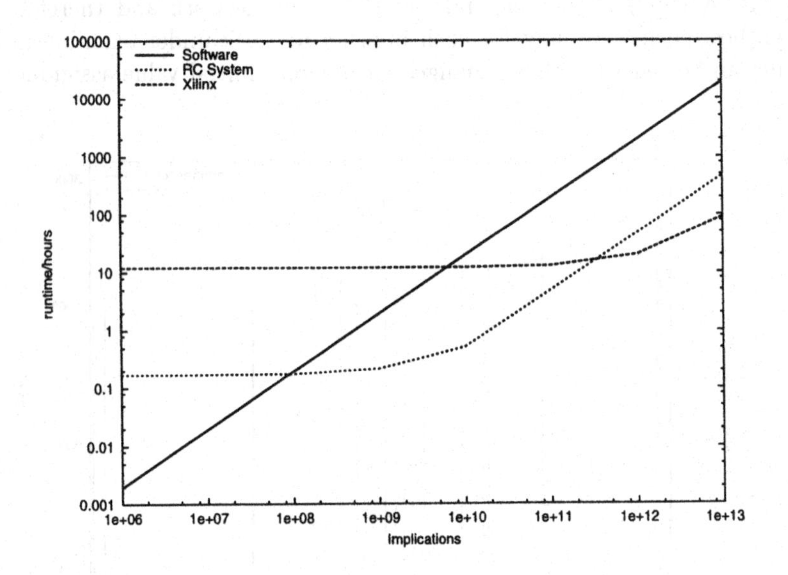

Figure6. Runtime by number of local implications.

Figure 6 shows the runtime including overhead of various implementations of recursive learning by the number of local implications for a 400 gate benchmark. Besides the measurements for the software version and our Xilinx implementation we included a rough estimate for an implementation of our approach on an innovative RC Architecture such as Berkeley's BRASS HSRA or MITs TSF-PGA, [2]. These numbers assume an architecture that allows synthesis of the circuit in 10 minutes but has a ten times lower performance than the highly optimized Xilinx implementation.

The software implementation beats our hardware implementation up to about 5 Giga implications which is about the number of implications usually performed in applications. However, the combined runtime of the hardware implementation does not increase significantly up to Tera implications. This means, that for typical application we can not improve on the runtime but we can provide a much more thorough search for implications in the same amount of time.

Fast synthesis for reconfigurable computing can potentially push the break even point below 100 Mega implications while still providing a very high peak acceleration. Therefore the development of very fast synthesis algorithms and reconfigurable architectures that support fast synthesis are essential for the presented approach and for reconfigurable computing in general. In the RC context, fast synthesis times are more important than performance optimization of the target circuit.

4 Conclusion

We successfully implemented recursive learning in FPGAs and achieved a very high speedup for typical cases. The high overhead limits the use to applications that spend multiple hours in computing implications in a single circuit, which is quite common in CAD. This limitation hopefully will be eliminated by innovative FPGA architectures and software.

References

1. Michael Chu, Kolja Sulimma, Nick Weaver, Andre DeHon and John Wawrzynek: "Object Oriented Circuit-Generators in Java"; *Proceedings of the IEEE Symposium on FPGAs for Custom Computing Machines, 1998*
`http://www.cs.berkeley.edu/projects/brass/documents/Generators_FCCM98.html`
2. Andre DeHon: "Reconfigurable Architechtures for General-Purpose Computing"; A.I. Technical Report No. 1586, Massachusets Institute of Technology, Artificial Intelligence Laboratory, 1996.
`http://www.ai.mit.edu/people/andre/phd.html`
3. Wolfgang Kunz and Dominik Stoffel: "Reasoning in Boolean Networks"; Kluwer Academic Publishers, 1997. ISBN 0-7923-9921-8
`http://www.wkap.nl/book.htm/0-7923-9921-8`
4. Wolfgang Kunz and Dhiraj K Pradhan: "Recursiv Learning: A New Implication Technique for Efficient Solutions to CAD Problems Test, Verification and Optimization"; IEEE Transactions on Computer Aided Design, Vol 13, No. 9, pp 1143-1158, Sep. 1994
`http://www.em.informatik.uni-frankfurt.de/forschung/pubs_arch/iccad94.html`
5. Kolja Sulimma: "Berechnung von Implikationen in Booleschen Netzen mit FPGAs", Universität Frankfurt, 1999. ISBN 3-933966-00-0
`http://verlag.prowokulta.org/isbn00/`
6. Peixing Zhong, Margaret Martonosi, Pranav Ashar and Sharad Malik: "Accelerating Boolean Satisfiability with Configurable Hardware"; *Proceedings of the IEEE Symposium on FPGAs for Custom Computing Machines*, 1998
`http://www.ee.princeton.edu/~mrm/pubs.html`

Author Index

Lecture Notes in Computer Science

For information about Vols. 1–1594
please contact your bookseller or Springer-Verlag

Vol. 1631: P. Narendran, M. Rusinowitch (Eds.), Rewriting Techniques and Applications. Proceedings, 1999. XI, 397 pages. 1999.

Vol. 1632: H. Ganzinger (Ed.), Automated Deduction – Cade-16. Proceedings, 1999. XIV, 429 pages. 1999. (Subseries LNAI).

Vol. 1633: N. Halbwachs, D. Peled (Eds.), Computer Aided Verification. Proceedings, 1999. XII, 506 pages. 1999.

Vol. 1634: S. Džeroski, P. Flach (Eds.), Inductive Logic Programming. Proceedings, 1999. VIII, 303 pages. 1999. (Subseries LNAI).

Vol. 1636: L. Knudsen (Ed.), Fast Software Encryption. Proceedings, 1999. VIII, 317 pages. 1999.

Vol. 1637: J.P. Walser, Integer Optimization by Local Search. XIX, 137 pages. 1999. (Subseries LNAI).

Vol. 1638: A. Hunter, S. Parsons (Eds.), Symbolic and Quantitative Approaches to Reasoning and Uncertainty. Proceedings, 1999. IX, 397 pages. 1999. (Subseries LNAI).

Vol. 1639: S. Donatelli, J. Kleijn (Eds.), Application and Theory of Petri Nets 1999. Proceedings, 1999. VIII, 425 pages. 1999.

Vol. 1640: W. Tepfenhart, W. Cyre (Eds.), Conceptual Structures: Standards and Practices. Proceedings, 1999. XII, 515 pages. 1999. (Subseries LNAI).

Vol. 1641: D. Hutter, W. Stephan, P. Traverso, M. Ullmann (Eds.), Applied Formal Methods – FM-Trends 98. Proceedings, 1998. XI, 377 pages. 1999.

Vol. 1642: D.J. Hand, J.N. Kok, M.R. Berthold (Eds.), Advances in Intelligent Data Analysis. Proceedings, 1999. XII, 538 pages. 1999.

Vol. 1643: J. Nešetřil (Ed.), Algorithms – ESA '99. Proceedings, 1999. XII, 552 pages. 1999.

Vol. 1644: J. Wiedermann, P. van Emde Boas, M. Nielsen (Eds.), Automata, Languages, and Programming. Proceedings, 1999. XIV, 720 pages. 1999.

Vol. 1645: M. Crochemore, M. Paterson (Eds.), Combinatorial Pattern Matching. Proceedings, 1999. VIII, 295 pages. 1999.

Vol. 1647: F.J. Garijo, M. Boman (Eds.), Multi-Agent System Engineering. Proceedings, 1999. X, 233 pages. 1999. (Subseries LNAI).

Vol. 1648: M. Franklin (Ed.), Financial Cryptography. Proceedings, 1999. VIII, 269 pages. 1999.

Vol. 1649: R.Y. Pinter, S. Tsur (Eds.), Next Generation Information Technologies and Systems. Proceedings, 1999. IX, 327 pages. 1999.

Vol. 1650: K.-D. Althoff, R. Bergmann, L.K. Branting (Eds.), Case-Based Reasoning Research and Development. Proceedings, 1999. XII, 598 pages. 1999. (Subseries LNAI).

Vol. 1651: R.H. Güting, D. Papadias, F. Lochovsky (Eds.), Advances in Spatial Databases. Proceedings, 1999. XI, 371 pages. 1999.

Vol. 1652: M. Klusch, O.M. Shehory, G. Weiss (Eds.), Cooperative Information Agents III. Proceedings, 1999. XI, 404 pages. 1999. (Subseries LNAI).

Vol. 1653: S. Covaci (Ed.), Active Networks. Proceedings, 1999. XIII, 346 pages. 1999.

Vol. 1654: E.R. Hancock, M. Pelillo (Eds.), Energy Minimization Methods in Computer Vision and Pattern Recognition. Proceedings, 1999. IX, 331 pages. 1999.

Vol. 1656: S. Chatterjee, J.F. Prins, L. Carter, J. Ferrante, Z. Li, D. Sehr, P.-C. Yew (Eds.), Languages and Compilers for Parallel Computing. Proceedings, 1998. XI, 384 pages. 1999.

Vol. 1661: C. Freksa, D.M. Mark (Eds.), Spatial Information Theory. Proceedings, 1999. XIII, 477 pages. 1999.

Vol. 1662: V. Malyshkin (Ed.), Parallel Computing Technologies. Proceedings, 1999. XIX, 510 pages. 1999.

Vol. 1663: F. Dehne, A. Gupta. J.-R. Sack, R. Tamassia (Eds.), Algorithms and Data Structures. Proceedings, 1999. IX, 366 pages. 1999.

Vol. 1664: J.C.M. Baeten, S. Mauw (Eds.), CONCUR'99. Concurrency Theory. Proceedings, 1999. XI, 573 pages. 1999.

Vol. 1666: M. Wiener (Ed.), Advances in Cryptology – CRYPTO '99. Proceedings, 1999. XII, 639 pages. 1999.

Vol. 1668: J.S. Vitter, C.D. Zaroliagis (Eds.), Algorithm Engineering. Proceedings, 1999. VIII, 361 pages. 1999.

Vol. 1671: D. Hochbaum, K. Jansen, J.D.P. Rolim, A. Sinclair (Eds.), Randomization, Approximation, and Combinatorial Optimization. Proceedings, 1999. IX, 289 pages. 1999.

Vol. 1672: M. Kutyłowski, L. Pacholski, T. Wierzbicki (Eds.), Mathematical Foundations of Computer Science 1999. Proceedings, 1999. XII, 455 pages. 1999.

Vol. 1673: P. Lysaght, J. Irvine, R. Hartenstein (Eds.), Field Programmable Logic and Applications. Proceedings, 1999. XI, 541 pages. 1999.

Vol. 1674: D. Floreano, J.-D. Nicoud, F. Mondad a(Eds.), Advances in Artificial Life. Proceedings, 1999. XVI, 737 pages. 1999. (Subseries LNAI).

Vol. 1677: T. Bench-Capon, G. Soda, A M. Tjoa (Eds.), Database and Expert Systems Applications. Proceedings, 1999. XVIII, 1105 pages. 1999.

Vol. 1678: M.H. Böhlen, C.S. Jensen, M.O. Scholl (Eds.), Spatio-Temporal Database Management. Proceedings, 1999. X, 243 pages. 1999.

Vol. 1684: G. Ciobanu, G. Păun (Eds.), Fundamentals of Computation Theory. Proceedings, 1999. XI, 570 pages. 1999.

Vol. 1685: P. Amestoy, P. Berger, M. Daydé, I. Duff, V. Frayssé, L. Giraud, D. Ruiz (Eds.), Euro-Par'99. Parallel Processing. Proceedings, 1999. XXXII, 1503 pages. 1999.

Vol. 1688: P. Bouquet, L. Serafini, P. Brézillon, M. Benerecetti, F. Castellani (Eds.), Modeling and Using Context. Proceedings, 1999. XII, 528 pages. 1999. (Subseries LNAI).

Vol. 1689: F. Solina, A. Leonardis (Eds.), Computer Analysis of Images and Patterns. Proceedings, 1999. XIV, 650 pages. 1999.

Vol. 1690: Y. Bertot, G. Dowek, A. Hirschowitz, C. Paulin, L. Théry (Eds.), Theorem Proving in Higher Order Logics. Proceedings, 1999. VIII, 359 pages. 1999.

Vol. 1694: A. Cortesi, G. Filé (Eds.), Static Analysis. Proceedings, 1999. VIII, 357 pages. 1999.